In *Unifying the Universe*, Hasan Padamsee has provided a beautiful disct
of our understanding of the physical world and universe. The author follow.
ence from the time of the ancient Greeks and Egyptians, then discusses the contributions of scientists in the Middle East, the catastrophe of the Dark Ages, and the ultimate revival and huge success of European science. Each new scientific discovery is described clearly and understandably with virtually no mathematics. It also gives the general reader an appreciation of the enormous contributions made by scientists from Archimedes and Pythagoras to Galileo, Newton, Einstein, and Bohr. Finally, he brings us to the ideas of the origin of the universe (the Big Bang), the internal structure of the proton and neutron (quarks), and the modern approach to symmetry. The book is ideally suited to provide a scientific background for university students not majoring in science.

David Lee, Nobel Laureate in Physics, Cornell University

Unifying the Universe presents a fascinating description of the role of symmetries in the development of scientific thought over the centuries. The text is written in a lively style of prose and makes frequent connections between contemporaneous scientific and humanistic themes. The careful reader will be rewarded with a renewed appreciation for the strong coupling between scientific and cultural development from the times of ancient Greece to the present-day frontiers of science.

Gerry Dugan, Cornell University, Physics Department

Unifying the Universe is an extremely readable introduction not just to the history of science but to the history of ideas generally. Taking the study of the world around and above us as its starting point, it leads the reader through the evolution of scientific thought from the Ancient Egyptians to the Enlightenment. In doing so, the book shows how, far from being a simple matter of progress from discovery to greater discovery, the creation of scientific truth is inextricably bound up with other cultural questions, such as the imagined relationship between human and divine, the role of art in society, and the desire to control the natural world. *Unifying the Universe* thus manages not only to communicate certain important facts about physics and astronomy, but also to suggest that such facts are only part of the picture; understanding how we know is at least as important as what we know in the process of learning to think about science. *Unifying the Universe* will be useful in a wide variety of classroom contexts, from a general introduction to physics, to a history of scientific thought, to a course on science and humanities. In addition, I expect that it will be interesting to a broad general audience interested in the history of ideas or the evolution of scientific thought.

Kristina Milnor, Classics Department, Columbia University

This book will inspire undergraduate students like no dry recital of the principles of physics can. It links the very inspirational history of scientific discoveries by the ancient philosophers with the very exciting new symmetries and principles of the most modern physics of today. As a practicing physicist I have found the presentation to be no less entrancing and I have recommended Padamsee's course at Cornell (from which this book is derived) to everyone old and young.

Nari Mistry, Cornell University, Physics Department

Padamsee vividly portrays physics as a human endeavour inextricably linked to other human endeavours. His graceful prose thrillingly describes the frustrations and relentless struggle to discover, to grapple with nature's mysteries. In the end, the reader is left with a sense of wisdom about the meaning, value, and visionary patterns of modern physics rooted in humanistic, cultural, and historical contexts.

Phat Vu, Soka University, Physics Department

Science does not develop in a vacuum. Padamsee's book is an excellent resource for anyone interested in the history and culture associated with the concepts and ideas we take for granted in elementary physics. It contains many valuable lessons from the history of physics and helps to bridge the unfortunate gap between science and the humanities.

Rich Galik, Cornell University, Physics Department

Unifying the Universe

The Physics of Heaven and Earth

Second Edition

Unifying the Universe

The Physics of Heaven and Earth
Second Edition

Hasan S Padamsee
Cornell University

CRC Press
Taylor & Francis Group
Boca Raton London New York

CRC Press is an imprint of the
Taylor & Francis Group, an **informa** business

CRC Press
Taylor & Francis Group
6000 Broken Sound Parkway NW, Suite 300
Boca Raton, FL 33487-2742

First issued in paperback 2021

ISBN-13: 978-1-138-38868-0 (hbk)
ISBN-13: 978-1-03-217481-5 (pbk)
DOI: 10.1201/9780429424410

Library of Congress Cataloging-in-Publication Data

Names: Padamsee, Hasan, 1945- author.
Title: Unifying the universe : the physics of heaven and earth / Hasan S.
Padamsee, Cornell University.
Description: Second edition. | Boca Raton : CRC Press, [2019] | Includes
bibliographical references and index.
Identifiers: LCCN 2019050825 | ISBN 9781138388680 (hardback) | ISBN
9780429424410 (ebook)
Subjects: LCSH: Physics--History. | Cosmology--History.
Classification: LCC QC7 .P26 2019 | DDC 530--dc23
LC record available at https://lccn.loc.gov/2019050825

Visit the Taylor & Francis Web site at
http://www.taylorandfrancis.com

and the CRC Press Web site at
http://www.crcpress.com

To my wife, Irma,
and my family, Tasleem, Xiomara, Kelly, Melissa, Zain, and Lexi

If nature were not beautiful, it would not be worth knowing

Henri Poincaré

Contents

PART 2   The Heavens

PART 3 *Synthesis*

Preface to Second Edition

As with the first edition, the goal of this book is to appeal to the lay-science enthusiast, and to professors who aim to present science requirement fulfilment courses to non-science students. Universities offer increasing numbers of such courses with various approaches to bridge the gap between science and humanities. The overall approach here remains to cross boundaries between disciplines by connecting science with other realms of human endeavour, such as art, philosophy, history, and poetry, along with captivating stories about the discoveries and personalities of the discoverers.

The second edition has a larger variety of modern updates to select from with many new topics such as exoplanets, is anyone out there, the secret of life, dark matter, dark energy, cosmic microwave background radiation, Higgs boson, string theory, and gravity waves. An entire new chapter on quantum physics has been added with sections on wave–particle duality, probability wave-functions, the uncertainty principle, quantum entanglement, the infamous Schroedinger's cat, and the exciting application of quantum computing. As with the first edition, the new modern topics and the new chapter are explored with human interest connections, and fascinating stories.

There is now a *Study Guide to Unifying the Universe* that professors can order from *Linus Publications*. The guide covers further exploration topics, numerical topics, detailed questions for thought, questions for home-works and tests, and multiple-choice questions for each chapter. The Study Guide makes it attractive to adopt this book for courses.

I am also happy to provide my lecture slides, exam questions, and other help if requested. Videos of my lectures with demonstration experiments are also available on YouTube: https://www.youtube. com/watch?v=IU5y0wn4QBM&list=PL6AO1HagveH_Kcq8Tf5aQBOXWJWb4MvnK&index=2.

Preface to First Edition

Our culture certainly values science for its stream of inventions. Aeroplanes routinely and speedily take us to distant lands. Television and compact disk players bring entertainers to our homes. Personal computers and cell phones are an integral part of daily life, connecting us through electronic-mail and the World Wide Web. With cell phones communicating through satellites, we talk instantly across the globe.

But science is important for much more than just its technological consequences. It brings us a great measure of intellectual satisfaction, opening new vistas that emancipate us from our provincial outlook. The universe is immense, full of planets like our earth, stars like our sun, and galaxies like our Milky Way. We are intrigued by a host of questions: where did the earth, moon, planets, sun, and stars come from? What is the origin of the universe? What are we and our world made of?

The deep longing to answer such questions is part of human nature. When we expand our thinking to the universe and our place in it, we wonder: is nature inherently complex and chaotic as it may appear at first glance? Or is there underlying order? What brings about the infinite variety that surrounds us? What is the cause of the incessant changes that envelop our environment? Is nature governed by natural laws? Can we understand the operation of the universe? Can we hope to unveil the laws? Can we fathom the cosmic mysteries of the origin and fate of the universe? Such questions lie at the very heart of scientific thought.

My aim in this book is to explore the first arena of science which emerged to reveal that the universe is indeed fundamentally orderly and governed by natural laws. We strive to find order and meaningful laws with our creative minds. This was the triumph of the science of mechanics, culminating in the laws of motion and gravitation, first understood clearly and succinctly by Isaac Newton in the 17th century.

Mysteries about terrestrial and celestial movements were some of the first questions to engage the earliest thinkers. Based on 2,000 years of observation and thought by his predecessors, Newton successfully explained the celestial drama, and in quantitative detail. What was truly astonishing about Newton's breakthrough was that he could explain celestial and terrestrial motion with the same laws. Before Newton's triumphant synthesis, it was generally thought that laws which operate in the heavens are quite distinct from laws valid in our earthly domain. The idea that heaven and earth are two separate realms with separate laws permeated our early science as thoroughly as it did our literature and art (see Figure 0.1). In the story of Creation, as told in *Genesis*, on the second day:[1]

> God said let there be a firmament in the midst of the waters, and let it divide the waters from the waters. And God made the firmament, and divided the waters which were under the firmament from the waters which were above the firmament.

Newton's synthesis of heaven and earth fired scientific imagination, generating intellectual optimism as never before. It is within our power to grasp the universe. Newton's approach became the prototype for future scientific inquiry. Science turned into one of the most important aspects of modern society, and a primary force in the advancement of civilization. Newton himself was surprisingly modest about his accomplishment. One of his most remarkable statements was:[2]

> If I have seen further than other men,
> it is because I stood on the shoulders of giants.

Who were these giants? What did they discover? How did they do it? Each discovery is a significant episode in intellectual history; each comes with its special story. Understanding how Newton and his predecessors reached the insights into nature's operations, we hope to appreciate not only their

FIGURE 0.1   Artist JS von Carolsfeld clearly depicts how God at the very beginning must have separated heaven and earth.[3]

science but also the processes that propel scientific progress in general. What matters most is not the facts, but how these were discovered, and how to think about them.

Aesthetic principles have been a strong guiding force throughout the development of science. In the quest for order and structure underlying nature's operations, the lamp of symmetry often lights the way. We look for underlying simplicity in the behaviour of nature. Humanity yearns for beauty as well as simplicity. We seek elegant explanations. One of the most important guiding principles is unity. Is there one fundamental substance out of which the universe is made? Is there one over-arching theory that explains all physical phenomena, as Newton succeeded with heaven and earth? When different areas of science come together with the weaving of general principles there is a dramatic leap in our understanding of nature.

At the end of every chapter in the story of the first synthesis of heaven and earth, updates with later ideas and developments link to contemporary physics. Here again, I plan to emphasize "how do we know?" By viewing exotic topics of modern physics in the context of the grand traditions they continue, I hope to make the abstract ideas more germane and meaningful. Some readers may wish to stay strictly on the path to the first synthesis, and choose to bypass these update sections. They should suffer no loss of continuity. But they may return later to follow the path to a grander unification that aims for an ultimate synthesis, between the microcosmos and the macrocosmos, between the subnuclear world and the universe on the largest physical scales. As laid out, the tales of the two syntheses entwine together. Through the course of the update sections I aim to cover modern answers to the fascinating questions relating to the origin and unity of the universe and its contents.

This non-technical approach to physics evolved from a conceptual course intended for non-science students. (Lectures are available as PowerPoint presentations from the author.) One of my primary aims was to cross boundaries, between physics and other disciplines. With the aim of addressing a general audience, I attempt to connect science with other realms of human activity. Scientific concepts are products of human imagination. Just as much as art and poetry are avenues for creative expression, our efforts to understand nature are also creative activities. Each in their own way, science, poetry, and art desire to see beyond the seen by teaching us to think, and by opening our minds to new ways of looking at the world. Thus the process of scientific thought is inextricably linked with humanistic, cultural, creative, and aesthetic aspects. Accordingly, I designed this book to appeal to the educated lay-science enthusiast.

Through the story of how heaven and earth come together, I hope to show that science is a cultural activity, a grand human endeavour to fathom the connection between us and the universe. From *The Tears of the Muses*, poet Edmund Spenser beckons us:[4]

> Through knowledge we behold the world's creation,
> How in his cradle first he fostered was;
> And judge of Nature's cunning operation,
> How things she formed of a formless mass.

Acknowledgements

I am immensely grateful to my editors of the second edition from CRC Press/Taylor and Francis, Kirsten Barr and Rebecca Davies.

I am grateful to Kelly Garrett and Greg Werner for kind assistance with the cover design and to Elizabeth Andrews for much help with the art research. To my readers I am ever grateful for their reactions. My lifelong friend Hoshang Dastoor carefully read an earlier version of the manuscript to provide insightful reactions and commentary, stylistic suggestions, and, above all, continued encouragement. Kristina Milnor of Columbia University kept a close eye on the classical content. My colleagues in the Cornell Physics Department, Rich Galik and Gerry Dugan, checked many chapters with great care to give valuable feedback on the physics and culture. Nari Mistry read the modern update segments and provided continuous support as I taught the course. A special thanks to Brian Greene of Columbia University for his generous encouragement all through the project and for looking over many chapters. I am immensely grateful to David Lee of Cornell for taking precious time to go through the manuscript and capture its essence in a nutshell. This effort would never have started without the early faith in my course initiative shown by the former Chairman, Doug Fitchen. I am deeply grateful to Phat Vu, Chairman of the Soka University Physics Department, for many stimulating discussions, for going through several chapters, and for offering the course at his university, as well as earlier at Wellesley and Holy Cross. My many teaching assistants in the Cornell Physics Department made innumerable suggestions and corrections over the years. And to Cornell students, I am ever grateful for their enthusiastic reception of my ambition to cross boundaries between disciplines.

Author

Hasan Padamsee has taught the "Physics of Heaven and Earth" to non-science students for the last 20 years. At Cornell he was the project leader of the Superconducting Radio Frequency Group pushing the advancement of accelerator technology for particle physics at the high energy and luminosity frontiers. Conducting research and development in the field for nearly 30 years, he collaborated with particle accelerator laboratories around the world, including Fermilab, Jefferson Lab, SLAC, CERN, and DESY, and has served as head of the Technical Division at Fermilab and as Chair of the Tesla Technology Collaboration. Among his many publications are two textbooks in the field of RF Superconductivity and review articles in encyclopaedias.

In 1990, he launched the TeV Energy Superconducting Linear Accelerator (TESLA) which morphed into the TESLA collaboration headed by DESY, and subsequently into the International Linear Collider (ILC). He was elected Fellow of the American Physical Society in 1993. He received one of the highest awards in accelerator physics, the 2015 American Physical Society Robert R. Wilson Prize for Achievement in the Physics of Particle Accelerators.

Part 1

The Earth

1 The Shapes of Nature: Beginnings of Scientific Thought

We see more of things themselves when we see more of their origin.

GK Chesterton[1]

NATURE IS NATURAL

The process of gaining knowledge about our world and the skies above was an integral part of every culture. For Western science, there was a crucial juncture in the development of thought: a transition from gathering knowledge about nature to gaining a genuine understanding. Many historians of science hold that the first such turning point took place within Greek culture. Human thought rose to a new level when Greek philosophers, starting around 650 BC, began to look for rational explanations for natural events in the world around us. As with most cultures, Greeks devised imaginative supernatural explanations for the daily motion of the sun, for monthly changes in the shape of the moon, or for terrifying lightning and thunder in a storm. Mystical accounts mingled with colourful myths about invisible gods in faraway places – on tops of mountains, beneath the earth, or in the heavens.

In the pivotal transformation from knowledge to understanding, Greek thinkers rejected supernatural explanations. Nature is not subject to the whims of gods and goddesses. It would seem at first glance, from the unpredictability of the wind, the timing of rainfall, or the size of the harvest, that nature, like the human being, is inherently capricious. But the familiar rising and setting of the sun, the faithful phases of the moon, the unchanging pattern of the stars, and the regular motion of the heavenly bodies, all provide a soothing reassurance. Nature is not intrinsically chaotic. There is design.

Greeks searched for reasoned alternatives, *rational explanations*. Proceeding to speculate about reasons underlying the infinite variety of natural phenomena, they based their first explanations on familiar mechanisms derived from commonly recognized physical relationships within their habitat. Surrounded by an apparently limitless ocean, it was tempting for creative-thinking inhabitants of Ionian islands to guess that the world must be a giant rock floating on a universal sea. They sought to explain nature by drawing from nature, the foundation for eventual understanding based on *natural laws*. Taking such steps, they were the first to launch what we call *scientific inquiry*.

This unique intellectual turn was just one aspect of a broader cultural development, as evidenced by their activities in trade, politics, sports, literature, and art. Politically and economically, Greeks were an adventurous people, a rugged nation of sailors and traders. Expert mariners and energetic merchants established small independent city-states, which traded abundantly with one another. Even though they frequently went to war with each other, they shared close cultural bonds, spoke a common language, and assembled together to compete in sports. The first Olympic Games were held in 776 BC. The roots of our political framework can be found in Greek culture. Our word "politics" comes from the Greek word *polis* for city-state. In the city-states of Athens, Corinth, and Thebes, the progressive idea of democracy was born. *Demos* means people, and *crates* means rule. When the rulers of a city-state called upon their citizens to fight or row warships, the people demanded and won the right to vote and participate in the government.

Turning to the literature of the 8th century BC, the blind poet Homer provided the Greek world with language and poetry through the epics of the *Iliad* and *Odyssey*. Most especially, he forged an attitude that pervaded Hellenic civilization for 1,000 years: people can be their own masters. Greek educators strived to instil the values that Homer imparted to his heroes. Homer's characters inspired a new self-consciousness, maturing over time into a supreme self-confidence.

Greeks modelled their gods after humans. Grand as they were, Greek gods exhibited human frailty; they made love and fought battles. Aphrodite was goddess of love, fertility, and beauty. Dionysus showed mortals how to cultivate grapes and make wine. But the gods who stood for reason and order towered above those who sponsored passion or brought chaos from drunkenness. Only disaster awaited those ruled by emotion. Passion only obscured truth. King among gods, Zeus stood strong, dignified, and in control (Figure 1.1). By exalting humanity, sculptors expressed faith in human potential and capacities. In architecture, Greek columns stood tall and erect. Dignity and poise in human form similarly pervaded Homer's literary attitude: we can be in control of our own destiny, although we may have to struggle against severe odds. Art, poetry, and drama were not the only vehicles for the expression of ideas. Dialogue and debate were critical components of Greek culture, destined to play a crucial role in the evolution of scientific thought. Homer gives his highest praise to characters who excel both in war and debate.[2]

Then among them spake Thoas, son of Andraemon, far the best of the Aetolians, well-skilled in throwing the javelin, but a good man too in close fight, and in the *place of assembly* could but few of the Achaeans surpass him, when the young men were *striving in debate.*

FIGURE 1.1   Bronze *Zeus*, ca. 460–450 BC. Poised and majestic, Zeus was the god of gods, ruler of the heavens. As god of thunder, Zeus prepares to strike with a lightning bolt, which he holds in his right hand.[3]

As Homer so often describes in the *Iliad*, even gods held assembly to debate their moves and ponder the fate of human protégés. Down on earth, Greek citizens in Athens congregated daily at the *agora* (open market) to discuss political developments and philosophic ideas.

It is tempting to suppose that seafaring traders debated the shape of the earth in the same spirit. Is it a circular disc surrounded by the ocean? Is it square? Is it flat or round? Clearly it is not infinite, since the sun and moon rise on one side and set on the other. The heavenly bodies are surely located beyond the earth; earth must have a finite extent. What supports the moon, the sun? What supports the stars?

DRAWING FROM THE WELL

Analysing the reasons for their broad cultural development, we can identify two important factors. First, Greek exploration of new frontiers challenged and expanded their habits of thought. Colonists encountered new environments as they enlarged their dominion to distant lands, to Syracuse in Sicily, to Masilia (today Marseilles in France). Homer's *Odyssey* overflows with a spirit of adventure that characterized maritime culture. The frontier also offered opportunity, and most often, prosperity. But exploiting possibilities demanded hard work as well as creativity. Seafarers had to be resilient and resourceful. New approaches led to improvements necessary to thrive in unfamiliar territory.

A second important factor was contact with Egyptian and Babylonian knowledge. Located near two advanced civilizations (Figure 1.2), Ionians absorbed their knowledge for more than

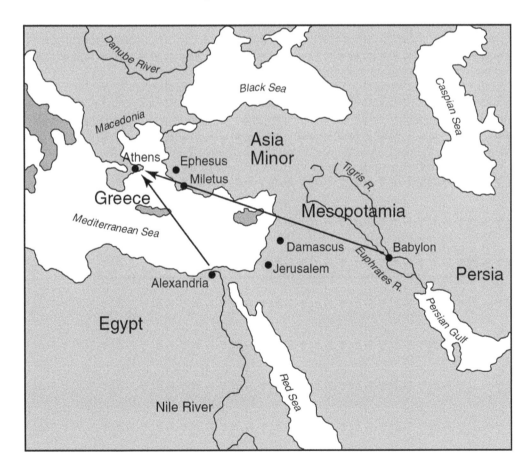

FIGURE 1.2   The Greek islands were situated near two ancient civilizations: Egypt, the land of the river Nile, and Mesopotamia, the land of the rivers Tigris and Euphrates. Knowledge flowed from ancient cultures into the Greek islands.

a millennium. In charming anecdotes, Greek historian Herodotus offers richly diverse accounts of neighbouring empires.

Egypt was the land of the river Nile, kingdom of the pharaohs, home of the grand pyramids, temples, and colonnades with majestic columns, built thousands of years prior. Situated between the rivers Tigris and Euphrates (now in Iraq and Syria), Babylon was the land of the ziggurat temple, and the Hanging Gardens. Here Emperor Hammurabi drew up a uniform code of laws from the conflicting rules of the many Sumerian city-states which his empire absorbed.

Culturally, intellectually, and technologically, Egyptians and Babylonians were thousands of years more advanced than the Greeks of Homer's time (for a timeline see Figure 1.3). For example, writing developed in Egypt as early as 3000 BC, probably out of the need to identify the ownership of land, or of granaries that stocked food. While the priests of Egypt routinely kept written records, even 2,000 years later, Greeks circulated Homer's epic poems by word-of-mouth. Egyptians were far ahead in medical knowledge and practice, as demonstrated by their highly developed art of embalming, for instance.

Rudiments of geometry evolved in Egypt, probably from the need to determine the length and width of canals they built to irrigate the Nile delta, guiding the water of the life-giving river from its central artery. The very name geometry means "to measure the earth." Every year in spring, the Nile would flood and wash out established property boundaries. Surveyors had to re-measure the land to lay down new property boundaries. Such activities led to important rules. One in particular was a powerful relationship between the sides of a right-angled triangle. Knowing that a 3–4–5 triangle defines a 90-degree angle, surveyors parcelled out land into well-defined rectangles using a right triangle of ropes. Such a device evolved into the set-square, one of our common drafting tools. We know that Egyptians had procedures to calculate areas of rectangles, triangles, and circles as well as prescriptions for the volume of cylindrical-shaped granaries.

Egyptian temple priests were master engineers, inventing dykes and canals to subdue mighty tributaries that ravaged villages with frequent floods. The tools and techniques eventually fed into construction of gigantic pyramids commanded by pharaohs who sought immortality in massive monuments. Standing in silent tribute to their vast engineering skills, the Great Pyramid of Giza covers an immense area of 13 acres. The enterprise demanded 2 million limestone blocks, each weighing more than 2 tons. Some of the blocks used in constructing the Great Pyramid at Giza are 27 feet long, 4 feet thick, and weigh 54 tons. Each granite obelisk weighed more than 1,000 tons.

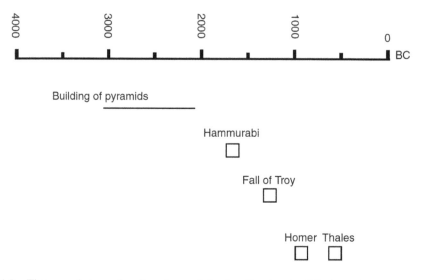

FIGURE 1.3   Time span between key Egyptian and Greek cultural and political events.

We still do not completely understand how they extracted such colossal pieces intact and moved them over land to transport the massive cargo along the Nile from stone quarries hundreds of miles away. At the pyramid site, workers used elementary mechanical devices, such as sledges and ropes to drag stones, ramps to raise hefty pieces, levers to accurately position megaliths into their final locations, and water levels to lay them straight. Architect-priests, well-versed in mathematics and astronomy, supervised construction. No doubt these monumental projects stimulated further advances in mathematical techniques.

While Egyptians were master surveyors and reckoners, Babylonians excelled in arithmetic, a skill which probably originated from a need to keep track of the calendar as well as lunar eclipse cycles for astrological forecasts. Expert calculators, they developed tables for multiplication, reciprocals, squares, and even square roots. Sumerians before them invented reckoning in the number system based on 60, instead of ours based on 10. The sexagesimal system probably started from an early estimate of the length of the solar cycle as 12 30-day months, $12 \times 30 = 360$. Accordingly, they divided the angle around a circle into a perfect cycle of 360 degrees, the fractioning of 1 degree into 60 minutes, and 1 minute into 60 seconds. We have inherited their system for the division of angles, as well as for hours into minutes and seconds.

FROM KNOWLEDGE TO UNDERSTANDING

Despite the advanced nature and wide range of Egyptian and Babylonian foundational knowledge, we would hesitate to classify this type of information as fully "scientific" in the modern sense. It was more an important collection of facts derived from individual experiences by trial and error. From generation to generation, sages handed down rules as workable prescriptions, established without question. Precious knowledge was oriented primarily towards practical use. But no underlying theory, no system, no abstract framework bound together the myriad separate rules and procedures.

When they first came to Egypt, the towering pyramids dazzled the Greeks. It was a sight even more spectacular than at present. According to some accounts, polished limestone covered the immense sides shimmering like marble in the brilliant sun. The Greeks eagerly absorbed Egyptian know-how and techniques. But as sailors on a vast open ocean they could not blindly accept arbitrary rules they encountered. They had to know if the rules for distance reckoning were correct, or they would be lost at sea. With inquisitiveness, they took a radically different approach. Greeks sought to expand the knowledge they assimilated by applying it to new circumstances. They debated the validity of standard prescriptions. As a result, Greeks did not merely inherit the vast store of Egyptian and Babylonian knowledge or pass it down unchanged from generation to generation. They asked: "What is the proof for the rules?" For example, Babylonians used the formula $3R^2$ for the area of a circle of radius R, while Egyptians asserted the area to be $(8/9) (2R)^2$. Of course, they did not use our modern algebraic representations. How were curious newcomers to determine the correct recipe? By attempting to prove the rules for themselves!

In seeking answers, Greek intellectuals created fundamentally new ways of thinking that eventually opened up new kinds of questions. Did any overall order or rationale underlie so many scattered observations? Were there any basic principles tying together the variety of geometrical rules? Going beyond the realm of mathematical rules, they began to ask similar questions about the nature of the world. They wondered whether any fundamental organizing principles existed beyond the baffling variety of behaviour observed in nature. Such questions lie at the heart of scientific thinking.

THE FIRST PHILOSOPHER'S GEOMETRIC INSIGHTS

Located at the mouth of a busy river, Miletus (in present-day Turkey) was one of the greatest Greek cities of the time, main harbour and rich market, prosperous from commerce and the colonization of other lands. Thales of Miletus (620–546 BC) was the original rationalist, the first to search for causes of natural events within nature itself, but not before drinking at the fountain of ancient know-how.

After his pilgrimages to the seats of learning in Egypt and Babylon, Thales returned to Miletus with a vast array of knowledge. By his time, Greeks were literate; they wrote on pottery and tablets. With new-found engineering skills, Thales supervised dyke-building to prevent flooding in Miletus. He extended geometric rules by demonstrating an easy method to measure the height of inaccessibly tall objects, such as pyramids and palm trees. Wait for the sun to reach a point in the sky so that the length of the shadow cast by a small stick is the same as the height of the stick. At that very moment, the length of the shadow made by the tree (easily measured on the ground) is also equal to the height of the tree (Figure 1.4).

It was a scheme based on *similar-triangle geometry*. When parallel rays of sunlight from the distant sun hit the treetop and the stick, the small right triangle formed by the stick and its shadow is *similar* to the large right triangle formed by the tree and its shadow. Because the corresponding angles of the two triangles are equal, the small triangle is a scaled replica of the big triangle. If the shadow of the stick is smaller than the shadow of the tree by a certain factor (say ten) then the stick itself is ten times smaller than the tree. Knowing the dimension of the stick determines the height of the tree. Starting with such particular useful applications, Thales progressed from the world of technique to the world of thought, from prescribed Egyptian know-how to abstract general principles.

Remarkable benefits emerged from fresh understanding of the abstract principle of similar-triangle geometry. Thales could apply it to new circumstances, for example, to determine the distance of ships from the shore, providing great benefits to a maritime culture (Figure 1.5). Centuries later, Aristarchus of Alexandria used similar-triangle geometry to estimate lunar and solar diameters as compared to earth's diameter, as well as the relative earth–sun and earth–moon distances. Already at these early stages, a deeper understanding of abstract principles could wield enormous intellectual power to open up brand-new directions.

Thales was not alone in his foray into the world of geometry. Geometrical forms prevailed in Greek art and architecture of his time and well before. Consider the example vases from the Greek geometric period, amply decorated with triangles, squares, and circles (Figure 1.6). Artists employed compass-like devices to draw precise, concentric circles as we know from tiny holes at

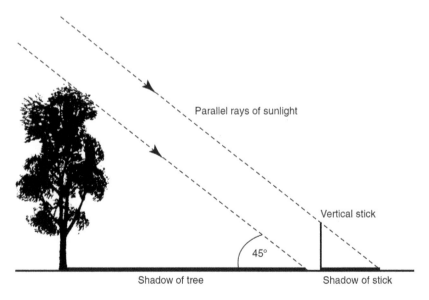

FIGURE 1.4   The use of similar triangles to determine the height of a tree. At a time of day when the height of a vertical stick on the right equals the length of its shadow, the height of a tall tree on the left will also be equal to the length of the tree's shadow, which can easily be measured on the ground. The right triangle formed by the tree and its shadow is similar to the right triangle formed by the stick and its shadow.

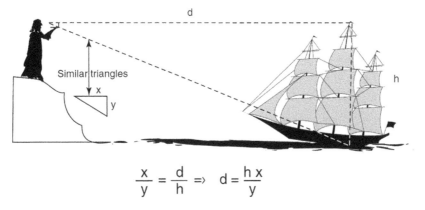

$$\frac{x}{y} = \frac{d}{h} \;\Rightarrow\; d = \frac{h\,x}{y}$$

FIGURE 1.5   Thales applied similar-triangle geometry to determine the ship-to-shore distance. He did not, of course, use modern algebraic notation. Assume the mast-height of the ship is a standard, known quantity. The trick is to form, for example with sticks, a small right triangle whose base and *hypotenuse* (largest side) visually line up with the corresponding sides of the large triangle over the ocean. One leg of the large triangle is the *known* mast-height of the ship, while the other is the *unknown* ship-to-shore distance. With the small-scale triangle in hand, determine the scaling factor between the mast-height and the short vertical leg of the stick-triangle. Applying the same scale factor to the horizontal side of the stick-triangle gives the corresponding ship-to-shore distance.

FIGURE 1.6   Vases from the geometric period, ca. 760–750 BC.[4]

the centre. But the emergence of new ideas in science was not directly influenced by artists or vice-versa. The commonality is that artistic thinking and scientific thinking were products of a shared culture, with different manifestations, like different fruits from a common soil.

Thales was not satisfied with extending ancient know-how to new applications. He went on to bridge the gap between the world of technique and the world of thought by converting myriad geometrical rules into a broader, more abstract study. Originating the idea of logical inferences and general proofs, he initiated deductive mathematics, proceeding to emphasize the value of making predictions. We owe some of the most elementary geometry we learn in school today to Thales. Applying logic, he proved the most basic theorems: angles at the base of an *isosceles* triangle are equal. (*Isosceles* comes from the Greek words *iso* meaning equal and *skeles* meaning legs.)

Opposite angles formed by intersecting straight lines are equal. The sum of the angles in a triangle is equal to the sum of two right angles (180°). Later these theorems will become transparent through symmetry principles recognized by Thales' successors.

UNITY BEHIND DIVERSITY

Most significantly for the evolution of scientific thought, Thales ventured beyond the realm of mathematics to apply rational thinking to a new quest: what is the nature of the substance of the universe, the *physis* of the *cosmos*? Along with the literal incorporation of the word *physis*, the meaning of physics has always encompassed Thales' original question: what is the universe made of? What is the nature of matter, the stuff we hold between our fingers? Contemplating the make-up of the universe Thales developed the first cosmology – one of the oldest subjects to captivate humanity. What supports the sun, moon, and stars in the sky? What supports the earth?

In searching for answers to the physics of the cosmos, Thales was guided by a profound aesthetic principle: there must be unity underlying the apparent diversity of nature. Entities, which at first glance appear quite different, may actually be alternate manifestations of the same essence. Perhaps one fundamental substance is at the basis of everything in the universe. We call it the primary *element*. Thales chose water because of its abundance, universal presence, and other good reasons. A vast billowing ocean surrounds the Greek isles. Water embodies two of the most fundamental attributes of nature: change and motion. Wherever there is water, there is motion, as evident from the flowing rivers and streams, and the ever-constant ebb and flow of ocean-tides. Most important, an essential property of water impressed Thales. Water can assume all the three distinct forms for matter: solid, liquid, and gas; ice, water, and steam. Perhaps, thought Thales, the earth is a solid water-disc floating on a vast ocean, a most natural method of support (Figure 1.7). Earthquakes occur when vapour bubbles rising from the oceans bob the floating earth plate. A single idea could account for several different aspects of the world. The sky is a hemispherical canopy containing water vapour, which explains why rain falls from the sky. The sun and the moon are disks like warriors' shields attached to the sky canopy. The stars are like lanterns pinned to the canopy. Thales' cosmology had a rational basis, influenced by direct observations.

Natural explanations for natural phenomena were far more appealing to Thales' scientific imagination than myths and fables spun by priests and poets. Choosing water as the primary element of the universe, Thales provided a natural link between substance, form, and change. Some may

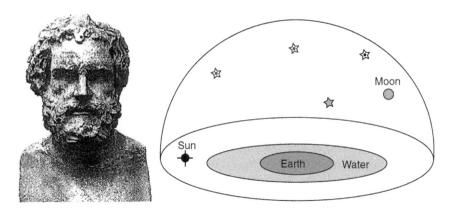

FIGURE 1.7   Thales and his model of the cosmos.

describe his approach as pure *speculation*, others may call it *insight*. The history of science is replete with examples of insightful guesswork, the creative part of the scientific process. There is always some insightful guesswork throughout the progress of science. But it is not random conjecture. It is guided by aesthetic principles. Does the proposed explanation unify or simplify the apparent complexity? Does the explanation provide a sensible ordering? Even though Thales' answers ultimately proved wrong, his efforts represented a deep engagement with the cosmos and with fundamental questions of all time.

Thales was also quite a colourful character, besides a pioneer mathematician, philosopher, and scientist. He was closely involved in the politics of Miletus and neighbouring islands. Miletus gained political importance from its position on essential trade routes. Commerce with Egypt, Babylon, and India brought caravans loaded with textiles, pottery, spices, wheat, gold, silver, ivory, and gems. But the future of Miletus and its neighbouring islands was under grave threat from the growing military power and westward expansion of the Persian empire. Always aiming to unify, Thales tried to persuade the small, independent city-states of Ionia to unite into one large entity to better resist the widening empire. Although he was unsuccessful during his time, Ionians did finally come together with mainland city-states such as Athens and Sparta to stem the tide of Persian expansion.

There is a popular story[5] about how Thales was taunted about being supremely wise but not very practical. While gazing intensely at the stars one night, he fell into a ditch, whereupon a servant girl mocked him for failing to see the obvious at his feet while trying to penetrate the mysteries of the heavens. People dismissed him as an idle philosopher who could not make enough money to support himself. A man had to be rich to be respected. Wisdom was not enough. On one occasion, Thales decided to teach his critics a lesson. Understanding the influence of weather on crop cycles, he guessed that a bumper olive crop was soon forthcoming. Quietly purchasing all available olive presses, he was able to charge exorbitant prices for their use at harvest time. In just one season he grew fabulously rich. Satisfied that acquiring riches was no great challenge to the mind, he abandoned business and returned to philosophy. Philosophers could easily acquire wealth if they put their minds to it. But thinkers had to have a higher ambition. Thales' attitude set a tone for Greek intellectual pursuit.

As intellectual fervour spread across the city-states, Thales' successors continued his philosophical quest for the universal element. But they did not immediately accept their master's dictum that all things must be composed of water. With love for intellectual contest, some chose air, earth, or fire as essential components of nature. Another added a primary element, called ether, as the fundamental essence of heavenly bodies. Thus the essential substance of the heavens was different from the elements found on earth.

Some challenged the notion that one element could suffice to explain the rich variety of substances present in nature. They claimed instead that all matter was composed from a mixture of rudimentary constituents: water, air, earth, and fire. Others argued that the universe originated from an infinite number of elements, growing like a forest from primary seeds. Such competition of philosophical ideas among independently minded thinkers was intellectually healthy. It was to become a crucial factor in the progress of science.

Thales' approach to addressing the composition of the universe led to a new style of thinking. The desire to find unity guided him to propose that all substances in the universe spring from a primary element. Nature's apparent complexity is not truly bottomless. It was a bold and captivating idea that has continued to motivate the progress of science ever since. Physicists of our time are still pursuing the prospect of unity among the constituents of the universe.

Present understanding of the elements has evolved far beyond Thales' simple but essential beginnings, topics for future chapters. Very briefly for now, water, earth, air, and fire are not elements; they decompose into the "chemical elements." Water breaks up into hydrogen and oxygen. More than 100 chemical elements populate the ubiquitous Periodic Table. Each element is composed of individual atoms, at first thought to be indivisible. By the turn of the 20th century, the perpetual hunt for basic constituents revealed that the atoms of chemical elements are also not elementary.

There is a central portion, called the nucleus, surrounded by electrons. While the electron continues to retain its elementary status, the nucleus consists of protons and neutrons, which in turn break down into components, called *quarks*. Electrons are just one among a broader family of *lepton* particles. The other leptons are called the *muon* and *tau*. There are three types of *neutrinos* which are also leptons. According to present thinking, six quarks and six leptons form the basic constituents of matter. But important loose strings remain to haunt us. Perhaps some will lead to more elementary forms in the future. We will return to these subjects in modern update sections through our journey toward the ultimate synthesis.

SWEET SYMMETRIES

In the generation following Thales, Pythagoras (570–495 BC) of Samos introduced pioneering ideas that turned into strong guiding principles for scientific thought. The first was a creative and powerful insight. Elegant principles must govern nature. One of the most elegant aspects that captivated Pythagoras was *symmetry*. His second crucial idea was that scientific understanding of nature must always incorporate mathematical descriptions. Symmetry and proportion in the mathematical realm have counterparts in the natural shapes and make-up of the universe.

Upon advice from Thales, Pythagoras, spent a good part of his young life in Egypt. Later he was a captive in Babylon. From both cultures he absorbed mathematics. We now remember Pythagoras best for his discovery of the famous theorem for right triangles (Figure 1.8). It was a powerful generalization of one of the geometrical rules discovered by the ancients. A triangle with sides of length 3 units, 4 units, and 5 units is always a right-angled triangle. A simple extension of similar-triangle geometry reveals that any triangle whose sides are scaled from the 3–4–5 triangle will also be a

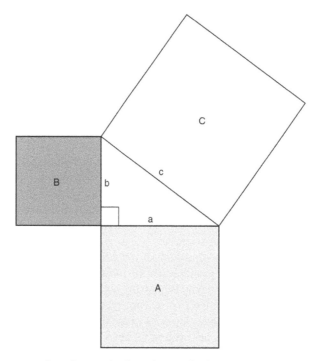

Area of square A + Area of square B = Area of square C

$$a^2 + b^2 = c^2$$

FIGURE 1.8   Pythagoras' theorem in geometric and algebraic forms.

right triangle, for example, the 6–8–10 triangle. But a triangle with sides 5, 12, 13, which is also a right triangle, cannot be obtained by scaling the 3–4–5 triangle. What is the magical relationship between the general sides (a, b, c) that will guarantee a right triangle? Expressed geometrically, if squares are drawn on all three sides of a right triangle, the area of the largest square is equal to the sum of the areas of the two smaller squares.

Expressed in modern algebraic form: $a^2 + b^2 = c^2$. Pythagoras not only recognized the abstract relationship, he also *demonstrated* its validity. An oft-told story, worth retelling, is how he offered 100 oxen to the temple of the Muses in thanks for the inspiration leading to the "proof." But the first rigorous proof of the famous theorem came from Euclid, two centuries later. A shorter, cleverer proof, based on similar triangles and some algebra, came from 12-year old Albert Einstein, as presented in a later section.

Like his predecessor Thales, Pythagoras progressed beyond geometry to creative insights about nature's operations, adopting the word *kosmos* for "universe" to encompass the idea that everything in our universe is beautifully ordered. Indeed, the words *cosmetic* and *cosmos* share a common root that embraces the idea of beauty. Pythagoras also coined the word "philosopher," which means "lover of wisdom." His enduring visions and theories continue to permeate our thinking.

To Pythagoras, symmetry is one of the most appealing aspects about the way nature is shaped and ordered. In symmetry there is simplicity; relationships do not change. Any constancy amidst change is deeply reassuring. No doubt Pythagoras was alert to symmetrical patterns that occur in nature, such as in fruits, flowers, butterflies, and palm trees. Another influence was Pythagoras' father, a jeweller[6], whose collection of precious stones fascinated the youngster and led him to recognize mathematical symmetries of solids in naturally occurring gems (Figure 1.9).

A symmetry exists whenever there is a quantity (example, shape) that remains unchanged after an alteration (example, rotation). The alteration is called transformation. More simply, a symmetry exists when there is "no change" after a "change." One type of alteration can be translation in space. Another is a change in orientation by rotation, or an interchange between identical elements of a composition.

Besides emerging as a way to understand nature, symmetry shaped expression in Greek art. Both front and back, the human exterior has bilateral (mirror) symmetry about the mid-plane, emphatically demonstrated by the Greek *Kouros of Anavyssos* (Figure 1.10). In this archaic statue of a youth, who died heroically in battle, the artist simultaneously expresses order, symmetry, and proportion.

The foot-soldiers in the painting of Figure 1.11 exhibit discrete translational symmetry; they appear nearly the same if the picture shifts by a certain distance. The cavaliers on horse-back manifest reflection symmetry. Many Greek vase paintings look the same when rotated by a finite angle. Such vases exhibit discrete rotational symmetry. An equilateral triangle manifests discrete rotational symmetry when rotated by 120°.

FIGURE 1.9   Natural gems.

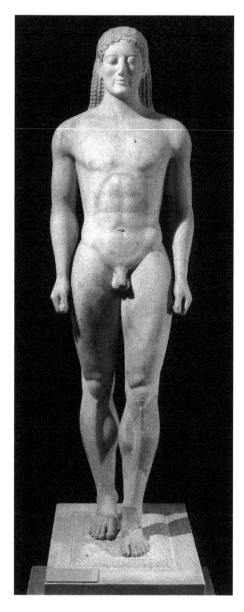

FIGURE 1.10   *Kouros of Anavyssos* ca. 525 BC.[7]

Although the soldiers of Figure 1.11 repeat, there are subtle variations in the warriors' shields. Shapes and rhythms do not repeat with monotonous rigidity. To an artist, such delicate shifts expressed a flowing sense of life and nature. For similar reasons, builders of antiquity purposely introduced minute variations from absolute symmetry in their columnar structures. Symmetry provided an overall governing form, but perfect symmetry would embody a frozen death. Being true to nature, artists were sensitive enough to avoid rigidity.

Rituals in life's everyday patterns, or in religious worship, also manifest symmetry. As relief from total chaos, people crave the comfort of structure that imposes rhythm upon their pace of activity. Music has underlying symmetries expressed by rhythm and tempo. Just as repeating patterns mould form in sculpture or painting, rhythm in music drives the natural flow of melody. Repeating at regular intervals, rhythm assures the continuous progression of music. But repetition

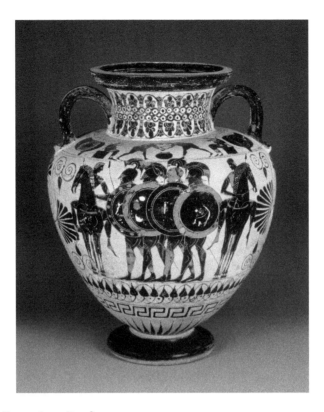

FIGURE 1.11   Hoplites and cavaliers.[8]

in music is also never exact; that would become quickly tiresome. Talented composers insert skilful pattern variations to melody, or clever rhythm alterations, such as syncopation. Conventional poetry has symmetry in rhyme and metre. As with art, music, and poetry, nature hardly exhibits overtly perfect symmetry. Although patterns never repeat exactly, there is an underlying structure present in nature. It is the objective of science to look for the underlying order.

Mathematical symmetry is exact. A square appears unaltered after rotation by 90 degrees, but it will look different at any other angle. Therefore a square manifests 90-degree rotation symmetry. In solid geometry, a cube also exhibits 90-degree rotational symmetry about an axis that passes through the centre of a face. There is only one marvellous solid – the sphere – that shows perfect angular symmetry because it always appears the same upon rotation by any angle.

Symmetry enlarged Pythagoras' sense of beauty. Just look at the bright blue sky, thought Pythagoras! No, it cannot be a hemispherical dome, as Thales envisioned. That would defy symmetry. It would be ugly! In any direction, the heavens look the same, judging from the apparently "random" distribution of stars. Beauty commands that the heavens must be spherical. Motivated by the perfect symmetry of the sphere, Pythagoras claimed that the moon and sun, which appear like discs, could not really be flat discs. They just appear to be discs from a distance. Symmetry dictates that heavenly bodies must be spherical in shape. If they were discs, they would appear different from different parts of the cosmos. Rotating a flat plate, it is easy to see that a disc will resemble a line when viewed edge-on. Thinking about heavenly bodies in universal terms, Pythagoras was a pioneer who freed himself from the restricted view of our unique position on earth. His powerful intuition about nature's intrinsic beauty made it imperative that the heavens, the sun, and moon must all be spheres with perfect symmetry (Figure 1.12). Pythagoras used symmetry to move toward truth.

Even the earth must be a sphere, claimed the sage. It is like a ball, floating free in space. Though it appears that the world stretches on forever in every direction, the earth is not flat, as appearances

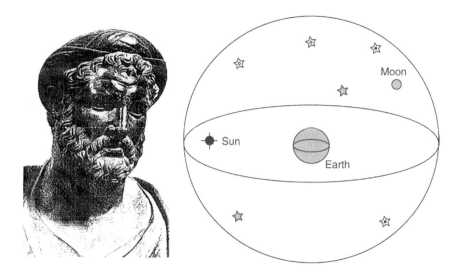

FIGURE 1.12   Symmetry motivated Pythagoras to think about the heavens and the earth as spherical.

might suggest. Likewise, the ocean is not an infinite, flat plane of water; it covers the surface of our spherical earth, like the skin of an orange. Symmetry was Pythagoras' trusted guide toward new insights.

It is a matter for speculation, but we can imagine that many of Pythagoras' contemporaries protested against the apparently absurd conclusion of a spherical earth. What supports the earth? Commonplace observations show nothing floats freely. On the other side of a round earth, would not rivers and oceans gush away into the emptiness below earth? If earth is indeed a sphere, is the other hemisphere inhabited? And if it is inhabited, what keeps people from falling into the endless abyss below? Would they have to live on trees, clinging to branches, so as not to fall off the face of the earth? Clearly a spherical earth flouts the most common observation; all things must "fall down." The spherical shape of earth demanded the idea of gravity, to be born two centuries later.

To think of earth as a sphere, Pythagoras defied commonplace notions based on common-sense observations. With the courage of his convictions about intrinsic beauty and symmetry, he was prepared to buck common sense. Despite appearances, he could "see" the spherical earth with his mind's eye. In his time, the idea was as radical as later ideas that our earth spins on its axis, or that earth orbits the sun, or that stars move, or that our planet may not be the only world to be populated by living, intelligent beings.

As time would show, Pythagoras' hypothesis about earth's shape, guided by the aesthetic preference of symmetry, was a brilliant success. Without invoking the power of symmetry, progress in physics would often be stuck. But as future developments will show, beauty does not necessarily make a theory right. To explain nature's behaviour by simple, elegant, and symmetric principles *alone* can sometimes lead to oversimplifying models that fail in their ability to describe and understand nature. Pure theoretical principles by themselves can often be deceiving. For example, invoking principles of elegance and symmetry, Pythagoras claimed that heavenly bodies must move in simple and symmetric paths, perfect circles. The sun and the moon do not go to bed in the west at night, to wake up miraculously the next morning over the east. They move in perfect circles around the earth through the spherical heavens. Aesthetic prejudice guided Pythagoras to put earth at the centre of the universe because the star-studded heavens look nearly the same in all directions. By symmetry, we on earth *must* be at the centre of the cosmos, another conclusion that later proved false. Earth-centric theories based on perfect circular orbits misguided astronomy for 2,000 years. Science cannot rely on creative speculation alone, or theoretical symmetry alone. It needs much more.

THE MYSTIC ELEVATES MATHEMATICS

Pythagoras' other inspiration was: nature must conform to mathematical description. How did he come to assign such far-reaching importance to mathematics? An indestructible and delightful legend exalts Pythagoras' insight. One day, while passing by a smith shop where workers were banging metals into useful shapes like pots and pans, Pythagoras noticed how different hammers produced different musical notes on striking metal. He persuaded the smiths to strike their hammers in an orderly sequence by weight to produce a pleasant series of notes of increasing pitch, a melody we would call a scale. Soon he came to a far more important realization. If the weight of one hammer was double the first, the note played by the second sounded similar to the first, but with a higher pitch. When struck together or in rapid sequence, the combined chord or interval sounded quite pleasant. There was music behind the apparent noise.

Returning home, he continued to look for numerical relations, now in the music of the lyre, to find a similar correlation between vibrating string-lengths and pitch. Shorter strings exuded a higher pitch. And once again, when string-lengths were in a 2:1 ratio, the combined sound was harmonious. Pythagoras was on the way toward another wonderful discovery. Whenever a pair of plucked strings produced a pleasant chord, their lengths showed a *simple ratio of natural numbers*, such as 4:3, 3:2, or 2:1. When ratios were not simple whole numbers, the sound had a grating quality. Even with lyre strings made from a different material, or out of a thicker stock, the musical interval would remain the same, as long as the proportion between the lengths was the same. If Pythagoras had used reeds or pipes instead of strings, he would have discovered harmonic ratios among pipe lengths. With the discovery that ratios produced music, Pythagoras married musical harmony to mathematical proportion. Music is number. Just as musical tones are pleasant to the ear, mathematical relationships are pleasant to the mind.

PATTERNS IN THE SAND

While the language of numbers expressed musical harmony, the language of geometry introduced visual patterns to bring order into physical space. Looking out over the clear blue ocean, the horizon formed a right-angle with the vertical line of free fall. Geometry organized space in another essential way. Right-angle rotations linked cardinal directions, north, east, south, and west, established from astronomical observations. The North Star defined north, while sunrise and sunset positions marked exact east and west (during the equinox, as discussed in Chapter 5). Pythagoras placed a 3–4–5 triangle flat on the ground so that the 4-side lined up with the north–south line (see Figure 1.13).

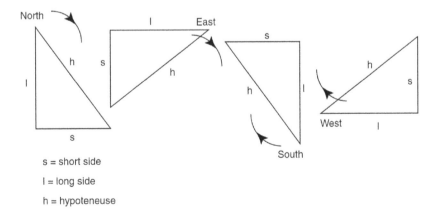

s = short side
l = long side
h = hypoteneuse

FIGURE 1.13 Pythagoras recognizes the importance of right-angle geometry in organizing the dimensions of space. Rotating a 3–4–5 right triangle repeatedly clockwise, the triangle points toward north, east, south, and west, the cardinal directions. For future reference, the lengths of the sides of the triangle are *l* for long side, *s* for short side, and *h* for hypotenuse.

In consecutive operations, he rotated the right triangle clockwise by 90 degrees to generate four right triangles. For each triangle, the vertex pointed north, east, south, and west. In the next step, he arranged the four right triangles into two different patterns to demonstrate the validity of the theorem that made him famous. To follow his proof, see Figure 1.14. The proof is not what we call rigorous, but it is a delightful graphical demonstration.

Inspired by Pythagoras' emphasis on symmetry, a very simple demonstration of his theorem results by symmetrizing the right triangle into an isosceles figure, i.e., with two equal sides (Figure 1.15) As 19th-century Russian poet Alexander Pushkin reminds us:[9] "Inspiration is needed in geometry, just as much as in poetry." Inspiration is also necessary for science.

Symmetry is a powerful tool to simplify solutions of mathematical problems as well as to understand natural behaviour. Rotation symmetry helps to prove Thales' theorem that vertically opposite angles are equal when two straight lines intersect (Figure 1.16) or the equality of alternate angles formed when a line intersects two parallel lines. Translation symmetry elegantly proves the equality of the corresponding angles formed when a line intersects two parallel lines (Figure 1.16).

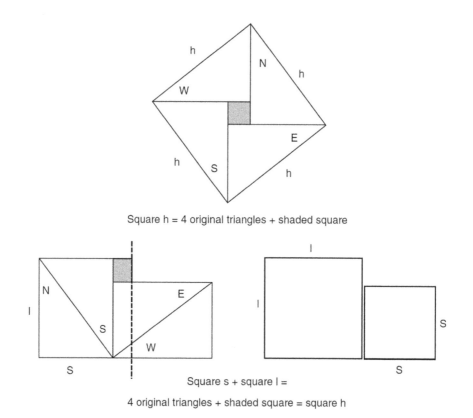

Square h = 4 original triangles + shaded square

Square s + square l =

4 original triangles + shaded square = square h

FIGURE 1.14 Pythagoras' demonstration of his famous theorem. (Top) in the first step, assemble four right triangles called N, E, S, and W to form a large square that includes a small shaded gap in the centre, also a square. The total area of the big square is equal to the square of the hypotenuse (h^2). (Lower left) in the next step, rearrange the four triangles and the *small shaded square in the gap* to form two new squares, demarcated by the dashed line. One of the new squares has the long side (*l*) of the original right-angled triangle. The area of that square is l^2. The other new square is formed by the short side (*s*) and has area s^2. (Lower right) eliminating (for clarity) the lines that define right triangles and the small square, new lines define two squares with areas l^2 and s^2, together equal to h^2.

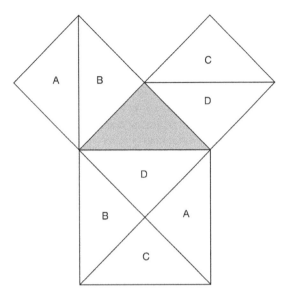

FIGURE 1.15   Symmetry greatly simplifies the proof of Pythagoras' theorem if the right triangle becomes isosceles. By drawing one diagonal across each of the small squares, we generate four triangles that fit exactly into the four triangles generated by two main diagonals across the large square. Note that each triangle A ... D also has the same area as the original right triangle.

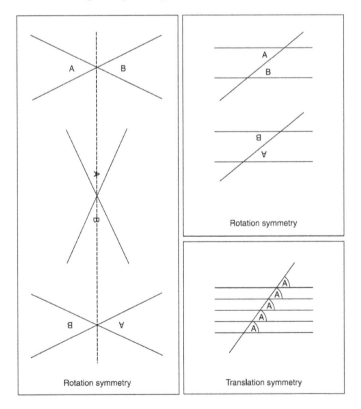

FIGURE 1.16   (Left) *rotation symmetry* shows that opposite angles (A and B) formed by a pair of intersecting lines are equal. Two successive rotations by 90 degrees leaves the pattern of intersecting lines unchanged, but angle A interchanges with angle B. (Right top) rotation symmetry demonstrates that alternate angles A and B are equal when formed by a line intersecting two parallel lines. (Right, bottom) *translation symmetry* dictates that all the corresponding angles (A) between parallel lines are equal.

ALL IS NUMBER

Symmetry, geometry, and numerical relationships gave special joy to Pythagoras. You could say, Pythagoras was drunk on math! Leaping to generalization, Pythagoras asserted: *"All is number."* What he probably meant was that mathematics is everywhere in nature. Words describe only a limited picture of the universe; the structure of nature is more aptly described with numbers. Could the divine mind have constructed our universe according to the principles of geometry and numbers? Indeed our faith that nature is "rational" can be traced back to the original significance Pythagoras attached to "ratios" of whole numbers. Mathematics helps in comprehending the order of nature because the regularities of nature are intrinsically mathematical. Pythagoras' basic intuition was on the mark; the language of mathematics provides a mirror for the physical world. Today mathematics permeates all of physics, although the field has progressed far beyond right-angle geometry and whole number ratios. Every natural phenomenon, every attempt to understand the forces of nature, or to control and engineer them to suit our needs, must be cast in the language of mathematics in order to be truly successful in revealing its full power.

In a bolt of mathematical reasoning, Pythagoras discovered another important connection between number properties and geometry: *the sum of the first n odd numbers is* n^2. Once again, he provided an exquisite geometrical demonstration (Figure 1.17). After the rediscovery of classical thought during the Renaissance, Galileo Galilei (1564–1642), father of modern science, recognized Pythagoras' mathematical pattern in free-fall motion to make one of the first quantitative breakthroughs in the laws of motion. Distance intervals covered by an object in free-fall grow with time

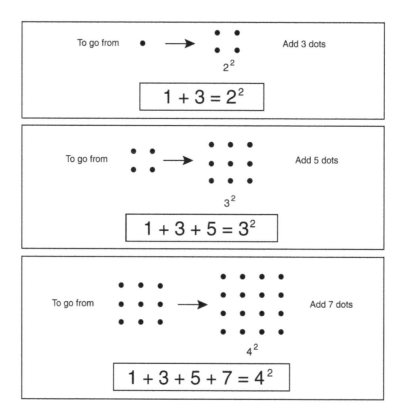

FIGURE 1.17   Start from a 'square' of one pebble. To make a square of four pebbles, add three pebbles ($1 + 3 = 2^2$). To make a square of nine pebbles, add five pebbles ($1 + 3 + 5 = 3^2$). To make a square of 16 pebbles, add seven pebbles ($1 + 3 + 5 + 7 = 4^2$). The sum of the first five odd numbers will be 25 and so on.

like the odd number series. As a result, the total distance an object covers in free-fall increases as the square of the elapsed time. Pythagoras and Galileo illuminated the regularities that reflect an underlying law of nature.

Pythagoras left behind a large cult of staunch followers. True or not, the story is often told that after one of his powerful addresses, 600 joined the Pythagorean commune without even going home to take leave of their families. Some pursued his mathematical and scientific ideas. Mystical teachings captivated others. Pythagoreans devoted to the study of number ratios discovered a severe limitation of integers. There is no simple ratio of whole numbers (m/n), which when multiplied by itself gives the whole number 2. Therefore $\sqrt{2}$ is an "irrational" number, unlike $\sqrt{25} = 5$, or $\sqrt{9} = 3$, which are rational numbers. In modern terms, when we express numbers as decimals, a rational number can always be expressed as a finite or a repeating decimal, e.g., 3/4 = 0.75, and 4/3 = 1.333 repeating. But an irrational number has an infinite decimal representation with no repeating sequences, e.g., $\sqrt{2} = 1.414213562...$.

The existence of irrational numbers upset the elegantly simple view that pure whole numbers encompassed all mathematics. Pythagoreans were too embarrassed to accept that something so simply realized as the diagonal of a square with sides of unit length was not describable in terms of whole numbers. It was so totally incomprehensible, so irrational! Moreover, being related to Pythagoras' theorem, the crisis tarnished the master's name. It challenged his teachings that the beauty of natural numbers exemplified the purest mathematics and expressed basic harmonies of nature. If the existence of irrationals became known, serious doubt would upset their master's doctrine that whole number ratios could explain the universe. Pythagoreans tried to keep the "evil" discovery a secret. One disciple was drowned for letting out the secret. With perspicacity, the remarkable incident provoked the modern cognitive scientist, Douglas Hofstadter, to wittily declare:[10] "Irrationality is the square root of all evil!"

SENSE OR REASON

Thales, Pythagoras, and their successors were pioneers, asking penetrating questions. What is the fundamental constituent of matter? What is the underlying cause of nature's vast diversity? Where is the order and beauty underlying nature's apparent complexity? What is the origin of the cosmos? But they were speculating about the answers. Dissatisfied with pure speculation, Parmeneides of Elea (515–450? BC) raised a most important issue: how can we reach proper conclusions about such fundamental questions, and ultimately about nature? How can we decipher the right answers from the morass of creative possibilities? He considered two distinct ways. We can make observations about the world with our senses and cross-check these against our conjectures. Or we can use the power of reason. Thales and Pythagoras showed how, in geometry. But the two thinkers took different approaches in understanding nature. Thales made observations about the world through the senses; Pythagoras used the power of reason with aesthetics as his guide.

Parmenides pointed out a serious flaw with the first method. To make observations one has to rely on the *accuracy* of the senses. When we look down upon the surface of a calm pond we see an identical twin staring at us through the water. But we know it is an illusory image; we would find no one when we dive into the pond. Our senses can betray and play tricks, leading us to erroneous conclusions. A modern painting by MC Escher (Figure 1.18) provides a striking illustration of Parmenides' concern about misleading perceptions. With one swift glance, many of us tend to see only white fish swimming to the left. But a closer look reveals how the black "formless" background is really a second, interweaving motif of black birds flying to the right, previously invisible to many. With limited sensory capacities, our mind gets used to looking at the world in a certain way. By jolting viewers, artists try to teach us new aspects of the world.

What the senses reveal is quite distinct from the "real world," argued Parmenides. We must question the "reality" of perceptions. Truth must be deeper. The astounding diversity of nature, its

FIGURE 1.18   *Fish and Birds*, MC Escher.[11] Translational symmetries in both horizontal and vertical directions rule the painting.

infinitely changing forms, the amazing variety of motion are all just superficial manifestations arising from sensory deception:[12]

> 'Do not let custom, born of everyday experience, tempt your eyes to be aimless, your ear and tongue to be echoes. Let reason be your judge', he advised.

Concerned about errors and sensory limitations, Parmenides felt that observation was a flawed path for seeking knowledge and understanding about the world. Information from the senses can be partial. Sensory perception can be marred by distractions and noise. He preferred to trust the process of rational thought and purely logical methods. As we follow the evolution of science, we will see that Parmenides was certainly right to stress the importance of reason. It is essential to "make sense" of observations using reason and logic. The eye of the mind reveals surprising conclusions that defy the information we gather from the senses alone. We can see much further to grasp nature at a deeper level through thought. But Parmenides' extreme position to relegate sensory information to little value turned out to be a serious mistake. It was all too restrictive. Science surely needs empirical content, information from nature.

Democritus was Parmenides' student, most famous for conceiving the notion of atoms. In pursuit of the eternal and unchanging laws underlying the infinite variety apparent in nature, he proposed that by continually subdividing matter, one must eventually reach the very smallest portion, which must be indivisible – to be the smallest. This he called "atomos," meaning "uncuttable." An atom is the smallest particle that you can divide an element into so that part still remains the element.

If the atom is to be the ultimate constituent of matter, it must also be "immutable." It cannot change by any process. If, on the other hand, you can subdivide each substance forever, everything should ultimately become indistinguishable from everything else. By making the atom indivisible, a portion (the atom of the element) would be fundamentally different from the smallest portion of another element.

While atoms are the primary, unchangeable, and indestructible entities, the relationship between atoms is in a constant state of flux. Different clusters of atoms provide the rich diversity of natural phenomena and substances we perceive. Wood, for example, is not an elementary substance. Wood atoms are combinations of ash-like earth atoms plus fire atoms plus water atoms. Drive away water atoms, and dry wood is ever so likely to catch fire. Drive away fire by burning the wood, and retrieve the constituent ash. How do clusters of atoms already formed undergo change? Because of their incessant motion. Like buzzing bees, atoms detach from one arrangement to re-congregate and form into another collection. To provide the space in which atoms can move, Democritus postulated the existence of void. "By convention sweet is sweet, bitter is bitter, hot is hot, cold is cold, colour is colour; but in truth there are only atoms and the void."[13] Democritus' introduction of the atomic hypothesis pioneered the idea that the familiar world of matter results from numerous combinations of a relatively few entities.

Democritus' conjectures about atoms and the void turned out to be an eerily insightful image bequeathed to posterity. Even though the Greek elements disappeared over the development of science, Democritus' atom survived. Well taught by Parmenides, Democritus did not need to directly *see* the infinitesimal discreteness of matter, nor its incessant motion. Nor did he have to actually experience the effects of vacuum (void). Just as Pythagoras deduced the spherical earth from the guiding principle of symmetry, pure reasoning about constancy underlying change led Democritus to the powerful concept of infinitesimal atoms in motion, darting about in a void like dust in a sunbeam.

Yet even Democritus could not avoid falling victim to the strong trap of perception. Attributing to atoms properties derived directly from the senses, he pictured water atoms as smooth and round, so water flowed easily and had no permanent shape. Fire atoms were thorny, so burns were painful. Earth was composed of jagged atoms so they held together in a variety of forms to produce the inexhaustible variety of substances. Shape and size became his key differentiators.

A GOLDEN AGE

By 500 BC the competing Greek city-states, most notably Sparta and Athens, united under the leadership of Athens to overcome a common threat: the advance of the Persians. Crucial to their defence was the mighty navy supplied by Athens. In a pivotal battle at Marathon (490 BC), the Greeks repelled the Persians. To bring the exciting news of their triumph, a courier ran 26 miles non-stop to Athens. After exclaiming: "Rejoice … we conquer," he fell dead of sheer exhaustion. We commemorate everywhere his run at the 26-mile modern marathon.

Greek victory heralded a Golden Age. Under Pericles (494–429 BC), Athens led the united city-states to become the most beautiful city of the period, the enlightened centre of the Greek world, nucleus of Western culture. Philosophy, poetry, plastic arts, and architecture flourished as never before. Through the Golden Age of Greece thrived legendary thinkers: Anaxagoras, Socrates, Plato, and Aristotle (see timeline, Figure 1.19). The Athenian government attempted the first democracy, but only free men were allowed to vote. Slaves, women, and foreign-born were excluded. They elected Pericles as chief general for 30 years in succession. It was fortunate that Pericles had philosophers Anaxagoras and Zeno as tutors during his youth. Because of his intellectual upbringing, he fostered a climate in which a glorious culture flourished. In celebration of his great victory, he commissioned the Parthenon temple and lavish monuments which still stand at the Acropolis as relics of an inspiring period, echoing his stirring words.[14] "We have forced every sea and land to be the highway of our daring and everywhere … have left imperishable monuments behind us." Even though the temple (Figure 1.20) is in ruins, we can appreciate its timeless beauty in the order of majestic, marble columns, the symmetry of its structure, and the balance of its proportion. The front and back are identical; so are the two sides, as if reflected in a mirror. A comforting rhythm prevails through the repetition of columns. Through its commanding presence, architects express geometric principles that rule the exploration of spatial dimensions. Knowing that a geometrically straight

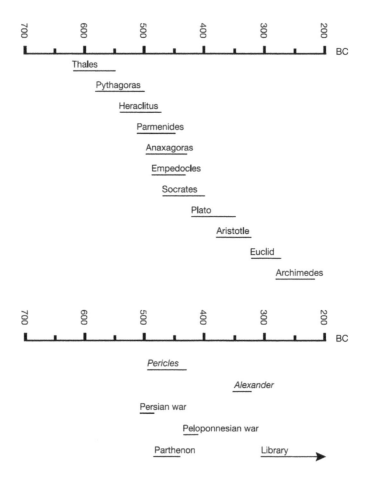

FIGURE 1.19   The Greek thinkers and the events of their time.

FIGURE 1.20   The Parthenon (built 448–432 BC), symbol of Greek triumph and cultural supremacy, standing majestic as it appeared before its destruction.[15] Including the pediment, the front face lies inside a golden rectangle.

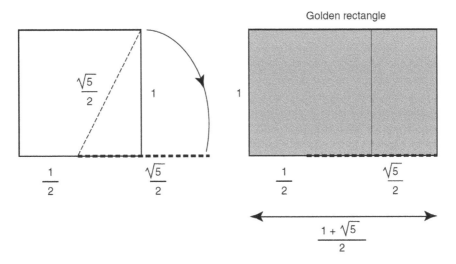

FIGURE 1.21 The construction of a golden rectangle from a square. Take the mid-point of the base and draw a line to an opposite corner. Rotate that line down toward the base. The end of the pivoted line extends to form the base of the golden rectangle. With the help of Pythagoras' theorem, it is easy to show that the base is $\left(1+\sqrt{5}\right)/2 = 1.618...$, a magical number called the golden mean, which we will meet again in an update section. Alternatively, the golden mean results from a simple prescription. In a well-proportioned rectangle, the ratio of the long side (l) to the short side (s) should be equal to the ratio of the sum of the two sides ($l + s$) to the long side (l): $l/s = (l + s)/l$. This equation simplifies to $l/s = 1 + s/l$. If we say $l/s = g$ (for golden mean) the equation becomes $g = 1 + 1/g$ or $g^2 - g - 1 = 0$. The quadratic equation yields the positive solution $g = \left(1+\sqrt{5}\right)/2 = 1.618$.

column appears slightly concave from afar due to an illusion of optical perception, they deliberately incorporated a slight bulge in the middle of each pillar to give the impression of perfectly straight and strong columns.

Architect Callicrates proportioned the front of the Parthenon, but not in a square outline. That would be perfectly symmetric, too rigid and unlifelike. Instead, he extended the square by an elegant geometrical construction to provide an aesthetically proportioned, golden rectangle (Figure 1.21), signature for the golden age.

A METHOD OF THOUGHT

Anaxagoras (500–428 BC) disputed Democritus' atoms:[16] "There is no smallest among the small and no largest among the large; but always something still smaller and something still larger." He posed some troubling questions that displayed a startlingly modern touch. If atoms exist, what are they made of? Does an atom have smaller parts with still smaller parts, like chains of seeds within seeds?

One of Anaxagoras' very controversial ideas was that the sun and moon are composed of the same materials as on earth. When news came to Athens that an enormous fiery stone (probably a meteor) had fallen from the sky, Anaxagoras declared the flaming rock to be a piece of the sun. For his impudent assertion that the heavenly sun is just a fiery rock, vastly bigger than any Greek island, Athenians prosecuted Anaxagoras on a charge of impiety and eventually drove him out of the glorious city. The charge was really an indirect political attack on his friend Pericles, who, as he grew more powerful, began to make enemies. Although Pericles tried hard, he could not obtain his teacher's acquittal, but he did manage to arrange for exile over death.

Greek philosophers who came before the Golden Age asked penetrating questions and conjured up creative explanations for nature's behaviour. Even though each theory sprang from observations and ideas about nature, every philosopher developed his own hypothesis about the cosmos. Socrates (469–399 BC) returned to the key question Parmenides asked: how to decide among so many creative conjectures? He tried to develop what he felt was missing: a critical *method* of thought. Astute questions and speculative answers were preliminary steps in launching an investigation. To make further progress, Socrates stressed the need for a proper form of deliberation. He proposed *deductive reasoning* as the most fertile method of analysis. One starts with a hypothesis. For example, the earth is spherical. Hypotheses can originate in a number of ways. Creative insight guided by aesthetics suggests a spherical shape for earth. Or a hypothesis may arise from an innovative attempt to organize scattered observations; for example, four fundamental elements underlie the chaotic variety of substances. From the hypothesis, one examines the consequences that follow. If the consequences prove true, the hypothesis is provisionally confirmed. But this can never be ultimate proof that the hypothesis is true. Only after sufficient confirmation of the conclusions drawn, can it be universally accepted as true. Any conclusion logically deduced from it can also be accepted as true.

But what are the universally accepted hypotheses, or "facts," as we may be tempted to call them? Socrates was not as concerned with facts as he was with logical methods of deriving reasoned consequences from facts. His chief approach was to question methods to arrive at conclusions, and so to address the validity of analysis. Socrates' emphasis on the need for a proper method of thought stimulated subsequent thinkers. They grew more critical in their approach to philosophical questions.

On the political scene, Athens continued to collect taxes from the city-states to support its powerful navy, marring the spirit of unity that brought victory and glory. Pericles began to abuse his power. As the Athenian Empire began to unravel, a ruinous war broke out between the two main protagonists, Sparta and Athens. Greece would never recover from the power struggle between the fierce contenders.

Reacting to the erosion of values amidst the ensuing chaos, Socrates channelled his intellect into pursuits of moral philosophy. He was not interested in the study of nature as his predecessors were, because such studies did not show how people ought to approach the ideals of "good" and "just." Rather, he directed his thought towards ideal character and conduct. A man of legendary ugliness, he talked in the streets to politicians and poets alike, making them examine their notions of right and wrong, teaching them the credo of Apollo: "Know thyself." It was important to define higher ideals to which leaders must aspire.

With the victory of Sparta and the downfall of their empire, Athenian statesmen indicted Socrates for corrupting the young, in particular those blamed for lost battles. They found great disfavour with his method of questioning, because his pupils began to dispute the validity of all they had been taught, and posed dangerous challenges. Instead of just accepting the "truth" from their teachers, the youth of Athens sought to debate the merit of "truth." Political leaders preferred not to be awakened. Believing that Athens would be better off with uncritical patriotism, they accused Socrates of impiety in neglecting the gods worshipped by the city. Speaking in his own defence, Socrates angered his accusers with biting irony. Finding him guilty in a court trial, prosecutors sentenced him to death. His allies offered him an opportunity to escape, but he refused. Because a recognized legal body delivered the judgement, he insisted that it must be obeyed – as a matter of higher principle. He drank the fatal hemlock. Vividly recreating the tragic incident, the Baroque artist Jacques Louis David (1748–1825) shows Socrates in a famous painting [not shown] pointing his finger high up to emphasize, with his last breath, the value of sticking to high ideals, despite the horror and agony of his colleagues who watch his suicide. Socrates' martyrdom consecrated his teachings. Today, his thought survives mainly through the writings of his most famous disciple, Plato, who immortalized the master's didactic method in a series of question-and-answer dialogues, destined to exert a powerful influence on all thinkers who followed him.

FOR GEOMETERS ONLY

Plato (427–347 BC) was just a nickname, given to the accomplished athlete by his wrestling coach because of his broad shoulders. His real name was Aristokles. Plato's creation of the Academy of Athens became the next pivotal event in the advancement of science. It was important to spread knowledge in an open society. At this first university of Western civilization, Plato aimed to educate the elite pupils of Athens to become future leaders of Greece, teaching them mathematics, philosophy, ethics, and principles that would guide them through any political climate. He used the Socratic form of teaching, a philosophic dialogue of questions and answers. Plato and his students systematically pursued rational inquiry in both humane and natural sciences, thereby originating the concept of a liberal education. New ideas and philosophies emerged. All were debated, many discarded. Like a modern university, the campus was beautiful, lined with olive trees, instead of the now familiar ivy of our universities. Buildings were open structures. Teachers became known as the "peripatetic" philosophers, or "walkers around," because of their habit of pacing to and fro while lecturing in open hallways. Isolated from city crowds, the school stood on grounds set aside in honour of a beloved Greek, named Academus, who helped take Helen of Troy back to Greece after the epic Trojan War, as legends claimed.

The Academy was a unique place of education. Teachers and students could pursue knowledge without worrying about popular opinion or external politics. Here originated the modern idea of academic freedom. Plato's Academy remained a vibrant centre of learning for nearly 1,000 years, through the rise and fall of Greek and Roman cultures, with reincarnations appearing over Europe through the Renaissance (Chapter 3). The Academy concept is now a permanent fixture in Western civilization.

Nearly 2,000 years after its founding, Renaissance artist Raphael Sanzo brought the academic tradition back to life in 1508 with a famous painting, *The School of Athens* (Figure 1.22).[17] In his homage to the greatness of classical Greece, Raphael paints Plato and his most famous student, Aristotle, entering through the open archway, engaged in healthy debate. Each character conveys a mood that reflects the philosophy he is most famous for. In the front left corner, Pythagoras scrutinizes a manuscript. He is surrounded by a group of curious students from all ages. Parmenides

FIGURE 1.22 *The School of Athens by Raphael* (1510–11), copyright Scala/Art Resource, NY, Stanza della Segnatura, Vatican Palace, Vatican State.

holds up a harmonic scale in front of Pythagoras. Facing Parmenides, a youth responds with two fingers alluding to octave doubling. Behind them, Socrates lectures a unified group, probably on the importance of questioning what they know. With the persuasive force of the dialectic method, he marks off point and counterpoint on his fingers. In the lower right corner, Ptolemy and Hipparchus discuss the spherical earth and celestial sphere, better visible on the book cover. Appropriately, Ptolemy wears a crown; he ruled astronomy for 2,000 years (Chapter 6). We will often return to the characters of Raphael's masterpiece as well as its engaging themes.

Aimed at knowledge of the eternal, geometry was a principal activity that Plato stressed at the Academy. Important mathematical works accomplished here during the 4th century BC were primarily contributions from friends and pupils of Plato. A favourite topic was the geometry of five perfect solids: the cube, tetrahedron, octahedron, icosahedron, and dodecahedron. Sensitive to symmetries in nature, and their mathematical counterparts, Pythagoras was the first to recognize a correspondence between these special solids and alluring gems: the quartz, beryl, and garnet. What is so perfect and magical about these solids? On a two-dimensional surface, such as a flat piece of paper, it is possible to draw *any number* of *regular* geometrical figures, such as an equilateral triangle, a square, a pentagon, and so on (see Figure 1.23). Each of these figures is regular because all sides and all interior angles are equal. In three-dimensional space, regular solids are those with identical faces of equilateral figures, such as squares, triangles, pentagons, and so on. For each face of a regular solid, all sides are equal and all angles are equal. An outstanding property of the perfect solids is that there are five – *and only five* – such solids.

Fascinated by their perfection and geometrical elegance, Plato believed that there existed only five regular solids for the special reason that nature has only five elements. More than a century

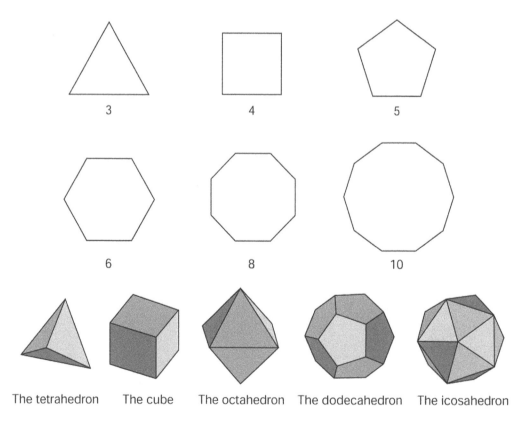

3	4	5
6	8	10

The tetrahedron   The cube   The octahedron   The dodecahedron   The icosahedron

FIGURE 1.23   Perfect geometric figures in two dimensions and three dimensions. There can be any number of two-dimensional regular figures but only five perfect solids.

later, Euclid provided a rigorous mathematical proof of why there are only five perfect solids. As we shall see, 2,000 years later, the five symmetrical solids and the perfect sphere shaped the thinking of a young Renaissance astronomer, Johannes Kepler. With perfect solids, he believed he had successfully penetrated the divine architecture of the heavens (the solar system). There are six planets because each rides a sphere that revolves in the space between the five regular solids, nested inside each other. Through symmetry in Platonic solids, Kepler thought he had penetrated a deep secret of the universe. Although absolutely wrong, it was not a wasted inspiration. His long quest for a pure geometrical scheme eventually led him to elliptical planetary orbits and mathematical laws which became the seeds of Newton's overarching theory of gravitation.

FORM VERSUS SUBSTANCE

Proposing a new approach to the study of nature, Plato split the cosmos into two distinct realms, the world of "form" and the world of "substance." The first is the domain of pure ideals emphasized by Socrates, such as justice and virtue, and perfect forms imagined by mathematicians, such as spheres, circles, and perfect solids. Forms are *idealizations* of what we sense exists in nature, but none actually exist in the real world. The second realm is the familiar, imperfect, and changing physical world, composed of shadowy copies of perfect forms of the realm of ideas. On the sand, the geometer can draw a shape resembling a square or a circle. No matter how sharp he makes the point of the stick or how steady his hand or what instruments he uses to guide construction, the figures he makes are always imperfect copies of the perfect circle or square, true subjects of a geometer's logic and proofs. Only the ideal circle has a precise and universal definition: a circle is a set of points that are equidistant from a fixed point called the centre of the circle. Even the point is an idealization. The ideal square has four exactly equal sides. Ideal geometric figures are endowed with eternal and necessary properties which only the mind can discover. Similarly, even though exact symmetry does not appear to exist in nature's substances or in natural behaviour, it can be contemplated by the mind. The realm of ideals is the deepest and eternal reality.

Now Plato went one step too far. He proclaimed that it is only the domain of universal ideas that is truly meaningful, while the familiar material world is just an illusion. To our eyes the earth appears flat, but in reality it is round. Senses lie. True reality can be grasped only by the mind. Observations are just inadequate imitations of the real essence. Echoing Parmenides, Plato claimed:[18]

> If we are ever to know anything absolutely, we must be free from the body and behold the actual realities with the eye of the soul.

And in the *Republic*, Plato exhorts:[19]

> So, if we mean to study astronomy in a way which makes proper use of the soul's intellect, we shall proceed as we do in geometry, and not waste time observing the heavens.

A sad point of view! Plato's philosophical emphasis on a "higher ideal" was hardly a singular trend. Similar movements dominated other aspects of Greek culture, well before Plato. Even with his last breath, Socrates emphasized how leaders should strive toward higher ideals of behaviour, rarely seen in real political figures. To create the perfect individual became the Greek ideal. Built as an altar on high sacred ground, the Parthenon temple in the Acropolis stood remote from common people, symbolic in its elevated status, reigning over Athens, like Zeus from Mt Olympus. The very word "Acropolis" meant "high-city." Here Greek culture focused its spiritual and ceremonial aspects, separated from the humdrum of daily life. Even earlier, while depicting the "ideal man" or the "ideal woman," Greek artists of the 6th and 5th centuries BC portrayed the essential elements of the human face, leaving out "unimportant" details, such as moles and wrinkles. The *Kouros of Anavyssos* (Figure 1.10) and the *Kritos Boy* (Figure 1.24) stress a calm strength in the athletic ideal

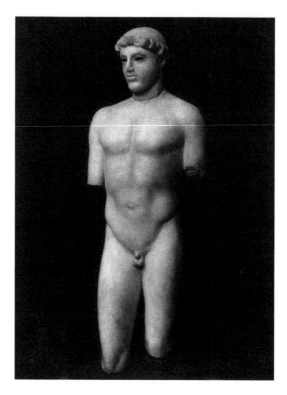

FIGURE 1.24   *The Kritos boy*, a young victorious athlete.[20]

of the human torso, without details such as bulging veins and rippling muscles normally expected in an athlete's body. Using abstract, idealized forms and ignoring "irrelevant" particulars, artists underscored universal human attributes. Only in their perfection are all things truly beautiful. With far-reaching insight, artists were emphasizing the ideal in sculptures nearly a century before Plato stressed idealism in his view of nature. These were the cultural origins of Plato's schism of the cosmos: a higher province of ideal forms and a lower domain of base, imperfect substances.

In search of knowledge and understanding about nature, Plato emphasized Socrates' deductive method as supreme. Geometry was its potent example. It showed clearly how to leap from the mundane world of everyday life to the supreme province of ideas. How wonderful that geometry could reveal so many truths from so few original propositions. Its study became essential to train the mind to recognize pure forms. The way to understand the cosmos was to think, and, through pure reason, to deduce nature's ideal laws.

Parmenides and Plato deserve credit for stressing the importance of making logical and reasoned abstractions. Their emphasis on idealizations played a key role in the development of scientific ideas. We will see how Galileo masterfully extracted ideal laws of motion by separating out the complications of the real world, such as friction and air resistance. But Plato's extreme position, that empirical investigation has little value, turned out to be a grave mistake. In *Timaeus*, Plato audaciously proclaims:[21]

> To apply an experimental test would be to show ignorance of the difference between human nature and divine.

For such arrogance, Plato has been severely chastised by modern scientists who have come to appreciate the high value of empirical investigation. Even though senses can deceive, it is a mistake to reject them outright. Nature is too vast, and its secrets are too subtle to be successfully penetrated by thought alone. Plato's bias had a lasting effect on later thinkers. Carried to the extreme, the doctrine became a great impediment to the progress of science because it discouraged scholars of the early Middle Ages from making fresh observations about nature.

One of the most famous narratives about Plato's thinking is his allegory of the cave. In a general form, Plato claimed that our views of the world are as limited as those of cave-men confined to looking at shadows on the wall projected by real objects outside the cave. From inside this cave of ignorant sensation, shadow is all we know. The "reality" we experience is but a glimmer of a grander reality. The world of our experience is only an imperfect copy of the world of eternal ideas. We should reject the limitations of the senses and rather struggle to view nature in different ways to seek the realm of ideals which will lead us to the deepest reality.

THE FIRST EMPIRICIST

Plato's most renowned student was Aristotle (384–322 BC), revered by succeeding generations. Plato nicknamed Aristotle the *nous*, meaning brain. In one story, when Plato arrived at the Academy to teach his class, he looked around and did not see Aristotle, whereupon he exclaimed:[22] "The room is empty, the *nous* is not here." It did not take long for the student to surpass the teacher. Plato himself had occasion to complain that his best pupil frequently acted like a young colt kicking the mother. After all, academic tradition encouraged open challenge and debate.

After studying at the Academy for 20 years, Aristotle struck out on his own when his master died, being passed over for the leadership of the Academy. Soon, King Philip of Macedon invited Aristotle to tutor his 13-year-old son Alexander. Aristotle set up a small private school for his charge. To prepare him for his future role as the military leader of a united Greece, he aimed to enlighten Alexander with the best tradition of Greek philosophical thought. As conqueror of vast regions of Europe and Asia, Aristotle's illustrious pupil later became his benefactor and supporter; Alexander the Great ordered fishermen and hunters throughout his vast empire to inform the sage of any unusual matters.

Being passed over again for the directorship of the Academy, Aristotle founded a rival school, naming it the *Lyceum*, because it was located near the temple to Apollo Lykaios, god of the shepherds. Speaking with a lisp, the master lectured extensively on a wide range of subjects: natural philosophy, astronomy, biology, literary criticism, politics, and ethics. By setting up a classification system, he disentangled subjects, creating distinct academic disciplines. Bringing order to a chaotic scene turned out to be Aristotle's forte.

Best remembered today for his work in the physical sciences, Aristotle was also a competent biologist, spending a great deal of time (including his honeymoon!) collecting and studying specimens of marine life at the seashore. He observed accurately and reasoned logically from what he saw:[23]

> There is a kind of ... bee that builds a cone-shaped nest of clay against a stone ... besmearing the clay with something like spittle ... Here the insects lay their eggs, and white grubs are produced wrapped in a black membrane.

His biological classification was a thorough and successful effort. For example, he divided animal species into red-blooded and bloodless, vertebrate and invertebrate, fashioning an effective language. Carrying out animal dissections, he observed blood vessels emerging from the heart to draw logical conclusions. The heart is at the centre of animal life, and the heartbeat is the primary sign of life. Despite its surprising reach and validity, his work in biology was largely ignored. Ironically, his work in the physical sciences exerted an enormous impact on centuries of thought, but has been largely disproved.

When Alexander the Great died, an anti-Macedonian trend enveloped Athens; Aristotle became one of the prime targets. Like Socrates before him, politicians indicted Aristotle on a charge of impiety, but he left the great city and saved the Athenians from repeating their travesty against philosophers. Through Aristotle's encyclopaedic writings Greek philosophy survived to reach us, albeit through a tortuous path. After his death, his books were preserved in caves near his home, and later sold to the library of Alexandria. But only about 50 of the 400 volumes he wrote have survived. Among these are titles such as: *Physics*, *Heaven and Earth*, *Meteorology*, *Generation and Corruption*, and *On Animals*.

OBSERVE THE ROUND EARTH

Aristotle initiated a breakthrough in scientific method. Contrary to Parmenides and Plato, he believed, as his biological work would attest, that creative hypotheses and pure reason are quite insufficient. To gain insight into the natural world, it is important to muster evidence:[24]

> Credit must be given to observation rather than to theories, and to theories only insofar as they are confirmed by the observed facts.

Speculation on the basis of symmetry and cosmological principles that the earth must be round is not enough. Where is the evidence? Only evidence can provide solid food for thought:[25]

> The sphericity of the earth is proved by the evidence of our senses, for otherwise lunar eclipses would not take such forms … in eclipses the dividing line is always rounded … if the eclipse is due to the interposition of the earth, the rounded line results from its spherical shape.

A careful observer, Aristotle also recognized how the curved shape of the moon during an eclipse is characteristically different from the shape of the lunar phases. The edge of the dark part of the moon is always convex. But during the progression of phases the dark part of the moon can be convex (crescent moon), concave (gibbous moon), or straight (quarter-cycle moon). Anaxagoras had provided the natural explanation that lunar eclipses occur when earth comes between the moon and its source of illumination, our sun. At this time, the shadow of earth falls upon part of the lunar surface. During a full moon, the sun is on the opposite part of the sky to the moon, so that it can fully illuminate our pearly companion (Figure 1.25). If the earth now comes in between the two heavenly bodies, it casts a curved shadow betraying its round shape (Figure 1.26). Aristotle did more than emphasize careful observation. Babylonians regularly observed eclipses and made extensive tables of eclipse cycles. But they paid no attention to the shape of the eclipsed moon. Aristotle made good sense of careful observation to reach a momentous conclusion about the earth's shape.

Raphael captures the critical distinction between the disparate approaches of Plato and Aristotle in the *School of Athens*, as the detail shows (Figure 1.27). Emphasizing the primacy of reason and higher ideal principles, Plato points up, toward the world of ideas, perhaps toward the heavens. He learned much from his master Socrates, who, even with his last breath, emphasized higher ideals.

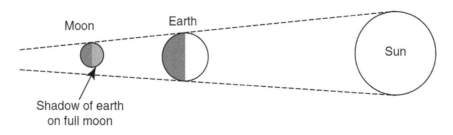

FIGURE 1.25   A lunar eclipse occurs when earth comes between the sun and the full moon.

FIGURE 1.26   During a full-moon night, the moon's shape changes gradually over the span of a few hours, and then returns to its original shape. From the earth's curved shadow on the moon's face, Aristotle could literally see the curvature of our world, direct evidence for the round earth. The usual reddish colour (not shown here) of the faintly glowing moon during totality is discussed later (Chapter 12).

FIGURE 1.27   Plato and Aristotle. Detail from the *School of Athens* by Raphael.[26]

But Aristotle gestures downward toward earth, stressing the importance of paying attention to the more mundane, but tangible, evidence that surrounds us.

Beside the shape of the moon during a lunar eclipse, Aristotle knew of other evidence for a round earth. Greek seafaring pilots had already learned from maritime experience to think of the earth as curved. They understood why the hull of a distant ship disappeared before the mast, as the ship sailed farther away from shore. No, the ships were not sinking in the ocean; they eventually returned from their trade missions. On a curved ocean, the hull would be the first to sink out of view below the horizon, and finally, the mast. Records of stargazers showed that as they travelled north, new constellations appeared on the northern horizon, while familiar ones in the south disappeared from view. The farther north you are, the more of the Northern Hemisphere stars you will be able to see on a round earth, but the horizon will block more of the Southern Hemisphere. If the earth were flat, no stars would be hidden from view. All stars in the night sky would always be visible from any location on earth, independent of the position of the traveller.

RECAPPING PIVOTAL EVENTS

Early Greek thinkers' accomplishments can be fairly cast as some of the first great events in the history of scientific thought. They tried to understand the world instead of shrouding it in mystery. They searched for reason in natural phenomena, rather than resorting to the supernatural, making the pivotal transformation from worship to science. Nature does make sense if we interpret it correctly. Some strived to find unity in diversity and extract the fundamental from the apparent variety. They were interested in identifying the substance of the universe, seeking the essential elements of its make-up. Collecting and correlating disconnected facts of Egyptian and Babylonian knowledge, they welded them into a coherent scheme, seeking harmony, order, balance, and symmetry. Some emphasized forms, patterns, and ratios found in nature. Natural phenomena must be subject to mathematical description. Believing that elegance and simplicity underlie nature's organizing principles, aesthetic ideals guided their efforts to understand nature.

Apart from the issues of form, substance, and organizing principles, early thinkers began to explore which paths lead to truthful conclusions. Logical deductions must be drawn from hypotheses and examined. They attached a high value to making idealizations. The ideal provided the unchanging standard of the best. Universals were far more important than particulars. Some preferred pure reason, while trying to avoid the confusing and conflicting information from the unreliable senses. Others stressed observation. Where is the evidence for a particular claim? Only with the culmination of two millennia of intellectual struggle can we appreciate that nature's secrets are far too subtle to unravel by limiting ourselves to a few select approaches. Creative speculation, logical reasoning, observation, and insightful interpretation are all essential components of the scientific process. And the progress of science relies on much more, as subsequent chapters will bear out.

UPDATES

SEARCH FOR SYMMETRIES

In his passion to demystify the world, Pythagoras recognized that symmetry plays a dominant role in the make-up of nature. His sharp insights about elegance, order, and symmetry have served as a guiding principle throughout the history of science. Poet Anna Wickham poignantly expresses symmetry's overwhelming appeal:[27]

> God, Thou great symmetry
> Who put a biting lust in me
> From whence my arrows spring,
> For all the frittered days

That I have spent in shapeless ways
Give me one perfect thing.

From outward appearances down to the deepest levels, symmetries govern the form and order of nature's substances. Consider the lustrous gems that fascinated Pythagoras. Each face of a beryl is a hexagon and each face of a garnet is a square because crystals build up by an orderly stacking of billiard-ball-like atoms. In perfect three-dimensional arrangements that repeat over and over, atoms stack themselves into cubic (Figure 1.28) and hexagonal crystals. Gold, silver, diamond, and garnet all form cubic crystals. The simplest stacking of atoms forms the primary building block which fills space by repetition to form an overall design. According to mathematical properties of stacking, just seven structures form the construction basis for all crystal solids to display cubic, tetragonal, hexagonal, and higher symmetries. No other regular solid constructions are possible because they leave empty spaces. As a two-dimensional analogy, a floor can be covered completely only by triangles, squares, rectangles, parallelograms, or hexagons. Tiling with other perfect figures such as circles or pentagons will leave empty spaces.

Snowflakes exhibit a perfect six-fold symmetry even though no two are the same (Figure 1.29). Every branch of a snowflake grows with the same exquisitely detailed features of the other five branches. Underlying the hexagonal symmetry is the triangular arrangement of the hydrogen and oxygen atoms in water molecules. Thus water molecules lock into orderly patterns.

Carbon atoms take on many different crystalline forms. Graphite is sheets of carbon atoms forming hexagons. Loosely bound together, the sheets can slide past each other to give good lubricating properties to graphite. When the hexagons of graphite intersperse with pentagons, the graphite sheets can curve, so that 12 pentagons and 20 hexagons bend all around to form a "sphere."

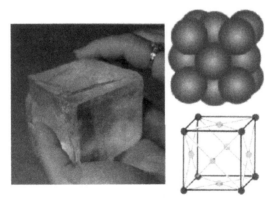

FIGURE 1.28   Stacking of atoms to form cubic crystals. (Left) salt crystal. (Right upper) face-centred cubic. (Right lower) skeleton arrangement of atoms in a face-centred cubic crystal.

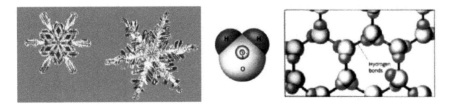

FIGURE 1.29   (Left) snowflakes exhibit six-fold symmetry due to (centre) the spatial arrangement of atoms in water molecules. (Right) when water solidifies into ice, there is a crystal structure with hexagonal symmetry.

The 1996 Nobel Prize in Chemistry went to the discovers of just such a structure which they named the Buckminsterfullerene because the structure resembles a dome created by the architect. Also called buckyballs, they make excellent lubricants because the balls can roll around between surfaces like ball-bearings. At conditions of high pressure and temperature that exist deep inside the earth carbon atoms bind together in tetrahedrons to form a cubic crystal. Such a structure makes diamond a hard substance that can abrade almost all other tough materials. The beauty of diamonds comes from the sharp angles of the crystal facets flashing rainbow colours from inside.

Through the bewildering array of different substances that exist in nature, symmetry relationships, most conveniently expressed in mathematical terms, help to unite substances into patterns of organization and order. In the words of chemist CA Coulson:[28]

> Man's sense of shape – his feeling for form – the fact that he exists in three dimensions – these must have conditioned his mind to thinking of structure, and sometimes encouraged him to dream about it … It is when symmetry interprets facts that it serves its purpose: and then it delights us because it links our study … with another world of the human spirit – the world of order, pattern, beauty and satisfaction.

Even the animate world is full of beautiful, symmetric forms, governed by mathematical patterns. With its mesmerizing spiral symmetry, the nautilus shell (Figure 1.30) exhibits mathematical proportions of the Greek golden mean. Aesthetic preferences that drove Greek architecture lurk inside the most common sea-shell. An overall perimeter around the shell is a golden rectangle like the one surrounding the Parthenon front. Separating out a perfect square leaves another golden rectangle. Repeating the process generates an infinite series of smaller squares and golden rectangles. At the core of the repeated structure lies the spiral shape of the shell's graceful arch obtained by drawing quarter circles inside each square. Minute life forms, such as radiolaria, protozoa (Figure 1.31), and viruses (Figure 1.32), exhibit spherically symmetric arrangements of amazing intricacy and beauty. Carbon atoms can join up with oxygen and hydrogen atoms to form a variety of fascinating crystal forms to become the fundamental element of life. The genetic molecule at the centre of all life is a long crystal with two counter spirals that can disassemble from each other to make copies. We discuss later the story of the discovery of the crystal structure of DNA (Chapter 2).

Through the delightful relation between musical harmony and whole number ratios, Pythagoras saw that natural behaviour can be ordered in stunningly simple ways. Casting his vision far ahead, he sought order in nature through the structure of numbers. Mathematical laws underlie apparent lawlessness. Along with mathematical patterns symmetry has always lit the way forward. These grand traditions of classical thought found surprising continuity in the exciting ideas of modern physics. As topics for the future, spatial symmetries governing the basic element, hydrogen, play a crucial role in the row and column organization of chemical elements into the Periodic Table. The lamp of symmetry and the quest for nature's harmonies continued to illuminate the way in the long

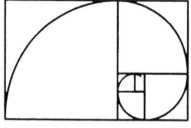

FIGURE 1.30   (Left) a nautilus shell. (Right) generating the spiral of a nautilus shell from golden rectangles.

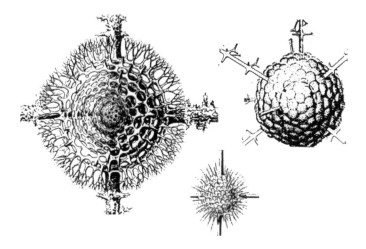

FIGURE 1.31   Radiolaria are protozoa found in lake water.[29]

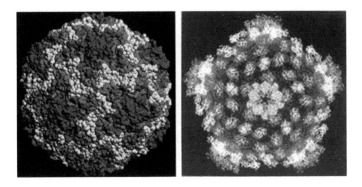

FIGURE 1.32   Molecular structures of certain viruses that exhibit spherical symmetry. Each X-ray map displays close to 50 million atoms.[30]

journey to uncover nature's irreducible elements, layer by layer, as the structure of matter unravelled from the atoms of chemical elements to electrons and nuclei, and finally to leptons and quarks, topics for Chapter 2 and Chapter 12.

FUN WITH FIBONACCI

Beauty, symmetry, and mathematical patterns frequently manifest in familiar life forms of fruits and vegetables. The seeds in the pods of a daisy or a sunflower are arranged in clockwise and anticlockwise spirals (Figure 1.33), 21 and 34 for the daisy and 55 and 89 for the sunflower. Similarly for the florets of a cauliflower. Careful examination of the outside of a pine-cone and pineapple reveals similar patterns, five spirals and eight anti-spirals for the pine-cone, eight and 13 for the pineapple. Arranging these numbers to increase sequentially forms an interesting series …5, 8, 13, 21, 34, 55, 89… Each term of the series is the sum of the previous two, the fundamental property of a famous mathematical series called the Fibonacci sequence 1, 1, 2, 3, 5, 8, 13, 21… In 1202, the Italian mathematician Fibonacci investigated the growth pattern of rabbit-pair populations that led him to the series that bears his name. Branches of a tree also follow a similar growth pattern. Fibonacci also provided a major benefit to Western civilization by importing the number system 0, 1, 2, 3 … 9 from the Arabs to replace the prevalent Roman numbering system 1, II, III, IV, V…

FIGURE 1.33   Clockwise and anti-clockwise pattern formation by seeds in a flower pod.

The Fibonacci series shows many fascinating mathematical properties which can be proved analytically. For example, the ratios of successive terms, 3/2, 8/5, 13/8, 34/21, 55/34, 89/55 which are 1.5, 1.625, 1.619, 1.6176, 1.6182…, oscillate around the converging number 1.618034… which is called the Fibonacci ratio (f), which is at first surprisingly identical to the golden mean $\left(1+\sqrt{5}\right)/2$ discovered by the Greek architects. The magical property about f is: $f = 1 + 1/f$, which leads to a quadratic equation $f^2 - f - 1 = 0$, with the positive solution $f = \left(1+\sqrt{5}\right)/2$, the same as the Greek golden mean.

DYNAMIC SYMMETRY

Symmetry principles simplify, as in the easier proof of Pythagoras' theorem obtained by replacing a general right triangle with a particular symmetric (isosceles) right triangle (Figure 1.15). When recognized in physical laws, symmetries also simplify by reducing the number of rules required to specify physical behaviour, bringing about economy of description. Symmetry laws impose order in the behaviour of physical systems by permitting only certain types of behaviour and forbidding others.

Basic laws that govern the universe reflect the principles of invariance symmetry. Physical behaviour remains invariant under changes, such as translation, rotation, or reflection. Consider reflection symmetry, as in a mirror. If we perform an experiment, and look at it either directly or in a mirror, the physical laws governing the two experiments are the same, even though any left–right asymmetry in the real-world will display as a right–left asymmetry in the mirror. For example, a left hand will look like a right hand. Even though the mirror image world may look different, the physical laws operative there are the same as in the real world. The crucial aspect of mirror symmetry is that mirror-image processes can also occur in the real world. The appearance of a clock-face is particularly intriguing. We are accustomed to associating time flow with the clockwise movement of the hands. But the mirror image of a clock in Figure 1.34 appears to run anti-clockwise, i.e., backward in time. Without number labels to help, the hour-hand pattern of 11 o'clock in the real world appears as 1 o'clock in the mirror. Both 11 o'clock and 1 o'clock are physically valid patterns.

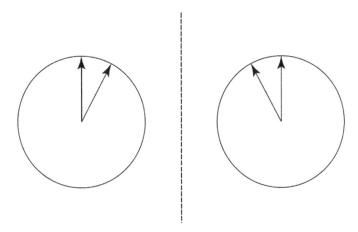

FIGURE 1.34   Mirror reflection reverses the sign of one spatial coordinate to make clockwise motion appear as anti-clockwise. But the mirror image case is possible in the real world.

The anti-clockwise movement of a clock mechanism does not violate any physical laws. A rearrangement of the gears inside a real clock can make the hands run in the anticlockwise direction. When the open right hand is held perpendicular to the mirror, the image looks like the open left hand. As another example, both L-amino acids and R-amino acids exist, although animals are made predominantly of L-hand amino acids, a pure accident of biological evolution. Digestive enzymes evolved to act on right-handed sugar molecules so that left-handed sugar passes right through the human body. Left-handed glucose tastes just like right-handed glucose, but will not be absorbed, a panacea for diabetics. However left-handed glucose is not a usual remedy because it is very expensive to make commercially, more so than gold!

Another important feature of mirror symmetry is that reflection reverses the sign of one spatial coordinate. If you walk north toward a mirror, your mirror image will move south toward you. (But east–west and up–down motion remains unaltered through the looking-glass.) The mirror image is space reversal in one dimension. As a result, mirror reflection also reverses velocities in that direction, i.e., toward the mirror plane. Notable exceptions to mirror symmetry encountered in modern physics occur in interactions involving the weak nuclear force, although the strong, electromagnetic, and gravitational forces all respect mirror symmetry. Important examples of symmetry violations will come up in future chapters.

Laws of physics are also symmetric with respect to time reversal. Of course, time never flows backward as far as we know, but thinking about time flowing backward is a helpful idealization to understand symmetries in motion. A video tape of a bouncing ball will look the same when played in reverse, provided the motion of the ball is ideal and not slowed down by air resistance or landing on the floor. But modern physics has fascinating cases where the weak interaction violates time-reversal symmetry, which prompted some modern physicists to jest that the weak interaction must be God's mistake in designing nature, or that God was slightly inebriated when inventing the weak interaction!

Modern treatments of physical laws show more general symmetries than time-reversal and mirror symmetries. Laws of physics do not depend upon a choice location in space. This is called translation symmetry. There is no special point in the universe that can be considered as the absolute reference point. Physical laws also do not change with respect to directional orientation, called rotation symmetry. When describing the behaviour of physical systems, the origin of time is also totally arbitrary. One moment is as good as the next to serve as the starting point. The same laws that applied in the past, apply now, and will apply in the future. From such general symmetry principles, physicists expect, and observations confirm, that all the fundamental laws of physics are symmetric with respect to time translation, space translation, and rotation through any angle. It is equivalent to the statement that the universe is homogeneous and isotropic in time and space.

Invariance symmetries lead to general *conservation laws* which are valuable in understanding physical interactions as well as in greatly simplifying descriptions of nature. In this context, "conservation" does not mean "avoiding waste," as in the common usage of the phrase for energy conservation. The principle that laws of physics are invariant under time translation leads to the *Law of Conservation of Energy*. Conservation here means that the total energy in a system remains unchanged despite any interactions that take place between different components of a system that change energy from one form to another. For example, electrical energy stored in a car battery converts to energy of motion of a starter motor. Albert Einstein (1879–1955) generalized the law to the *Conservation of Mass and Energy*, now a golden rule of physics, another topic for the future (Chapter 4 updates).

Invariance under spatial translation leads to the *Conservation of Momentum*. That nature looks the same in every direction is another symmetry principle. It gives rise to the *Conservation of Angular Momentum*. Because of their deep origins in symmetry, conservation principles hold true even when physical situations are not ideal, as for example when oscillations of a bouncing tennis ball attenuate due to interactions with the floor. But to make a proper account of the energy in such cases, it is also necessary to include energy lost to moving atoms of the floor and the ball.

It has become the "holy grail" of particle physics and cosmology to find the deepest symmetry at the very origin of all things: all forces, all particles, all space, and all time. Our final chapter will address the status of this on-going quest. A modern physicist, Steven Weinberg, echoes Plato, by expressing the idealizations of symmetry and unity, when he says:[31]

> The task of the physicist is to see through the appearances down to the underlying, very simple, symmetric reality.

Einstein's Proof of Pythagoras' Theorem[32]

We will see the full measure of Albert Einstein's intellect when we discuss his special relativity in Chapter 4 and his general relativity in Chapter 11. Einstein recalls that when he was 12 years old he thought very hard about the famous Pythagorean theorem, and came up with a breathtakingly simple proof using similar triangles (Figure 1.35). One could say it was Einstein's first masterpiece, and it is particularly germane for this chapter because it is based on Thales' similar triangles despite the use of some algebra that advanced over centuries.

Einstein Proves Atomic Motion

Even though Democritus proposed atoms dancing in the void more than 2,000 years ago, the first real proof for the existence of atoms in motion came from Einstein with his explanation for Brownian motion. In 1828, English Botanist Robert Brown was intrigued by the incessant random motion of microscopic bits of pollen particles suspended in a fluid he was observing under a microscope. At first, he wondered if the jerky motion arose from the pollen's origin from living plants. Perhaps the dancing pollen grains possessed a residue of some vital living spirit. He decided to try particles from other sources. He scraped fine fragments of petrified wood that had been dead for a long time. Again, the particles moved about frenetically in the fluid. Surely particles that were dead for such a long time should have no residual vitality that could make them move about on their own. Brown proceeded systematically to inanimate flecks, using shards from rocks, then from glass, volcanic ash, and even meteorites, presumably from outer space. He scratched powder from a piece of the Sphinx, unarguably lifeless. To his surprise everything jiggled about in the fluid. He never came up with a satisfying answer.

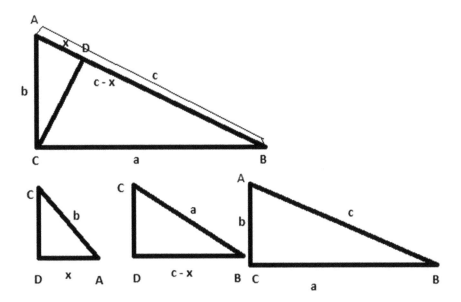

FIGURE 1.35 The aim is to prove $a^2 + b^2 = c^2$. Draw a perpendicular to side AB from vertex, C to divide the side AB into two parts with lengths: x and $c - x$. Re-draw below the three similar right triangles and label the side lengths. From the first and the third triangles similarity gives $b/x = c/b$, which can be re-written as $b^2 = cx$. From the second and third triangles we find $a/(c - x) = c/a$, which can be written as $a^2 = c^2 - cx$. Substituting b^2 for cx from the first equation into the second equation we get $a^2 = c^2 - b^2$, the theorem we are after: $a^2 + b^2 = c^2$.

Several investigators after Brown independently speculated that the random motion of the micron-size particles must arise from the many collisions of molecules of the fluid with the particles. But how can molecules of water, vastly smaller than pollen grains, cause such unpredictable staggering movement, large enough to be visible to the eye under an ordinary microscope? Clearly molecules of liquid are far too small to cause any observable motion of a large object with single impacts.

Besides, the existence of atoms and molecules was not universally accepted, even as late as 1905. Scientists belonged to one of two camps: "atomists" who accepted the existence of atoms and molecules from Democritus and Dalton (Chapter 2), and their opponents the "positivists" who claimed that knowledge should be confined to pure experience, blithely ignoring the admonitions of Plato and Parmenides. Since atoms could not be observed, or their size measured, the atomic hypothesis was considered a purely fictitious concept. Nevertheless, the idea was enormously useful for deriving many experimentally observable results about pressure and temperature in the field of heat.

Einstein proved the positivists wrong by providing a clear, quantitative explanation of Brownian motion with verifiable predictions. Brownian motion provided unequivocal manifestation of the existence of atoms. A 10-micron pollen particle is continuously bombarded by a vast number of molecules of fluid on all sides, but the frequent impacts on its different sides *vary* by large numbers to make the corresponding jitter large enough to be visible under a microscope. The particle is hit more on one side than another, leading to the seemingly random nature of the motion.

Einstein calculated how far and how fast the pollen particle should move. He showed that the mean of the square of the displacements of the particle depends on the size of the particle, the viscosity of the fluid, and the time over which the particle diffuses, along with Avogadro's number, the famous constant that gives the number of molecules in a given mass of liquid (Chapter 2). Thus Einstein could predict that the root-mean-square displacement of a 1-micron Brownian particle is of the order of a few microns when observed over a period of one minute. The displacement of a

Brownian particle is proportional to the square root of elapsed time, and not directly proportional to time as one may naively expect.

Experiments confirmed the predictions. Thus Einstein established the existence of atoms from the molecular mechanism of Brownian motion. His analysis also provided a way to determine the size of a molecule from observable and measurable quantities in Brownian motion. For the first time, a measurable quantity (diffusion lengths of pollen particles) probed the atomic realm.

2 Matter in Motion – An Elementary Quest: Beyond Reason and Observation

Where we cannot use the compass of mathematics or the torch of experience … it is certain we cannot take a single step forward.

François Voltaire[1]

BRIEF OVERVIEW

Motion of objects on earth and stars in the heavens were among the first questions to engage early thinkers. Aristotle reasoned that the nature of motion is intimately connected with the nature of the objects that move. Believing in an underlying unity of natural phenomena, he coupled the understanding of motion to the nature of substance. This fundamental connection led him to the idea of *natural motions* for earthly and heavenly bodies, and a scheme to classify and order the apparently inexhaustible complexity of matter and motion. Order was immensely reassuring. If he could find order in nature, then the behaviour of nature was predictable. Determinism was inherently comforting.

Early Greek views on the essential substances of nature and natural motion changed drastically as physics unravelled matter layer-by-layer, and penetrated the fundamental laws of motion. As it turned out, contrary to Aristotle's thinking, elements composing earthly substances are identical to elements composing planets and stars, a surprising unity between heaven and earth. Laws governing motion on earth are also the same as the laws of motion through the entire universe.

To fathom the modern view of the cosmic connections, it is helpful follow the quest for the elements through the many stages of fundamental reconstruction, from the elements of antiquity all the way to modern quarks and leptons. How the universe evolved from quarks and leptons into stars, galaxies, and planets will become the ultimate synthesis between the world of the very small and the universe on the largest physical scale. The synthesis of the laws of motion on earth and in the heavens will be the climactic conclusion for the book (Chapter 13).

As to the fate of Aristotle's natural motion, one key aspect did survive – the natural force of gravitation. Gravity is our main topic throughout the book, and comes to its pinnacle with the synthesis under Isaac Newton and Albert Einstein in Chapter 11.

Starting with Thales, Pythagoras, and Aristotle, the desire to find unity, order, and symmetry prevailed all through the voyage to the present understanding of the cosmos and its constituents. Beginning with Pythagoras, who taught that the regularities of nature are intrinsically mathematical, numerical patterns played a decisive role in comprehending order and structure at every stage of the elementary quest.

THE WALKING PHILOSOPHER'S NATURAL MOTION

Pythagoras deserves applause for his bold and creative spherical earth hypothesis, inspired by aesthetic judgement. Aristotle's careful observation of the earth's shadow on the face of the moon must also be celebrated, together with his creative and insightful interpretation. Creativity is crucial to the scientific process, both for the imagining of hypotheses as well as for judicious interpretation of observations. Contemplation after careful observation led Aristotle to validate the daring, round-earth theory. In fact, the word "theory" comes from *theoria*, meaning "to contemplate."

How the Greeks deduced the earth's sphericity is a brilliant success story of interplay between reason, observation, and insightful interpretation, all necessary for understanding nature. Yet, even together, these methods remained insufficient for the continued progress of science. They also led to some remarkable failures. For example, observation and contemplation led Aristotle to an interesting hypothesis about the organization of the cosmos, and to an intimately coupled theory about motion. Both turned out to be dead wrong. But science was to realize this only after new approaches to understanding nature germinated and matured. An examination of Aristotle's mistakes helps one to appreciate why some of the later scientists fell into similar traps. And even today, science is not completely free from such pitfalls.

Aristotle came to some remarkable conclusions about the substances composing heaven and earth, as well as about their motions. Notice, he might have said, how from day to day and from year to year the sun never changes its appearance. Year after year, it follows the exact same paths through the sky (Chapter 5). The phases of the moon repeat with certainty, month after month. The stars always stay in the same positions relative to each other, fixed in their constellations. Every night they rise from the exact same place on the horizon. On the other hand, everything on earth is in constant change. A beautiful piece of shiny sword-iron will grow dull with time and eventually become ugly with rust. Solid-white sparkling snow will melt into plain, clear, liquid water. Constant change, generation, and decay imbibe the character of matter on earth. Animals are born; plants grow out of the soil. And all life forms eventually wither, die, and decay. Change and decay seem to obviously distinguish the substance of earth from the substance of the heavens.

Now observe motion. Rocks and branches fall *down*, fire and air move *up*, water *spreads* out over the earth's surface. But all the heavenly bodies – the sun, the moon, the planets, and the stars – move only in *circles* forever and ever across the sky. Note how all motion on earth eventually comes to a stop. A rolling cart comes to a stop unless pulled by a horse. Ships need the thrust of the wind. Movement on earth decays, like plants and animals. But heavenly motion goes on forever. Why do earthly substances and heavenly bodies have such different features and such disparate motion?

Advancing from careful observation to insightful interpretation, Aristotle attempted to organize the evidence. Diversity of motion arises from innate properties of elements which compose substances. Four familiar elements, earth, water, air, and fire, compose objects on earth. In sharp contrast, a different element – ether – comprises the heavenly bodies.

Casting his vision further, Aristotle progressed to theory. The cosmos can be ordered in a stunningly simple way. To each of the four earthly elements he assigned a *natural place*, and so a natural order. Organization was Aristotle's strong suit. The successful categorization of biological species and classification of academic disciplines testify to his strength. By imposing a hierarchy among the elements, he sought to bring order into an imperfect and disordered cosmos. The first major challenge for any of the sciences is to bring some semblance of order into apparent chaos – a step known as taxonomy. Swedish botanist Carl Linnaeus created the classification system for biology to divide the animal kingdom into phyla, such as anthropoda, brachiopoda, chordata, and so on. Darwin classified a large variety of finches he studied on Galapagos Island to ultimately hit upon the idea of evolution. The 19th-century chemist, Dmitry Mendeleyev, ordered the chemical elements into the familiar Periodic Table. Guided by symmetry ideas from his predecessors, Aristotle envisioned a spherical architecture. The natural place for the element earth is at the centre of our spherical world, at the centre of the spherical universe. Then comes water. Its natural place is the spherical surface of the world. Above the world is the spherical shell of air, and above that is another shell of fire (Figure 2.1). Above the fiery shell come the seven heavenly bodies, including the moon, the sun, and five known planets. Finally, in the outermost region of the heavens comes the star-studded celestial sphere.

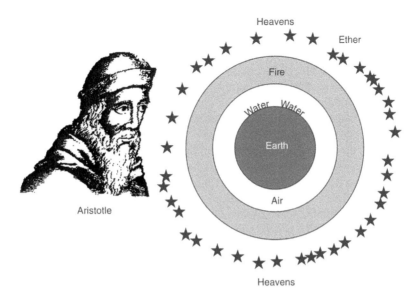

FIGURE 2.1   Aristotle and his theory of the natural place for each of the five elements.

Bringing order among the elements of the cosmos, Aristotle shared a guiding principle with art-ists and architects who designed orderly temples to decorate every Greek isle. Order and symmetry were dominant components of Greek culture. Aristotle believed that the structure of the universe exhibits the reasonableness, orderliness, and elegance underlying the symmetry of recurring spheri-cal shells. Echoes of Pythagoras' wisdom rang through Aristotle's structure. Indeed, the tenor's strain continues to resound all the way to our time. In *Troilus and Cressida*, Shakespeare, the immortal bard of the Renaissance, asserts:[2]

> Take but degree away, untune that string,
> And hark, what discord follows. Each thing meets

Aristotle's scheme divided the cosmos into two distinct realms, heaven and earth. The familiar elements – earth, air, water, and fire – have their natural places below the moon, in the sub-lunar, terrestrial region. The heavenly bodies, in the superlunary, celestial region, are composed of the fifth element, ether. Pure and unchangeable like the stars, ether is the singular *eternal* element. The origin and derivatives of the word *ether* are quite enlightening. Ether comes from the Greek word meaning "burn" or "glow." Our language still reflects Aristotle's idea of ether as heavenly when we refer to lofty feelings as "ethereal." Similarly, we call the purest essence of something the "quint-essential," which comes from the Latin word, "quintus," for fifth, as in fifth element. (In a related context, 19th-century physicists adopted the same word to name the space-pervading medium that transports light waves, just as air carries sound waves.) Aristotle and his followers believed that in an ideal world, such as the one that Plato envisioned, the "ethical" purity of the heavens would ultimately penetrate to the earth and reach humankind. No doubt Plato's crucial separation between higher ideal forms and base actualities influenced Aristotle's schism of the cosmos into distinct regions of heaven and earth.

Bifurcating the universe, Aristotle made it eminently reasonable that the properties of heavenly and earthly bodies appear starkly different. Made of the eternal element, heavenly bodies are natu-rally incorruptible. Of the two regions, only the exalted one is stable and permanent; the lowly earth is rife with change and decay.

With the hierarchical model of a natural place for each of the elements, Aristotle catapulted to a theoretical understanding for all motion. In the heavens, ether has its own natural movement, *circular motion*. Without beginning or end, circular motion is eternal motion without change, quite distinct from the motion of sublunary elements. In our imperfect world, composed of four base elements, natural motion is *down* for objects composed mostly of earth or water, and *up* for objects composed mostly of fire or air.

Made of earth, a stone seeks its natural place, moving downward to the centre of the universe. When thrown up, a stone's motion decays and comes to a stop so it can pursue its most *natural* motion down to the centre of earth. Thus the stone displays its *gravitas*, the sensory property of heaviness. A stone with more earth feels heavier; it has more gravity. Falling bodies speed up as they approach earth. The closer a body approaches its true place at the centre of the earth, the more eagerly it travels to its "home." Aristotle's idea of gravity as the natural tendency for rocks and heavy elements to seek the centre of earth offered a natural relief for the distress over the spherical shape of the earth (Figure 2.2). Heavy objects on the other half of the earth need not descend into the abyss below. Rather, by symmetry, every object on the earth seeks the centre of earth. Already at its true centre, the centre of the universe, earth needs no support. Symmetry commands that the spherical earth remain intact. Aristotle's theory provided a fundamental reason why the earth must be a sphere without need for support. All parts of the earth, made of the element earth, try to reach their natural place, the centre, which is also the centre of the universe. Aristotle was the first to propose an intimate connection between the shape of the world and the notion of gravity. This first coupling between gravity, motion, and geometry was to evolve in meaning all through the development of physics to a first synthesis under Newton's universal gravitation, and the ultimate synthesis under Einstein's unification of space, time, and matter.

But sensory observations deceived Aristotle, as Parmenides had warned. Since light objects such as leaves and feathers fall slowly, and heavy rocks fall rapidly to earth, Aristotle reasoned that gravity, the innate property of heaviness, manifests an object's natural eagerness to return to its proper place. A big stone, which contains more earth, must fall faster than a small stone.

How do other elements behave? Because water spreads out over the ground when spilled, water's natural place is a spherical shell on the surface of the earth: thus the oceans. It rains because water seeks to return to the surface of the world from the air, which is not its natural place. Water trapped

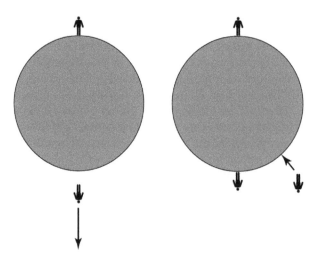

FIGURE 2.2   With the idea that all objects made of the element earth are attracted to the centre of the earth, Aristotle answered a serious difficulty concerning earth's spherical shape. It was the first step to earth's gravity as an attractive force.

beneath the earth rises to the surface in the form of springs and geysers. Fire and air tend to rise to their natural places, which are spherical shells above the earth. They show their inherent "levity" versus the "gravity" of a stone. Air bubbles rise up through water because air seeks to return to its natural shell, which we still refer to as "atmosphere." The normally invisible shell of fire sometimes becomes visible in the form of lightning. On striking a tree, lightning will set it on fire. Thus Aristotle provided a first rational explanation for lightning.

With the hierarchical model of a natural place for basic elements, Aristotle devised an overarching theory to encompass all types of motion. *An object moves in the way that is inherent to its nature. This is natural motion.* Through observation and theory Aristotle scored important successes, providing captivating explanations for many everyday observations. The diversity of motion surrounding us reflects the diversity of the world's material composition. When elements are mixed up and out of their natural places, natural motion restores order. A volcano erupts because earth and fire are mixed up under the mountain. An eruption throws burning lava up towards the sphere of fire. When lava rocks lose their fire, stones come crashing down to earth. Steam rises because of fire introduced into water. Hidden fire explains why steam burns. When cooled, steam gives up its fire. Condensed moisture, which is mostly cold water, moves home to the surface of earth. Aristotle was following Socrates' deductive method, deriving consequences from his grand organization hypothesis. Finding so many conclusions valid, it was tempting to accept the hypothesis as universally true, as succeeding generations came to believe.

VIOLENT MOTION

Aristotle's division of the cosmos into distinct realms of heaven and earth incorporated another crucial principle. Not only is the substance of heaven different from earth but also the laws governing terrestrial motion and celestial motion are quite distinct. The heavens are constant and unchanging because the stuff of the heavens is permanently in its natural place. Circular motion is natural for heaven; no agent is necessary to maintain eternal circular motion.

Down here on earth natural motion is down or up. Otherwise, everything will remain forever at rest unless moved by an external agent. Again, Aristotle relied on common-place observations. Objects on earth, once set in horizontal motion, soon come to a stop. For a cart to keep moving horizontally, a horse must pull it. Wind must propel ships. So it is also for circular motion on earth. An athlete must continually pull on a string to whirl a stone in a circle. Cut the string, and the stone sails for a short time in a straight path before it naturally falls to earth. On earth, one never sees objects moving naturally in horizontal straight lines or circles. Other than down or up, any motion on earth is *unnatural*, and will therefore deteriorate. Only *force* can maintain this type of motion, doing violence to natural behaviour. Aristotle classified it as *violent motion*. To move is to *be* moved.

Aristotle yearned to capture the whole, to encompass all of heaven and earth: the nature of substances, their arrangement, and their motion. It was an imposing philosophical framework. So persuasive was the emerging paradigm that whenever there was an obvious deviation, he and his followers used their creativity to devise rational explanations conforming to the overall model. If the moon is a heavenly body, made of pure ether, what about the obvious dark patches on its face? Aristotelians proposed that the moon could have some stains since it was positioned at the edge of the terrestrial–celestial divide. But the sun and stars must be perfect. If heavens are regular and unalterable, what is the cause of the sudden and erratic appearance of comets? These must be atmospheric phenomena. Where the spheres of fire and air meet, sometimes fire ignites dry pockets of air to form hairy-tailed comets. What about divers who fall toward the centre of the earth but then rise upward to the ocean's surface? Do humans move up or down? A reasonable Aristotelian interpretation for the observed sequence of motions might be: humans are made mostly of water, partly of earth, and partly of air. Like rain, we fall to the ground when we jump off a cliff. The forced motion carries us underwater. But eventually we come to a stop, then rise to float on the surface of the ocean. When an arrow leaves the bow it continues to fly freely with unabated horizontal motion. Why does it not immediately fall to the ground? Aristotle proposed that an arrow leaves behind a vacuum when it cuts through air. As air rushes in to fill that vacuum, it propels the arrow. "Nature abhors a vacuum" was another of Aristotle's tenets, and one he used effectively to dismiss Democritus' wild notions about atoms moving forever in a void. Aristotle and his followers worked hard to shore up his constructions.

Aristotle was a shrewd observer, but he relied too strongly on direct observations from the senses, ignoring the warnings of his teacher who advised that one must look beyond the senses, to search for the underlying ideal. The best student had missed a most important lesson. And his observations were purely passive. His reasoning was so cogent, and his comprehensive picture was so appealing, that he and his followers did not see any reason to put explanations to a test, such as to drop a large rock and a small stone from the same height. Do these actually fall at different rates? They do not! Aristotelians never progressed to the modern idea of making a controlled experiment to test theories or to challenge accepted understanding. It would take inordinate courage to defy the paradigm once it became firmly established in the Western intellect. The idea of conducting an experiment to check theory took another 2,000 years to fully germinate after many transformations of the Western intellect (to be covered in Chapter 3). With the clear hindsight of two millennia, we realize, perhaps too easily, that Aristotle made serious errors. Both his models and their conclusions suffered grievous flaws. His theories of matter and motion changed drastically over the subsequent development of physics. So did the associated cosmology.

Aristotle's experience shows that the development of science can be full of mistakes and errors, as well as triumphs. Being human creations, scientific theories evolve with imaginative perception. They can be fallible in their grasp of reality unless they are continually subject to challenge and test. Despite his notorious failure, we should not dismiss Aristotle lightly. Even though his models were abandoned and his laws totally supplanted, he made immensely valuable contributions through observations, reasoning, and modelling. He set up a provisional philosophical framework using methods and approaches that should be classified among the first breakthroughs in the development of science.

The fate of Greek ideas provides an excellent example of how science progresses. Greek thinkers started by making a model based on theoretical principles or on observations. For example, Pythagoras chose a spherical earth model guided by aesthetic prejudice. Aristotle made a model based on observations to put elements in order, and into their natural places. It was up to those who followed the trailblazers to challenge the models. Thinkers who came after Thales challenged the master. A single element could never suffice to explain the rich diversity of nature's substances. Aristotle challenged Pythagoras. Where was the evidence that the earth was a sphere? But serious challenges to Aristotle's shell model and the connected ideas of motion took 2,000 years due to intervening cultural and political transformations – topics for Chapter 3. Even today, if we cease to test and challenge models, scientific progress will also cease.

THE EGYPTIAN MELTING POT

As Greeks established colonies to the north in Italy and to the west as far as Marseilles, their culture spread widely and rapidly. Starting with a mission to avenge the century-old Persian invaders who razed Greek temples to the ground, Alexander the Great spread Greek conquests into Asia as far as India, and into Europe as far north as the Danube. In Egypt, he founded the city of Alexandria in 332 BC at a choice location with a seaport large enough to accept his mammoth fleet. With two natural harbours, one facing east and one facing west, ships could dock at this hospitable port no matter which way the winds blew. From here, heavy trade in goods criss-crossed the vast new empire.

Alexander died prematurely at age 33 without naming an heir. His generals grabbed parts of the empire, diffusing the remains of the great power. Ptolemy seized Egypt to establish his capital at Alexandria, where he founded a dynasty that lasted 250 years. Many of his followers adopted the name Ptolemy. From the bustling city of Alexandria, Greek civilization began to rekindle intellectual activity in Egypt. Greeks, Macedonians, Jews, Arabs, Egyptians, and Hindus converged in the dynamic city, bringing with them new knowledge and culture from remote points. Alexandria turned into a cosmopolitan city, a melting pot of Mediterranean culture. Although situated in Egypt, on the site of Egyptian-temple culture, Greek became the language of learned men. Under the powerful influence of Greek culture, Alexandria rapidly transformed into the intellectual capital of the world.

Even divided up, portions of Alexander's empire formed large political units when compared with any independent Greek city-state. As a result, the kings of Egypt commanded substantial financial resources. Alexander left the rulers another important legacy – a drive to encourage cultural activities. It was an attitude which Aristotle had instilled in his young pupil. Appropriately motivated and enormously rich from the spoils of their eastern conquests, the Ptolemies supported huge projects, such as the Museum, a temple to the muses, goddesses of the arts and sciences. The Museum rapidly developed into a research institute where scholars advanced learning in all subjects. Ptolemy-I (ca. 300 BC) also established plans for a major library on advice from the governor of Athens, another one of Aristotle's pupils, who later became the first director of the new library, built by Ptolemy-II.

As true patrons of the arts and sciences, the Ptolemies endowed the new library with all of the world's existing knowledge in a prodigious collection of 400,000 papyri, the ancient equivalent of books. Each papyrus was a foot wide, and unrolled to about 20 feet. It contained the equivalent of about 60 pages of a modern text book. The Greeks called a strip of papyrus "biblion," from which the word "Bible" and other words pertaining to books, such as "bibliography," are derived.

Alexandrian authorities required ships anchoring in the harbour of the busy port city to hand over their books so that the library could make copies for the great collection. Under the shadow of Athens' glory, the intellectual unification of the world at the Great Library of Alexandria surpassed Alexander's ambition to dominate the entire world with the military might of Greece. The repertoire of volumes spanned a wealth of subjects: philosophy, natural history, drama, poetry, cookbooks, and magic. The workmanship of the library staff was so good that occasionally they kept the original and sent back the copy. On one occasion, Ptolemy-II brought over 72 Jewish scholars to translate the Jewish Bible. The Alexandrian version became the Bible of the early Christian Church, and eventually the Old Testament. It is also referred to as the *Septuagint* (from Latin for 70).

THE ATHENIAN BRAIN DRAIN

Both cosmopolitan and complex, Alexandria remained a centre of vigorous intellectual activity for more than 500 years. Euclid, Erathosthenes, and Archimedes carried out monumental advances in geometry, geography, and physics. The Museum was the seat of astronomical research conducted by Hipparchus, Aristarchus, and Ptolemy (not one of the kings).

A Greek mathematician educated at Plato's Academy, Euclid was one of the earliest scholars attracted to the Museum. His move from Athens to Alexandria symbolized the shift in the seat of scientific pursuit and the brain drain that followed as the once-great Athenian culture slowly withered at the roots. Under the spell of true Platonic schooling, Euclid contemplated pure geometric forms without reference to tangible aspects of the real world. Synthesizing into one great work the geometry developed since Thales, he derived volumes of theorems from just a few basic postulates. For example, starting from the axiom that the distance between parallel lines remains constant, he proved that the sum of angles in a triangle equals 180 degrees. Symmetry principles simplify the derivation of some of his powerful theorems (Figure 1.16). In the lower right corner of *The School of Athens*, Raphael recreates the spirit of the classical geometer (Figure 1.22). Surrounded by eager students, Euclid bends over compass and board. One student has just grasped the construction and bends over to explain it to his kneeling companion who still appears mystified. A third looks up with surprise at an excited youth who nearly loses his balance in epiphany.

A student once asked Euclid to explain the usefulness of the knowledge he was acquiring. Without hesitation, Euclid rewarded the student with a coin to satisfy his utilitarian urge, and then dismissed him. For Euclid, knowledge and thought existed purely for the good of the soul, not so much for practical use. To cap that story, when one of the Ptolemaic kings asked Euclid if there was a faster way to learn the fascinating subject, Euclid spurned the impatient monarch with the remark:[3] "There is no royal road to geometry." Euclid called his *magnum opus The Elements of Geometry.* The everlasting classic is the most successful mathematics book of all time. Since the first printed edition produced in Venice in 1482, more than 1,000 editions have been published in a variety of languages.

TO CATCH A THIEF

Alexandrian culture fermented with new ideas. Coming a century after Aristotle and Plato, Archimedes (287–212 BC) was the trumpeter of a new era. Not satisfied with abstract principles and logical proofs, he wanted to know details. The son of an astronomer, he became an accomplished mathematician who calculated the most accurate value of π, the universal irrational number that binds together the diameter and the circumference of any circle (Figure 2.3). To him we owe the formulae for the volume and surface area of a sphere in terms of the ubiquitous π (Figure 2.4). All

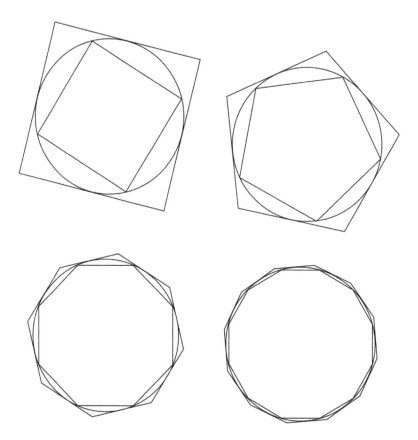

FIGURE 2.3   Archimedes determined an accurate value of π by placing its value between calculable perimeters of many-sided polygons drawn inside and around a circle. His approximations improved as he progressed from squares to pentagons to eight sides to 12 sides, capturing the circle ever more accurately with increasing numbers of sides.

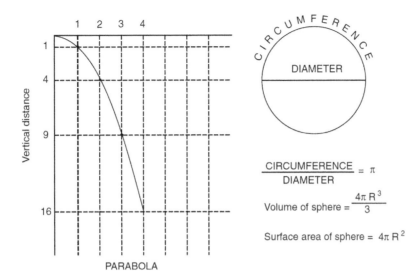

FIGURE 2.4   Some of the mathematical contributions of Archimedes that later proved essential to the development of physics. The parabola, the value of π, and the volume and area of sphere.

through mathematics and physics we encounter the symbol π, reminding us how the geometry of the perfect circle often underlies our attempts to idealize the world.

Archimedes was the first to depart from the pure geometry of straight lines and circles to introduce the parabola (Figure 2.4), a mathematical work to play a pivotal role in physics centuries later. Along such a figure, the vertical distance increases with the square of the horizontal distance. It was the same relationship that Galileo would later recognize in the path of water fountains (Chapter 4).

Unlike Plato and Euclid, Archimedes was not above tackling everyday problems. Rejecting Platonic blinkers, he was prepared to broaden the Athenian Greek's narrow focus on abstract idealism by paying close attention to detailed features of the real world. Becoming intimate with the complexity and diversity of natural phenomena is essential if one hopes to gain any real understanding. While trying to solve a practical problem, Archimedes made a marvellous discovery – the force of buoyancy – a celebrated story about Archimedes and his bath.

Archimedes' home was Syracuse (in present-day Sicily), where King Hiero commanded the court jeweller to make a gold crown. When the crown arrived, there was a nasty rumour that the artisan had substituted cheap silver for some of the gold. But the crown weighed exactly the same amount as the quantity of gold allotted to the craftsman. Not wishing to diminish the glory of the crown nor to be cheated, Hiero requested his relative, the wise Archimedes, to investigate.

Archimedes struggled with the challenge, thinking about it day and night. One day, while getting into a tub, he noticed what most people would consider a rather trivial detail. The more his body sank, the more water he displaced. Instantly, the solution to the crown problem became clear. He jumped out of the tub and rushed naked through the streets of Syracuse, screaming "*Eureka, Eureka*" ("I've got it"). Ever since that famous incident, sudden leaps of insight are called "eureka moments."

Obtaining chunks of pure gold and pure silver, each equal to the weight of the crown, he submerged them individually in a flask filled with water. The silver displaced a larger volume of water than pure gold, and so did the crown. The crown was in fact adulterated. Most importantly, Archimedes launched another major advance in scientific method: experimentation. To go beyond observations, he did not just wait for nature to do something, and then passively observe it to discover correlations, as Aristotle had done with the lunar eclipse. He initiated the practice of manipulating the objects under investigation, changing situations in a systematic way, and then comparing the results. He used a control, gold versus silver, to compare the results with deliberately introduced changes.

The scientific result from close observation, measurement, and experimentation is that the density of silver is less than the density of gold. Put another way, for the same *weight* of metal, the piece which displaces a higher *volume* of water has the lower density. The concepts of weight and volume were quite familiar to the Egyptians and Greeks for centuries through the needs of trade. Now Archimedes progressed to the key idea of *density*: *Density = weight/volume.*

What Archimedes determined next turned out to be far more valuable than Hiero's crown. He advanced to the idea of *buoyant force.* A crown weighs less when immersed in water than it weighs in air (Figure 2.5). The reason is that water exerts an upward force on the crown, the force of buoyancy. How do we know this? By an argument involving substitution symmetry. Imagine a "block of water" surrounded by water inside a beaker (Figure 2.5). The water block has weight, so why does it not fall? Because the water below it must exert an upward force of buoyancy to precisely balance the weight. If a block of gold replaces a block of water, leaving everything else the same, the water below the gold block should continue to exert the same upward buoyant force on the gold block. This upward force reduces the apparent weight of the gold block in water. The loss of weight is proportional to the volume of the block. Archimedes hit upon a quick method to determine density (more correctly *specific gravity*) by weight measurements alone, rather than by combining weight and volume measurements.

The specific gravity of an object compares its density with that of water. For example, iron's specific gravity is about eight; volume for volume, iron is eight times heavier than water. The human body's specific gravity is 0.98; indeed, we are composed mostly of water, as Aristotelians guessed.

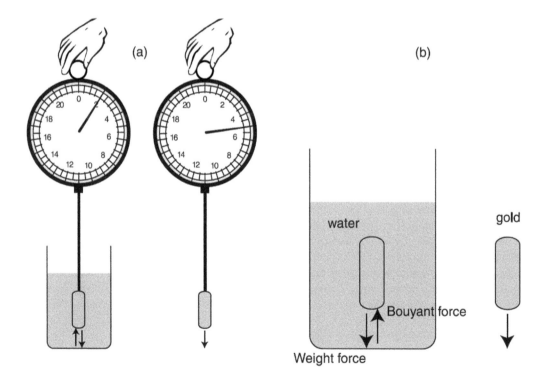

FIGURE 2.5 (a) An object immersed in water will weigh less than it weighs in air because water exerts an upward buoyant force. (b) An imaginary block of water of the same volume as the object remains stationary in water, even though the water block has weight. The upward buoyant force of the water exactly balances the weight.

Since our density is 2 per cent less than the density of water, 2 per cent of our bodies remains above water when we swim. A floating cork from a wine bottle has a specific gravity of about 0.2, and air is a minuscule 0.0013.

Archimedes' comparative densities and buoyant force explained the contrary motions of a wine cork in air and in water without invoking Aristotle's diatribe on separate motions specific to each of the elements, and elemental mixtures. A cork immersed in water will expeditiously rise *up* to the surface because the specific gravity of cork is five times less than water. Accordingly, the upward buoyant force of water is also five times greater than the weight of the cork. Moving through air, the cork will fall straight *down*. Because the specific gravity of cork is 200 times more than of air, the upward buoyant force of the air is very small compared to the weight of the cork.

Archimedes' success with densities and buoyancy showed that to make progress in understanding nature one must pay close attention to details that may at first seem irrelevant. Going further, it is important to make measurements. In his famous eureka moment, he took heed of the trivial displacement of water as he sank into his bath-tub. He carefully measured volumes of water displaced by gold and silver pieces. Interest in detail is similarly evident in the evolution of the artistic world of the Alexandrian era. Compare, for example, the faces, attire and jewellery of the pottery of Figure 2.6 with the abstract figures of the vases from the earlier periods in Figures 1.10 or 1.11. Fancy jewellery and elaborate hair styles start to adorn abstract figures. Examine the bulging muscles of the Discobolus (Figure 2.7) as compared to the simple and dignified Greek athlete of Figure 1.24. Protruding ribs, rippling veins, and a precarious poise convince us that action is imminent. Details of the world absorbed the kindred minds of scientists and artists.

Ignoring the early Greek intellectuals' distaste for details, Archimedes was poised to make new breakthroughs in *statics*, the mechanics of bodies in equilibrium. Unlike Aristotle's sweeping dynamics, Archimedes' careful advances remain valid. In explorations through mechanics, Archimedes understood the principles underlying the many potent devices which Egyptians invented to amplify muscle power. His nimble mind analysed the function of simple machines, such as pulley arrangements, the wheel-and-axle, and the lever. To impress the king with the power of these mechanical devices, he claimed he could single-handedly pull a giant ship from shore into the water. The king and his cohorts laughed in disbelief, whereupon Archimedes arranged a spectacular demonstration. With a combination of levers, pulleys, and ropes, he amazed the king and his courtiers.

FIGURE 2.6 *Vase of Maedias*, ca. 410 BC.[4]

FIGURE 2.7   *Discobolus*, 450 BC, Myron.[5]

Apart from advances in mathematics and mechanics, Archimedes was a talented inventor who developed new technologies based on a deeper understanding of scientific principles. Many devices were of immediate service. In the longer run, the advancing technological base was to benefit Alexandrian science. To ease the chore of pumping water out of wells, he devised a clever water pump, winding a pipe in a spiral around a rod and tilting the combination into a water reservoir, with the open end of the pipe in the water. As the rod rotated, the pipe scooped up water and transported it upwards and out of the reservoir. Archimedes' screw pump fulfilled agricultural needs through the heyday of the Roman Empire.

When Archimedes returned home to Syracuse from Alexandria, he helped defend his native city-state against Roman conquerors by creating mechanical war machines. Most of his devices were based on the principles of levers and pulleys. Catapults hurled huge stones at Roman battalions, and claws capsized their boats. After a while, Roman soldiers were so terrified that whenever they saw a piece of rope or a fragment of wood from the city wall, they retreated in terror, as if Archimedes was about to unleash some new, deadlier contraption upon them. Having to give up direct attacks on the city, Roman general Marcellus set siege to it for two long years. When the Romans finally captured Syracuse, Marcellus gave direct orders to capture alive the wily adversary. Where was this brilliant general who single-handedly held the mighty Roman army at bay? Little did they know he was just a mathematician! When a soldier broke into Archimedes' house and trod over his complex mathematical diagrams in the sand, Archimedes arrogantly warned: "Do not touch my drawings." Whereupon the ignorant soldier drove a spear through his body. Archimedes died as a martyr to mathematics.

Unfortunately, Archimedes was not as proud of his practical accomplishments as he was of his pure mathematics, probably a vestige of Plato's elite academic tradition. Accordingly, he never published much about his clever and immensely useful inventions. In keeping with his wishes,

Archimedes' tombstone was decorated, not with the lever or a pulley, but with a figure of a sphere that fits exactly within a cylinder, together with a formula that he derived showing the ratio of the volumes.

A LIBRARIAN MEASURES THE EARTH FROM SHADOWS

Eratosthenes (276–194 BC) advanced the new stage in scientific development launched by Archimedes to make the first successful *measurement of the size of the earth*. Although he was lowly keeper of the Great Library of Alexandria, he was no ordinary desk clerk. Highly educated at the Academy, he developed the key idea to mark locations on earth with parallel lines of *latitude* circling the globe, and *longitude* slices from north to south. It was an elegant method for elementary geometry on the surface of a sphere.

Being chief librarian, Eratosthenes had access to volumes of reports from different regions. A curious chronicle triggered Eratosthenes' thinking. From the city of Syenne (modern Aswan), 500 miles to the south of Alexandria, the inquisitive librarian found an interesting record. It noted that every year, on the longest day of the year (the summer solstice) and precisely at noon, the reflection of the sun was completely visible at the bottom of a deep well. Geometrically minded Eratosthenes recognized this could only happen if the sun's rays were vertical, which also meant that a vertical obelisk would cast no shadow at all, since rays from the sun arrive parallel to the vertical shaft. But he knew well that an obelisk in Alexandria would indeed cast a *finite* noon shadow, even on that auspicious day of summer solstice (Figure 2.8). The report from Syenne clearly conflicted with observations at Alexandria.

Why did such a dissimilarity exist between two locations on earth? It violated translation symmetry. Whenever we recognize an asymmetry in the behaviour of nature, it rings alarm bells, generally pointing to some gap in our understanding, which eventually leads to new revelations. Why

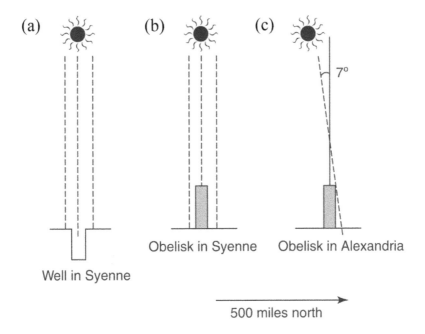

FIGURE 2.8   The conundrum which led Eratosthenes to the size of the earth. (a) In Syenne every year, on June 21, the day of the summer solstice, the image of the sun is visible at the bottom of a well exactly at noon. (b) This is equivalent to the situation that an obelisk at Syenne will have no shadow. (c) In Alexandria on that same day of the summer solstice and at noon, an Alexandrian could usually observe shadows. Alexandria is about 500 miles north of Syenne.

should parallel rays from the distant sun cast different length shadows at different places on the earth? In a flash of insight, Eratosthenes realized that the two observations are completely consistent because the earth is spherical in shape. Unequal shadows provided another piece of evidence to support the round earth. When the sun is directly overhead at Syenne, it is at a small angle from the zenith in Alexandria. Hence there is no shadow in Syenne, but a shadow at Alexandria (Figure 2.9). The apparent asymmetry between Alexandria and Syenne disappears for a spherical earth.

At this point Eratosthenes came to an intoxicating realization. By making two crucial measurements, he could determine the size of the earth. His quantitative approach opened a new door to understanding the world. Eratosthenes made a direct measurement of the shadow length of an obelisk in Alexandria. By triangle geometry, he calculated that the sun's rays were inclined by an angle of 7 degrees from a line pointing directly overhead. Two vertical obelisks, one at Syenne and another at Alexandria, extend downward right into the spherical earth, so that the two lines converge exactly at the centre. From Figure 2.10 he made a simple geometric deduction based on Euclid's alternate angle theorem. Radii drawn from the ends of the arc between Syenne and Alexandria to the centre of earth form an angle of 7 degrees (Figure 2.10). We say the arc between the two cities *subtends* an angle of 7 degrees to the centre. Since the complete angle subtended by a full circle to its centre is 360 degrees, the circumference of the earth must be 360/7 = 50 times larger than the distance between Syenne and Alexandria. All Eratosthenes needed now was an accurate measurement of the short distance from Alexander to Syenne. He accomplished this by finding out how long it took a battalion of soldiers to march from one city to the other. Knowing the distance between the two cities (486 miles), he calculated a value of 25,000 miles (40,000 km) as the circumference of the earth:

$$\frac{360}{7} \times 486 = 25,000$$

This was an enormous distance, judging from the estimated size of the known world. Eratosthenes' result was only off by 1.5 per cent from the modern value. It was an outstanding mathematical and technical advance over Thales, who had determined the height of a tree from the length of its shadow. Now Eratosthenes used shadows and geometry to set a scale for the size of the earth.

It is well worth emphasizing the important symmetries and simplifying assumptions underlying Eratosthenes' extraordinary deduction. The earth is a sphere, and the distance to the sun is much larger than the diameter of the earth. When the sun's rays reach the earth they are essentially

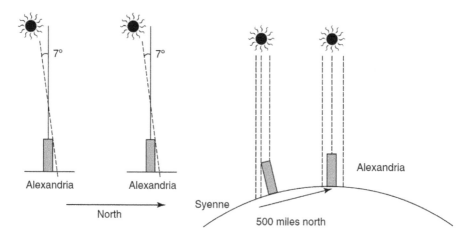

FIGURE 2.9 (Left) on a flat earth, translation symmetry demands that all obelisks of the same height should cast the same length shadows, since the rays from the distant sun are parallel lines. (Right) a spherical earth resolves the shadow paradox between Syenne and Alexandria.

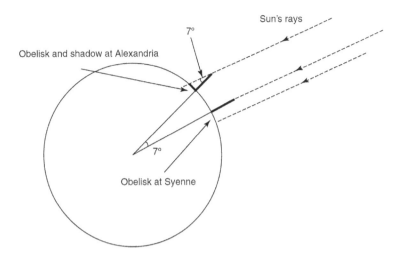

FIGURE 2.10 Eratosthenes determines earth's size. On a spherical earth, Alexandria is at latitude 31 degrees and Syenne is at latitude 23.5 degrees, a system of marking locations on a sphere developed by Eratosthenes. After measuring the angle between the shadow and the vertical, and the distance between the two cities, Eratosthenes determined that the latitude difference between the two cities must be 7 degrees. From this he deduced the circumference of the earth.

parallel lines, the same principle Thales relied on to obtain similar shadow triangles (Figure 1.4). The arc that joins Alexandria with Syenne runs along the north–south direction. In Eratosthenes' system of marking locations on the surface of a sphere, Alexandria and Syenne are on the same meridian, the same line of longitude. Like all longitude circles, the meridian line connects the two cities with *a great circle, with a circumference equal to that of the earth*. The circumference (*C*) of the circle joining the two cities is therefore directly related to the radius (*R*) of the earth through the well-known equation: $C = 2\pi R$.

Rather than idly dismissing shadow reports as curious anomalies, Eratosthenes paid close attention to details, proceeding to angle and distance measurements. The impact was phenomenal. He had extended the intellectual concern from the shape of our abode to a quantitative assessment of its size to find the earth to be surprisingly enormous. Parallel developments in observation and measurement took place in Alexandrian astronomy, topics for the future (Chapter 6). Aristarchus' preoccupation with details about shadows of the earth on the moon during a lunar eclipse, as well as the shadow of the moon on the earth during a solar eclipse, led him to a first estimate of the earth–moon and earth–sun relative distances. In the same stroke, Aristarchus would determine the relative sizes of the earth, moon, and sun.

Ever since Eratosthenes' triumph, quantitative measurements have played an increasing role in our efforts to understand nature. By setting a scale for our globe, Eratosthenes resolved the common-sense deduction that earth looks flat. Once we realize that the circumference of the earth is a million times larger than the human scale, it becomes transparent how our senses trick us into believing that the earth is flat (Figure 2.11).

In their zest to explore details of nature, scientists of Alexandria began to manipulate earth, air, fire, and water in practical experiments. Rather than rest with Aristotle's static model, they cooked up new substances. For example, they distilled alcohol, presumably to extract powerful liquors with potency to exceed wine. They thought that when alcohol evaporates, it turns into a spirit. From this belief comes our reference to alcoholic beverage as "spirits." Distillers developed new technology to make spirits re-materialize by condensation, using flasks, beakers, and condensers, familiar apparatuses of modern chemistry laboratories. In experimenting with spirits, Alexandrians launched research on gases. Philo of Byzantium (ca. 200 BC) watched air bubbles stream out of a

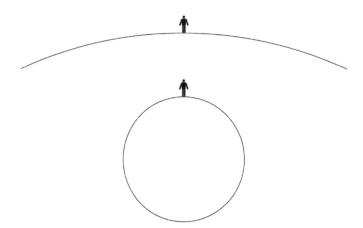

FIGURE 2.11   Showing that earth's circumference is a million times larger than human size, Eratosthenes explained why earth appears flat to the casual observer.

pipe immersed under water when he connected the other end of the pipe to a metal ball and heated it with sunlight. He guessed that heat must drive out some of the air from the ball. When the ball cooled down, water climbed into the pipe to replace the air that departed. Burning a candle in a closed air space confined over water, he noticed water creep up into the air space. Again he reasoned correctly that some of the air must disappear by the burning process that consumed the candle.

But Alexandrian science and technology stopped dead in their tracks with the rise of the Roman Empire and Christian spirituality, tragic developments for the continued progress of science, stories for Chapter 3. The progress of science finally continued after 1,800 years, with renewed questions about the element "air." Does air have weight? Does vacuum exist? Archimedes' methods of careful observation, measurement, and experimentation finally returned.

UPDATES

The Frustrated Fountain Builders of Florence

Digging for precious metals, miners in Joachimstahl (now in Czech Republic) hit one of the greatest silver strikes in history (1516). In commemoration of the great find, they named the silver coin minted in the region as the "stahler," from which comes our word, "dollar." As they dug deeper to unearth the treasure, ground water seeped into the mines, and eventually flooded them. To remove the water, a most effective method was a mechanical pump. The device was based on the ancient principle of the air pump, invented by Ctesibos (285222 BC) in Alexandria. One end of a long hollow pipe with a tight-fitting piston inside was stuck into the flooding water. As the piston was pulled up it sucked the water into the pipe till it was full, when a valve at the bottom of the pipe was closed and another valve at the top opened. On the descending piston stroke, the piston forced the water out of the tube via the open valve at the top.

Although extremely effective, the method suffered a mysterious limitation. It was impossible for the piston-pump to raise the water any higher than 34 feet above the mine. A similar problem arose from a different arena. Fountain builders of Florence, who wished to build increasingly spectacular displays, discovered to their chagrin that they could not pump a column of water to a height of more than 34 feet to provide higher pressure for even more spectacular fountains.

Perhaps the inability to raise water above 34 feet was connected to the weight of the liquid, thought Evangelista Torricelli (1608–47). He tried further experiments with mercury, which is about 14 times heavier than water. He filled glass tubes of different lengths with mercury and inverted them into a bowl of mercury. Now there seemed to be a problem filling tubes longer than only 30 inches. The mercury would run out until the remaining column reached a height of only 30". The ratio of height of the tallest possible column of mercury to that of tallest column of water was the same as the ratio of the densities of the two fluids in the same column.

Torricelli reached a monumental conclusion: a simple mechanical balance is in play, with the atmosphere on one side and the enclosed fluid on the other. Since we live at the bottom of an ocean of air, a short, 30-inch (76 cm) column of mercury can balance the same weight of air as a tall, 34-foot-high, column of water. But the mechanical balance explanation means that air has weight. Although it was commonly accepted that air is an elementary substance, no one had yet demonstrated that air has weight. And what could possibly reside in the "empty" space above the column of mercury in tube more than 30" tall? Could it be a vacuum? Can a vacuum exist? Can it be produced artificially? Many confidently denied these possibilities. Aristotle's frequently quoted maxim: "Nature abhors a vacuum" was the definitive word on the subject. The controversy over the existence of vacuum re-ignited.

In France, physicist and inventor, Blaise Pascal (1623–62) heard about Torricelli's experiments. Pascal proposed a definitive experiment for the weight of air by reasoning along the following line: if air has weight, the weight is surely finite. Therefore, there can only be a finite total height to the air above the earth. Hence the weight of the atmosphere above should decrease with increasing altitude. The height of the maximum column of quicksilver should be less at the top of a mountain than at the base. Because of his frail health, Pascal was unable to ascend a tall mountain peak.

A year later, Pascal's brother-in-law performed the experiment. As he scaled the 1.5 km high mountain, he verified the continuous decrease of atmospheric pressure. At the peak, they recorded a difference of nearly 100 mm of mercury out of 760 mm at the base. The effect was so large that Pascal could later check it for himself on top of a tall church steeple, finding a measurable difference of a few millimetres. From these results, Pascal guessed that the height of the atmosphere must be 15 km, about ten times the height of the mountain. The human body experiences about one ton of atmospheric weight over the body's areas, comparable to the weight of a car. But we don't collapse under this enormous weight because the body has evolved to be in equilibrium with the atmospheric pressure.

Contemplating the finite extent of the atmosphere, Pascal had a bolt of insight, one that seemed totally incredible for his time. Above the finite atmosphere, there must exist a vast vacuum. The utter silence of empty space was a frightening concept. These spaces till now had been abundantly populated with angels and spirits that cared for the world and for humanity.

ALL THE KING'S HORSES

Otto von Guericke (1602–86), mayor of Madgeburg in Germany, became interested in the philosophic controversy concerning the existence of a vacuum. He decided to settle the question by experimentally creating a vacuum, rather than entertain endless argument. Trained as a military engineer, he had several occasions to sharpen his mechanical skills during the Protestant/ Catholic wars which raged over Germany for 30 years. Perhaps air can be pumped out of a vessel just like water from a mine. Stimulated by the needs of miners and fountain builders, the technology of water pumps had advanced considerably through the Middle Ages and now could be applied to air.

Von Guericke's first efforts to produce an artificial vacuum were dismal failures. He filled an empty wine barrel with water, sealed all the openings tightly and tried to pump out the water with the water pump, hopefully leaving empty space (vacuum) behind. But the seals leaked, and air instantly rushed in to replace the exiting water. Even after improving the seals, he remained unsuccessful, as the air continued to leak into the cask through minute pores in the wood.

It was time to switch to a stronger, metal sphere. He also decided to pump out the air directly, skipping the water altogether. At first the piston of the air pump glided easily as it sucked out the air. But soon, a resistance began to build up against the motion of the piston. Even two strong men working together could hardly pull the piston out. Suddenly, with a loud bang, the metal sphere crumpled with a horrible clamour. The dramatic collapse was utterly astonishing.

Not one to give up easily, von Guericke made another sphere, this time with a thicker metal wall, taking care to shape a more perfect sphere and avoid flat regions. Only a truly spherical container could withstand the weight of the atmosphere and withhold a vacuum inside. They removed the air successfully without collapse, and closed the stop-cock. When they opened the stopcock, the air rushed into the sphere with such that it seemed as if it as if one of the assistants would be sucked into the sphere. von Guericke successfully created a vacuum inside the metal sphere. The question finally was settled.

Von Guericke had a theatrical flair. To show the atmospheric pressure effect dramatically, he rigged up the famous demonstration of the "Madgeburg hemispheres" before the Holy Roman Emperor, Ferdinand II. He placed together two copper bowls (half-spheres), 36 cm in diameter, with a good seal in between the two halves, to form a complete hollow sphere. After the air was pumped out, even a team of 16 horses, pulling on each sphere, was unable to tear them apart. But after he let the air in with a long, loud hiss, a single man could effortlessly take apart the spheres.

Later von Guericke demonstrated several other interesting properties of vacuum: sound does not travel in a vacuum, a candle cannot burn, and animals cannot live in a vacuum. But light can travel through a vacuum.

UPDATING THE ELEMENTS

One of the first breakthroughs concerning the fundamental nature of elements since the Greeks came during the turbulent times of the French Revolution. Despite the social and political turmoil brewing around them, French scientists, such as Antoine Lavoisier (1743–94), continued to flourish at the Academy of Paris, one of the many successors to the pioneer Academy of Athens. Armed with an analytic balance to determine precise weights, Lavoisier turned his keen mind to Philo's aborted investigations into combustion. Through his emphasis on precise measurement and close attention to details, Lavoisier continued the innovative approach to investigating nature pioneered by Alexandrian scientists.

From an early interest in Paris street lighting improvements, Lavoisier started to explore the nature of combustion. By now, the "experimental method" had become well-established, thanks to Galileo (see Chapters 3 and 4). Burning sulphur and phosphor in air, Lavoisier found the earthy ash-like products to weigh *more* than the starting substances. If the earthy substances left behind after combustion were supposed to be the "element earth" how could did these "elementary" substances weigh *more* than the starting substances? Clearly some material in air combined with the burning materials.

Something rather unusual was going on. Lavoisier was moving in directions sharply contradictory to prevailing ideas about combustion. Harking back to Aristotle, most chemists of the 18th century believed that metals were not elementary. When roasted in air metals gave up their "fire" element (they called it "phlogiston") to leave behind the more elementary, "earthy" substance, called "calx." Throwing earthy ore into a fiery furnace yielded a shiny metal (production of metal from ores). The very brilliance of a metal betrayed the presence of fire within. Dirty ore plus brilliant fire equalled shiny metal. But weight measurements told the analytically minded Lavoisier that earthy calx could not be more elementary than metal; calx *weighed more* than metal. The calx had to be a *combination* of metal and air. There was *no loss* of fiery phlogiston, but a *gain* of air.

Lavoisier's next experiment was decisive. When combustion took place in a fixed volume of air, trapped inside a tube turned over a bowl of water, the burning reaction lasted only for a brief period (Figure 2.12). When the combustion came to a halt, the level of water in the tube rose to fill part of the tube. Lavoisier concluded that only part of the air must be consumed by combustion, but not all the air, since the rising water did not fill the whole tube. Some gas still remained in the tube. And that remaining gas did not support combustion, which is why the candle flame went out. Therefore

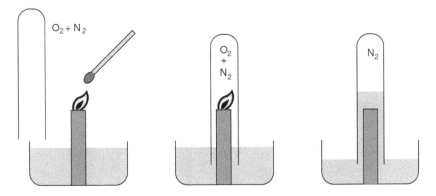

FIGURE 2.12   (Left) after lighting a candle, (centre) combustion proceeds in a closed volume of air, trapped inside the tube over water. (Right) when combustion consumes all the oxygen in the air, the candle extinguishes. The water level rises to replace any missing gas.

the original air must consist of at least two components, one that supports combustion, and one that does not. Lavoisier was headed toward another revolutionary conclusion. Air cannot be an element since it consists of at least two types of gases.

Other effects were also taking place in the candle-burning experiment. For example, the burning candle heated the air and expanded it. Some hot air escaped slowly as bubbles through the water below. As long as the candle burned, the remaining hot air inside the tube could keep the water out of the tube – due to its higher pressure. As soon as the candle was extinguished, the remaining gas cooled and contracted, to cause the water to enter the tube due to a drop in temperature and pressure of the remaining gas.

To be sure, he needed to isolate that "active component" of air involved in combustion. Perhaps he could obtain it by reversing the process of combustion, turning a calx back into its parent metal, releasing at the same time the active part of air that combined with the metal. But it was not so easy to disintegrate a calx in a controlled fashion to collect the products and to weigh those.

At this stage, Lavoisier ran into a bit of luck. English chemist Joseph Priestley (1733–1804) told him of a remarkable experience. When heating the red calx of mercury, Priestley liberated and collected a "new gas" that possessed some rather unusual properties. Combustibles, such as sulphur and phosphorus, burned more brilliantly in it than in plain air. Mice were particularly frisky in it. Priestley's head felt light and easy when he sniffed it. Mistakenly, Priestley thought that the "new air" was somehow deprived of phlogiston. In its eagerness to combine with phlogiston, it turned flames brilliant. The fiery element of antiquity continued to hold a strong grip over his thinking. Trapped in the old dogma, he missed the true significance of his seminal discovery. For the progress of science it is not enough to make a new discovery. It is crucial to have the insight to properly interpret the find, and courage to turn against the prevailing paradigm.

Fortunately, the seeds of Priestley's experience fell upon fertile soil. Lavoisier's experiments took a successful turn. Finally there was a calx that transformed directly into a gleaming metal simultaneously releasing a gas, which was a part of the air. Lavoisier decomposed mercury calx to collect the gas and the shiny metal left behind (Figure 2.13). Once again, the analytical balance proved decisive. The weight of the gas collected plus the weight of the shiny mercury metal left behind was equal to the weight of the starting calx. The gas showed all the combustible properties discovered by Priestley. He had successfully captured the combustible part of air.

Lavoisier was confident. *Air could not be an element.* It must consist of two gases, at least. One component, which he now successfully isolated, actively supports combustion and breathing, hence the frisky mice. This gas he named "oxygen," from *oxus* for acid and *genea* for produce. (Erroneously he thought he also isolated the active element for acids. But that was later discovered to be hydrogen.)

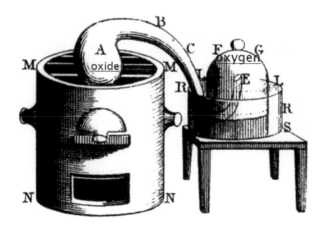

FIGURE 2.13   Decomposition of mercuric-oxide into mercury (metal) and oxygen gas. The metal is left behind in the hot flask (left), and the gas is collected in a container (right) over water.

The other constituent of air he called *azote*, meaning "no life" from Greek. While the French retain Lavoisier's *azote*, in English we call it "nitrogen." The gas left in the tube after candle combustion over water was mostly nitrogen. Breathing pure nitrogen will cause suffocation. A burning matchstick will quickly extinguish in nitrogen gas. With analytical balance and analytic mind, Lavoisier *decomposed* a stubborn element of antiquity into two basic substances: *air equals oxygen and nitrogen*. Air consists of 20 per cent oxygen. It was a watershed event separating the ancient from the modern world.

If one earthy calx decomposed, so could others. Lavoisier moved on to cast serious doubt upon the elementary nature of earthy substances. Air is not an element. Earth is not an element. And what would be the fate of fiery phlogiston? Lavoisier published a scathing attack. Combustion is not liberation of a fiery element. Rather, a heated substance combines with oxygen in air. Lavoisier jolted the scientific community by attacking their certitudes about earth and fire. His frontal blow on fire as an element started an intense quarrel that took on national overtones. Staunch English scientists, including Priestley, refused to abandon phlogiston. It was impossible to accept the cracks in established theory, much less the idea that the entire ancient construction would have to be razed to the ground.

One by one, elements of antiquity began to disintegrate. And what about water? Lavoisier synthesized water from two separate gases, the newly discovered oxygen, and another gas known for a long time as "flammable air." Flammable air (now known as hydrogen), was well known since the Middle Ages as the gas released when nitric or sulphuric acid attacked a metal. Lavoisier collected flammable air and burned it with regular air to form a clear liquid product that behaved in all respects like water. He was sure that oxygen in air burned with flammable air to produce water. Water was no element! It was time to give flammable air a new name, *hydrogen*, meaning (from Greek) "to give rise to water." Lavoisier's names started to catch on. He did not realize it, but Lavoisier had named the most basic of all chemical elements, and the first element to form after the creation of the universe in the Big Bang.

In the midst of these major challenges and breakthroughs, Lavoisier understood the limits of his own discoveries:[6]

> Nothing gives us the certainty that substances which we now believe to be simple are actually simple … this stuff presents today merely the limits of analytical chemistry, it cannot be further decomposed with our present knowledge and devices.

Prophetic words indeed! We know that hydrogen, oxygen, and nitrogen are also not elementary. They can be decomposed into protons, neutrons, and electrons. And protons and neutrons are not elementary either; they consist of quarks. Only electrons remain elementary.

Lavoisier launched the science of chemistry. But a terrible tragedy snuffed out the triumph of his breakthroughs. The brilliant mind became a casualty of the bloody French Revolution. Early in life, Lavoisier made a financial investment that later turned out to be a grievous political blunder. Being born of a wealthy family, he invested half-a-million francs in the hated tax-collecting agency of France. When the Jacobians unleashed their Reign of Terror during the Revolution, they rounded up "enemies of the people" to summarily try them and execute them. Lavoisier faced arrest as a member of the dreaded tax-collecting agency. In a trial that lasted less than a day, a fanatical horde condemned him to death and guillotined the aristocrat at the Place de la Revolution. Lavoisier's scientific colleagues, Joseph Lagrange and Charles Coulomb, stalwarts in the rapidly maturing sciences of mechanics and electricity, bid their colleague a final farewell. Lagrange mourned:[7]

> A moment was all that was necessary to strike off his head, and probably a hundred years will not be sufficient to produce another like it.

His body was thrown into a common grave. Only the tax collector died at the guillotine; the scientist lives on forever.

THE CRUSHING WEIGHT OF OUR ATMOSPHERE

If all the air were to somehow be expelled from Lavoisier's tube of Figure 2.12, the water would rise all the way to the top of the tube. For example, the air could be sucked out via a tube attached to a pump. Alternatively, the tube could be filled completely with water and then turned upside down over the bowl of water so that none of the water from the tube falls out. This can be accomplished with the help of a card at the end of the tube. The card is removed once the end of the tube is under water. Again, all the water would stay inside the tube, defying gravity, and not falling into the bowl. French mathematician, physicist, and inventor, Blaise Pascal (1623–62), carried out such experiments with tubes of various lengths to find there is a maximum length of tube from which the water will not fall out. Incredibly this length was 10.3 m (33.8 feet), taller than a house. For tubes longer than 10.3 m, the water would fall out to a length of 10.3 m, but no lower. Pascal figured out that the pressure exerted by the column of water in the tube must be balanced by the pressure exerted by the weight of the atmosphere. When the water pressure in a column taller than 10.3 m exceeded the atmospheric pressure, the water level would drop to 10.3 m to leave a "vacuum" in the space above the water (except for some water vapour). The weight of the atmosphere exerts a pressure of 14.7 lbs over every square inch of area. To express the pressure in metric units, the atmosphere has a weight of 1 kg for every square centimetre of area over the surface of the earth (at sea level). The human body experiences about one ton of atmospheric weight over the body's areas, comparable to the weight of a car. But we don't collapse under this enormous weight because the body has evolved to be in equilibrium with the atmospheric pressure.

Atmospheric pressure plays an important role in many arenas. At higher elevations, as on a high mountain top, or where the airplanes fly, the amount of air above is significantly less, so that the pressure decreases, and it would be hard to breathe because of the insufficient oxygen level. At even higher elevations air in our bodies would escape so that the ear drums would rupture, and blood gases would boil out.

THE DISCRETE CLUES TO THE ATOM

From Manchester, England, the world's largest industrial city, John Dalton (1766–1844) took the next crucial step by examining multiple compound *molecules* forming from the same elements. Carbon and oxygen form two gaseous compounds. One is deadly toxic. The other is a harmless gas that dissolves in water to make a pleasant effervescent, soda water drink, imparting a tingling tartness. There was a quantitative relationship between the weight-ratio of carbon and oxygen in the

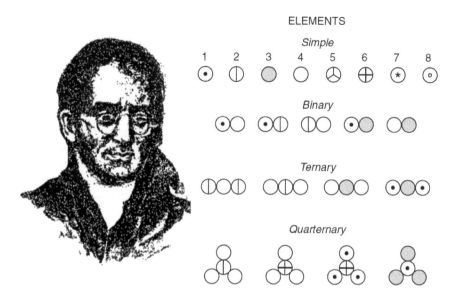

FIGURE 2.14   Dalton explains chemical compounds molecules with an atomic theory.[8]

first compound to the weight-ratio of the same elements in the second compound. Analysing the data, Dalton was thrilled to find a simple and striking numerical order. If 1 gram of carbon combines with 1.3 grams of oxygen in compound A, then 1 gram of carbon will combine with $1.3 \times n$ grams of oxygen to form compound B, where n is a small integer. Another fascinating example comes from nitrogen and oxygen compounds, which come in five different forms. The weights of oxygen combining with a unit weight of nitrogen in each compound are in the ratio 1:2:3:4:5. A simple integer order appeared in chemical combinations.

Pythagoras was right! At its roots nature does show regularity in simple mathematical patterns. Dalton announced his discovery in the form of *The Law of Multiple Proportions*. But he was hot on the trail of an even grander concept, the "atomic" nature of matter. When two elements A and B form compounds, there is a structural reason for the integral combinations. One atom of A can combine with one atom of B to form a binary compound. Or one atom of A can combine with two atoms of B to form a ternary compound. Weight relations follow suit. He even imagined simple geometric patterns to describe the bonds (Figure 2.14).

Contemplating the fundamental nature of matter in terms of atoms, Dalton was inspired by the Greek philosopher Democritus, whose concepts about the atomic nature of matter sprung straight from the mind, like Athena from the brow of Zeus. Democritus conjectured that atoms are indivisible and unchanging, darting about eternally and randomly in a vacuum, like particles of dust in a sunbeam. Dalton's atoms participated in chemical reactions by joining with other atoms. The richness of natural phenomena and the many types of substances all arise from the varied configurations of atoms. While Democritus' atomic theory was astute speculation, aesthetically motivated by the desire to find something constant amidst the bewildering change, Dalton resurrected atoms by applying quantitative comparison to systematic weight measurements on compounds and their constituents. Striking numerical patterns which emerged in compound formation gave Dalton the confidence to revive the concept of invisible atoms:[9] "The existence of these ultimate particles can scarce be doubted though they are probably much too small ever to be exhibited." With the rediscovery of fundamental units of matter, the alchemy of medieval times could finally turn into the science of chemistry.

Some of Dalton's contemporary scientists were greatly dissatisfied. Insisting on observational evidence over theories born out of pure aesthetic preferences, the famous chemist Sir Humphrey

Davy (1778–1829) warned against speculations about invisible entities, unfortunately expressing his prejudice on the same occasion as he was conferring upon Dalton the Royal Society's medal, a high honour for any British scientist. Professing similar caution, chemist Jean Dumas censured atomists who followed Dalton:[10]

> I would efface the word atoms from science, persuaded that it goes further than experience … In chemistry we should never go further than experience. Could there be any hope of ever identifying the minuscule entities?

Like Aristotle, they preferred to trust the senses, needing to see the atoms. In the hot dispute, the positivist chemists completely missed an overriding lesson from the classical thought of Pythagoras and Plato. Our perceptions of nature are limited by the senses, but the mind must be ready to transcend these limitations. Our sensory experiences are only a faint inkling of a richer reality that flickers beyond reach.

Democritus and Dalton could "see" atoms with the eye of the mind although atoms were far too small to visualize with any microscope available to them.

Only in the 1980s did it finally become possible to see individual atoms on a fine scale (Figure 2.15a) using a revolutionary new instrument called the scanning tunnelling microscope (STM). Today it is not only possible to visualize atoms but also to manipulate the placement of individual atoms using laser tweezers or the tip of the microscopic needle of the STM to write your name with atoms or with molecules (Figure 2.15b).

How Much Does an Atom Weigh?

For Dalton, the important difference between one atom and another was not shape, as originally conjured by Democritus, but *weight*. After Lavoisier's breakthroughs with the analytic balance, Dalton would naturally attach much significance to weight. With no hope of weighing individual atoms, Dalton proceeded to determine *relative weights* between atoms of different elements through his law of proportions governing the way atoms combine. He devised a strategy to guess the formula of chemical compounds. Motivated by simplicity and elegance, Dalton posited that when two elements, A and B, combine, the lightest compound forms by the *simplest* combination. One atom of A combines with one atom of B. For example, when hydrogen (H) and oxygen (O) combine to form water, as Lavoisier determined, the most obvious combination, according to Dalton, would be

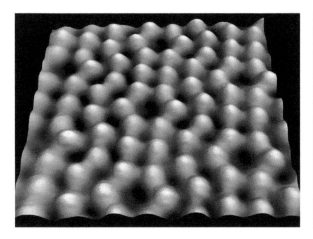
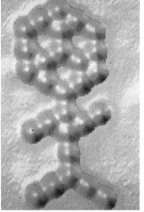

FIGURE 2.15   (a) Scanning tunnelling microscope (STM) image of silicon atoms arranged in a regular pattern in a silicon crystal.[11] (b) "Molecular man" built from 28 CO molecules.[12]

HO. Since experiments showed that 8 grams of oxygen combine with 1 gram of hydrogen to form 9 grams of water, he deduced that one atom of oxygen is eight times heavier than one atom of hydrogen. With this approach to analyse compounds, Dalton started to prepare a table of *atomic weights*, the relative weights of atoms of different elements.

His rule of simplicity was attractive, but off the mark. Aesthetic principles alone have their limitations. They are powerful guides, but their power can also be deceptively strong. We now know that the *simplest compound* of H and O is not HO, but H_2O. Also, the more complex H_2O_2 also exists. Between iron (Fe) and chlorine (Cl), the simplest compound is $FeCl_2$, not FeCl. When two elements form multiple compounds, their formulae are not given by simplest integer progressions. The three oxides of iron do not follow the obvious progression: FeO, Fe_2O, and FeO_2. Rather, nature prefers the progression: FeO, Fe_2O_3, and Fe_3O_4. Later discoveries of the electronic structure of atoms made this understandable. Nevertheless, in spite of the failure of his rule of simplicity, and the resulting errors in the first table of *atomic weights*, the framework which Dalton set up did much to spur the progress of chemistry. Dalton's inspiration gave chemists a badly needed starting point from which they could focus on a common goal to analyse data: derive atomic weights of elements to put them in order.

As the number of chemical elements increased with chemists' discoveries, the table of atomic weights continued to grow. Remarkably, hydrogen remained in its primary position with the lowest atomic weight. In 1815, William Prout proposed an intriguing hypothesis based on the astute observation that the atomic weights of most elements appeared to be close to integer numbers, barring a few notable exceptions like chlorine. Once again, a simple numerical pattern was appearing in nature's organization. Perhaps all elements are made from hydrogen. Wondering whether he hit upon the primary element, he conjured up the name "protyle," to capture the tantalizing possibility. Through the thick veil shrouding the mystery of substance, Prout had glimpsed an underlying structure inherently simpler than the many chemical elements. Indeed, as the story unfolds, the hydrogen atom plays a major role in the structure of matter. Thales would have been proud, although hydrogen, not water, turns out to be one of the basic constituents of matter.

When Dalton re-introduced atoms, many scientists resisted accepting their existence for a long time, until Einstein's atomic explanation of Brownian motion (see Chapter 1 update). Nevertheless, the idea of "fine-scale structure" to the world was slowly seeping into the intellectual culture. Artist Georges Seurat (1859–91) introduced *Pointillism*, where small dots of colour emerge from afar as an uninterrupted image as in Figure 2.16. From a distance the scene appears to be continuous, but at close range you may grasp the artistic atomism of individual dots of colour. Seurat was not directly influenced by Dalton's ideas, but the atomistic mindset had permeated the culture.

AVOGADRO'S GOT YOUR NUMBER

In 1809, a year after Dalton published his atomic revival, Gay-Lussac published another law about gas reactions, discussing combining volumes instead of Dalton's combining masses. When gases react they combine in simple *volume* ratios. For example, two litres of hydrogen gas react with one litre of oxygen gas to form *not three, but one litre of water vapour*. Recall Dalton's mass analysis: 8 grams of oxygen combine with 1 gram of hydrogen to form 9 grams of water, to conclude (erroneously) that the atom of O must have eight times the mass of the H atom and also (erroneously) that the combination molecule of water takes the simplest form, HO. Gay-Lussac's volume law showed that the *volume* of water vapour decreased in the reaction, whereas Dalton's law showed that *masses* combined to increase. Why the difference?

Amadeo Avogadro (1776–1856) based the explanation on his new idea that equal volumes of gas must carry the same number of particles (atoms or molecules). The explanation for the volume decrease was that the number of molecules after reaction decreased. This is known as Avogadro's Law, a marvellous deduction. For example, when 200 particles of H in two volumes of H combine with 100 particles of O in one volume of O, they only form one volume (or 100 particles) of water vapour. But by using HO for the water compound according to Dalton, 100 H particles (one volume)

FIGURE 2.16 *A Sunday Afternoon on the Island of La Grande Jatte* by Georges Seurat[13] using the new art form of pointillism. The foliage of the tree on the upper left brings out the technique.

would be left over, which was not observed. Hence HO could not be the correct combination for water!

Avogadro came to several striking conclusions. First, the correct formula for water must be H_2O, not HO, so that the number of particles is preserved in the reaction. Two hundred particles of H combine with 100 particles of O to give 100 particles of H_2O.

$$2H + O = H_2O$$

Before becoming the first physics professor in Italy, Avogadro was a lawyer and mathematician. He argued that since oxygen and hydrogen so readily combine with each other to form water, can free O really exist in mono-atomic form? Can free H really exist in mono-atomic form? An O atom should readily combine with another O atom, so that oxygen must exist as O_2 – a di-atomic molecule of oxygen. Similarly for N_2 and H_2. Chemically active gases such as hydrogen, nitrogen, and oxygen are made up of molecules of two atoms each, H_2, O_2, N_2, instead of individual atoms. By the same argument, inert helium gas must remain monatomic. Therefore he thought in terms of molecular weight of such gases instead of their atomic weight.

Another brilliant deduction was that atomic weights (or molecular weights) could be determined directly from measurement without recourse to Dalton's arbitrary rule of simplicity, and combination assumptions. If a certain volume of O_2 and the same volume of H_2 contain the same number of molecules (Avogadro's main hypothesis), the relative weights of the molecules of the two gases are the same as the ratios of the densities of these gases (under the same conditions of temperature and pressure.) Knowing the density of oxygen, he could assign a molecular weight of 32 for O_2, and therefore 16 for O, instead of Dalton's 8.

Avogadro corrected yet another Dalton conclusion. He altered the compound formation of water to include the diatomic molecular properties of oxygen and hydrogen:

$$2H_2 + O_2 = 2H_2O$$

Going one step further, Avogadro expressed his law in an amazingly simple way: one molecular weight (or mole) of a gas such as H_2, N_2, or O_2 contains a unique number (Avogadro's number) of molecules and occupies a unique volume (22.4 litres) at room temperature and pressure. One can generalize Avogadro's law to: one mole of any element or compound contains the same number of molecules, independent of its state, whether gas, liquid, or solid. But of course he did not know that enormous number of molecules in a mole, now called Avogadro's number: 6.02×10^{23}. You would need to spend a million dollars per second for more than the entire age of the universe to spend that many dollars! We will see how it is possible to determine this critical number from X-ray analysis of crystals. Einstein's analysis of Brownian motion (Chapter 1 – update) provided an early path to estimate Avogadro's number which led Jean Perrin to win the Nobel Prize for 1926. In a later thesis on the diffusion of sugar and viscosity of sugar solutions, Einstein also estimated Avogadro's number as well as the size of a sugar molecule as one-billionth of one metre (one nano-metre).

The Electric Atom and the Atom of Electricity

In Italy, Alessandro Volta (1745–1827), professor of physics at the University of Pavia, discovered that metals are inherent sources of electricity. When two dissimilar metals, such as zinc and copper, come in contact with each other through salty water, the metals charge up. If a wire connects the outer surfaces together, a charge flows from one metal to the other to provide flowing *electric current*. In the first form of the modern *battery*, Volta stacked together several pairs of discs into a pile. Each pair consisted of one zinc and one copper disc, separated by a piece of pasteboard, soaked in brine. In this soldier-line formation (hence the name battery) each copper–zinc pair was in direct contact with an adjacent pair so as to combine the voltages and provide a larger current. For the first time in history, there was a brand new source of electricity, truly distinct from familiar frictional electricity, obtained by rubbing fur against amber, or electricity snatched from the clouds, as through Benjamin Franklin's famous kite experiment. Volta's discovery began the age of electricity by providing a source of continuous and controllable electrical current with new and wonderful powers and applications. Summoned by Napoleon to Paris, Volta demonstrated the power of the electric pile to the future emperor of France.

Electric current flowing through fine filaments of metal turned into an intense source of heat, sufficient to melt metals. Flowing electricity opened up the entirely new field of electrochemistry. Within six weeks of Volta's report, Joseph Priestley passed electrical current through water to decompose it into hydrogen and oxygen gases. The new electrical magic was immensely more powerful than any dreamed of by alchemists, trying in vain for centuries to transform elements into each other. The amazingly simple method of dipping two wires connected to a voltaic pile effortlessly decomposed water into its basic elements.

With a powerful battery invented by Volta (Chapter 3), English chemist Davy (Chapter 3) unlocked new chemical elements from earthy ash, and transformed them into brilliant metals. Decomposing alkali earths, like soda and potash, Davy extracted the chemical elements sodium and potassium.

At the back of a bookbinding shop in London, a young apprentice boy named Michael Faraday (1791–1867) tinkered with electrical experiments he found described in books that he was supposed to bind. Faraday's name more aptly belongs to every chapter of electricity and magnetism, but his pioneering work in electrochemistry paved the way to the electric atom. His is truly a Cinderella story. One of ten children of a blacksmith who lived in a grimy section of London, he became a bookbinder's apprentice at age 13. Put to work as an errand boy, he delivered newspapers, dusted lodgers' rooms, and shined their shoes. Eventually he advanced to bookbinding. But Faraday did not stick to binding. In his spare time, he devoured the books he was supposed to bind and started to explore their contents. Using voltaic cells the size of a penny, he decomposed the sulphate of magnesia to cover wires with thick streams of gas bubbles.

Fate soon smiled on Faraday. The famous chemist Humphrey Davy was going to deliver a series of lectures. A customer of the bookbinding shop gave Faraday an admission ticket to the sold-out lectures. Through Davy's words, Faraday heard his true calling. Taking detailed notes on every lecture, he transcribed them with lavish, calligraphic writing and artistic illustrations. After binding them in the shop, he sent off the volume to Davy, along with a job application.

Intrigued by the resourcefulness of the young man, Davy asked one of his colleagues for advice on whether to hire Faraday. "Let him wash bottles," was the answer.[14] "If he is any good, he will accept the work. If he refuses, he is not good for anything." Davy hired him, but only as a bottle-washer and valet. Faraday was eager to help in the laboratory, but Davy tried to dissuade him from pursuing science. It was not fitting for a valet. When an accident injured one of Davy's laboratory assistants, Davy relented and offered the job to Faraday. Davy's epoch-making discoveries in chemistry are said to be surpassed only by his discovery of Faraday.

Harking back to his days of mechanical puttering in the bookshop, Faraday wondered about the mechanism underlying electrochemical decomposition. Chemists who were open to the idea of atoms visualized compound molecules in terms of hooks and eyes joining constituent atoms together. For Faraday, a natural explanation was that electricity somehow bound together the atoms in the water molecule, and electricity also played a role in tearing the molecule apart. But the nature and origin of electricity were still utter mysteries. Why did the unhooked atoms not simply turn into gas bubbles all over the solution? Why did gas atoms take form only at the terminals?

Continuing emphasis on careful measurement from Lavoisier and Dalton, and before that from Eratosthenes and Archimedes of Alexandria, Faraday accurately weighed the products of electrochemical decomposition. Through exquisite experiments followed by insightful interpretation, his thoughts crystallized two revealing relationships regarding electrical decomposition or electroplating of metals:

In a given time, the amount of material that evolves during electrochemical decomposition is proportional to:

(a) the amount of current flowing, and
(b) *a simple fraction* × the atomic weight of the constituents.

Once again, simple rational numbers were cropping up in the order of nature. Pythagoras would have been immensely proud. Apart from numerical patterns, the laws of electrolysis unveiled an exciting link between electricity and atoms. Bonds which hold compound molecules together are not mechanical eye-hooks, but electrical in nature. One positively charged part attracts a negatively charged part. In solution, the molecules break up into charged "fragments" which Faraday named "ions" (or wanderers) from the Greek, because they can move around freely, conducting electricity, carrying charge to one electrode or the other. Also from the Greek, he coined the word "electrodes" for the terminals, at the end of the "road" of electricity. The an-ode is the upper (+) route, and the ca-thode is the lower (−) route in the electrolytic cell. The positive charge fragment moves to the cathode while the negative ion moves to the anode. At the electrodes, the positive charge on an ion neutralizes with negative charge arriving from the flowing current, releasing a charge-free atom of gas which bubbles out during decomposition, or a charge-free atom of metal which plates out of the liquid during electroplating.

What is the reason behind the simple fractions? When a common electrical current flows through each solution, transporting the same amount of charge, the number of copper atoms deposited is half the number of silver atoms deposited, and the number of aluminium atoms is one-third. So copper ions must carry twice the charge of silver ions, and aluminium ions three times the charge of silver ions.

Just as simple ratios appearing in Dalton's Law of Multiple Proportion gave powerful testimony to the atomic nature of matter, simple fractions now pointed to the "atomic" nature of electricity. Faraday realized that all charges on ions are integral multiples of a single fundamental unit of charge without the existence of any fractional charge.

In one bold stroke, Faraday unlocked the intrinsic atomic nature of electricity and discovered there has to be a minimum quantity of electricity, an "atom" of electricity. Dramatic confirmation had to wait yet another 60 years for one Joseph Thomson and his discovery of the electron. Faraday had prepared the stage for Thomson. In yet another prescient experiment Faraday attempted to pass electricity through low-pressure gases to light an arc beginning at the negative electrode and ending at the positive. The colours of light from gas discharge carried powerful clues to the structure of the atom (Chapter 13).

We will continue the story of electricity and magnetism in Chapter 3 with other major discoverers.

SHUFFLING THE ELEMENTS

Under Dalton's elegant atomic framework, chemistry advanced so rapidly that when the Siberian-born chemist Dmitry Mendeleyev (1834–1907) wanted to offer a college course in chemistry, there was hardly a suitable textbook. Mendeleyev decided to write his own, assembling a new list for the known elements. Chemists of the day were occupied in adding to chemical facts rather than in contemplating relations between properties to enhance understanding. It was time to make some sense of the myriad properties in the hope of seeing some order by which to classify the chemical elements. As 20th-century mathematician Henri Poincaré eloquently put it:[15]

> The scientist must order. One makes science with facts as a house with stones; but an accumulation of facts is no more science than a pile of stones is a house.

Arranging the elements in order of atomic weight, Mendeleyev created a deck of cards. On each he wrote down the salient chemical properties of each known element. For instance, lithium, sodium, and potassium were active metals and helium, neon, and argon were inert gases.

In a jumbled deck of playing cards order can be found by rearranging cards by suits or numbers. Shuffling element-cards around in a game of chemical solitaire, trying to find patterns, Mendeleyev made an intriguing discovery. There was an uncanny regularity in the occurrence of similar chemical properties. For the first 18 elements, chemical properties began to repeat every eight cards. For example, helium, neon, and argon are all inert gases. We call them *noble gases* because no other elements will react with them, just as in the feudal society of the Middle Ages, no common person could approach a noble lord. Mendeleyev's card positions of noble gases are 2, 10, and 18, with a separation of eight. His card positions are now called *atomic number*. Lithium, sodium, and potassium are all silvery white, metals so violently reactive that they will burst into flames upon contact with water. Their card positions are 3, 11, and 19, again a separation of eight. After two rows of eight elements, properties began to repeat with a larger period of 18 elements (Figure 2.17 and Figure 2.18).

It took consummate skill and deep insight to group elements into rows and columns. Mendeleyev had to focus on the most relevant chemical properties with the courage to ignore inessential differences. Bold faith in striking periodicities stirred him to insert blank cards whenever there was no known element in the crossword puzzle, but where the periodic behaviour demanded there should be one. One empty spot he named eka-aluminium, and another eka-silicon. *Eka* is the Sanskrit word for "one," a mystic touch consistent with his otherwise wild character. Mendeleyev was the typical mad professor, prone to start dancing with rage or excitement, greatly popular with students for his electrifying chemistry lectures, as well as his unruly hair and beard which he had cut only once per year by a shepherd with wool shears!

Contemporary chemists laughed at Mendeleyev's folly in predicting new elements from a stack of playing cards. But five years later, a new element discovery vindicated Mendeleyev's leap of faith. With properties fitting the blank spot labelled eka-aluminium, it earned the name gallium. Eleven years later, germanium filled the eka-silicon spot. With the wild scientist's chemical prophesies fulfilled, the idea of periodic properties began to catch on.

More elements filled the table after Mendeleyev. After two rows of eight elements followed by two rows of 18 elements, the period changed to 32.

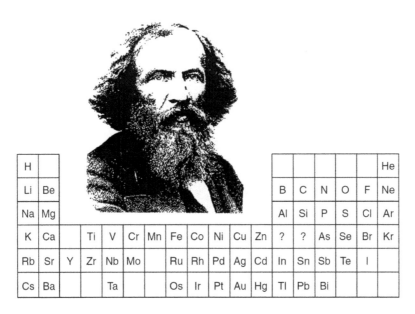

H																	He
Li	Be											B	C	N	O	F	Ne
Na	Mg											Al	Si	P	S	Cl	Ar
K	Ca		Ti	V	Cr	Mn	Fe	Co	Ni	Cu	Zn	?	?	As	Se	Br	Kr
Rb	Sr	Y	Zr	Nb	Mo		Ru	Rh	Pd	Ag	Cd	In	Sn	Sb	Te	I	
Cs	Ba			Ta			Os	Ir	Pt	Au	Hg	Tl	Pb	Bi			

FIGURE 2.17 Mendeleyev arranged 60 elements known by his time into a table where elements appearing in each column showed similar chemical properties.[16]

IUPAC Periodic Table of the Elements

FIGURE 2.18 Modern Periodic Table of chemical elements.[17]

The Periodic Table brought a semblance of order to the apparently inexhaustible variety of chemical elements, holding the potential of some deeper structural principles. Culminating an era of scientific thought and discovery, Periodic Tables decorating science classrooms everywhere now carry 118 elements (Figure 2.18). The last known element, artificially made, is called organessum.

In spite of the great advance that it represents, the table poses tantalizing questions. Why are there so many *elements*? Where did they come from? Is nature really so extravagant that it needs

FIGURE 2.19   Emission spectrum of excited hydrogen.[18] The wavelengths of the colour lines are discussed in Chapter 13.

more than 100 fundamental elements? Surely the Greeks were better off with just five? Perhaps the chemical elements are not so elementary after all. Could the mysterious regularity speak for a possible structure of the atom? If so, the atom is not elementary either! There is order in the table, but with so many columns and sub-columns the arrangement is neither simple nor elegant. Like gorgeous markings on a butterfly's wing, the regularity is fascinating. But what is the underlying meaning? What is the magic in the sequence of numbers of elements in the periods 2, 8, 18, and 32? The first period of the table has two entries (H, He). The second and third periods have eight entries (Li to Ne; Na to Ar). The fourth and fifth periods have 18 entries (K … Kr). The sixth and seventh periods have 32 entries.

Indeed the repeating patterns and the number sequence point toward an underlying symmetry. To penetrate that symmetry is to find principles which govern the construction of the atom. As we will see, the lightest and simplest element, hydrogen, holds the clue to the structure of matter as well as to the organizing principle behind the Periodic Table. As to the origin of elements, our story of the heavens will show that stars are the furnaces that forge elements from basic hydrogen (Chapter 8). To prepare the way, we need to delve into the structure and composition of atoms.

When a hydrogen atom is "excited" by heating with a flame or by an electrical discharge, it releases discrete amounts of energy in the form of coloured lines of light (Figure 2.19). Every element in the Periodic Table when excited also displays a unique series of colour lines which serve as a signature for that element. The line spectrum identifies the presence of an element in familiar substances, such as ores from the bowels of the earth, or elements in the sun and the stars. Indeed, the mathematical pattern governing the spacing of characteristic lines for excited hydrogen led to the model for atomic structure. We will continue the story of the structure of the atom in Chapter 13 on the quantum.

TEARING THE ATOM APART

The journey to the heart of matter continued another pioneering effort by Faraday with the effort to push electric current through gases, lighting up the gas in a brilliant arc discharge. Bright gas discharge tubes eventually led to the neon signs of different colours that blaze on city streets with brilliant advertisements. With the help of improved vacuum technology, pumps could lower the gas pressure in a sealed glass tube to one-millionth of an atmosphere after hours of pumping. With most of the gas gone, the glow of light inside the discharge tube also disappeared, yet electricity could still flow through the space between metal plates (Figure 2.20) connected to the electrodes. A layer of fluorescent material coating on the far end of the tube past the positive plate lights up like a screen struck by some powerful rays. Objects placed inside the tube cast sharp shadows on the fluorescent screen. The bright spot on the screen can be manipulated by an external magnet to make entertaining patterns on the screen.

The discharge tube provided the first and most boring television show, but it was profound intellectual stimulation for Joseph Thomson (1856–1940). What kind of invisible rays were flying past the positive anode to cast bright spots on the screen? Objects inside the tube could block the rays and cast shadows. Speculation about the nature of the rays grew intense: could the rays be atoms, molecules, or matter in a still finer state of subdivision than gas-atoms?

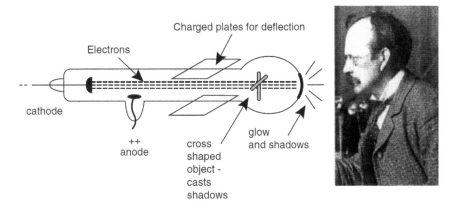

FIGURE 2.20   Thomson and his pioneer apparatus for producing, detecting, and deflecting electrons.[19]

In a decisive experiment, Thomson deflected the light spot on the fluorescent screen with electrical force, using a pair of metal plates placed above and below the tube, each connected to the terminals of a voltaic cell that charged the plates. The rays turned from their straight path toward the positively charged plate, judging from the movement of the spot. Clearly the rays were comprised of particles carrying a negative electric charge. Thomson could cancel the electrical deflection by a second, magnetic force. Through a delicate balancing act between electric and magnetic deflections, followed by careful, quantitative comparison of electric and magnetic forces, he discovered a shocking new feature of the stream of negative particles: their mass-to-charge ratio (m/e) was extremely small, when compared to the mass- to-charge ratio of hydrogen ions moving through Faraday's electrolytes. To date hydrogen ions showed the smallest mass-to-charge ratio known for charge-carrying particles moving through any fluid. Indeed the ratio for the mysterious new ray particles was *nearly 2,000 times smaller* than the smallest value known. If the enigmatic charged particles flying through the discharge tube carried the smallest unit of charge that hydrogen ions carry, the only way their m/e ratio could be 2,000 times smaller was if the particles were 2,000 times *less massive* than hydrogen atoms. But how could that be? Did not hydrogen have the smallest atomic mass among all the elements? Was the minuscule m/e ratio of the negative particles powerful new evidence for the first speck of subatomic matter? It was an intoxicating discovery.

Thomson took a daring view. Particles in the discharge tube rays must be tiny negative charge fragments detached from atoms. From the Greek word for the precious stone, amber, he coined the name "electron." Indeed, it was Thales of Miletus who took special notice of the tiny electrical sparks shooting out when he rubbed a stick of yellow amber briskly with sheep wool.

In the discharge tube, the negatively charged electrons come from gas atoms. Electricity is not a fluid as many pictured, but a flow of electron particles. Just as moving ions transport electric current in Faraday's liquid electrolyte, electrons carry current through the gas tube. Upon collisions with the gas atoms remaining in the tube, they stimulate the atoms to give out an eerie glow. Starting with other gases in the discharge tube, and pumping out the gas to very low pressure to obtain the same rays always showed the same value of m/e, independent of the starting gas species present in the tube. Clearly the electron is a particle common to all types of atoms, to all matter.

Naturally there was much scepticism. Chemists who were finally prepared to accept atoms were now appalled. How could Democritus' indivisible atom have smaller parts? Even the bold and imaginative Mendeleyev was uncomfortable. There was something even more troublesome. If bits of atoms could be broken off, the bits could stick to other atoms. It would then be possible to change one atom into another, violating another essential property of atoms, their immutability. Mendeleyev poked fun at such nonsense. To accept new ideas about the divisibility and mutability of atoms was painful, demanding the sacrifice of fundamental convictions.

If electrons are real, Dalton's atoms could not possibly be elementary. They must have substructure. Of the monumental discovery, Thomson's student AA Robb celebrates with:[20]

> All preconceived notions he sets at defiance
> By means of some neat and ingenious appliance
> By which he discovers a new law of science
> Which no one had ever suspected before.
> All the chemists went off into fits,
> Some of them thought they were losing their wits,
> When quite without warning
> (Their theories scorning)
> The atom one morning
> He broke into bits.

Dismantling the atom, Thomson initiated the relentless descent into the heart of matter. Thomson was not interested in applications for the electron. He often raised a toast to the electron:[21] "May it never be of use to anyone!" The future proved him dead wrong.

It was the beginning of the electron century. Humanity was poised to reap the benefits of the rudimentary discharge tubes which eventually turned into television tubes for every living room, and later into computer monitors on every desktop. Instead of ink and paper people now prefer to write with electrons on phosphorescent screens. Electrons carry the current through keyboards, electrical appliances, transistors, integrated circuits, and computer microchips. Electrons flow through the central nervous system to maintain consciousness. The motion of electrons from atom to atom, from molecule to molecule is the basis of all chemistry, biology, and life itself. The electron has thoroughly permeated not only our technology but also our culture through electronic mail (e-mail), the electronic network, and electronic commerce. Electrons flow through the core of modern life. Whenever you read email, remember the *e* stands for the electron.

If the negatively charged electron is one of the crucial building blocks of the atom, the rest of the atom must be positively charged, since the atom, as a whole, is electrically neutral. What is the arrangement of the negative and positive electric charges within the atom? Since the negative electron is 2,000 times lighter than the hydrogen atom, most of the atomic mass must be positively charged. Some proposed that the positive charge must be distributed over a spherical atom, with negative charges embedded within the sphere, like raisins in pudding, *the raisin-pudding model*. Spherical symmetry was an obvious and aesthetically pleasing choice for the model of the atom. Others pictured electrons to be smeared out like negative icing over a positively charged spherical cake. Another appealing analogy was that lightweight electrons orbit a massive, central, positive charge at the centre, like light planets revolving our massive sun. To resolve atomic structure demanded a look *inside* the minute atoms, a daunting challenge, considering that no one had yet realized an image of a complete atom.

PENETRATING RAYS

Near the turn of the 20th century, the treasure of *radioactivity* provided rays powerful enough to peer into the atom. Remarkably, just a few weeks before radioactivity appeared on the scene, the discovery of X-rays provided a major tool to probe inside the human body. Both finds were intimately related.

X-rays were yet another surprise to emerge from the prolific discharge tube. Paying close attention to minor, but bothersome, details, Wilhelm Roentgen (1845–1923) first noticed how photographic plates placed near the tube became annoyingly fogged. Even when he tightly wrapped the plates in black paper to exclude stray light, the vexing problem persisted, foggy plates. What were these invisible, penetrating rays ensuing from the discharge tube? One fateful night, after experiments were complete and the gas tube shrouded over by a black mantle, a marvellous surprise greeted Roentgen. There was a distinct glow on a phosphorescent paper sitting far away on the other side of the room. When later asked "What did you think?" about this crucial moment of discovery, Roentgen reacted:[22] "I did not think, I investigated." But think and act and think he did. Putting pieces of wood, rubber, tin foil, decks of cards, and thick books between the tube and the phosphor-paper, he found that some rays went right through to light up the remote paper. Imagine his astonishment when his hand also failed to block the rays but cast an eerie shadow of his bones instead. The flesh of his hand seemingly melted away in the path of the rays. With their identity very much in question, he thought to call them X-rays. After weeks of secret work in a frenzy of mental activity and excitement, he finally introduced the discovery to his wife by X-raying her hand just a few days before Christmas, 1895. The X-rays penetrated skin and tissue, but her bones and her gold ring blocked the phantom light. A sharp shadow skeleton on the photographic plate gave the world one of the most famous pictures in the history of science (Figure 2.21). The X-ray was to be a Christmas present for humanity – 20th-century physics began a few years early.

Few discoveries have captured the popular imagination more quickly. Discharge tubes had been making X-rays for years, but no one realized it until Roentgen. After his first public lecture, students continued thunderous applause for one full hour. Monarchs summoned scientists, including Roentgen, to give X-ray shows. Laboratories around the world promptly repeated the result. Thomas Edison boasted that it was child's play to reproduce photographs of skeleton hands. By spring of the following year, dentists were X-raying teeth for cavities. Doctors were examining the injured and sick to find bone fractures, kidney stones, and gall stones. Police detectives were searching for bullets inside the brains of a murdered woman. Humourists warned against the intrusive power of the new rays:

X-ACTLY SO! (appears in)[23]
The Roentgen Rays, the Roentgen Rays
What is this craze?
The town's ablaze
With the new phase
of X-ray's ways.

 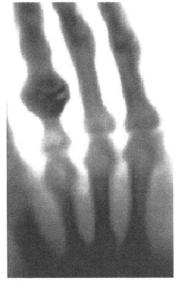

FIGURE 2.21   Roentgen and the first X-ray photograph.[24]

I'm full of daze
Shock and amaze
For now-a-days
I hear they'll gaze
Thro' cloak and gown – and even stays,
These naughty, naughty Roentgen Rays.

Today X-ray machines are routine at dental clinics, hospitals, and airports. Roentgen received the first Nobel Prize in physics for the monumental find. Ever since that time, a Nobel Prize has elevated the recipient to a lofty position.

Avogadro's Number Revealed

Avogadro had his moment in history with his law that equal volumes of gas carry the same number of molecules, which is also the number of molecules in one mole of a compound. But he did not know the value of the famous number that bears his name. Today it is possible to determine this number in many different ways. Perrin determined the number from Einstein's analysis of Brownian motion, and Einstein made an independent estimate from the diffusion of sugar molecules and the viscosity of sugar solutions. Here we use the molecular dimensions of salt molecules. X-ray analysis of the cubic salt crystal (NaCl) reveals the structure of the crystal. If you shine light at a cube at various angles the shadows form different two-dimensional images, which can be reconstructed with mathematics into the three-dimensional model of the cube. X-ray diffraction uses similar principles – two-dimensional ordered images formed by scattering X-rays off a crystal can be transformed into a model of the crystal's molecular structure.

X-ray modelling of the crystal shows us that the side of the cube of the salt crystal is 5.64×10^{-10} m with four atoms of Na and four atoms of Cl inside the crystal. From that information we determine that the volume of one NaCl crystal is 1.8×10^{-28} m^3. Simple scaling yields 5.55×10^{27} unit crystals in one cubic metre, with four molecules of NaCl each. Therefore, there are 2.22×10^{28} molecules of NaCl in one m^3. The density of NaCl is independently known via measurements (à la Archimedes) to be 2.2×10^6 gm/m^3. Therefore 2.22×10^{28} molecules have a mass of 2.2×10^6 gm. Since the molecular weight of NaCl is 58.4 (adding together the atomic weights of Na and Cl) we find the number of molecules of NaCl 58.4 gm (one mole of NaCl). This will be Avogadro's number, N_A. Thus the famous number is:

$$N_A = \left(58.4 \text{ gm}/2.2 \times 10^6 \text{ gm/m}^3\right) \times 2.22 \times 10^{28} \text{ molecules/m}^3 \sim 6 \times 10^{23} \text{ molecules/mole}$$

Photograph 51 – The Secret of Life

X-ray crystallography played a seminal role in solving the three-dimensional molecular structure of DNA (deoxyribonucleic acid), the secret of life. Unlike crystal gems, biological and chemical molecules in the liquid or the gas form dance around randomly, to give only a hazy shadow when illuminated by X-rays. To get a clean image of the underlying structure, the technique is to form a crystal from a solution of molecules so that the atoms of the giant molecule are locked into position. Now X-ray diffraction patterns become regular as for the salt crystal; the structure in three-dimensional space can be deciphered. Linus Pauling and Robert Corey won the 1954 Nobel Prize by solving the molecular structure of several proteins using X-ray diffraction. Ten years later Dorothy Hodgkin at Oxford won a Nobel Prize for solving the crystal structure of penicillin.

In our main story, the key players in the discovery of the structure of DNA were James Watson, Francis Crick, Maurice Wilkins, and Rosalind Franklin. Franklin was recruited to Wilkins laboratory by Wilkins' boss for her famous work on X-ray crystallography of coals. Despite her boss' suggestion to collaborate with Wilkins on the on-going effort to crack the structure of DNA, the fiercely independent female scientist, thrown into a world dominated by men, defiantly would not agree. Besides, Franklin aimed to make her own ground-breaking discoveries, and so had little desire to work as anyone's assistant.

Crystallized DNA material presented great challenges to photograph clearly. Wilkins found that the crystallized molecules switched configurations from "dry" to "wet" generating fuzzy photographs. Franklin found a way to stabilize the wet configuration by increasing the humidity of their environment. In a few weeks of working on the project she generated the most beautiful X-ray photographs, the best in the world. Franklin took a massive number of such photographs with increasing clarity, but she was not equipped with the advanced mathematics to transform those images into a 3D model. Wilkins agreed to work on the dry form of DNA to keep the peace with Franklin.

Enter young James Watson, bird biologist, who had developed a burning desire to unravel the secret of life. But he was totally ignorant of X-ray diffraction techniques. He was going to solve the structure of DNA by building molecular models out of sticks and spheres. Watson teamed up with a bombastic, 23-year-old Francis Crick, who had switched fields from physics to biology, to master the complex mathematical theory of crystallography. Crick's dream was to decipher the structure of the molecule of life that carries hereditary information.

After attending a seminar by Franklin where she showed some of her crisp X-rays, an excited Watson turned to Crick to begin building models. They realized they had to assemble several strands of long and complex molecules. The race was on to decipher the code of the messages hidden in the DNA. When they showed one of their first attempts to Wilkins and Franklin, Wilkins was not impressed, but politely remained quiet. Franklin on the other hand tore the model apart piece by piece.

The competition intensified. Soon after the failed model incident she was successful in taking a most perfectly crisp image (Figure 2.22) which she labelled "Photograph 51." When Watson visited Wilkins, Wilkins went into Franklin's laboratory and showed Watson the masterpiece, without asking for Franklin's permission. Watson was stupefied by the clarity of the photography and the message it bore.[25]

> The instant I saw the picture my mouth fell open and my pulse started to race. The pattern was unbelievably simpler than those obtained previously … The black cross could arise only from a helical structure … After only a few minutes' calculations, the number of chains in the molecule could be fixed. DNA had to be made of two intertwined, helical chains.

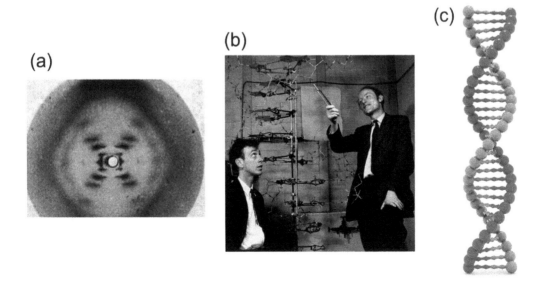

FIGURE 2.22 (a) Famous X-ray *Photograph 51* taken by Rosalind Franklin[26] revealing to James Watson the double-helix structure of DNA. (b) James Watson (left) and Francis Crick (right) discussing their successful 3D model of DNA.[27] (c) 3D rendering of DNA showing the double helix.[28]

Rushing the news to Crick the model-building moved swiftly into a new electrifying phase. Two helical strands of the double helix of DNA locked together, as in a zipper. The structure manifested the function of duplication for genetic material. As Watson recalled, Crick could not resist boasting they had found the secret of life. They invited Wilkins to look at their newest model.[29]

> [It] had a life of its own—rather like looking at a baby that had just been born … The model seemed to speak for itself, saying—'I don't care what you think—I know I am right.'

Franklin saw the model few days later, and she too was quickly convinced. The beauty of the structure testified to its truth. On April 25, 1953, Watson and Crick published a paper in *Nature* magazine: "Molecular Structure of Nucleic Acids: A Structure for Deoxyribose Nucleic Acid." Accompanying the article was another, by Gosling and Franklin, providing strong crystallographic evidence for the double-helical structure. A third article, from Wilkins, corroborated the evidence with experimental data from DNA crystals. The historic collaboration had revealed the molecule of life.

In 1962, Watson, Crick, and Wilkins won the Nobel Prize for their discovery. Franklin was not included in the prize. She had died in 1958, at the age of 37, from ovarian cancer. Crick later remarked,[30] "It is the idea of what it does." As the blue-print for life, the double-helical molecule carries the instructions to build, run, repair, and reproduce life in all its forms. Once we learn to manipulate this chemical, diseases would be cured, fates changed, futures reconfigured. The complete DNA sequences of many organisms have been solved including the human genome. In the next decades the cracking of the DNA code will lead to phenomenal benefits in medicine and agriculture.

RADIOACTIVITY

From where in the discharge tube did the magical X-rays emerge? When electrons bombarded the coated screen at the end of the tube, they made the coating fluoresce to emit light and X-rays. Could fluorescence be the source of the penetrating rays? Many materials were known to naturally fluoresce after exposure to sunlight. Perhaps natural fluorescent materials harboured the same magic. Was there any connection between X-rays and exposure to sunlight? The race was on to find a natural source for X-rays.

In Paris, Henri Becquerel (1852–1908) was familiar with a uranium salt that glowed mysteriously after brief exposure to sunlight. Could it also be emitting invisible X-rays? It was easy to find out. Wrapping a photographic plate in black paper to exclude light, he placed the precious salt on it. In between the salt and black paper he put a metal cross which he hoped would cast a shadow. Alas, the day was cloudy; there would hardly be enough sunlight to induce fluorescence in the uranium. Becquerel safely tucked the package away in his drawer to wait for the sun. After three successive gloomy days, an impatient Becquerel developed the plates anyway. What a delightful shock it was to find a clear shadow image of the metal cross on the plate. Becquerel had discovered spontaneous emissions from uranium. "Becquerel rays" continued to pour out from the salt for weeks on end without any attenuation. Making air conductive, the rays discharged electrified bodies. Becquerel named the uranium emission *radioactive*.

How can a lump of inert substance sitting on a laboratory bench emit such powerful rays? What is the underlying cause of the emissions? Were there other more powerful radioactive materials? The hunt was on. The prize would go to another Cinderella figure in the eternal quest for the structure of matter. Escaping from oppressive Russian-dominated Poland to Paris, Marie Sklodowska (1867–1934) longed for freedom and education in the West. Paris was more hospitable to female students than any other European city. She was busy eking out a living by cleaning laboratory apparatus at the Sorbonne when she fell in love with Pierre Curie, and married him after finishing her degree. By this time, radioactivity had moved to the centre of the scientific scene. For her doctoral thesis, she started work in her husband's laboratory when she made a baffling discovery. While examining dirty ores instead of pure metals she found a remarkable ore called pitchblende emitting more radiation than the pure uranium, even though pitchblende contained no uranium. Others would have dismissed this as a minor detail. But Curie guessed there must be a trace of something more powerful inside the ore. Could they extract it? Carloads of pitchblende from Bohemia went through the Curies' kitchen.

With Herculean effort, the husband and wife team crushed, filtered, and chemically separated tons of ore by manual labour. In five months, Curie purified a scant few milligrams to identify two new elements, new additions to the Periodic Table with mysterious and alarming powers. The first was 300 times more powerful in radioactive emanations than uranium. She named it polonium, in honour of her home country, proud patriot that she was. The second element emitted rays with nearly a million times the intensity of uranium. As the quintessence of radiation, she gave it the name "radium." Trail-blazer of the first order, Marie Curie became the first woman in France to earn a doctorate in science, the first woman professor in the world, the first woman to win a Nobel Prize, and the first scientist to win two Nobel Prizes, one in chemistry and one in physics. Despite her illustrious career and amazing successes, Curie was rejected induction by the all-male French Academy of Sciences.

Overnight, radium turned from a laboratory "Curie-osity" into a sensational discovery with glistening promise. It glowed so strongly in the dark you could read by its light. Pierre and Marie held the purified radium near their closed eyes and saw flashes and meteors inside their eyeballs. Radioactivity from radium released enough energy to boil water. Diluted in a liquid, it became a popular phosphorescent paint for chorus girls' costumes, gleaming casino chips, and luminous watch dials. But the rays are dangerous. The "Radium girls" who diligently painted watch dials with luminous paint in watch factories contracted radiation poisoning and cancer. They were instructed to bring the paint-brushes to a fine point by using their lips, so ingesting the powdered radium.

Rays from the perpetually glowing solid could burn off warts. Some touted radium as a new wonder cure for skin cancer, gout, and rheumatism. In St Petersburg a doctor claimed to have cured blind mice with radium emanations. Indeed, modern medical diagnostic procedures rely on trace quantities of radioactive materials. But the Curies realized the dangers too late. Both Marie Curie and her husband suffered severe skin burns. Her laboratory notebooks were so dangerously contaminated that they must be kept inside lead-lined boxes. Marie eventually died of cancer from cellular mutations inflicted by the deadly rays. As special tribute, element number 96 earned her name.

Radioactivity remained a subject of profound astonishment for some time. How could a lump of apparently inert substance emit such powerful, invisible rays? The spontaneity of the radiation was truly an enigma. What was the underlying cause behind the emission? How was it possible for radium atoms to manufacture energy seemingly out of nothing? It was as astonishing as perpetual motion. The answers will emerge from the developing story of the structure of the atom. Atoms of the heavier elements of the Periodic Table are unstable and spontaneously break up due to the action of the weak force, releasing energy.

TO THE HEART OF MATTER

Although science stood at the brink of the atomic era in the first decade of the 1900s, civilization was stuck in the horse and buggy age. Even bicycles were a novelty. Although rich homes had electric lights, there were no refrigerators, and telephones were rare. The first cinematographs, primitive motion pictures, opened in just big cities. Television and aeroplanes were decades in the future.

In England, Thomson's student Ernest Rutherford (1871–1937) was ready to use radioactive emanations to penetrate the mysteries of atomic structure. Growing up on a New Zealand sheep farm he joined Cambridge at a thrilling moment with so many fundamental discoveries swirling around. How were positive and negative charges distributed inside the atom? With lively curiosity, he began to investigate the nature of emissions from radium, polonium, and thorium. How could radium produce such energy, as if from nothing? There were clearly several types of rays, judging by the different thickness of metal foils required to stop them. Using electric forces, he succeeded in separating and identifying three types of rays, appropriately naming them *alpha*, *beta*, and *gamma* rays. Gamma rays carry no charge he deduced, since electric forces could not deflect them from their straight path. Like X-rays, they penetrated metal foils, but they were far more energetic than X-rays, going through as many as half a dozen pennies. (Gamma rays and X-rays are both light rays with higher energy.) Beta rays were akin to Thomson's electron rays, betraying their negative charge from the polarity of their deflection by charged metal plates. The alpha rays were new. Easily stopped by strips of foil, they appeared uncharged at first. But when Rutherford deployed stronger electric forces, alpha rays turned to the opposite side as electron rays. Alpha rays were positively charged particles, and significantly more massive than electrons, even more massive than hydrogen atoms. Rutherford placed small samples of radium and polonium outside a sealed glass tube to find helium gas accumulate in the tube where none existed before. Alpha emanations must be charged helium atoms speeding out of radium like atomic bullets.

Energetic thoroughness was Rutherford's characteristic trait. If helium atoms were flying out from radioactive substances, what did they leave behind? To find out, he teamed up with chemist Frederick Soddy when he became professor at McGill University in Canada. Together they identified a new element, now called radon, emerging from radioactive thorium and radium. It appeared as if a *radium atom was breaking up* into separate pieces. One was a radon atom, with atomic number 86, and the other a helium atom, atomic number 2. Both were noble gases. Rutherford could now explain radioactivity to arise from the break-up of atoms.

What a rude shock for the atomists! Atoms of one kind were disintegrating into atoms of another kind! Radium atoms were crumbling into helium and radon atoms. One element from the Periodic Table was breaking up into two other elements. Soddy was overcome with excitement to be[31] "chosen from all chemists of all ages to discover natural transmutation." Yet Rutherford was cautious about such medieval fantasies.[32] "Don't call it transmutation," he warned. "They'll have our heads off as alchemists." For the discovery of transmutation, Rutherford received the Nobel Prize in chemistry. Later, he artificially transformed nitrogen (atomic number 7) into oxygen (atomic number 8) by bombarding nitrogen with alpha particles. New atoms with higher atomic number could be constructed from atoms with lower numbers. These were troublesome discoveries about the atom of the elements, as Mendeleyev had feared. If the atom has sub-components, how would it ever be possible to reconcile the mutation of matter with the eternal stability of its elementary constituents? So much for the word "elements." The very definition of *elementary* as a property of the atom was in question. The bedrock of chemistry was under threat.

Rutherford's most important work was yet to come. He was after the structure of an atom. Where do the positive and negative charges reside on the atom? With a mass four times that of a hydrogen atom, and a speed of more than 1 million metres per second, the positively charged alpha particle atomic bullets proved ideal ammunition to strike at the heart of matter. His idea was to shoot alpha rays emerging from polonium or radium through a thin foil of gold, just a few atoms thick. He planned to study the deflection of the emerging particles by catching them as spots of light when they struck a surrounding screen of phosphorescent zinc sulphide. Each scintillation was a single alpha particle collision with the screen. By counting how many particles emerged, and at what angle, he hoped to learn about the structure of atoms. But the target material had to be as thin as possible to avoid masking the effect of single interactions with individual atoms. Gold has the convenient property that a foil can be rolled out into an extremely fine thickness of about 400 atoms, just 20-millionths of an inch.

As if there was nothing inside the atom, most of the alpha-bullets sailed right through the foil to emerge nearly undeflected, and light up a fluorescent screen placed behind the foil (Figure 2.23). Occasionally the alphas changed their direction slightly. Gold atoms are completely transparent to

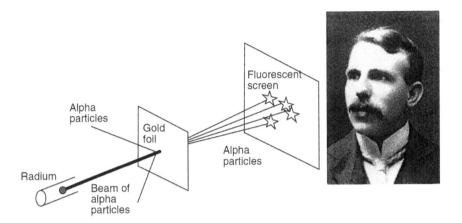

FIGURE 2.23 Rutherford's famous experiment that revealed the structure of the atom. A beam of fast-moving alpha particles emerging from a radium source bombards a very thin gold foil. Most of the alpha particles traverse right through the foil with little or no deflection. But 1 out of every 10,000 alpha particles "bounces back" with a large angle.[33]

bombarding particles. Proponents of the raisin-bread atomic structure could claim victory. Massive, fast-moving particles ran through a uniformly distributed ball of positive charge like William Tell's arrow through an apple, scarcely deflected by the tiny electron-like apple seeds.

Demanding thoroughness once again, Rutherford suggested to a young colleague, Robert Marsden, who had just joined his research group, to look at all angles around the foil, not just behind the foil. Marsden sat in a dark room for hours watching for flashes of light as alpha-particles struck the scintillator. Much to Rutherford's surprise, Marsden succeeded in finding a few energetic alpha particles occasionally emerging *backwards*! The rare find was a tribute to Marsden's acuity and precision, since only about one in every 10,000 alphas scattered backwards. It was a baffling result, but to Rutherford's insightful mind it was a stunning revelation:[34]

It was almost as incredible as if you fired a 15 inch [cannon] shell at a piece of tissue paper and it came back and hit you.

Reflecting on the results for two long years, Rutherford emerged with a breakthrough planetary model for atomic structure. Alphas sail right through matter because atoms are mostly empty space. Occasionally a positively charged alpha particle bullet comes near the very small positively charged *nucleus* of the atom, and reflects off the hard core like a ball bouncing off a hard wall. The nucleus diameter is about 100,000 times smaller than the entire atom. Lightweight electrons buzz around the nucleus like planets around the sun. The raisin-bread model was dead.

It was an epoch-making discovery. To approach the inner sanctum of the nucleus, Rutherford's fast-moving alpha particles provided a million times more energy than Thomson needed to rip electrons off an atom in his discharge tube. Once again, reflecting upon the atomic breakthrough, Robb delights us with:[35]

What's in an atom, the innermost substratum
That's the problem he is working at today.
He lately did discover how to shoot them down like plover,
And the poor little things can't get away.
He uses as munitions on his hunting expeditions
Alpha particle that out of Radium spring.
It's really surprising, and it needed some devising

How to shoot down an atom on the wing.

Rutherford brought us the modern atom. For our civilization, the impact of Rutherford's eureka moment, the discovery of the nucleus, was tantamount to the discovery of fire by prehistoric humans. Through the unprecedented power it confers upon humanity, the nucleus has become both the talisman and the terror of our age. The ability to manipulate the nucleus provides 20 per cent of the world's energy through nuclear fission. Sad to say, as with fire, humanity has found ways to use the same power for mass destruction, threatening the very survival of the human species on planet earth, perhaps threatening all life.

Chemical elements are not elementary. Their atoms comprise many electrons buzzing around a central positive nucleus. The number of electrons (atomic number) determines the location of chemical elements in the Periodic Table. Hydrogen consists of one electron. Iron has 26 electrons. Near the bottom of the table, radium has 88 electrons. Familiar chemical properties, such as an element's inertness or its readiness to form chemical compounds, arise from its electronic structure, i.e., how electrons arrange themselves around the nucleus. The story of electronic structure emerges from quantum theory which we will discuss in Chapter 12.

With the atom dismantled, Rutherford's nucleus and planetary model for the electrons raised a constellation of troubling questions. How are the electrons distributed in the empty space around the nucleus? Why do the negatively charged electrons not simply fall into the positively charged centre? As we will see from Chapter 3 *Updates on Electricity and Magnetism*, a circulating charge should lose copious amounts of electromagnetic energy as it moves in the circle. If the electron indeed circles around the nucleus, it radiates its energy. Rutherford's picturesque analogy of electrons circulating around the nucleus like planets around the sun fails, as the electron spirals into the centre. As we will see in the chapter on the quantum (Chapter 12), clues from the nature of light and emission of light from the discharge in gases will provide fascinating answers to the many puzzles about where to place the electrons around the nucleus. The answers will open new doors to nature's wonders.

At the summit of his atomic enterprise, Rutherford wished to cast his vision even further. When he transmuted nitrogen (atomic weight 14) into oxygen (atomic weight 16) by bombarding nitrogen with alpha particles (atomic weight 4), hydrogen nuclei came flying out, as if there was remaining debris flying out. In a flash of insight, he realized that since hydrogen is the lightest atom, the hydrogen nucleus must be an integral part of nitrogen, oxygen, and helium nuclei. Perhaps all nuclei are conglomerates of hydrogen nuclei. What better name to give these primary nuclear particles than *protons*, recalling Prout's protyle. Thus was born the proton, the first sub-nuclear particle.

If the nucleus is so very small and positively charged, what holds the protons together, so close to each other? Why do they not fly apart from the ferocious mutual repulsion within the positively charged core? Is there a force much stronger than electrical operating within the nucleus?

POLTERGEISTS

There has to be another constituent to the nucleus. Helium is element number two, so it has two negative electrons and two positive protons inside the nucleus. But if the nucleus of helium has two protons, why does it have atomic weight four, equivalent to a mass of four protons? Indeed, a glance at the Periodic Table reveals that the nuclei of most elements have more than twice the atomic weight as the number of protons. For example, chlorine with atomic number 17 has atomic weight 35. Rutherford conjectured the existence of electrically neutral particles inside the nucleus, the *neutrons*, perhaps comprising an electron and proton, more tightly bound together than in the hydrogen atom. The idea of another sub-nuclear particle was born.

In 1932, James Chadwick identified a mysterious form of radiation emerging from beryllium metal bombarded with alpha particles. He was studying charged particle tracks formed in a *cloud chamber* where saturated water vapour in a sealed container formed minute droplets revealing the paths of fast-moving charged particles. When he exposed a nitrogen filled cloud chamber to "beryllium-radiation," Chadwick saw visible tracks fly off from the middle of his chamber without any incoming tracks. Invisible rays seemed to hit nitrogen atoms, charge them up, and propel them. With shrewd insight, he realized that the strange beryllium-radiation must consist of uncharged particles, which formed no tracks. Without seeing its tracks, Chadwick had discovered the neutron. Moving like a ghost though the cloud chamber, a neutron projectile struck a nitrogen atom, knocked off an electron, and propelled the departing charged nitrogen ion to form a visible track. From the sudden recoil of different gas atoms inside the detection chamber, he determined the mass of invisible neutrons to be nearly the same as proton mass. The neutron was confirmed. A strong nuclear force holds together the protons and the neutrons so the positively charged protons do not fly apart.

Neutrons introduce fascinating new properties to atoms and chemical elements. Bombarding atoms with neutrons can increase their atomic weight, producing new elements. Thus the Periodic Table can be artificially expanded. Technetium (Tc) was the first element to be produced artificially by bombardment with a deuteron (proton plus neutron). Its existence was another one predicted by Mendeleyev who named it eka-manganese.

Atoms of elements come in different versions, called isotopes, all with the same number of electrons, but with different numbers of neutrons in the nucleus. Deuterium above, also called heavy hydrogen, is a stable isotope of hydrogen. It is naturally present in oceans in a trace quantity (0.02%). Water with a high concentration of deuterium instead of hydrogen is called "heavy water." Some isotopes are stable, lasting essentially forever; other isotopes are radioactive and unstable, which means they can decay spontaneously over a period of time. Thus Curie's radium decays into mostly radon over a half-life of 1,600 years. Half-life means if you keep track of 1 gram of radium, half of that substance will be left as the original radium after 1,600 years.

Carbon has three naturally occurring isotopes, C-12, C-13, and C-14. C-12 is the familiar stable form and has the same number (six) of protons and neutrons. C-14 has two extra neutrons and decays into C-12 with a precise half-life of 5,730 years. The well-known decay rate can be used as a radioactive clock. Radioactive dating is used in geology to determine the archaeological age of layers of rocks in the different eras of evolution of life forms (Cretacious, Jurassic, Devonian...). Radioactive isotopes have an amazing range of half-lives, for example rubidium decays into strontium with a half-life over 50 billion years.

Most of the carbon in the carbon dioxide of the atmosphere is C-12, with a very small fraction (one in one trillion atoms) of C-14. This fraction stays constant over centuries in spite of the decay of C-14 into C-12. The reason is that cosmic rays from the universe constantly convert a small amount of N-14 into C-14, keeping the C-14 content essentially unchanged. N-14 has seven protons and seven neutrons. When high energy cosmic rays (Chapter 13) bombard a proton of the N nucleus they convert it to a neutron, reducing the atomic number from seven (nitrogen) to six (carbon), but keeping the atomic mass at 14. Over centuries the rate of cosmic bombardment is about the same (small changes can be corrected for by other means).

All *living* creatures from plants to animals to humans possess the same ratio of C-12 to C-14 because living creatures are dependent on the carbon content of the atmosphere. But when a plant or animal dies, it is disconnected from the atmosphere. There is no more C-14 that enters the organism, and the amount C-14 starts to decay slowly to N-14. Therefore the measured ratio of C-14 to C-12 in a fossil or skeleton can be used to tell the length of time since the organism died.

A most famous example of carbon dating is the dating for the shroud of Turin, which is reported to have an imprinted image of Jesus Christ crucified. The Vatican permitted a tiny piece of the shroud to be dated by three independent laboratories, all of which determined the shroud's date of origin to be between 1200 and 1300 AD. Radioactive clocks using other isotopes are used to determine the age of rocks, which can be then be tracked to the age of the earth as about 4.6 billion years.

Inside the nucleus, the neutrons are stably bound to protons by nuclear force (strong force). But a free neutron will live for only 15 minutes before spontaneous decay into a proton by emission of an electron. Because emerging electrons are the same as Rutherford's beta-rays, neutron decay earned the name *beta-decay*. A disturbing fact cropped up about beta-decay; it appeared to violate the law of energy conservation, by now sacrosanct in physics. During a beta-decay, the sum of masses and energies of the final particles (proton and electron) did not always add up to the mass and energy sum of the original neutron. Beta-decays also violated other general symmetries underlying the laws of conservation of linear and angular momentum (Chapter 1). For example, when a stationary nucleus suffered beta-decay, the emerging electron and the recoiling nucleus did not always fly off in opposite directions. Puzzling over the mystery of beta-decay, reputable physicists, including Curie, grew sceptical about whether conservation laws continued to hold in the micro-world. With unshakeable faith in conservation, Pauli in 1931 proposed[36] a "desperate remedy." An invisible, charge-neutral particle must carry off the missing energy. Being neutral, it was immune to the electromagnetic force and thereby very difficult to detect. With minuscule, or possibly non-existent, mass, Enrico Fermi later jokingly referred to it as the "little neutral one" or *neutrino* in Italian. It was a cute way to distinguish it from the neutron, the other neutral particle just discovered by Chadwick.

The neutrino appeared to possess the most bizarre properties. Fermi proposed that the neutrino would be very hard to detect because it ignored both the electromagnetic and strong forces. The neutrino was only sensitive to a new type of force: Fermi called it the *weak-force*. Thus energy conservation, which spawned the idea of the neutrino, led indirectly to the hypothesis of a brand new force. Fermi's weak-force theory also explained the process of beta-decay. But journal referees rejected his landmark paper as too speculative, which made him turn from a brilliant theorist into a superb experimentalist. The new force was so weak that neutrinos, moving at nearly the velocity of light, could on average traverse through the thickness of 1 billion earths, stacked end to end, before they could interact with matter. But how to accept ephemeral particles that hardly ever interacted?

For 26 years the neutrino remained a ghost until in 1956 Clyde Cowan and Frederick Reines snared a few from a mighty torrent of neutrinos emerging from a nuclear reactor used to generate nuclear energy. One of the hardest experiments to do at the time, they dubbed the effort Project Poltergeist, in deference to the elusive character of the new speck of matter. Project Poltergeist successfully detected a few neutrinos because a nuclear reactor produces 50 trillion neutrinos per square centimetre.

Immensely more powerful than nuclear reactors, the sun and stars produce such copious quantities of neutrinos that 1 trillion neutrinos from the cosmos rain day or night on every square inch of earth every second. Incredibly, all pass right through, except one every few days. At most one neutrino would be absorbed by a human body over a lifetime. Ambitious scientists harvest 100 per year in giant neutrino detectors. These are located inside mines deep below the earth's surface to avoid confusing signals from other cosmic rays. Contemporary literary figure John Updike makes light of the phantom speck in *Cosmic Gall*:[37]

> Neutrinos, they are very small.
> They have no charge and have no mass
> And do not interact at all.
> The earth is just a silly ball
> To them, through which they simply pass,
> Like dustmaids down a drafty hall
> Or photons through a sheet of glass.
> They snub the most exquisite gas,
> Ignore the most substantial wall,
> Cold-shoulder steel and sounding brass,
> Insult the stallion in his stall,
> And, scorning barriers of class,
> Infiltrate you and me! Like tall
> And painless guillotines, they fall
> Down through our heads into the grass.
> At night they enter at Nepal
> And pierce the lover and his lass

From underneath the bed – you call
It wonderful; I call it crass.

We return to neutrinos in the story of how the sun and stars shine by fusing four hydrogen nuclei (protons) together for form the nucleus of helium, and release massive amounts of energy in the reaction (Chapter 8). When the star runs out of hydrogen it starts fusing helium nuclei inside its energy-generating furnace to form the nucleus of lithium, the next element of the Periodic Table, and so on to forge many elements of the Periodic Table.

At the turn of the 21st century, physicists discovered that neutrinos do have a small mass, a few millionth times the mass of the electron. But neutrinos are so abundant in the universe that their total mass is significant compared to the mass of all ordinary matter which makes up stars and galaxies. There are 100 million neutrinos for every proton.

If atoms are not basic entities, are protons and electrons the new building blocks of matter? Or can we shoot at protons to find even smaller constituents?

When elementary particle physics experimenters emulated Rutherford 30 years later to determine the structure of the proton by bombarding it with energetic projectiles of electrons and other protons, a veritable riot of new particles emerged. The proton could never be a conglomeration of such a large variety of particles. Some were even more massive than protons! It was as if physicists opened Pandora's box to reveal a frightening number of new entities. We continue the story of the structure of the proton in Chapter 13 on quarks, leptons, and strings to the present view of the ultimate constituents of matter.

From Thales' bubbling imagination of water as the primary cosmic element to Rutherford's protons, the scientific odyssey through the structure of matter saw stage after stage of fundamental reconstruction with many loose ends to open leads to new frontiers. All through the journey to comprehend the substance of the universe, the process of understanding was the same. Creative speculation, insight, and aesthetic guiding principles lead to the formation of hypotheses. Observation, experiment, measurement, and precision found evidence to test hypotheses. Reason, logic, idealization, extrapolation, model building, and mathematical theory formed a comprehensive picture.

Always, the aesthetic principle of symmetry, illuminated by the language of mathematics, expanded our understanding of matter beyond that attainable through the senses alone. Pythagoras' mind envisioned the spherical earth through symmetry; Aristotle saw the curved shadow of the earth on the face of the moon. Democritus and Dalton imagined atoms, now clearly imaged by the scanning tunnelling microscope. All of chemistry stems from the elements of the Periodic Table. The best scientific advances blended inspiration, insight, and reason with concrete observation strengthened by close attention to details and precise measurements. Thomson could not see electrons, but he witnessed the shadows they cast on the phosphor-coated face of the discharge tube. Today ephemeral electrons dominate our culture as they light up television windows to other worlds and computer screens connecting minds. Rutherford fathomed the depth of the nucleus to glimpse the very heart of matter through the scintillations of bouncing radioactive emanations. Today, 20 per cent of the world's energy supply stems from the nucleus. Pythagoras, Mendeleyev, and Dalton, each in their own way grasped reality through pure thought, as Einstein echoed in his 1933 lecture, pointing up like Plato:[38]

> It is my conviction that pure mathematical construction enables us to discover the concepts and the laws connecting them, which gives us the key to understanding nature ... In a certain sense, therefore, I hold it true that *pure thought can grasp reality,* as the ancients dreamed.

But then, in a 1950 *Scientific American* article, Einstein pointed down like Aristotle:[39]

> The sceptic will say: 'it may well be true that this system of equations is reasonable from a logical standpoint. But this does not prove that it corresponds to nature.' You are right, dear sceptic. Experience alone can decide on truth.

3 Science Lost, Science Regained: The Rise of Empiricism

If at every turn we had to construct science anew … without the guidance of style and knowledge in their widest sense, how could we hope to catch this complex and infinitely fascinating world with our minds at all?

Gerald Holton[1]

THE LIBRARY IS BURNED, THE ACADEMY IS CLOSED

We return to the tragic loss of Greek and Alexandrian science where we left off in Chapter 2 before the modern updates. Over a period of two centuries (starting from 200 BC), the Romans became the dominant political power, assembling a vast empire that spread in all directions. When the illustrious city of Alexandria stood at the brink of military defeat in 31 BC, the romantic and ruthless Queen Cleopatra, seventh and last monarch of the Ptolemaic dynasty, used all her wiles to hold on to her kingdom as she fought to keep her dynasty alive. Even though she courted Julius Caesar and Mark Antony before the final battle with Rome, she could not stem the tide of Roman expansion. Alas, one battle between Egypt and Rome destroyed more than 100,000 scrolls of the great library of Alexandria. Later, Mark Antony plundered a large library at Pergamon and presented 200,000 books to his lover Cleopatra to compensate for this terrible loss. Books were still valued treasure.

Despite the steady decline of their political power, Greeks continued to be teachers in a changing world dominated by Rome's ascendance. Although Latin became the master language of the new Empire, Greek culture remained its principal component. Romans derived important deities from Greek gods and goddesses, changing only their names. For example, Zeus became Jupiter; Aphrodite became Venus. Conquering generals hauled over shiploads of Greek marble statues to decorate their mansions. After they stripped Greek islands bare, they hired artists to make copies. Without such faithful and prolific facsimiles, much of Greek art would have been lost forever. Influenced by classical Greece, symmetry was pervasive in Roman art and architecture. But there was new interest in realism, in parallel with the Alexandrians. While Greeks had sought the ideal, Romans preferred the everyday reality. Roman artists were not loath to explore commonplace subjects in ordinary settings, including mundane topics such as feeding farm animals. Their human portraits showed warts and all.

Through the glory days of the Roman Empire, Athens remained the seat of wisdom, and Alexandria the city of high learning. Even though Romans became masters of the land and the Greek people, they remained servants of their parent culture. Greeks provided engineering and medical skills needed by the Roman army. Greek slaves served as tutors of rich Roman boys. Greek academies educated Roman youth. By installing extensive irrigation systems based on the Archimedes' screw pump, Romans made the water pump popular throughout their vast empire. As they expanded westward, they imprinted Greek intellectual and artistic inheritance all over Europe.

Pragmatic and militaristic, Romans were obsessed with power and territorial expansion, focusing less interest on mathematical or scientific advances. Some notable exceptions were the literary master/philosopher, Marcus Cicero (106–46 BC), who introduced Romans to the schools of Greek philosophy, and playwright/philosopher Lucius Seneca (4 BC–AD 65) who compiled a collection of meteorological facts. Another poet/philosopher was Lucretius who wrote the poem *De Rerum Natura* (On the Nature of Things) reviving and expounding Greek atomism.

Glorifying the greatness of Rome, Virgil (around 30 BC) declared in the *Aeneid*:[2]

> Let others better mould the running mass
> Of metals, and inform the breathing brass.
> And soften into flesh a marble face;
> Plead better at the bar; describe the skies,
> And when the stars ascend, and when they rise.
> But Rome! 'tis thine alone, with awful sway,
> To rule mankind, and make the world obey.

In select places, such as Alexandria, an erudite but secluded few kept the old flames burning. Archimedes' death at the hand of a common soldier was at once symbolic of Roman lust for power and of their disdain for science. Roman intelligentsia made excellent compendia of knowledge. Comprehensive treatises on military art, architecture, agriculture, and geography were of great value to heads of Roman legions who preferred practical writings to original investigations. They wanted to consult an encyclopaedia to find immediate answers to questions about military technique, medicine, or civil engineering. An inquisitive Roman sailor, Pliny the Elder, compiled 37 volumes of facts into an encyclopaedia called *Historia Naturalis*. Vitruvius wrote a ten-volume treatise on architecture and mechanics serving as a source of inspiration to later Renaissance builders. Strabo collected geography in 17 volumes. Galen compiled 20 tracts on medicine. In these works, they threw together observations of ancient scientists, often mingled with empirical prescriptions, unverified anecdotes, and stories from sorcerers. For example, Galen, who practiced medicine in Rome around AD 130, based his descriptions of human anatomy upon animal dissections, indiscriminately reporting crocodile's blood, mouse dung, and viper's flesh as medicines. Nor did he hesitate to pass on unchecked remedies such as treatment of warts by touching ugly protrusions with beautiful gems.

One area of mathematics in which Romans took an active interest was surveying, essential to establish the boundaries of their vast empire. To control their territories, they constructed impressive networks of carefully laid out, well-paved roads and bridges over which their legions could march rapidly to any military crises. And they made important advances in state organizations, government, military institutions, and law.

While Rome prospered, Christian religion spread through Europe. Eventually, Christianity became the official religion of Rome (see timeline, Figure 3.1). Even the Roman Emperor Constantine converted to Christianity in AD 312. But in its early stages, when Church fathers were fighting for the precarious existence of their new religion, they leaned heavily on authority and doctrine. Christian theology rejected intellectual activities and Greek learning. Spirituality and after-life were considered far more valuable. St Ambrose wrote:[3]

> To discuss the nature and position of the earth does not help us in our hope of the life to come.

In spurning Greek culture, Christian thinkers attached a derogatory connotation to everything pagan in origin, including Greek thought. Not being followers of Christ, the Greeks and their accomplishments had no connection with teachings of the Bible. Translations from Greek were forbidden. Only Catholic faith contained knowledge necessary for salvation. In this mood, St Bonaventure wrote:[4]

> The tree of science cheats many of the tree of life, or exposes them to the severest pains of purgatory.

St Jerome fought a lifelong battle against the temptation to read pagan classics:[5]

> Lord, if ever again I possess worldly books, or if ever again I read such, I have denied Thee.

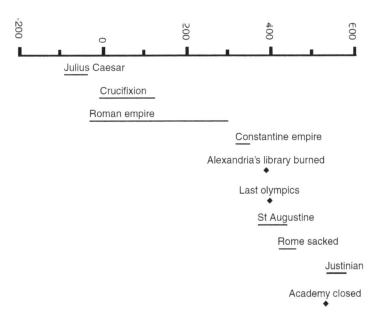

FIGURE 3.1   The rise and fall of the Roman Empire.

As the decline of Greek culture accelerated, the value of scientific thought continued to diminish to near extinction. A tragic result of Roman and Christian attitudes toward Greek learning was a growing condemnation of continuing scholarly activities at Alexandria. A remarkable case was the attack on the beautiful and talented Hypatia. Philosopher, mathematician, astronomer, and head librarian, she attracted many students to her eloquent lectures. The participation of women in the advancement of science was a unique development of the complex and cosmopolitan Alexandrian culture. Donning her philosopher's cloak, she would walk through the town, like a peripatetic philosopher, giving lectures interpreting Plato and Aristotle. Her commentaries made Apollonius' conic sections intelligible to students. But Christians of Alexandria had a different agenda, demanding an end to scientific activities. After a serious conflict between the learned Hypatia and Cyril, the Bishop of Alexandria, a riotous mob pulled the rebellious woman from her carriage and ruthlessly tore her body to pieces.

At the Museum, scholars locked themselves in their tower, growing increasingly isolated from the world at large. They no longer tried to bring their knowledge to the people, nor did they seek ways to apply it. Their isolation accelerated the approaching crisis of culture. Gone was the openness that encourages discussion, debate, and challenge. Gone was the spirit of the Academy and the Lyceum where Plato and Aristotle sought to disseminate knowledge, encouraging students to learn from their predecessors, and to challenge what they learned. Eventually the spread of Greek knowledge came to a complete halt when the Roman Empire started to crumble under repeated attacks of the Goths, Huns, and Vandal tribes. As the flood of invasions washed away the ancient world, the power of the disintegrating Empire shifted to the east, and the capital moved to Byzantium, later renamed Constantinople (now Istanbul, Turkey).

Denounced as a symbol of pagan culture, waves of invaders set fire to the library of Alexandria several times. A million scrolls went up in flame. It was a cosmic tragedy. Fortunately, well before the devastating conflagration of AD 389, many Alexandrian scholars fled east, carrying with them as many precious scholarly works as they could, saving the treasures of Greek knowledge from total oblivion. Some settled in Byzantium, the eastern part of the Roman Empire, remote from the ravages of massive invasions. Perpetuating the literature and traditions of Hellenic civilization, Byzantium survived as the bastion of the Roman Empire, enduring for 1,000 years after the collapse of the western half. Here, scholars kept the spirit of Greek science alive through the Dark Age that followed. They scrutinized Aristotle's sweeping theories, and even hinted at shortcomings.

Wiping out the last remaining symbols of Greek cultural heritage, Roman Emperor Theodosius ordered the Olympic Games to shut down (AD 394). Sixteen years later, Alaric, the Visigoth, marched through the gates of Rome, sacking and pillaging the heart of the glorious Empire.

All through the sprawling Empire, bridges crumbled, roadways disintegrated, and aqueducts dried up. In AD 529, one of the final few Roman Emperors, Justinian, continued to rail against the spread of paganism. Issuing a decree forbidding pagans to teach, he rubbed out Plato's stubborn legacy by shutting down forever the Athens Academy, temple of ancient philosophy. It took Europe 1,000 years to recover from the disasters. A Dark Age descended.

How intrinsically precarious is the collective human intellect! How quickly knowledge can evaporate from human consciousness. In spite of 1,000 years of intellectual advances, painstakingly accumulated at the great library, only a paltry fraction of that massive collection has passed to us. Over the next 1,000 years there was a drastic reduction, even a regression in the pace of scientific activities. A few tenuous threads of continuity remained and some meagre progress as well. Yet it is truly a wonder that the Western intellect ever survived. Even more remarkable is the fact that Europe ultimately redeemed a significant part of its treasured heritage so that intellectual activity eventually recovered to thrive again.

RETURN TO A FLAT EARTH

After the collapse of the Western Roman Empire, Europe became a purely Latin-speaking region, shutting down access to Greek culture. During the period of disintegration which followed the cultural decline, Christianity and the Church were the only forces left to assimilate the civilization of mighty Rome and to transmit it to the Middle Ages. The fragmented Roman mosaic would have shattered irreparably and dispersed forever had it not been for the unifying bond of Christianity that held the disparate pieces together. With the fall of Rome, and disintegration of their august empire into small kingdoms, the Pope became the only remaining authority. Ruler of Rome, the Pontiff became the stable focus of a rapidly transforming society. In the spirit of their founding fathers and Roman masters, Latin priests of the Christian Church remained uninterested in scientific inquiry. They shunned Greek philosophy and science as a retreat to pagan learning. Dangerous ideas could pollute the mind and desecrate Christian souls. Instead of scientific issues, Christian thinkers turned to religious dogma and questions of faith, predestination, and free will.

Father Augustine (later made a saint), who personally witnessed the onset of the collapse of the Roman Empire, warned in his handbook for Christians that neither philosophy, science, nor technical arts were of any use to a Christian:[6]

> It is not necessary to probe into the nature of things as was done by ... the Greeks ... nor need we be in alarm lest the Christian should be ignorant of the force and number of the elements – the motion and order of the eclipses of the heavenly bodies; the form of the heavens; the species and the nature of animals, plants, stones, fountains, rivers, mountains ... and a thousand other things which those philosophers either have found out or think they have found out ... It is enough for the Christian to believe that the only cause of all created things ... is the goodness of the Creator.

From this passage it is clear that Augustine himself was quite familiar with the issues and progress of Greek science. Yet in one sweeping stroke, the influential priest dismissed philosophy, physics, astronomy, biology, and geology. Christianity preached the essential unimportance, even unreality, of worldly things. This world was not only transitory, but also full of evil. Only in the invisible world of the "after-life" would followers of Christ ultimately find eternal salvation, provided they moulded their lives to prepare for Heaven. Sinful by nature, humans could only hope to achieve immortal bliss if they rejected the material of this world, refused the body and its pleasures, and devoted themselves instead to cultivating the spiritual realm.

FIGURE 3.2   *Madonna and Child in Majesty Surrounded by Angels*, 1270, Cimabue.

Painters abandoned the human body and the natural world, becoming singularly obsessed with pious themes and angelic subjects. We can appreciate the mystical, otherworldly character of their art in the conceptual compositions of Cimabue's *Madonna and Child in Majesty Surrounded by Angels* (Figure 3.2).[7] All of the figures have nearly identical features. It is even hard to find any natural distinction in Mary's visage from those of surrounding angels. With emphasis on the inward vision of the soul, human figures are mere prototypes, far from real people. Yet the power of symmetry prevails. This most basic of human instinct continues to mould expression.

The retreat of scientific culture flattened the collective mental picture of our globe into a rectangle, with Christian Hell in one corner, and the Garden of Eden in another. Lactantius, tutor to the son of Emperor Constantine, revived the antipodes problem to reject the sphericity of the earth. On the other side of the earth, up would become down, and down up!

ARABIAN NIGHTS AND THE HOUSE OF WISDOM

The negative bias of early Church leaders continued to drive scientists and philosophers toward the east, where the rise of Islam provided in many places a more welcoming climate for the survival and growth of scientific activities. As Arabia, Iraq, Syria, Lebanon, Palestine, Egypt, much of North Africa, Central Asia, and Spain, as well as parts of China and India came under Arab dominance, language and political barriers dissolved to allow scholars from different regions and backgrounds to travel and interact. The Abbasid empire united Baghdad, Basra, and Kufa to build a network of caravan roads into capitals, with vast palaces, glorious mosques, schools, and hospitals.

United under the sign of a crescent moon and stars, Islam shunned neither astronomy nor Greek science. An intense translation movement from Greek to Arabic provided the foundation for inquiry in the sciences. Besides serving their needs for timekeeping and calendar reckoning, stars provided a reliable method to navigate through vast tracts of barren deserts. Mullahs relied on heavenly motions to fix the precise hours of prayer, the direction of Mecca, and periods of devotional fasting. Along with regulating every aspect of daily life with religious laws, prayers, hymns, poems, moral principles, and social and political instructions, the Koran encouraged the study of nature, *taffakur*, as well as the mastery of nature through technology, *tashkeer*. According to Mohammed, Islam was not to be a religion of miracles, an attitude more open to Greek rationalist ideas.

From nomads, the Bedouin tribes of Arabia advanced to become the primary heirs and practitioners of ancient wisdom. In Baghdad, Haroun Al-Rashid (AD 766–809), the Abbasid king from the fable *A Thousand and One Nights*, founded a school of science that came to be known popularly as the "House of Wisdom." Gathering an army of learned Arabs together with Christian scholars whose ancestors had fled from Alexandria to Byzantium, Al-Rashid commissioned translations of Greek manuscripts salvaged from burnt-down libraries. Surviving works of Aristotle, Archimedes, Euclid, and Ptolemy became available in Arabic. More than just transmitting Greek and Alexandrian science, Arabs condensed these subjects, making volumes of commentaries while adding original contributions. Through the residue of ancient Greek stimulus, scholars coupled Persian and Indian learning, creating at Baghdad a lively bazaar of scientific knowledge. From this mixture emerged an attitude of tolerance toward the constituent cultures of a vigorous, cosmopolitan society. Baghdad became the most intellectually vigorous city in the world for 200 years, a true successor to Alexandria (see timeline, Figure 3.3).

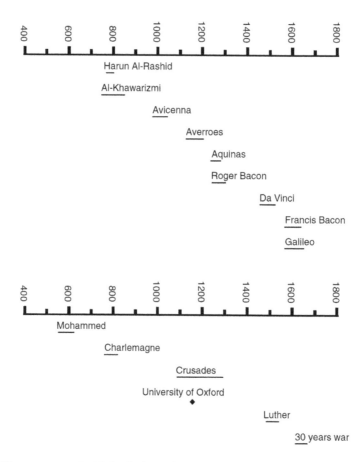

FIGURE 3.3   The recovery and rebirth of science through Islamic culture, and the re-awakening of Europe.

Islamic art imbibed Greek ideas of symmetry, expanding to multi fold symmetry, exhibited by the decorative patterns of the design in Figure 3.4. In this progressive and creative climate, Arabs made important progress in mathematics. Al-Khwarizmi wrote a treatise entitled *ilm al-jabr wa'l muqabalah*, the science of transposition and cancellation. The word *al-jabr*, which later became "algebra" in Latin, means restoration of equilibrium by adding, subtracting, multiplying, or dividing the same amounts on both sides of an equation to preserve symmetry. *Al-muqabalah* means the simplification of equations by combining equivalent terms into a single expression, removing complications, synthesizing to achieve simplicity. Thus symmetry and simplicity expanded from the Greek province of geometry to abstract manipulation of numbers. Borrowing the Hindu concept of zero, Al-Khwarizmi used it as a place holder to simplify writing of numbers, and operations of arithmetic. By converting Hindu numerals into an easily workable code, he made arithmetic so simple that even a child today can handle what would then have been considered a hopelessly complex computation.

As the empire and culture of Islam spread westward along the coast of Africa, all the way to Spain, the city of Cordoba evolved into another Muslim cultural and intellectual centre. Arabs built observatories and compiled tables of astronomical measurements to make accurate calendars. No doubt there was also a desire to make astrological predictions. Muslim doctors experimented with new drugs. Continuing Alexandrian ways, Arabs enthusiastically practised alchemy, offering recipes for converting base metals into silver and gold.

In a few centuries, Spain transformed miraculously through cross-fertilization between Muslim, Jewish, and Christian cultures. At Cordoba, the mosque was larger than any cathedral in medieval Europe. Here the Emir organized a palace library of 400,000 volumes, placing agents in cultural centres, Baghdad, Damascus, and Cairo, to scout for the latest works. Often he paid huge sums for first copies so that a new work could become known in Islamic Spain well before it started to circulate anywhere else. Arab rulers in Seville, Valencia, and Toledo emulated the shining example set by Cordoba.

FIGURE 3.4   An example of multi-fold symmetry pervasive in Islamic art.[8]

RELIGHTING THE LAMP OF KNOWLEDGE

While Islam flourished, Europe recovered sluggishly from 500 years of anarchy and confusion. Shreds of the Roman Empire evolved into the new Holy Roman Empire, extending over Italy, Prussia, and Austria. Prosperity increased through agricultural advances, such as crop rotation and substitution of animal labour for human. Christianity continued as the dominating, unifying influence. Shards of the Roman Empire evolved into the Holy Roman Empire. Cathedrals to praise the Lord began to dot the landscape. Some of the most splendid structures took centuries to erect. Moulding the physical space to express spiritual goals of Christianity, the architects' imagination soared to the infinite vaults of the Gothic arches. Their vision expanded upward to heaven. Entering the dark cavernous space of a cathedral, a radiance bursts forth from a high stained-glass window through which a rose of divine light penetrates from the purity of Heaven to reach worshippers. No windows at eye level distract visitors with concern for the dismal affairs of the real world.

In AD 784, Charlemagne, ruler of the Holy Roman Empire, ordered that all abbeys and cathedrals should have schools attached. Besides monks and clergy, children of noble lords should benefit from formal education. Although the Emperor himself could neither read nor write, he encouraged literacy, so that people would read the Bible to better understand Christian doctrine. He personally developed an interest in astronomy. An appreciative abbot praised the enlightened ruler, effusing with optimism:[9]

> If many people became imbued with your ideas a new Athens would be established in France – nay, an Athens fairer than the Athens of old, for it would be ennobled by the teachings of Christ, and ours would surpass all the wisdom of the ancient academy.

With emphasis on literacy for Bible study, Europe reawakened to the value of learning. By 1100, European monarchs founded the universities of Paris, Bologna, Oxford, and Cambridge. Supervising all instruction, the Church provided a basic curriculum. The first stage *trivium* (hence our word "trivia") comprised Latin grammar, rhetoric, and logic. It was followed by the *quadrivium*, which included arithmetic, geometry, music, and astronomy. Mathematics and science began to creep into the upper level of feudal society. But masters of the medieval schools had access only to a meagre portion of ancient thought from Greece and Alexandria. Unaware of the rich culture of Islam, which thrived right outside the border of Christendom, monks relied on distorted versions and on encyclopaedias of Greek knowledge compiled by Romans such as Pliny. Only a few of Plato's philosophical works managed to drift into Europe through Byzantium. Scrutinizing the paltry collection, scholars at cathedral schools developed a worshipful respect for ancient masters. From a knowledge of obscure titles, they became aware that the giants of antiquity had written many great works. If only they could gain access to those precious manuscripts.

With rapid expansion of the Islamic Empire, a serious religious conflict erupted between Christians and Muslims, starting in the 11th century. When Muslim Turks conquered Jerusalem, they posed a dire threat to the Byzantine Empire in the East, jeopardizing at the same time the cherished Christian tradition of making regular pilgrimages to the tomb of Jesus, the Holy Sepulchre in Jerusalem. Responding to the appeal of the Byzantine Emperor to protect his kingdom from total oblivion, Pope Urban II incited lords and knights of Western Europe with offers of Papal indulgences to launch a crusade to rescue the Holy Land from advancing Muslims.

As they took the East, zealous crusaders learned about the luxuries of perfumes, silks, and spices, unknown at home. Where did such stupendous riches come from? How could Europeans gain access? Such curiosity stimulated a new interest in geography and navigation. German chieftains and Swedish kings eagerly sought to buy Arabic weapons, tools, and jewellery. By the end of the 11th century, crusades turned to the western front. When they beat back Muslims in Spain, they discovered that the Arabs, whom they regarded as heathens and barbarians, had a higher state of culture than did the feudal lords and knights of Europe.

Falling to Christians in the year 1085, Toledo's Islamic libraries disclosed a staggering treasure of classics. Rather than discovering infidels and barbarians, Christians found a devout and learned culture. The real prize of the crusader's victories was the precious wisdom of Greek culture. What a treat for medieval scholars who trickled into Spain from the abbeys and cathedrals of Europe! As word spread about the new-found fortune, monks began to descend upon Spain in droves. One of them, Gerard of Cremona, spent more than 30 years translating nearly 100 works from Arabic into Latin. He uncovered the depth and breadth of ancient knowledge: Aristotle's works, Euclid's elements, Al-Khwarizmi's algebra, and manuscripts on medicine, optics, acoustics, chemical substances, geology, physics, mathematics, and mechanics. It was the second time that Greek works were translated. Because Latin had no words to describe many of the ancient findings, translators freely introduced Arabic words. Thus we inherited: alcohol, alcove, alkali, zero, cipher, azimuth, sofa, zenith, nadir, saffron, and nitre. As new learning enriched the meagre curriculum of cathedral schools, it eventually outstripped traditional trivium and quadrivium.

Through Islamic translators and commentators, Europeans learned medicine by studying the writings of Greek physicians, such as Hippocrates and Galen. Leonardo Fibonacci (1170–1240) discovered Arab mathematics directly from North Africa. Returning to Pisa, he wrote *Liber Abace*, the new math, where he showed systematic use of Arabic numerals and place value. Thinkers were growing alert once again to patterns in nature. One of Fibonacci's remarkable discoveries was a simple number series that emulates population growth sequence.

NIGHTMARES FOR THE CHURCH

Despite the twists and turns, the rich legacy of the past finally found its way back to Europe. Slowly, Latin translations of Aristotle spread through the monasteries. Now abbeys and universities began to scrutinize and discuss the re-discovered masters, reviving the pioneering spirit of Plato's Academy and Aristotle's Lyceum. From the new outpost of Christendom in Spain, Islam tutored Europe in the accumulated and diversified wisdom of the Greeks. Along with Aristotle's works, medieval scholars discovered insightful commentaries on Aristotle by scholars, such as ibn-Sina, the Iranian philosopher-physician, and Abu'l-Walid ibn-Rushid, philosopher from Cordoba. Unable to pronounce the Arabic, monks liberally Latinized their names to Avicenna and Averroes. Medieval scholars found Arab commentaries on classical works quite provocative. Promoting the thought of Islam, Al-Rashid affirmed that all knowledge stems from the word of Allah, through teachings of the Koran. Such ideas Muslims shared with their Christian brethren, who also believed that everything people ever needed to know comes from the word of Christ and the Bible. But Al-Rashid (Averroes) incorporated a brand-new position: teachings of faith can be supported by rational arguments. An extremely progressive idea for the time, it strongly influenced later European Scholastics. Another of Averroes' ideas was that human thought should be free from the restrictions of theological doctrine to pursue a rational path. In his *Decisive Treatise on the Harmony of Religion and Philosophy (Fasl al-maqal)*, he asserts:[10]

> It is evident that the study of books of the ancient is obligatory ... and whoever forbids the study of them to anyone who is fit to study them ... is blocking people from the door by which the law summons them to the knowledge of God, the door of theoretical study which leads to the truest knowledge of Him.

Later, Thomas Aquinas supported Averroes' position to wield a pivotal influence on the subsequent direction of medieval Christian thought.

Discovery of Greek thought and Islamic commentary presented a thorny problem to Christianity, causing nightmares for the Church. Aristotle electrified medieval Europe. Educated in the universal importance of faith and knowledge through scriptures, Christian theologians were now forced to deal with demands of rationalism. Which was the right path for humanity: faith or reason, divine revelation or reasoned insight, theology or philosophy? Christian thinkers faced a crisis.

At first, the Church reacted with vehement opposition to Aristotle's ideas. For example, one of Aristotle's primary tenets was that the cosmos is eternal. To the Church, this was a blatant denial of God's creative act. To believe that nature's processes are regular and unalterable was to deny the possibility of miracles and saints. Under penalty of excommunication, Church authorities forbid the writings of Aristotle. When a few radical students and faculty at the University of Paris invoked the name of Averroes in defiance, entrenched powers easily crushed the isolated rebellion. As before, the pagan origins of Greek philosophy posed a dangerous challenge to the central idea of Christianity: divine revelation is the sole source of truth. Anyone spreading Greek ideas faced excommunication and damnation.

A SAINT RESTORES REASON

Europe stood once again at the brink of intellectual catastrophe, a disaster comparable in dimension to the destruction of Alexandria following the rejection of Greek knowledge and culture. Thomas Aquinas (1225–74) averted the calamity by attempting to reconcile newly discovered Greek ideas with the teachings of Christianity. In commentaries he argued that reason and faith can co-exist in harmony. Science can operate according to its own principles without endangering doctrines of faith. Physical nature has determined laws, which the rational approach can reveal. There need be no compulsion to explain the operations of nature through miracles or the Providence of God. All knowledge does not have to come from divine illumination. The study of nature through rational thought is an acceptable path, and should not be feared. It must eventually lead to God, for God is the creator both of nature *and* its laws:[11]

> It should be said that the gifts of grace are so added to nature that they do not destroy it, but rather perfect it; so too the light of faith, which is infused in us by grace, does not destroy the light of natural reason divinely placed within us. And although the natural light of the human mind is insufficient to manifest things made manifest by faith, still it is impossible that those things which have been divinely taught us through faith should be contrary to what has been placed in us through nature.

Later, Galileo invoked the same line of argument to defend his investigations into the system of the world even though they appeared contrary to the literal passages in the Bible. Nature is a book written by God. But on the sensitive issue of creation, Aquinas was prudently cautious:[12]

> With respect to creatures, the philosopher and the believer are concerned with different things, since the philosopher considers what belongs to them given their proper natures, for example, that fire leaps up. The believer however considers only those things in creatures which belong to them insofar as they are related to God, for example, that they are created by God, are subject to God.

Reason alone is insufficient to explain creation. Revealed knowledge from scriptures is imperative to this key topic as it is to the subject of salvation, and cardinal questions connected with the human soul. By advocating rationalism with restrictions, Aquinas made a crucial breakthrough. Through treacherous thickets of Christian theology, he provided a path to continue rationalist philosophy by endorsing logical reasoning as a respected method for expanding knowledge. It was permitted to study Aristotle; there was no danger to the soul. Aquinas' teachings began to make science respectable in Christian Europe after the vast intellectual desert of the Dark Ages. Philosophy was no longer an undesirable, pagan activity.

It became suitable once again to study the heavens and the earth. Aquinas' teachings eventually became fully incorporated into Christian philosophy. As a new symbol of co-existence, medieval artists portrayed Pythagoras and Aristotle together with Christ and the saints on the door panels of Gothic cathedrals.

Gradually, natural science and mathematics became important elements in education. Ignorance of these subjects even became a reason for embarrassment. Eventually, the study of Aristotle became part of the standard curriculum. By the 13th century, Aristotle and Christian theology blended so fully together that monks studied the master with total devotion, but little criticism. Ignoring flaws, they put Aristotle on par with the scriptures. In their zeal to recapture the Golden Age, they never aspired to surpass it. By sanctifying Aristotle's ideas, they failed to appreciate his methods, completely missing how Greeks had advanced their knowledge and understanding by regular debate among themselves, and continuous challenge to ideas of their predecessors. Aristotle gained the stature of the dominant authority in natural philosophy.

Teachers who followed Aquinas worked myopically through Greek masters, translating assiduously, writing volumes of commentaries. Treating knowledge as an inheritance, they sought to purify and return it to its original form before passing it on to the next generation. For the central dogma of faith, they aimed to arrive at proofs by pure logical arguments. With renewed enthusiasm for logical exposition, they lost sight of how intellectuals of antiquity had used both observations as well as logical reasoning to advance knowledge and understanding about nature. Like their Roman predecessors, Scholastics eagerly compiled encyclopaedias. Anything needing explanation could be found in works of ancient sages or in the Bible. Aquinas' reconciliation of Aristotle was so effective that Scholastics, by now in virtual control of all academic establishments, held that there was no greater authority on nature's operations. If there were matters Aristotle had not talked about, such matters were not even worth considering.

True or not, there is a popular anecdote about a dispute among a group of medieval Scholastics who gather together to answer how many teeth there are in the mouth of a horse. Each one offers long and weighty arguments, why the number of teeth should be so and so. A resolution of the question seems impossible. Now if a copy of Aristotle's appropriate work was at hand they could just look up the answer. It so happens that a child nearby is listening to their arguments. "Why don't you just pry open the mouth of the horse, and count the teeth," asks the innocent child. Hearing this impertinent proposal, the elders get very angry. Scolding and ridiculing the ignorant child, they drive the boy away.

A MEDIEVAL MONK CUTS A NEW PATH

As Plato and Aristotle took a choke-hold over the medieval mind, resurrected ideas turned into dogma. There was hardly any drive for new knowledge, except for the daring comments of a few, in particular an upstart English monk named Roger Bacon (1214–92). He started out as a typical Scholastic of the age. Born into a wealthy family, and well-versed in trivium and quadrivium, he became a popular lecturer on works of Aristotle as well as on Greek rationalist and deductive methods. Following the norm, he wrote volumes of numbing commentaries on Greek masters. However, before reaching age 30, he started to doubt whether the Scholastic approach, or even Greek methods of pure rational deductions, could be the most productive routes to expanding knowledge:[13]

> Of the three ways in which men think and acquire knowledge – authority, reasoning, and experience – only the last is effective and able to bring peace to the intellect.

It was true that rediscovery of Plato sharpened the medieval mind by exposing it to logical, rational thought. But Plato's emphasis on the *ideal-world* of universal forms over the tarnished *real-world* of experience had dulled the eyes of scholars. Bacon criticized his contemporaries for wasting paper and ink with volumes of trivial Scholastic summaries and stupefying, logic-chopping expositions that paralysed the mind. He proposed to lay bare the secrets of nature by positive study. Instead of being cooped up in school rooms, scholars should get out into the open air to make fresh observations. Bacon might have said: "Consider a fragment of wisdom set down by Pliny. Goat's blood can destroy diamonds. But did anyone bother to try it?"

As a Franciscan monk, Bacon was strongly influenced by the teachings of the founder of his order. St Francis (1182–1226) rejected rote learning as the most virtuous way to practise Christian faith. Instead, he praised the spiritual value and the beauty of nature, urging fellow monks to abandon their monastery sanctuaries, go out into the world, appreciate nature, and minister to the poor and needy. Breaking free from ritual and dogma, Francis strived to re-humanize religion by preaching the original message of Christ.

Bacon sought a way to go beyond the Greeks. It was more important to observe natural phenomena than to speculate on metaphysical causes, as was the rage of his time. Crying out for a reform of the universities, he proposed investigations into every corner of knowledge. Hypotheses should be tested by experiment.

While most scholars were busy making amalgams of theology with Aristotelian philosophy, Bacon dabbled with curved bits of glass, called "burning glass," to focus the rays of the sun and light candles. Later, these glass artefacts took on the name of *lens*, after their lentil-like, convex shape with outward curving glass surfaces. The convex lens also served as a magnifying glass. When held near the page it helped the aged to read. Bacon shaped lenses into spectacles to improve the vision of the elderly. Held close to the near-sighted eye, a concave lens spectacle, where glass surfaces are curved inward, helped adolescents to see distant objects clearer.

Optical studies raised strong objections from fellow monks. Should artificial instruments aid the natural eye? God gave humans the gift of eyes to recognize the true shape, size, and colour of objects. Devices such as mirrors and lenses were only instruments of trickery distorting the truth. Indeed, primitive mirrors produced contorted reflections, and the first lenses generated blurred images, soiled by unrealistic colours. Such deficiencies only strengthened superstitious claims. The university ordered Bacon's lenses destroyed because students were wasting too much time lighting candles with burning glass, instead of studying holy texts.

Despite resistance from many quarters, practical eyeglasses prevailed. Monks with failing eyesight no longer had to forsake reading scriptures. Tailors could continue their profession well past middle age. Spectacles prolonged the working life of ageing scribes. What a welcome relief to visions strained by years of assiduous translations! No doubt the wide availability of books after the proliferation of the printing press made eyeglasses even more popular. Eminent Europeans were no longer embarrassed to pose for portraits wearing spectacles. Every major town in Europe was proud to have its optician's shop.

With their magical properties, lenses stimulated Bacon to prophesize that it may one day become possible to make a device to see letters from incredible distances, perhaps to bring the sun, moon, and stars closer to the human eye. Bacon's prescience of a scientific method, yet to mature, was only surpassed by his prophetic vision of its revolutionary possibilities. Discussing the sphericity of earth, he imagined it would be possible to sail to Asia going west from Spain. When he argued that science could become a tool for clarifying truths of Christian religion, his enemies accused him of regarding reason and philosophy as superior to faith and revealed knowledge. Imprisoning him for 14 years, they successfully suppressed his upstart ideas.

For a man who argued strongly for the value of fresh experience, it is surprising how few original investigations Bacon himself carried out. Apart from his isolated efforts in optics, he also encouraged some of his contemporaries to probe magnetic phenomena.

MAGNETS AND THE FINGER OF REASON

Records about magnetism are found in various cultures. In one account, the word, "magnetism" is ascribed to the shepherd Magnes who, while pasturing his flocks, discovered to his chagrin that the nails of his shoes and the tip of his staff stuck fast in a rock. He struggled greatly to free himself. When he regained his calm, he went back and started digging to find out what in the ground was responsible. He found a stone that attracted iron nails. Another possible origin is from the district of Magnesia, a Greek province in Asia minor where the magnetic ore magnetite (Fe_3O_4) was first found.

Magnetism was the epitome of occult forces. The strange and remote influence violated all intuition to become the subject of many colourful stories. Magnetic mountains jut out from the sea to tear nails off ships that sail by. Magnets provide protection against the power of witches. Taken internally as powder, they cure ailments from toothaches to headaches to gout. A magnet under the pillow drives an adulteress from her bed. Its attractive power can bring peace between man and wife. Diamonds have the power to magnetize iron, and garlic to destroy magnetism. Washing a destroyed magnet with goat's blood restores its power. A wound made by a magnetic needle is painless.

Supernatural powers aside, rationalists wondered: from where does the force that tugs on a piece of iron originate? Familiar forces that push and pull all work through contact, not via empty space. Gravity was still not thought of as a physical force, just an inherent tendency for heavy objects to seek the centre of earth. But magnetism clearly manifested as a force between nearby objects, a force that defied the need for contact. And it had another amazing property.

About 3,000 years ago, the Chinese discovered that magnets rotate and orient themselves in a particular direction. Short for "leading stone," the name "lodestone" arose in connection with the magnet's primary application, the compass. A lodestone in the shape of a needle when floated on cork over water always set in a definite north–south direction. It became an astonishingly simple tool for navigation. The mystery of the magnetic compass was deep and its power far-reaching. From China, Muslim sailors imported the use of a magnet as a compass to Arabia, and thence to Europe, sometime before the 13th century. Along with the printing press and gunpowder, magnets turned out to be one of the key inventions that played a dominant role in the technological transformation of European civilization.

By repeatedly stroking a thin rod of iron in the same direction, lodestone imparted its magnetism into the iron needle. Subsequently, the magnetized needle naturally found the north–south orientation. What was the reason for the magical property? For those resorting to the supernatural, the compass was the finger of God, showing north and south as real directions of the universe. Rationalists preferred to invoke the pull of a giant magnetic mountain near the North Pole of earth.

Encouraged by philosopher Roger Bacon, French crusader Peter Peregrinus (1200–70) was one of the few medieval scholars who practiced experimentation in the modern sense. After proving the attractive power of a magnet to be strongest near the ends, referred to as north and south poles, Peregrinus went on to discover that opposite poles (north and south) attract each another, while the same poles (north and north) repel. If a magnet breaks into two pieces, a new pair of opposite poles appear at the ends, but disappear upon re-joining the two fragments. To make the compass usable for voyagers, he placed a magnetic needle on a sharp pivot, rather than floating it on a piece of cork, as was customary. Pinning the needle was a major step toward making the compass practical for navigation on the open seas. We will continue advances in magnetism in later sections.

RAZOR-SHARP LOGIC

Roger Bacon left his mark upon other medieval thinkers at the Universities of Oxford and Paris. William of Ockham (1285–1349) and Jean Buridan (1300–58) grappled with logical problems created by Aristotle's physics. Aristotle's comprehensive picture of order among the elements and their natural motion duly impressed most medieval scholars. It appeared to fit many "facts" of everyday observation. But the English scholar Ockham was bothered by an important anomaly. An arrow when released from the bow does not immediately fall to the ground. Once it leaves the bow, it continues to fly. What propels it to preserve Aristotle's "violent motion"?

Projectile motion was particularly interesting because of its military consequences. To understand violent motion, Ockham shifted attention from mover to moving object. During flight, an arrow carries a "cargo" with it, which ensures the progress of its motion. Once the stretched bow releases and delivers its cargo to the arrow, the cargo remains with the arrow, like heat in a red-hot, fireplace poker, gradually wearing out as the arrow continues to fly. The modern term for Ockham's cargo of motion is *momentum*.

Ockham is more famous for the principle of "Ockham's razor," an important outcome of his work on logic, which he pursued at a Franciscan monastery. Like Aquinas and Bacon, he tried to make distinctions between knowledge that comes from faith and knowledge that derives from reason and experience. What is the genius behind the principle of Ockham's razor? Industrious Scholastics of Ockham's time were busy deriving all kinds of entities to explain phenomena. Linking with Pythagoras' desire to find the simplest, most elegant explanations for the operations of nature, Ockham promoted the idea that one must make as few assumptions as possible because assumptions are the weakest points of argument. An explanation which uses fewer assumptions must be closer to the truth. In arguing with fellow Scholastics, he applied the principle so frequently, and with such sharp logic, that it came to be known as "Ockham's razor." Occam's razor cuts away many cumbersome possibilities.

Applying his razor, Ockham severed Aristotle's explanation for the continuing motion of the arrow after it leaves the bow. The ancient master invoked the role of air into the flight. Air rushing in to fill the void created by the moving arrow propels the arrow along. Aristotle's explanation involved unnecessary devices. Appealing to observation, Ockham dismissed the extraneous agency. If air was responsible, then a thread tied to the back of a moving object ought to be blown ahead of it, not behind it, as observed. Challenging Aristotle and thinking independently, Ockham opened a wedge; but he did not push hard enough to crumble the edifice. That was for Galileo.

At the University of Paris, Ockham's disciple, Jean Buridan, continued his master's quest to understand the cargo of motion (*momentum*). The young university was a spring of knowledge. Scholars pored over newly translated works of Greek masters. Buridan built on Ockham's concept. A massive cannon ball inflicts severe damage upon an enemy's fortress wall even though it barely moves when compared to a small rock hurled fast with a catapult. Along with speed, weight must be an important component of the cargo. The heavier a body, the more freight it carries. Cargo of motion must be a product of weight and speed. (The modern formulation is momentum = mass × velocity. Momentum also has the property of direction, as does velocity.) For the same reasons, it was more difficult to propel a big stone to a given speed than a little one. The big stone needed more cargo to change from its state of rest to the state of motion. Buridan caught a glimpse of the modern concepts of *momentum* and *inertia*, topics for future chapters.

CULTURAL TRANSFORMATIONS

Roger Bacon and his followers initiated fresh methods of actively probing nature to reveal her secrets. They were ahead of their time by centuries. Such ideas took hold very slowly, and only after fundamental transformations in the mental outlook of European society as it evolved from the Middle Ages into the Renaissance. Broad social and political developments swept through Europe over these intervening centuries. A new cultural spirit was simmering, one that would eventually boil over into major upheavals in political, social, and religious life. It is impossible to do justice to these engaging developments in our present scope. We can only view these as a lively backdrop for the emerging transformations in intellectual activity and scientific thought. William Gilbert (1544–1603) was one among many to be swayed by the strong cultural currents in his investigations on magnetism (below), and von Guericke (1602–86) in his investigations on static electricity phenomena (below). Galileo's breakthroughs in the science of motion (Chapter 4) will show how strongly this innovator was influenced by new patterns in the emerging cultural fabric. Many of the same cultural trends influenced the ideas and approaches of trailblazer astronomers Copernicus, Brahe, and Kepler (Chapters 7 and 8).

Fundamental economic shifts were ripping through the crusty layers of feudal society. Along with surplus agricultural output during the 15th and 16th centuries, increased trade led to increased prosperity. Via extensive fleets, merchants of Pisa, Genoa, and Venice controlled European commerce with Islam and the East. Wealth was in motion. A middle class was forming. Along with God and salvation, people sought new ways to acquire success and fortune. Merchants prayed just as fervently in church, but with brocaded gold purses at their sides. The new commercial class threatened the entrenched hierarchy of the Church and the land-holding nobles. Like Greek sailors and traders of 6th century BC, entrepreneurs failed to see the relevance and applicability of arcane rules and prescriptions handed down through the authority of ages. With enterprising spirit, they invited trouble by questioning orthodox rules and restrictions which hampered their freedom to trade. Medieval society was poised to topple the hierarchical structure which had put peasants, nobles, clergy, kings, and queens in their rightful places. Feudalism was slowly crumbling, diminishing the power of nobles and lords.

Under the authority of Pope and Emperor, the Holy Roman Empire long embraced an assortment of ethnicities united in Christianity. But at their roots, people were diverse in origin and language. It was impossible to hold them together. Christianity's unifying force was under serious threat. With increasing prosperity, the clergy grew materialistic, accumulating enormous wealth and power. The ensuing corruption of the Church coupled with its total dominion led to widespread abuses. Betraying Christian ideals, the clergy tarnished religious orders and abused fiscal machinery. Adopting a variation of an ancient German tribal custom of commuting penalties for crimes by monetary payment, Church organizations attached a price tag for forgiveness of each sin: one toll for blasphemy, a higher charge for adultery. For the ultimate prize, the salvation of the soul, the rich could purchase a "papal indulgence" at the steepest price. Through such blatant extortion, the papacy began to lose prestige.

With the break-up of feudalism, widespread abuses of the clergy, and increased corruption of the Church, uncertainty arose concerning the entrenched power structures. Responding to growing anxiety, the Humanism movement led by writers, poets, educators, and philosophers turned to emphasizing individual human values, as opposed to religious beliefs imposed by the Church, or the archaic customs dictated by the disintegrating feudal hierarchy.

Seeking to establish a new cultural framework, Humanists such as Francesco Petrarca (Petrarch) (1304–74), Desiderus Erasmus (1466–1536), Marsilio Ficino (1433–99), and Pico della Mirandola (1463–94), promoted a return to classical ideals. Cults sprang up to re-translate original Greek texts directly into Latin and avoid loose interpretations. With lessons drawn from classical Greece, Humanists stressed the central importance of basic values, as opposed to ordained religious beliefs. They glorified Homeric epics, ethical ideas in works of Plato, and the importance of human dignity and honour spelled out by the Greek tragedies of Aeschylus, Sophocles, and Euripides. Such dramas addressed major themes that envelop the world: the conflict between original thought and traditional belief, the power of hope against fatality, the belief in a higher moral order, and that justice will triumph at the end of struggle.

Striving to redress the eroding morals of their time, the rebels appealed to the literary beauty and moral content of Greek classics. While remaining devout Christians, Humanists sought to apply the values and lessons of pagan antiquity to questions of Christian faith. And they were prepared to challenge established religious authority. In biting satire and well-argued dialogues, they launched severe criticisms against the hypocrisies of their age. With merciless wit, Erasmus attacked the increasing corruption, superstition, and foolishness. Later, Galileo was to emulate the Humanists' biting sarcasm and compose eloquent dialogues to challenge the Church's insistence on the centrality and immobility of earth. Erasmus spared no one. Among his targets were grammarians, lawyers, logicians, theologians, clerics, and those who called themselves "scientists."

Most important, rebels promoted the idea of a humanistic education, one that would inculcate not only Christian doctrine but also ancient secular values consistent with Christian teachings. The post-medieval mind was not only rebellious but also intensely curious. Navigators sailed the uncharted seas, pushing back frontiers, opening the Age of Exploration. News of extraordinary discoveries overseas excited both imagination and avarice. With the recovery of classical texts, the spherical earth was at last re-accepted. Geographers dared once again to show the world as a sphere, sketching maps with topographical features that needed accurate representation based on detailed measurement. Explorers charted new territory. The Age of Exploration ushered in a new age of astronomy demanding increased precision in observation (Chapter 7).

Technologically innovative, people of the enterprising new class harnessed water power and wind power to enhance productivity. Inventors left their mark upon the course of history with the printing press, a major technological breakthrough. Before its invention, books were painstakingly copied by hand. Only the privileged few had access to precious volumes. An alternative method was stamping, but this was not yet widespread. To stamp out a 100-page book demanded the exorbitant labour and investment of 100 full-page stamps. In a brilliant invention that merged many technologies, a goldsmith from Mainz, by the name of Johann Gutenberg (1398–1468), devised a procedure that far outstripped the one-shot stamp. To form a page, he assembled a series of small carved stamps to represent each letter. He then rearranged the same letter-stamps to compose a different page. It was the first moveable type. Being a goldsmith by trade, he knew that stamps made out of metal would not wear out as easily as wood. Gutenberg had learned techniques of working metals from his father who used to work in the bishop's mint. To secure a firm and even print, he adapted commonly available screw presses for wine-making.

From concept to product, the technical development took Gutenberg 20 years, during which he worked in complete secrecy. As his very first business venture, he produced 300 copies of the Bible. Surviving Gutenberg Bibles are the most valuable books in the world today. Unfortunately, despite his labours, creativity, and secrecy, Gutenberg's printing business failed. Debtors took over his presses. But Western civilization owes an unpayable debt to the ingenious inventor.

Within two decades, there were more than 100 presses in towns all over Europe. The printing capital of the world was Venice, and the most prolific printer of all was Manutius Aldus (1449–1515), a man who also specialized in both translating and printing Greek classics. A sign outside one his shops said[14] "If you wish to speak to Aldus, hurry – time presses." One of his most popular innovations was small and cheap pocket books. "How-to" books proliferated in subjects from metallurgy to good manners. Printers began to issue books in the vernacular, disseminating knowledge among the common people. Within 30 years of the invention, presses printed some 6 million books.

Mechanical printing helped to spread new ideas and discoveries. No longer the exclusive province of Scholastics, knowledge became accessible to all who desired. With new enthusiasm for all forms of learning, the democratization of knowledge spelled an end to mental ossification. After reading newly printed books for themselves, a new generation of scholars could go beyond the endless hair-splitting commentaries of their teachers. By reading poems of Homer and works of Archimedes, they could directly receive the wisdom of 2,000 years. The revolutionary Copernicus benefited directly from Greek astronomy texts available at Italian universities, while Galileo studied fresh translations of Archimedes' mathematical works. The pulse of European intellectual life was quickening. Centuries of restriction on intellectual activity began to lift. A Renaissance, literally a rebirth, had begun.

ADMIRING CREATION

As the mind of Europe awakened from hibernation, artists awoke to the wonders of nature, rejecting the dominant theme of the spiritual era that our world is mere preparation for the next. Instead of focusing exclusively on figures receiving divine inspiration, painters flirted with a new approach, to explore the natural world. No longer would they restrict their creativity to purely spiritual themes. Instead of always glorifying the Creator, it was time to glorify his Creation.

Art historians[15] have drawn explicit connections between common forces that propelled creators and scientists of the time. Like artists and explorers, Gilbert and Galileo began to make fresh observations, probe nature, and devise experiments to draw from the rich diversity of sensory experience. Like the Humanists, Galileo's thinking was strongly influenced at first by the immense classical revival, especially the works of Pythagoras and Archimedes. But Galileo was prepared to challenge and demolish orthodox views on motion. Casting off the ropes of Aristotelian dogma rotting under medieval preservation, Galileo sailed free, pushing thought in new directions. We devote Chapter 4 to him after examining here the cultural and creative climate in which his luxuriant imagination flourished.

Artists began to paint faces of human characters in the facades of cathedrals. In the frescoes, angels and pious subjects turned into real people in streets and houses. Growing prosperity increased the appeal of the natural world. The importance of life here-and-now replaced the obsession of the here-after. Wealthy patrons of the new middle class commissioned artists to preserve their faces for posterity. Thousands of lively portraits fill our museums with galleries of unique faces. Raphael painted contemporary portraits of leading Renaissance figures into *The School of Athens*. Thus Leonardo da Vinci is Plato, Michelangelo Buonarroti models the lonely, brooding Heraclitus, and the famous Vatican architect Donato Bramante stands for Euclid. Many subjects proudly hold books. Plato clutches *Timaeus* in which he discusses the laws of cosmic harmony, while Aristotle carries *Ethics*, the laws of moral behaviour.

Adopting St Francis as a key figure, artists began to appreciate God's handwork in the beauty of nature. In *St. Francis in Ecstasy*, Giovanni Bellini portrays how the saint gives up a monk's ascetic life of isolation to emerge from the cave and gaze in rapture at the sun, marvelling at the surrounding world of plants and animals (Figure 3.5).[16]

A layer of spiritual meaning is evident from the stigmata on his outstretched hands, pierced by rays of divine light from the glory of Christ. Yet the artist's forceful realism allows us to deduce, from the direction of the shadow behind him, that St Francis looks straight up not at the divine light of Christ, but at the glorious sun. Bellini paints rolling hills in the background, trees in realistic detail, and a blue sky with billowing white clouds.

Compare this distinctive Renaissance landscape with the medieval treatment of Buoninsegna's *Flight Into Egypt* (Figure 3.6).[17] The older scene depicts the sky as heavenly gold, trees and rocks are figurative, while haloes dominate the faces of Joseph and Mary. In the spiritual theme underpinning his topic, the natural world is unimportant. There are no shadows. Ignoring gravity, baby Jesus floats in the hands of the holy mother.

FIGURE 3.5   *St. Francis in Ecstasy*, 1480-5, Giovanni Bellini.

FIGURE 3.6   *Flight Into Egypt,* 1308-11, Duccio Di Buoninsegna.

Renaissance artists grew interested in representing the human body as it actually looks. They painted Jesus as a real child, not a cherub. It became essential to display the inescapable consequence of gravity through the weight of subjects. Archetype of the period, German painter Albrecht Dürer expressed a newfound love for the world with his passion for reproducing its exact features. "Nature holds the beautiful for him who has the insight to extract it," claimed Dürer.[18] He painted *Hare* and *The Large Turf* (Figure 3.7) with such excruciating detail and realism that they could

FIGURE 3.7   *Hare* and *The Large Turf,* 1502, Albrecht Dürer,[19]

almost be photographs! It was important to imitate nature faithfully to reveal the truths hidden within nature. With the new approach, artists became excellent observers, as did the scientists of the period.

While the medieval artists' focus was purely on the soul, they ignored the human body. In stark contrast, the nude body became a favourite vehicle for the Renaissance artist. To accept nudity was to accept the dignity of the individual. With better knowledge of human anatomy emerging from rebellious physicians, artists could portray their subjects with new-found realism. In the *Theatro Anatomico* at the University of Padua, professors dared to conduct dissections of human cadavers against rigid sanctions imposed by the Church. Surely the inviolable sanctity of the body had to be preserved if it was to be successfully united with the soul on Judgment Day? Why else would early Christians go through elaborate procedures to save bodies in complex mazes of Roman catacombs? To help defiant professors of Padua conduct dissection of human cadavers, a conspiring look-out kept watch for Church police. At his warning, the surgeon would quickly flip over the operating table so that the body of an animal affixed to the reverse side would appear on top. Hired agents scurried away the human body to a safe hiding place via an underground river. Against such odds Vesalius, the visiting Belgian physician, gathered accurate anatomical information (Figure 3.8) to fill a richly illustrated and comprehensive textbook. Like Vesalius, both da Vinci and Michelangelo

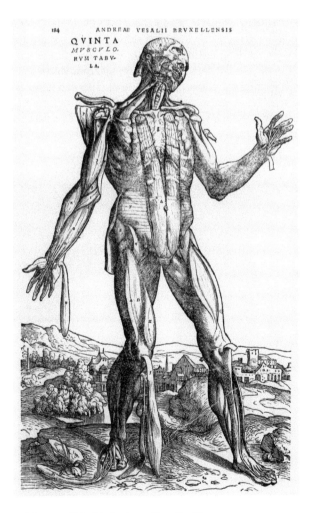

FIGURE 3.8   *Anatomical Study*, 1514–64, Andreas Vesalius.[20]

carried out dissections under abominable conditions but which were essential to making faithful anatomical drawings.

Slowly the Western intellect diminished its emphasis on the afterlife, so deeply rooted in the European mind since the fall of the Roman Empire. The value of human dignity and self-worth returned. Greek statues inspired Italian sculptors who revived and even rivalled antiquity. As a universal figure representing the youthful vitality of the Renaissance, David (of the legend of David and Goliath) became a favourite subject (Figure 3.9). Michelangelo sculpted an 18-foot marble David with a piercing gaze that seeks to penetrate the challenge before him. It is a defiant David who prepares to face the giant, with a sling over one shoulder and a stone in the other hand. The human spirit is once again self-confident. In choosing to portray the moment *before* the legendary battle, Michelangelo departs from the tradition of his Renaissance predecessors, who always showed a victorious David standing triumphantly over the severed head of the giant. Michelangelo sternly fixes David's gaze upon the approaching foe, ready for action. We can see latent energy bound up in the muscular strength of his body, and determination in the calm and ideal beauty of his face. Michelangelo claimed to have extracted the idea of David by carving away the excess marble from a giant block left over from a previously aborted work. With newfound pride in heroic human potential, Florentines proudly installed the superhuman structure in the city centre.

Raphael's *The School of Athens* is a complete and forceful statement of the vitality of classical heritage and the value of humanist education, painted with masterful realism incorporating techniques of perspective at their pinnacle. Even though Pythagoras, Socrates, Plato, Aristotle, Euclid, and Ptolemy lived centuries apart, it makes sense that they share the same canvas; their ideas were the foundation of the long-lasting academic tradition. To the left of Plato's entourage stands Socrates, engaged in dialectic discourse with a group from various professions including a disgruntled Scholastic and a young, attractive soldier. A hotly disputed topic must engage the crowd, since another Scholastic tries to weigh in with authoritative books carried by an assistant. But a close disciple of Socrates waves them away. Taught to think for themselves, they have no need for knowledge from authority.

FIGURE 3.9   *David*, ca. 1501–4, Michelangelo.[21]

With revived interest in the surrounding world, the individual full of curiosity became the new ideal of humanity. Confident of their genius, artists proudly imprinted their individuality by painting themselves into the picture. Raphael paints a self-portrait into *The School of Athens* (Figure 1.22). His juvenile face gazes at us on the right from the detail on the cover of this book. Unlike the unknown creators of Gothic cathedrals of the Middle Ages, Renaissance artists were no longer satisfied to be anonymous craftsmen in the service of God. The egotistical Dürer even went so far as to paint self-portraits in the image of Christ.

In a crucial innovation, artists developed techniques of linear perspective to project three-dimensional spatial relationships onto the two-dimensional canvas. The effort to describe nature through geometry and numbers had returned through art. In *Christ Giving Keys to St Peter* (Figure 3.10), Perugino aims to capture the holy event upon which the papacy based its claim to infallibility and total dominance over Christianity. To create the illusion of depth, he paints distant objects proportionally smaller in size. Perugino draws parallel lines receding from the viewer as they diminish to a "vanishing point" inside the central building on the horizon. Orthogonals to the parallels also meet at the same point. At the corners, he places triumphal arches in an architectural background as a tool to enhance the three-dimensionality of space. Accordingly, he integrates the chief characters, Peter and Christ, with the central axis passing through the vanishing point. For the most realistic effect, proportions are mathematically precise, with lines in correct convergence. Perugino's student, Raphael, employs both techniques of architectural settings and perspective to impart a vivid three-dimensional realism to the hallowed halls of ancient wisdom in *The School of Athens* (Figure 1.22). Architectural columns housing Apollo and Minerva amplify the depth. Just below the arch of the ceiling, seemingly parallel lines on tops of columns converge accurately to the vanishing point below Plato's wrist. As the steps of the hallway fade into the distance, their parallel edges vanish to the same point. Artists had to make careful measurements and calculations to arrange their observations of the physical world and bring out the third dimension.

Italy, the most prosperous and cultured region of Europe, bred the universal genius Leonardo da Vinci (1452–1519), who attempted to define the structure of nature through realism based on careful observation, through perspective and proportion based on measure, and through the mechanics of bodily motion based on anatomy. In his *Canons of Proportion* (Figure 3.11), da Vinci puts humanity

FIGURE 3.10   *Christ Giving Keys to St Peter*, 1481–3, Perugino.[22]

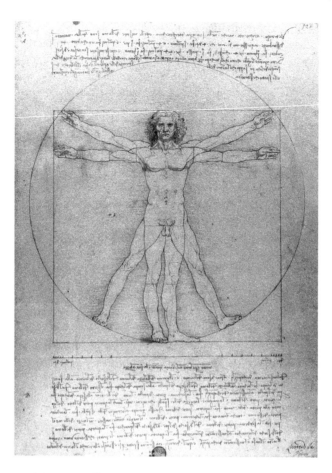

FIGURE 3.11   *Canons of Proportion.*[23]

at the centre of the rational world order. Through simple geometry, he relates the human body to the mathematical systems of the universe. By placing man symmetrically inside the perfect circle and the perfect square, he connects the microcosm of humanity to the macrocosm of the universe.

Although the painter in da Vinci eclipsed the scientist and inventor, his private notebooks, lost for centuries, show a spirit of inquiry far ahead of his time. They touch upon an incredible range of topics: botany, zoology, anatomy, perspective, light, optics, colour, mechanics, hydraulics, and military engineering. Secretly he worked on concepts for a bicycle and a flying machine, made drawings for elaborate trains of gears and patterns of flow in liquids. Each scientific drawing is a work of art in itself. From his accurate sketches (Figure 3.12) it is clear that da Vinci carefully observed and recognized parabolic trajectories of water fountains and cannon balls. But he could not connect the graceful arches with the mathematical advances of Archimedes. That was for Galileo. And da Vinci never shared scientific insights with contemporaries, nor did he exploit any of his marvellous ideas. Fortunately, he did record them for posterity in notebooks filling 3,500 pages. One was only discovered as late as 1960. Even though his impact on art is without parallel, he could exert little influence on the subsequent course of scientific development. As always, progress of science depends heavily on the wide transmission of ideas.

Realism was manifest not just in the artist segment of human creativity. Renaissance writers spun tales about people in the hustle and bustle of a real city. Familiar with the whole world of human nature, Shakespeare (1564–1616) embraced both the sorrows and joys of everyday people, as well as the grandeur and anguish of princes and kings. In *Hamlet,* the bard sums up the spirit

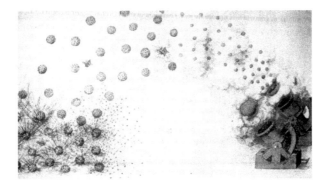

FIGURE 3.12   Da Vinci's sketch of a projectile shower shows parabolic paths.[24]

of the Renaissance with its love for nature, its resurgence of intellectual activity, and most of all its renewed pride and individualism.[25]

> This goodly frame, the earth …
> This most excellent canopy, the air …
> Look you,
> this brave o'erhanging firmament,
> This majestical roof fretted with golden fire …
> What a piece of work is a man!
> How noble in reason!
> How infinite in faculty!
> In form and moving
> How express and admirable!
> In action how like an angel!
> In apprehension how like a god!

REASON'S FINGER

Growing navigational interest in the compass triggered the English sailor Robert Norman to investigate (in 1581) a troubling property called "declination." The compass did not point exactly towards north, as defined by the north star. The declination, or deviation from north, was known by the time of Columbus. In his third voyage to the Americas, Columbus discovered that the declination varies at different points on the earth's surface, a fact which greatly upset his crew. Investors in future sea voyages grew worried. If the needle could not be trusted, their fortunes could drift into oblivion.

Such concerns triggered the experiments of Norman. Why did the needle err? He did not find the answer, but he made an important new discovery, For Norman, experiment was "reason's finger, pointing to truth." Paying close attention to details, he discovered another intriguing aspect of compass behaviour. Holding down a magnetized steel needle carefully balanced on a pivot, he found the needle to swing north–south as expected. But the needle also tipped downward, as if attracted to the earth. The phenomena, now called "magnetic dip," provided the crucial next clue to understand the innate directionality of the compass.

Why did a magnetic needle always point north? The first decisive explanation came from Renaissance experimentalist William Gilbert (1544–1603), a contemporary of Shakespeare and court physician to Queen Elizabeth I. Rather than indulge in hunting and other pleasures of the English court, Gilbert became absorbed in scientific questions, such as the gnawing riddle of the compass needle, and its fluctuations on voyages. Resonating with Bacon's teachings about the empirical route to knowledge, Gilbert dismissed preposterous ideas about the magical power of magnets. Before witnesses he demonstrated how diamonds, garlic, and goat's blood had none of the advertised power over magnets. Cutting through thickets of explanatory hypotheses medieval Scholastics provided on so many subjects, Gilbert denounced their "vast ocean of books."

Norman's investigations on magnetic dip spurred Gilbert to a most intriguing hypothesis: the entire earth must behave as one enormous magnet (Figure 3.13)! From a giant piece of lodestone, he sculpted a spherical model for the earth. Marking magnetic poles on the ball, he showed how a compass needle placed anywhere on the surface always pointed towards one pole. He could even simulate compass dip by the slight tilt of the test needle, as the poles of his giant sphere pulled on the tips of the needle towards the body of the sphere. In the same bold stroke, the reason for declination became apparent. The magnetic poles of the earth magnet are slightly offset from the geographic poles. (Today we know that the magnetic poles also change location on earth and with time.) It was a compelling demonstration. One theory embraced a host of disparate phenomena: north–south orientation, magnetic declination, and magnetic dip; the economy of the explanation was a sure sign that Gilbert was on the right track. Ockham's razor cut away past profuse and colourful explanations.

Proud of his explanations for the mysterious properties of the compass needle, he put together demonstrations of the earth-magnet for his queen, whose interest in science encompassed a broad

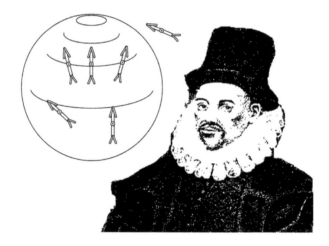

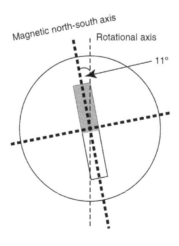

FIGURE 3.13   (Top) Gilbert's model of the earth as a magnet. (Bottom) the magnetic axis is tilted by 11 degrees relative to the polar, rotational axis, which accounts for the declination of the compass needle.[26] The earth is a weak magnet, about 1,000 times weaker than a common horseshoe, refrigerator magnet.

outlook for her country's intensifying exploration and empire-building activities. England had just conquered Spain's invincible Armada and was on the way to becoming the world's greatest naval power. With advances in navigation, the seas were open for commerce and colonist expansion. Voyages of the great navigators inspired the literature of Elizabethan England. Queen Elizabeth conferred knighthood upon the scientist.

Over 15 years, Gilbert spent a considerable amount of his personal fortune on innovative magnetic experiments, many with results now included in standard physics textbooks. For example, when he dipped a magnet into a cup of iron filings, the slivers clustered strongly near the ends, showing that the force of attraction is strongest near the poles. Iron filings sprinkled around a magnet tend to distribute themselves in a very definite and remarkable pattern around the poles, which Michael Faraday later attributed to "lines of magnetic force." Published in the year 1600, Gilbert's treatise, De Magnete, was a giant stride in scientific methodology, serving as a beacon for the scientific method to Kepler (Chapter 8) and Galileo. Chiding Scholastic tradition, and heralding the empirical method, Gilbert wrote in his preface:[27]

To you alone … who not only in books but in things themselves look for knowledge, have I dedicated these foundations of magnetic science … a new style of philosophizing.

Seeded by Roger Bacon's dabbling with lenses, the empirical method flowered into impactful new discoveries. Gilbert opened another new door, from magnetism to electricity.

THE SOUL OF AMBER

The nature and origin of electricity and electrical phenomena trailed far behind advances in magnetism. Other than lightning, nature does not provide a convenient source of electricity. In experiments with bodies electrically charged by rubbing amber with fur, Gilbert sprinkled bits of thread and sawdust around electrified pieces of amber to make comparable patterns to iron filings around magnetic poles. Gilbert coined the word "electricity" from Thales' word "electron" for the precious stone, amber. Amber is a yellow, hard, fossil resin of ancient pines, found on seashores, and used mostly for jewellery. A stick of yellow amber when rubbed briskly with wool attracts sawdust as discovered centuries earlier by Thales. When rubbed vigorously the stick gives off tiny, crackling sparks.

It was easy to confuse the attractive powers of charged amber with the similar attractive properties of magnetite. Gilbert distinguished magnetite from amber as definitively separate. A sheet of paper has no effect on magnetic attraction, but can significantly impede electrical attraction.

Where do magnetic and electric forces originate? Familiar forces all work through contact, not empty space. Yet electric and magnetic force defy the need for contact. Fascinated by their remote attractive power, an ancient explanation touted that electric and magnetic objects move other objects by moving the air in between. Contact of some kind has to be involved. By holding a candle flame near a piece of charged-up amber, Gilbert proved that the air is not involved at all, because he could deliberately disrupt air currents by heat, and observe no effect whatsoever on either of the attractions.

Gilbert found at least 20 other substances which acquire electricity when rubbed, calling them "electrics." However, a metal (e.g., brass) did not exhibit any electrical properties after rubbing, the first hint of a classification of substances between conductors and insulators.

In these early stages, the scope of discovery in electrical phenomena was limited to small frictional charges, easily obtained by rubbing glass or amber. But ordinary humidity rendered the remarkable results irreproducible and sometimes confusing. By inventing a machine to consistently produce a strong charge of static electricity, Otto von Guericke (1602–86) opened the road to a variety of fascinating electrical experiments revealing brand new features. As we saw in Chapter 2, the flamboyant mayor of Madgeburg, Otto von Guericke, was already famous for spectacular experiments on vacuum. If the earth is a magnet perhaps the sun holds the earth in orbit through the force of magnetism. Gilbert had assembled a magnetic model earth from magnetite. But von Guericke wished to re-create the magnetic attraction with a sphere of earthy soil. By coincidence, his finished sphere had a substantial amount of sulphur in it. On rapidly rotating the sphere suspended by an insulated metal rod, and rubbing it with fur, he drew a healthy stream of sparks. He had created a highly charged electrical sphere.

Touching the sphere against his bare hands, he felt his hair rise and received a strong shock when he touched the supporting metal rod. Bits of paper, cloth, sawdust, or straw, lying on a nearby wooden table jumped up towards the sphere. In one experiment, he hung scraps of paper and cloth from the ceiling with silk strings. As before, these were drawn to the charged ball, but von Guericke spotted another marvellous phenomenon. When any non-metal object touched the ball, it swung away from the ball, as if charged with a different type of electricity. The explanation came later. When objects, like bits of paper, came in proximity to the charged ball, they were charged by induction. The charge within the paper separated to leave the opposite charge on the end of the paper close to the ball; hence the paper was attracted to the sulphur ball. If the paper touched the ball, it acquired the same charge as the sphere and would therefore be repelled.

A UNIVERSAL PROPERTY OF ALL MATTER

Von Guericke's experiments established that all non-metallic objects could acquire charge. Charles Dufay (1698–1739) found that even metals could be electrified if properly isolated (insulated) from the earth by a piece of glass. Electrification must be a universal property of all matter. Dufay suggested that there are different kinds of electrical fluid, one type when glass is rubbed with silk, and another when amber (resin) is rubbed with fur. The charge from electrified glass could be transferred to corks by contact, and so also the charge from electrified resin could be passed to corks. But all electrified corks did not behave the same. All glass-electrified corks repel each other, and similarly for all resin-electrified corks. But a glass-electrified cork *attracts* a resin-electrified cork. Dufay deduced there must be two kinds of electricity, which he named: resinous (from amber) and vitreous (from glass).

Stephen Gray (1666–1736) discovered that electric charge can travel from place to place, an essential precursor to the discovery of electrical current. When he electrified a long glass tube at one end, he found that the cork which plugged the other end was also electrified. But he had not touched the cork at the far end by any charged object. Like a fluid, electricity was flowing along the tube. To determine how far electricity could flow, he tied long silk strings to the rod to find that charge could find its way to the end of longer and longer strings.

Another crucial electrical advance came from John Canton (1718–72) who identified a new method of electrification, one which involves no rubbing at all. He supported a metal rod by silk threads and brought a charged body near it, but did not touch the rod with it. Both ends of the rod showed charge (but of the opposite sign that Canton was not aware of). Canton had discovered charging by induction. For the first time, one object could impart electricity to another through empty space, without contact.

SHOCKED BY A JAR

Unlike magnetism which lasts "forever" in magnetite, electric charge on objects would always leak away in a short time, especially in humid weather. Was there a better way to store charge? While experimenting with the electrical properties of water in 1746, Peter van Musschenbroek (1692–1761), a professor of mathematics and natural science at the University of Leyden, received a very nasty shock that stunned him. Musschenbroek used von Guericke's static electricity generator to charge up a beaker of water. A gun barrel, suspended by silk threads, rubbed against the rotating sphere (Figure 3.14). At the end of the barrel was a wire, which he immersed into the jar of water to transfer the charge. Holding the beaker in one hand, he curiously touched the barrel with the other hand, and received a severe shock which intimidated him so much that he swore that nothing in the

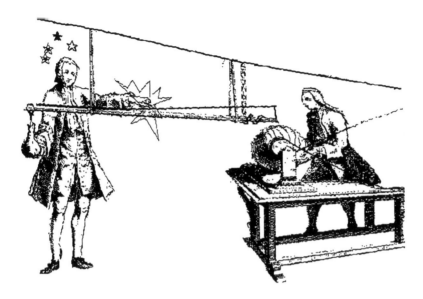

FIGURE 3.14   Invention of the "Leyden jar." After storing up a large quantity of charge in a jar, Musschenbroek gets a nasty shock.

world would tempt him to try such experiments again. (The metal rod acquired one type of charge while the glass beaker, hand, and body acquired another type.) It appeared that a simple water-filled jar could hold far more charge than any object of the same size.

Fortunately, Musschenbroek's students were much braver. Together with their teacher, they refined the reservoir idea into the "Leyden jar." They covered the outside of a water-filled jar with a metal foil (one terminal), and closed it with a wooden lid through which the metal rod (opposite terminal) was inserted. The bright sparks from the discharge across the terminals of a Leyden jar jumped over larger gaps of air than any seen before. People started to quote the length of the spark in air to gauge the amount of stored charge.

Electric Kisses

Along with von Guericke's electrostatic machine, Leyden jars soon became common electrical show apparatus for amusing court spectacles. In one demonstration, 180 soldiers of the King Louis XIV's guard when probed with a charged Leyden jar jumped in unison, with a precision exceeding any of their military manoeuvres. Enthusiasm was only slightly tempered when people realized that more than a dozen Leyden jars hooked up in series proved lethal! (The metal rod of one jar was connected by a wire to the foil covering the next jar.)

Electrical experiments became the rage. New discoveries were hurriedly reported in the popular press. Electric shows travelled around Europe with cart-loads of equipment to please growing crowds. In a most popular demonstration, a beautiful girl was suspended horizontally from the ceiling with insulating silk chords and then electrified by a hidden machine. Volunteers from the audience were invited to come up to the stage to steal a kiss from the prone lady, with a shocking result, of course. University students complained that they could hardly find seats at electrical lecture demonstrations as the general public crowded them out. Quacks abounded everywhere ready to swindle eager believers in the therapeutic powers of electricity. In one of the most bizarre cults, a Temple of Health in London offered the use of an electrical bed for a childless couple to reverse their fate under the therapy of an electrical charge while they engaged in sex!

UPDATES

A SINGLE FIRE

By Benjamin Franklin's (1706–90) time, electricity had grown into a novel and fascinating subject. Electrical demonstrations spread to colonial America, where Franklin witnessed his share. But what was this electrical fire? How many different kinds of electrical charge exist? Such questions fascinated the American. The point of view by his time (from Dufay) was that there are two different kinds of charge: vitreous and resinous. Franklin repeated popular electrical experiments to conceive a brand-new idea. Since all substances, including metals, can be electrified, electric charge must be present in all matter. Electrical charge is not created, only re-distributed. All kinds of electricity can be explained by transfers of electric charge. If an object loses some charge, it behaves negatively charged. If it gains charge, he called it positively charged. At once, he unified the jumble of notions. A charge of vitreous electricity can neutralize a charge of resinous electricity, leaving no charge behind. The two types of charge are not different; they are just opposite.

Franklin's simplified + and − description was a major breakthrough, the obvious explanation for why electricity flows from one place to another; negative charge moves towards the positive to restore natural balance. Franklin demonstrated his idea in a dramatic display. Two volunteers stood on separate insulated platforms. After electrifying a glass tube by rubbing with a piece of fur, he charged up one participant with the glass, and the other with the fur. When their fingers came together, a spark flew between them, and both were neutralized.

ST ELMO'S FIRE

Lightning has always been one of nature's mysterious and frightening phenomena. In his divine wrath, Zeus hurled thunderbolts to punish. When Columbus set out on his historic voyage seeking the fabled spice islands, he sailed the open treacherous ocean for two long months without any sign of land. His crew was about to mutiny. It was time to return home. On the same day, an ugly storm was brewing. As the sky blackened, a miraculous sight appeared. A glow of purple streamers filled the air from the tips of the masts, shooting up to heaven. While the crew stood awe-struck, Columbus seized the moment. St Elmo, the patron saint of sailors, sent his holy fire as a blessing for their voyage. God commanded that the search for land should not be abandoned.

A popular superstition of the time was that lightning is a diabolical agency acting in storms. The devout rang church bells to drive away the evil destructive forces and prevent harm. Many church bells bore proud inscriptions testifying to the power of the bell in successfully dissipating the effects of thunder and lightning. In the light of these "miracles," the faithful easily forgot how many churches were struck by lightning bolts, and how many bell ringers were killed during storms. Recognizing that more than 100 bell ringers were struck dead by lightning over the past 30 years, the Parliament of Paris passed an edict in 1786 to ban the practice of ringing tower bells in a thunderstorm.

CATCHING LIGHTNING IN A BOTTLE

Franklin suggested the first experimental proof for the electrical nature of lightning. One of his early remarkable discoveries was a special way to drain charge from objects. A grounded metal rod sharpened to a point at one end when brought near a charged object could discharge the object without even touching the object (by induction). The pointy rod could drain off the charge to ground far more easily than a blunt rod.

If there is indeed electricity in the clouds, Franklin wished to attract it with a pointed conductor. He proposed the "sentry box experiment" to tap into the electrical fire of the clouds with a pointed iron rod emerging from a sentry box on top of a tall building, such as a church steeple. An experimenter could stand in the sentry box on top of an insulated stand, and[28]

> bring near to the rod a loop of wire that has one end fastened to the ground, holding the wire by a wax handle so the sparks will fly from the rod to the wire and not effect him.

Franklin sent his suggestion as a paper to the Royal Society in England. But they did not consider the American's idea worthy of publication. Franklin tried to acquire a 100-ft iron rod to erect over Christ Church in Philadelphia. But it took a long time to acquire such a tall rod. Meanwhile, at Marly (near Paris), Dailbard (1709–78), an avid reader of Franklin's work, ventured to test the hypothesis. He set up a modest, but more readily available, 40-ft iron rod, 1 inch in diameter, insulating it from the ground with a wooden plank resting on three wine bottles. It did not take long for nature to provide the desired thunderstorm. The rod gave off brilliant electrical sparks and made crackling noises. He drew off the electricity into a small Leyden jar.

For the first time in history, humans harnessed electricity from the clouds. Here was the definitive proof that lightning involves electricity. The crucial Leyden jar captured nature's most awe-inspiring force. As word spread of the successful attempt, many in Germany and England rushed to repeat the experiment. Franklin became famous in Europe.

Before news of the successful sentry box experiment in Paris reached him in America, Franklin grew impatient to test his idea. Abandoning hopes for delivery of the 100-ft rod he devised the famous kite experiment instead. Near the ground, at the bottom end of the hemp twine lifting the kite, he attached a silk ribbon to hold on to the flying kite, thus safely insulated from the electricity in the cloud. At the junction of the twine and the silk ribbon he attached a metal key. He stood with his son under cover from rain, so that the protective silk ribbon would not become wet, and so lose its insulation. Franklin did not have to wait long for a violent thunderstorm to launch the kite. Soon the fibres of the hemp cord began to crackle sharply and spread apart they repelled one another upon charging with electricity. Franklin received an encouraging spark when he cautiously touched the metal key with his knuckles. He used the key to charge a Leyden jar as he described in the sentry box experiment. Franklin tamed nature's most baffling and terrifying phenomena.

ELECTROCUTIONS

The atmosphere develops static electricity on a grand scale. The earth and the sky form a giant Leyden jar, highly charged with electricity, until lightning and thunder accompany their violent discharge. Both the kite and the sentry box experiments are very dangerous, especially if the precautions that Franklin observed are neglected. If the silk ribbon had been soaked to become conducting, Franklin or his son may have been electrocuted. In Saint Petersburg, a scientist by the name of Richman electrocuted himself when he improperly performed the sentry box experiment. On that ill-fated occasion, Richman stood on the floor, and not on an insulated stand. Even his assistant, standing nearby, was knocked unconscious. News of the electrical accident spread quickly around the world. Scientific publications described the hideous state of his electrocuted organs. The faint-hearted duly warned that instead of interfering with nature and conducting intrusive experiments, nature's mysteries must be treated with humility and respect.

PIERCING THE HEAVENS WITH SHARP POINTS

Having confirmed that lightning is a thunderous discharge of electricity from the clouds, Franklin leapt to the idea of using lightning rods to divert potential damage from lightning strikes. In time, the lightning rod stood over many roofs as a symbol of progress with knowledge, a magnificent example of Francis Bacon's (1561–1626) early perspective that knowledge and understanding of nature will ultimately lead to control over nature's forces. Fashionable Parisian hat makers provided lightning-proof hats for ladies, and lightning-proof umbrellas for men with pointed rods above and a trailing discharge wire that ran down to make contact with the ground.

The sudden appearance of so many sharp points piercing the heavens caused great anxiety among the superstitious. A popular satirist of the time, Antoine de Rivarol wrote[29]:

> You may distinguish the learned and the superstitious … when it thunders. One seeks protection in sacred relics, the other in a lightning rod.

It took a long time to convince religious enthusiasts that the glory of God is not diminished by erecting a thin metal rod on the roof. Eventually the religious objections dissipated, and even ministers placed lightning rods on their houses. Nearly two centuries earlier, Shakespeare captured the changing times in *All's Well That Ends Well*[30]:

> They say miracles are past;
> And we have our philosophical persons,
> to make modern and familiar,
> things supernatural and causeless.
> Hence it is that we make trifles of terrors,
> Ensconcing ourselves into seeming knowledge,
> When we should submit ourselves to an unknown fear.

At one point a notorious political feud erupted over the exact shape that a lightning rod should have. During the struggle for American independence, King George III reacted against the American's rebel tendencies by insisting that lightning conductors at the Kew Palace in England should have rounded knobs instead of Franklin's sharp points. Sir John Pringle, the president of the Royal Society refused to carry out the unscientific decree, preferring to resign. A contemporary captured the rival scientific and political exchange[31]:

> While you Great George, for safety hunt,
> And Sharp conductors change for blunt,
> The nation's out of joint.
> Franklin a wiser course pursues
> And all your thunder fearless views.

REVOLUTIONARY, SCIENTIST, AND INVENTOR

Franklin began his scientific work at the age of 40, having previously been too busy earning a living. By the time of the American Revolution, he had become one of the world's most distinguished scientists. His book on electricity was published in ten editions in four languages. The Royal Society, which at first rebuffed his sentry box experiment proposal, awarded him a membership and their highest honour, the Copley medal. As a scientist, Franklin is best remembered today for his research into the nature of lightning with the kite experiment, but it was his penetrating work on the nature of electricity that was far more influential on the progress of science. Louis XV was so fascinated he ordered Franklin's experiments to be performed in his presence. Fame from his political activities came well after his scientific prominence. When he died, the French, who were in the throes of their own revolution, eulogized him as symbol of freedom and enlightenment. His epitaph aptly describes the man who:[32]

> snatched lightning from the heavens and scepters from Kings.

Twitching Frog's Legs

As professor of anatomy at the University of Bologna, Luigi Galvani (1737–98) spent most of his time researching the anatomy and the nervous system of frogs. Being also interested in the fascinating curiosities of static electricity, several electrical machines occupied his dissecting room. By chance, during a routine frog dissection experiment, one of his assistants took a spark from an electrical machine at exactly the same time as he touched a nerve in the frog's leg with his scalpel. They were astonished to see the animal's muscle go into convulsions even though the frog had long been dead.

The event made a deep impression. Familiar with the connection between electricity and lightning, Galvani wondered what would happen to a frog's leg in a thunderstorm. Hanging some recently dissected moist frog's legs outside (Figure 3.15), he waited for a storm to study the reaction. To his delight, every time there was a flash of lightning, the legs twitched violently. Perhaps lightning brought some form of life back to the animal with electricity.

But the most significant discovery was yet to come. One day, with perfectly fine weather, Galvani clasped a freshly prepared frog specimen to hang the legs on the iron lattice outside. By accident he bumped the legs hung on brass hooks to the iron trellis. Without the presence of any electrical machine, and not a sign of lightning in the sky, there were surprising leg convulsions. This was clearly a new phenomenon. Why did the legs convulse?

He tried several variations. Wrapping the frog's muscles with a metal foil, he fashioned a sort of Leyden jar. Now he observed intense contractions when he touched a part of the muscle with another metal. When he stretched out a leg on an iron plate and touched it with an iron rod, the contractions disappeared. But if the probing rod was brass, there was a strong response. To Galvani, ever the physiologist at heart, the reproducible twitching meant that electricity was really stored in the animal, like in a Leyden jar. Perhaps the nerves and muscles contain the subtle electric fluid, which he hoped to eventually isolate.

Galvani's experiments stimulating a dead frog with electricity inspired Mary Shelley in her novel, *Frankenstein*. She imagined a hideous mixture of body parts from dead people which her hero Doctor Frankenstein brought to life using electricity from a lightning storm. The phrase "galvanized to action" has become part of our language.

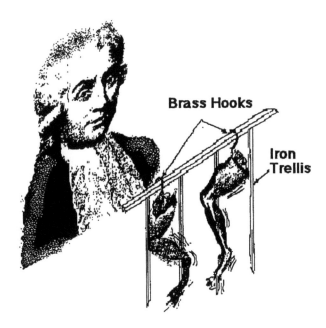

FIGURE 3.15   Galvani makes frogs' legs twitch by poking them with different metals.

But what Galvani revealed was the tip of the iceberg. Alessandro Volta, professor of physics at the University of Pavia uncovered the true nature of his friend and rival Galvani's discoveries. Now the term "volts" has more aptly permeated our scientific and engineering vocabularies. Volta was sceptical about Galvani's animal electricity. He repeated Galvani's experiments to search for the underlying source of electricity. Could the convulsing muscle merely be responding to electrical charge flowing through the fluid between the dissimilar metals? Which was more important, the animal muscle, or just the fluid in the muscles? What if he left out the muscle altogether, replacing it with a piece of cloth soaked with body fluid or even brine (salt water)? By placing disks of dissimilar metals in contact with his tongue, Volta felt a tingle and also a bitter taste. Could such sensations be similarly due to electricity flowing through his tongue between different metals?

For a definitive experiment to dispute Galvani, he brought a small silver disk into contact with a zinc disc through a wet cloth, and detected a surplus of electric charge on one metal from a slight needle deflection of an electrometer (a sensitive device to detect charge). Similarly a shortage of charge on the other conductor also registered on the electrometer. If he closed the circuit by putting the two conductors in contact, electricity would flow. The twitching of the frog's legs was just a detector of the flowing current produced by the different metals in contact through the muscle's fluid. The conflicting interpretations of the two Italian scientists sparked a lively controversy that divided the scientific community for some time.

Later (around 1800) Volta delivered the final blow against Galvani by amplifying the effect and connecting a pile of 20 or more pairs of silver/zinc discs through wet pasteboard, to produce strong shocks and sparks comparable to those from charged Leyden jars.

For the first time in history, there was a brand new source of electricity, truly distinct from previous sources, all based on static electricity through friction. Volta's discovery launched the age of electricity by providing a source of continuous and controllable electrical current.

UNLEASHING MORE ELEMENTS

In the first form of our ubiquitous battery, he stacked together many pairs of discs into a pile, each pair consisting of one zinc and one copper (or silver) disk, separated by a piece of pasteboard, soaked in brine. A marvel of the ages supplied a continuous flow of electrical current. It was a calm current, with none of the violence of a Leyden jar discharge. In one remarkable application, the continuous current flowing through fine filaments of metal turned into an intense source of heat, sufficient to melt metal.

Electric current opened up the field of electrochemistry. Within six weeks of Volta's report, Joseph Priestley (Chapter 2) passed an electrical current through water and decomposed it into hydrogen and oxygen gas.

With the calm, continuous current, English chemist, Humphrey Davy, who we met in Chapter 2, transformed earthy ash into a brilliant burning metal. Electrical currents could unlock new chemical elements from many ores of the so-called "element" earth. The new electrical magic was immensely more powerful than any dreamed of by the alchemists, who tried in vain for centuries to transform elements into each other. Davy used a pile of 274 plates to electrically decompose the alkali earths, potash and soda, to extract the new elements, sodium (Na) and potassium (K). Davy collected the gas liberated during decomposition as Lavoisier's oxygen.

By 1808, the voltaic cells had grown to a giant size and the number of pairs climbed from the hundreds to the thousands. The power delivered by a large voltaic pile was 10 kilowatts, compared to the typical 1 watt of power released in the discharge of static electricity from a Leyden jar. Using an assembly of 2,000 pairs of voltaic cells, Davy went on to extract barium (Ba), strontium (Sr), and magnesium (Mg) from other alkaline earths, filling out the Periodic Table. Then he turned to the electrical decomposition of muriatic (hydrochloric) acid, which, at the time, was believed to be the strongest possible acid because of its virulent attack on the strongest metal, iron. Two wonderful surprises greeted the persistent chemist. Electric current decomposed muriatic acid into hydrogen and a new green gas. Davy named the new chemical element, "chlorine." Another astonishing revelation was that acids, when decomposed, release hydrogen. So hydrogen is the active element in acids, not oxygen, as Lavoisier believed.

UNIFICATION OF MAGNETISM WITH ELECTRICITY

Electricity and magnetism were developing independently, with no apparent connection between the two powerful forces. The experimental method came to full force under Faraday who revealed the first crucial link between two disparate forces of nature. We met Faraday in Chapter 2 when he discovered the discreteness of electric charge, preparing the way for the discovery of the "electron."

Even after Gilbert's breakthroughs on magnetism, the origin of the strange force remained a deep mystery until the beginning of the 19th century. With random reports of how a bolt of lightning magically rendered a box full of knives and forks magnetic, or how lightning altered the polarity of compass needles, many suspected a link between electric and magnetic forces. Hoping for a simple analogy, two French investigators in 1805 made a straightforward but naive attempt to determine whether a freely suspended battery would align itself like a compass needle. Their hopes were dashed. Conjecturing on similarities between electricity and magnetism, Danish physicist Hans Christian Oersted (1777–1851), set out with systematic attempts to determine if electricity had any direct action on a magnet.

Living at a time when a Romantic movement was blossoming in the aftermath of the French Revolution, Oersted felt great affinity with the ideas of German philosopher Immanuel Kant (1724–1804) and his followers, the Nature Philosophers. They were greatly dissatisfied with the purely mechanistic philosophy of nature, which dreamed of reducing all phenomena to the motion of atoms, explained by laws of motion and rigid mathematics. Nature Philosophers wished to replace a dry mechanical universe by one filled with potent forces, yet to be understood. Ardent admirer of Kant, Friedrich Schelling (1775–1854) believed that the very essence of nature was the activity of forces in opposition. Optical, thermal, electric, and magnetic effects would all turn out to be manifestations of one force, but under different circumstances. They were exuberant when electricity decomposed water into its elements, hydrogen and oxygen. Here was an identity between patently diverse arenas: chemistry and electricity. But the ideas of the Nature Philosophers were nebulous.

Oersted, the scientist, became a disciple of Kant. In his doctoral dissertation, he evaluated the importance of Kant's philosophy to science. Later he served for a short period as the editor of a philosophy journal. In travels over the continent to study the new-found relationship between electricity and chemistry, he met with Schelling. Now Oersted came to believe that one power produced all phenomena, a vision which guided his 20-year search for an effect of electricity on magnetism.

In first attempts, Oersted always laid wires conducting electricity across the compass needle, instead of parallel to it, consistently failing to find any effect on the magnet. During a private lecture before advanced students in 1820, Oersted detected a first glimmer of a link. The event is often referred to as an accident. But as Louis Pasteur once said:[33] "in the field of experimentation, accidents favour the prepared mind." Oersted was busy demonstrating how electric current produces heat when he caught the flicker of a nearby compass needle (Figure 3.16). Appreciating its deeper significance, he mentally tucked away for later investigation the possible influence of electric currents on magnets.

The decisive discovery eventually conferred immortality upon Oersted, but it lay dormant for another three months. He needed to be absolutely sure. More experiments with higher current demanded a powerful voltaic cell, which took a long time to prepare. Systematic attempts followed to carefully check for other possible effects. Could air currents transporting heat dissipated in the wire deflect the magnet? A piece of cardboard between the current-carrying wire and the magnetic needle eliminated air motion, but the deflection remained the same. After months of cross-checking, Oersted was finally ready to declare success.

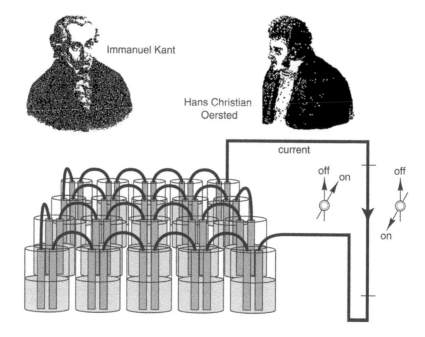

FIGURE 3.16 Influenced by the ideas of Kant and the Nature Philosophers, Oersted found a first link between electricity and magnetism.[34] A large voltaic pile (battery) produces a strong current through a wire which in turn deflects a parallel place magnetic needle.

A current-carrying wire always deflects a magnetic needle oriented *parallel* to the wire (Figure 3.16). The needle veers 90 degrees to become tangent to a circle around the wire. If on one side of the wire, the north pole of a compass needle points in one direction, then on the opposite side, the north pole would point in the opposite direction. When current reverses direction, the needle turns in the opposite direction. An electric current travelling through a conductor creates a circular magnetic force around the conductor to deflect the magnetic needle. Oersted had brought together two disjoint pieces of a giant puzzle. Electricity generates magnetism. It was the first small step towards a grander synthesis.

There was yet another remarkable effect Oersted found. A closed current-carrying loop of wire behaves like a bar-magnet with a north and south pole, the first one-loop electromagnet. Thanks to his powerful voltaic cell, he could detect the magnetic influence 10 feet away. It was magnetism without magnets. Electric currents create magnetism and motion (of magnets).

In England, Michael Faraday quickly devised a simple scheme for electric current to continuously rotate a magnet, a simple electromagnetic motor, progenitor of powerful industries. Casting his vision beyond magnetism from electricity, Faraday wondered about the converse effect. Could a magnet generate electric current? Ten years of experiment after experiment ended with repeated failure. Through his persistent quest with thousands of experiments, he devised ingenious techniques to increase the strength of electromagnets. Powerful voltaic cells raised the current. Stacking layers upon layers of wire loops increased the strength of the electromagnet. But no magnet or electromagnet could create electrical current.

Finally, in 1831, came a decisive success. Working with 400 feet of copper wire and a battery with 100 pairs of plates, he wound two coils of insulated wire around opposite sides of a thick iron ring. One coil he connected to a battery, and the other to a *galvanometer*, a device which detects weak electric currents by deflecting a magnetic needle (Figure 3.17). On closing the first circuit to energize the magnet, he was overjoyed to see a flicker in the galvanometer needle on the other side. It was only a fleeting response, far from the solid connection he was after. But then came another

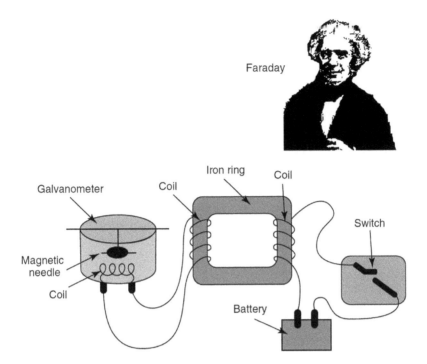

FIGURE 3.17   Opening and closing the current loop of the electromagnet, Faraday induced electric current in a second loop, and detected it with a galvanometer.[35]

surprise. Opening the circuit again to stop the flow of current, the galvanometer jumped again, this time in the opposite direction. In a sudden flash of insight, Faraday realized that the key to generating electric current from magnetism lay not in the mere presence of a strong current or magnet; it was in the *changing current*. A moving magnet induces a flow of electric current in a wire, an effect now called *induction*.

Still Faraday was not completely satisfied. To clinch the discovery, he needed a direct effect from a pure magnet, without any electrical current at all. Now on the right track, it took him only a few months. Repeatedly moving a permanent magnet in and out of a coil of wire connected to a galvanometer, he detected an oscillating current.

Faraday discovered a deep symmetry between electricity and magnetism. As Oersted found, a moving charge (electrical current) exerts force on a stationary magnet. Conversely, a moving magnet induces an equal and opposite force on a charge. What Faraday missed in early failed attempts was the importance of relative motion between charge and magnet.

Faraday's two crucial discoveries, electromagnetic rotation and electromagnetic induction, were forerunners of new industries to come, powerful motors to turn the wheels of industry and gigantic turbines impelled by gushing waterfalls to generate abundant quantities of electricity. Before his discovery of induction, Faraday had already gained fame for the electromagnetic motor and electrochemical discoveries. When British Prime Minister Sir Robert Peel visited his laboratory, Faraday proudly displayed his brand-new invention, a rudimentary generator of electricity from a moving magnet. Gaping in total disbelief at the worthless pile of wires and iron, the Prime Minister inquired,[36] "Of what use is it?" Confidently Faraday replied: "I know not, but I wager that one day your government will tax it."

With the understanding of the underlying identity between electricity and magnetism came amazing technical possibilities: to communicate over long distances, operate mammoth machines spinning gigantic wheels, generate prodigious quantities of power, and to transport that power over enormous distances.

SPACE IS NOT NOTHING

As his crowning achievement, Faraday set the stage for the complete unification of electricity and magnetism with a radically new approach to integrating space and force. For Faraday, a magnet's remote power, acting mysteriously across empty space, was unphysical. To move an object demanded contact of some kind. Visualizing mechanical forces between electric charges and magnets, he wished to fill the space between them with linkages to pull or push. In his picture, space is not empty; "lines of force" surround a charge or a magnet (Figure 3.18). Iron powder sprinkled around a magnet forms a characteristic pattern, as Gilbert first discovered, to reveal "lines of magnetic force." Sawdust sprinkled around a charged rod reveals "lines of electric force." More than an artistic picture, lines of force possess a physical reality, thought Faraday. Perhaps the properties of a force line are similar to a vibrating string, like tension and motion. Near a magnet pole, where the force becomes stronger, the lines of force grow dense. When lines of force are far apart the force is weak. Magnets and charges create invisible "somethings" that permeate the surrounding space.

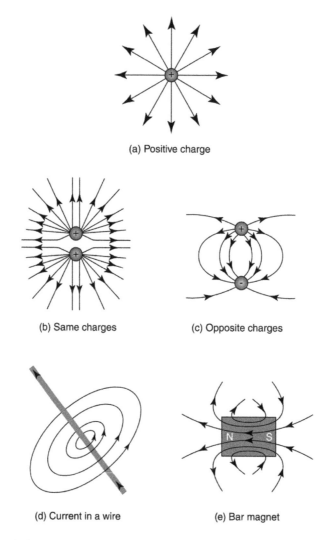

(a) Positive charge

(b) Same charges   (c) Opposite charges

(d) Current in a wire   (e) Bar magnet

FIGURE 3.18 Faraday's electric lines of force surrounding (a) a single positive charge (b) two like charges (c) two unlike charges (d) magnetic lines of force surrounding a current-carrying wire (e) a bar magnet.

Space is not empty; it is a medium filled with lines of electric and magnetic force. Real electrical and mechanical energy reside in the space around the poles of a magnet.

Mathematically inclined physicists of the time ignored Faraday's lines of force as the ramblings of a tinkering chemist. They refused to take the new ideas seriously because Faraday, who lacked formal training in mathematics, could not express his intuitive notions in precise mathematical terms. Fortunately, younger scientists were open to the new ideas.

A Temporary Scaffolding

James Clerk Maxwell (1831–79) was born in Scotland in the same year as Faraday announced his discovery of induction of electricity from magnetism. His parents were wealthy, and his family distinguished. When James arrived at school wearing clothes designed by his father, a dozen boys surrounded him, poking fun at his attire and his accent, then beat him up, tearing his fine apparel to shreds. Taking him for a rustic halfwit, they called him Daffy, a nickname that stuck through school. Within a year, Maxwell summoned the courage to confront the gang leader and pummel him into silence.

Developing an early interest in mechanical models and geometry, Maxwell made physical models of the five regular solids. At 14, he published his first scientific paper in which he described a generalized series of oval curves, such as ellipses, which he could trace with pins and strings. These early fascinations had a strong influence on his theoretical work in electromagnetism, which started out with elaborate mechanical models and mathematical relationships. He thought of electrical charge as a pump, continuously churning out an electrical "fluid." Spinning vortices in the fluid were agents of magnetic force. In his diagrams, he represented cylindrical vortices with a hexagonal cross-section for ease of drawing (Figure 3.19). Mechanical effects arising from whirling vortices produced magnetic force. Then he added strings of spherical particles of electric charge between vortices. The physical movement of charge-carrying spheres represented electric current. But the

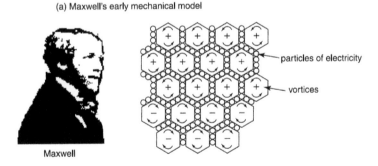

(a) Maxwell's early mechanical model

particles of electricity

vortices

Maxwell

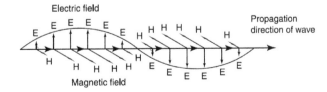

(b) Electromagnetic theory

Electric field

Propagation direction of wave

Magnetic field

FIGURE 3.19 Maxwell started out with mechanical imagery. Electric charge churns out electric fluid. Vortices in the fluid provide magnetic force. Electric particles act as ball-bearings to couple vortices. Eventually Maxwell abandoned these mechanical crutches to progress to vector electric (E) and vector magnetic (H) *fields* carrying electromagnetic waves. E and H arrows along the wave represent fields.[37]

spheres also acted like spinning gears or idle wheels. Their translation produced changes in vortex rotation. Thus he provided a mechanical model for electric currents to generate magnetic force. Similarly moving magnets could change vortex spin causing the spheres to move and generate electric current. The mechanical model covered both Oersted and Faraday effects.

Communications became the first beneficiary of rapidly evolving electromagnetic science and technology. With an undersea Atlantic cable connecting Europe and America, Morse-code clicks generating dots and dashes in Ireland moved needles and mirrors in Newfoundland. Queen Victoria exchanged instant telegraph greetings with President Buchanan in 1866. Fascinated by the possibilities of long-distance communication via electricity and magnetism, amateur poet Maxwell wrote this ode to pioneers of electrical and magnetic science who paved the way to the historic moment of transatlantic communications:[38]

> Valentine by a telegraph clerk to a telegraph clerk
> The tendrils of my soul are twined
> With thine, though many a mile apart,
> And thine in closed coiled circuits wind
> Around the needle of my heart…
> O tell me when along the line
> From my full heart the message flows,
> What currents are induced in thine?
> One click from thee
> Through many an Ohm and Weber flows
> And clicked this answer back to me –
> I am thy Farad staunch and true
> Charged to a Volt with love for thee.

UNIFYING LIGHT

Eventually Maxwell stripped away the temporary conceptual scaffolding to completely replace all mechanical models by pure concepts of *electric and magnetic fields*, vectors with magnitude and direction. Maxwell continued to follow Faraday's siren in search of deeper harmonies. He expressed Faraday's quaint geometrical trellis of force lines into precise mathematical form. At every point in space, an electric charge gives rise to a *field* of "mathematical arrows," an invisible electrical mist, which provides an electric force on any charge placed at the point. An arrow's orientation gives the direction, while its size gives the force magnitude. Similarly, surrounding any magnetic pole, or current-carrying wire, there is a magnetic mist, a field of magnetic arrows at every point in space which provide directed magnetic forces acting on any magnet present. Fields carry energy and exist independent of any charges or magnets used to probe it. Empty space is more than a vacant framework in which matter exists and moves. It has an underlying structure capable of transmitting electric and magnetic force.

The real power behind the mathematical treatment was a thoroughly novel prediction. Faraday showed that a moving magnet generates electric current. In Maxwell's language, a changing magnetic field produces an electric field, which moves charges to make current. For complete symmetry between electricity and magnetism, Maxwell's formalism demanded that a changing electric field should also produce a magnetic field. Adding a new term to his powerful equations to make the theory symmetric with respect to electric and magnetic fields, Maxwell completed the unification of electricity and magnetism. Only pure fields and connecting equations remained. With a few potent lines, Maxwell captured centuries of research into a comprehensive and elegant form. Maxwell's equations unified electricity and magnetism.

But the best part was yet to come. The elegant mathematical formulation boldly predicted that oscillating electric and magnetic fields can detach from charges and currents that generate those fields, and propagate freely into space as electromagnetic waves. In modern terms, electrons oscillating inside metal antennae generate electromagnetic waves, intimately familiar to us as radio and television signals, as well as microwaves in the kitchen or microwaves between cell-phones and relaying towers. Electric fields move up and down like the surface of a water wave; magnetic fields move side to side as the wave travels forward. Along the wave, the direction of the electric field is perpendicular to the direction of the magnetic field, and both electric and magnetic fields are perpendicular to the direction of propagation (Figure 3.19). Maxwell could calculate the speed of wave propagation directly from electrical and magnetic force constants measurable by laboratory instruments.

There was a stunning new result. To Maxwell's surprise and wonder, the speed of propagation of the new waves came out to be the same as the speed of light, 300 million metres per second (3×10^8 m/sec). (The speed of light was known from astronomical measurements.) The identity of the speed of electromagnetic waves with the speed of light revealed to Maxwell that the true nature of light is the same as waves of electromagnetism. In 1865, Maxwell wrote:[39]

> I have … an electromagnetic theory of light which … I hold to be great guns. All the properties of light in solids, crystals, empty space follow without further ado directly from mathematics. And the speed of light is predicted.

It was one of the boldest projections in the history of science – a grand synthesis. Before Maxwell, there was no obvious connection between light and electricity or magnetism. From purely theoretical considerations, Maxwell showed that light is an electromagnetic wave. Electric and magnetic fields live an independent existence, intertwined with each other, barrelling down empty space at the velocity of light. And that velocity falls out of the theory as a constant of the natural forces of electricity and magnetism.

Space is not empty; it is a cosmic ocean pulsating with electromagnetic waves. Our visual experience does not directly track the undulating fields, unless the wavelengths fall within a certain range to become the visible, coloured spectrum of light. Maxwell's imagination became the fountain of modern technology. Understanding electromagnetic waves predicted new kinds of radiation with a wide range of wavelengths and applications from radio waves (tens of metres) and microwaves (centimetres to a metre) for communicating across vast distances, to infra-red waves (micro-metres) for sensing heat or picking infra-red images of warm bodies in the dark, to the visible light spectrum (sub-micrometres), to X-rays (nano-metres) for penetrating the structure of crystals and DNA, to gamma-rays (pico-metres) for probing the structure of the atom. We already encountered X-rays and gamma-rays in Chapter 2.

New Ways of Looking

Reviewing the centuries of developments in this chapter, the ascendancy of Rome aborted the march of Greek scientific thought and Alexandria empiricism. Emphasizing spirituality, Christian thought set the mind of Europe on a different path – to salvation. With the dissolution of the mighty Empire of Rome, Europe fell into the Dark Ages with Christianity holding together the shards of the fallen empire. Recovering the intellectual heritage of the Greeks through Islam, Europe awoke from a long intellectual stupor. But accepting and absorbing the wisdom of ancient cultures was a long struggle. To surpass the early Greeks and continue on the road to empiricism paved by the Alexandrians, a few radical medieval scholars emphasized the value of empiricism. Nature cannot be deduced from pure logic. Science needs empirical content.

Scientific advances during the Renaissance took place within broad cultural advances. As the medieval fog began to lift, artists awoke to the wonders of nature. Painters expressed their love of the world by reproducing its exact features. To do so they became excellent observers of nature. Anatomists dared to probe the human body by dissection. Just as much as art and poetry were avenues for creative expression, efforts to understand magnetism and electricity were creative activities that started slowly with Gilbert and others to reach their pinnacle under Oersted, Faraday, and finally Maxwell.

Each in their own way, scientists and artists desired to see beyond the seen. They opened the human mind to new ways of looking at the world. Oersted looked for a unity among disparate forces. Electricity generates magnetism. Symmetry propelled Faraday to find the reciprocal link: moving magnets generate electricity. Faraday's imagination created invisible lines of force to fill space. Symmetry guided the experiments of Oersted and Faraday, as much as the mathematics of Maxwell. Abandoning concrete mechanical models, Maxwell imagined invisible fields. With field theory, he unified electromagnetism. In a dramatic leap to understanding nature, he penetrated the electromagnetic wave nature of light.

4 Terrestrial Motion: Dynamic Symmetries

> Philosophy is written in this grand book, the universe, which stands continually open to our gaze. But the book cannot be understood unless one first learns to comprehend the language and reads the letters in which it is composed. It is written in the language of mathematics.
>
> **Galileo Galilei**[1]

FALLING FROM THE LEANING TOWER

As we saw in Chapter 3, Renaissance Europe was teeming with creative activity, probing in all directions with a fresh outlook, seething with rebellion. Upon this turbulent scene burst an outspoken, upstart personality, a young man with wavy red hair and a feisty demeanour. Galileo Galilei dared to openly criticize Aristotle and his laws of motion, not with Scholastic argument or logical expositions, but with fresh observations and the inventive approach launched by Archimedes and Gilbert – controlled experiments.

Galileo had a dispute with the Greek master's conclusion about falling bodies. Remember how Aristotle had reasoned that a large rock, with more of the element earth, seeks more eagerly to return to its natural resting place, to the centre of the earth, to the centre of the universe. True or not, according to one famous story that deserves to be told, Galileo announced to the University of Pisa professors that he would perform a public demonstration to show how the indisputable authority was mistaken in his conclusion about falling bodies. A heavy ball of iron and a light ball of wood when dropped from the tower of Pisa fall to the earth at the same time. Several faculty and students attended the spectacle. The experiment was a success.

Galileo's challenge to Aristotle was real. But the faculty at the University of Pisa were unimpressed! Aristotle's system of truth was firmly entrenched in the "learned mind." One contrary isolated phenomenon could hardly topple an overall philosophical system that encompassed a wide meaning. They were not interested in a young upstart's mischievous contradictions of their age-old mentor. To start questioning even one doctrine would be to cast doubt on them all. To them, Galileo's experiment was absurd and idle tinkering – even dangerous to Aristotle's now sacred doctrines.

Born in Pisa in 1564, the same year Michelangelo died and Shakespeare was born, Galileo grew up in a family that valued the arts. His father made seminal contributions to the development of music. After sending his son to a Jesuit monastery at age 11, he withdrew the boy at age 15 so Galileo would not end up a monk – a popular occupation of the Scholastic times. Pointing him in the direction of medicine, he enrolled young Galileo at the University of Pisa. A physician of the time earned a comfortable living, 30 times as much as a mathematician. Like all other students, Galileo had to study the revered master, Aristotle.

By mistake one day the young man walked into the wrong class, where a discussion was in progress on problems of advanced geometry. It was an accident that changed the course of history. Mathematical demonstrations aroused his curiosity. Galileo began to study Euclid's geometry in his spare time, without formally enrolling in classes. Seeds of classical revival found fertile ground in the eager mind. Delighted by Pythagoras' proof that the sum of the first n odd numbers is n^2, and fascinated by geometrical curves, such as the parabola of Archimedes, Galileo was destined to make good use of these re-discoveries in the analysis of motion soon to engage his attention.

wheel

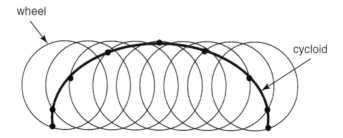

cycloid

FIGURE 4.1 When a wheel rolls on the floor, the path followed by the nail stuck on the rim is a new mathematical figure, the cycloid, discovered by Galileo.

A lively curiosity of the Renaissance instilled in Galileo a passion for careful observation of the world around him. Galileo discovered his own mathematical curve, the cycloid. A nail attached to a wheel rolling on the ground describes a cycloid path (Figure 4.1). Perchance the attentive student caught a fleeting glimpse of the flight of a nail stuck on the wheel of a passing horse-carriage. Perhaps he fell to wondering about the precise form of the curve traced by the nail. Just as da Vinci and Dürer had been keen scientific observers of nature, Galileo was fascinated with the detailed aspects of the surrounding world.

Inspired by his virgin encounters with Alexandrian science, Galileo delved deep into Archimedes' works, only recently translated into Latin from Arabic. No doubt the inventor in Archimedes captivated Galileo's imagination. Fascinated by the Alexandrian's seminal work on density, Galileo invented his own hydrostatic balance – an instrument to measure densities by determining the loss of weight of objects in water due to buoyancy. With such consuming distractions, he put aside medical studies. Eventually, commercial interest and the success of the hydrostatic balance landed him a teaching post at the University of Pisa.

THE PULSE AND THE PENDULUM

How did Galileo come to realize such a profound contradiction about free fall? On a fateful day in 1583, while the 19-year-old Galileo attended mass at the Cathedral of Pisa, he caught the interesting behaviour of a swinging candelabrum. Clearly, he was distracted from the litany but fascinated by a very natural event that most would never pause to contemplate. As the swings became smaller, Galileo wondered whether the time for each cycle was also getting shorter, a common-sense expectation. All he needed was to time the duration of each oscillation. Of course he could not reach into his pocket for a stopwatch, yet to be invented. But training in medicine inspired him to use the regular beat of his pulse as a timepiece. To his surprise, he found that the duration of the chandelier swings remained the same, independent of the amplitude of the to-and-fro motion. This is what we now call the *pendulum principle*. It was an epoch-making discovery.

It occurred to Galileo that, because of its constant property, a pendulum would provide an excellent way to regulate mechanical clocks, notorious in his time for their inaccuracy. A generation later, the Dutch scientist Christian Huygens would successfully realize Galileo's dream to use the pendulum principle as the controlling mechanism for the ticks of a mechanical clock, turning it into an accurate timepiece. As word of the discovery spread, physicians used the pendulum to monitor the pulse, which tends to change during a serious illness. Galileo's first discovery in physics became a valuable contribution to the field of medicine he had just abandoned.

Returning home from church, Galileo was eager to try new experiments, this time with stones of various weights dangling from separate strings (Figure 4.2) equal in length. Now he made another surprising discovery. All the different stones fell to the bottom of the swing in the same time. Reasoning by extrapolation, an approach he used effectively on many occasions, Galileo began to conjecture. Light and heavy stones swinging from a string take the same time to travel to the

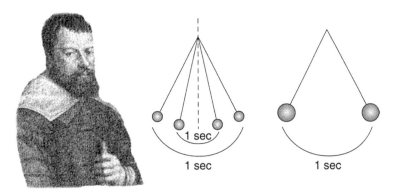

FIGURE 4.2   Galileo[2] discovered that the period of a pendulum is independent of the extent of its swing (for small swings). The period is also independent of the mass of the bob.

bottom of the arc. Extending that behaviour to objects free from strings, he reached a startling conclusion with far-reaching consequences. When released in free fall, a light stone and a heavy stone must descend to the bottom in the same time. But the conclusion contradicted a core lesson from Aristotle's teaching, which was that a heavy object, with more gravitas, should fall faster than a light one. It was time to demonstrate the pivotal result to the world from the legendary tower.

Why would the haughty university professors blithely ignore the evidence of their own eyes? Perhaps their Scholastic upbringing trained them to believe only in purely logical arguments. Galileo offered a stunningly simple one to show how the Aristotelian view leads to a paradox. Suppose (Figure 4.3) an extremely fine separation divides a large slab of marble into two half-size pieces. The earth continues to attract both halves. Why should such a minuscule change abruptly make the dissected pieces fall half as slowly as the whole slab? It would violate symmetry. Aristotle's reasoning led to a logical contradiction. To Galileo it was an outright absurdity. Whenever a paradox occurs it points to a serious flaw in the way we think about nature. For Galileo, the resolution of the contradiction was that the large slab and the two half-slabs must all fall at the same rate. And this is what the experiment clearly showed.

ROCKS AND FEATHERS

What about Aristotle's compelling *observation*, so plain for all to see? A leaf does fall slower than a stone. That, explained Galileo, is how our senses mislead us to obscure the underlying basic laws of motion. "Where the sense fails us, reason must step in," became Galileo's credo.[2] The universe is a master of disguise. We must unmask the cosmos by sophisticated inferences with our mind. A feather falls slower than a stone because the air through which it falls offers resistance to motion. By first identifying and then thinking away the role of air resistance – Galileo made the crucial *idealization* to penetrate the underlying ideal law of motion. Resolution of the paradox led to a radical new vision.

There is an easy experiment to eliminate air resistance and demonstrate that even a feathery light object such as a sheet of paper will fall to the ground at the same time as a heavy book. Sure enough, when released separately, the paper will float like a feather compared to the rapidly descending book. Now just lay the sheet directly on top of the book before they are released. They will both reach the ground at the same time. (If you conduct the experiment, be sure to choose a sheet of paper with dimensions that are no larger than those of the book. Otherwise, air resistance will slow down the paper's fall. It is a simple and delightful experiment to startle colleagues unfamiliar with the subtle underlying laws of free-fall motion.)

Heavy reliance on the senses had led Aristotle astray, as Parmenides had warned. Aristotle had missed the underlying *ideal aspect of motion*. By placing undue emphasis on sensory evidence, a

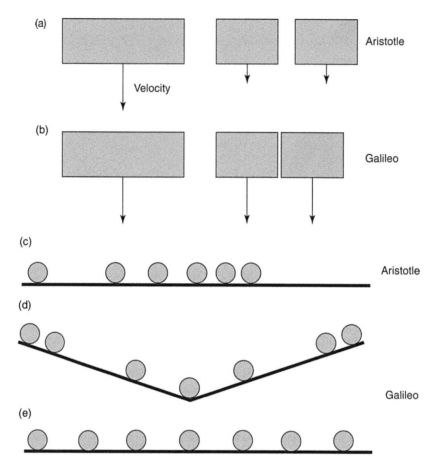

FIGURE 4.3 (a) Asymmetry in Aristotle's doctrine that two equal segments of a marble slab fall slower than the original piece. (b) Galileo expected that the two segments should fall at equal rates from a symmetry argument. Aristotle's idea that horizontal motion must naturally decay also harboured a bothersome asymmetry. (d) Galileo recognized the symmetry between the speeding up of a body sliding down a slope, and the slowing down of a body climbing a slope. (e) From the symmetry, he deduced that without any slope there should be no slowing down. On a horizontal plane, a body should ideally move forever at constant speed.

universal truth had escaped recognition, as Plato had feared. Yet this failure of Aristotle to properly interpret evidence could hardly be sufficient reason for Plato's summary rejection of all sensory information. Guided by careful observations of natural behaviour, Galileo's insight and creativity took him beyond sensory observations. Galileo brought Plato and Aristotle together, to harmonize the ideal and the real. His mind escaped from Plato's cave to grasp the richer reality that flickers beyond the reach of the senses. The shadows on the cave's wall hide the essence of the ideal laws and their inherent beauty.

In another crucial departure from Aristotle's approach, Galileo rejected grand ambitions to provide complete explanations of all aspects of motion from the very beginning of an inquiry. Believing that truth is to be found at the end of exploration, he limited his scope of inquiry to win one step of understanding at a time. To make progress, he would avoid dealing with phenomena in all their detail and complexity from the very first step, resisting the urge to immediately jump to answer the predominant concern *why*, without at first patiently exploring the limited *how*. Only after examining one part at a time could he harbour the hope of eventually piecing it all together.

Galileo first simplified the analysis of free fall motion by idealization. All objects fall to the ground at the same rate, independent of weight or composition or shape. The next step was to introduce the role of the medium. Dropping objects through water, he showed that the medium does play an important role in free-fall motion. A flat piece of sheet metal (e.g., a coin) will fall like a rock through air but float like a feather through a tall pitcher of water. But the shape of the falling object is also important. Drop the coin edgewise and it will descend rapidly, as if water offers little resistance. Weight, shape, surface area, density, all play a role in determining rate of fall. Take two identical sheets of paper, fold one in half repeatedly a few times and leave the other one spread open; the folded package will fall much faster. The two pieces of paper still have the same density *and* the same weight. But their shapes are different; the areas exposed to air and the corresponding resistances are different. A rectangular piece of cloth tethered to the rock by strings at the corners will slow down the fall of the rock – a principle now quite familiar to us as the parachute. If the parachute fails to open to intercept the air beneath, the passenger and parachute fall at the same rate with unfortunate consequences.

Equally important is the density of the falling object compared to the density of the medium through which it drops. Archimedes' work is enlightening here. Immerse a piece of cork under water and release it; it will *rise* toward the surface. Since the specific gravity of cork is about 0.2, water exerts a buoyancy force which exceeds the downward force of gravity. Therefore, the cork will ascend through water even though it will descend through air. Air bubbles rise up for the same reason. So does the swimmer who dives deep. Two centuries after Galileo, humans figured out how to defeat the force of gravity by rising in hot-air balloons using the fundamental principle of buoyancy: the density of hot air in the balloon is less than the density of the air through which it ascends.

BEYOND THE CLASSICS

Galileo's free-fall experiments through air and water demonstrated the value of fresh observations and experiments over arid argument. Yet die-hard professors preferred to ignore the force of evidence. Even direct experience can often fail to shake deep convictions built up by longstanding tradition and belief. Escaping the paradigm demanded a mental breakthrough. Eventually, the light of evidence broke through the thick intellectual fog. With lectures that fired the imagination, young Galileo filled the auditorium with enthusiastic young students, while those of fellow professors ran empty. Naturally they grew envious of the upstart who, with his challenging demeanour, upset staid beliefs. With new ideas, the irascible fellow aroused the animosity of influential people. Even worse, he used witty barbs to make fun of their thinking. Once he penned a humorous poem, poking fun at Scholastics and their habit of walking around like wax Aristotles. Surely he was influenced by the arrogance of Humanists (like Erasmus) who had ridiculed with biting satire the stock absurdities of an orthodox clergy.

A deeper rebellion was brewing in Galileo's mental outlook. Galileo was not exclusively challenging Aristotle. He was upsetting the whole world-view of the Renaissance, ruled firmly by classical principles. It was not enough to just gather knowledge from masters of antiquity, as the medieval monks had acquired from the Greek classics or, in their turn, as Greeks had derived from Egyptians and Babylonians. Accumulating knowledge does not win understanding. In his enigmatic work *Melancholia* (Figure 4.4), Dürer bemoans the fate of just amassing knowledge. The dejected goddess has wings but she cannot fly. Without divine inspiration, without the light of understanding, tools of art and science just lie scattered about in pure confusion. For Galileo, it was not enough to think about the cosmos in limited terms of geometric order and harmony. Transcending the longing to recreate glories of the ancient past, Galileo investigated motion with imaginative freedom. By challenging old habits with an inquisitive bent of mind, by devising fresh experiments with bubbling imagination, Galileo turned an historical corner into a brand-new era.

FIGURE 4.4   *Melancholia*, Dürer (1514).[3]

We can see a parallel development in the artistic world during the Baroque era dawning at the culmination of the Renaissance. Coined by art critics to characterize the rebellious new style, the word "baroque" signifies "absurd" and "wilful" – a wanton defiance of classical rules. Instead of calm, poised subjects set within a static, symmetric background, as exemplified by Michelangelo's *The Delphic Sibyl* (Figure 4.5), Baroque art teemed with motion. Exuberance and frenetic energy charge the works of Rubens (Figure 4.6).[4] Celebrating the arrival of the new wife of the French king after a long sea voyage from Italy, winged trumpeters welcome her as they swoop overhead and shower her with a cascade of gold coins. In *The Adoration of the Shepherds* (Figure 4.7),[5] El Greco's admiring angels flap their wings to hover above the miracle of Christ's birth. After two centuries of classical revival, the organized form, symmetry, and geometric clarity of the late Renaissance was ready to yield to a new dynamic. Both Rubens and El Greco studied in Venice, a city in constant motion, floating on a sparkling sea. Day and night, bustling gondolas ferried their way to and fro across canals. Witness in *Aurora* (Figure 4.8) the eternal horizontal motion in Guercino's celestial parade of the goddess of dawn and mother of the west wind. Led out by the flying victories, she rides triumphant in her chariot, as clouds, birds, and angels effortlessly glide by her. Even though the theme is classical, Guercino's dynamics go far beyond a revival of realism to add a brand-new sense of vitality.

In their zeal to portray vivacious movement, Baroque artists captured the flow of time during action. Bernini's *David* convincingly depicts the moment at which the legendary hero (Figure 4.9) is about to unleash the stone with lethal energy. But Bernini's work moves beyond Michelangelo's static, introspective champion. The contorted face and the pent-up stance clearly display power about to be released. For Baroque artists, it was time to renounce the bedrock of classicism to

FIGURE 4.5   Detail from *The Delphic Sibyl*, 1508–12, Michelangelo, Sistine Chapel, Vatican City, Rome, Italy. Possessed by the spirit of God, the Sibyl foretold the coming of Christ to Roman Emperor Caesar Augustus.[6]

FIGURE 4.6   Detail from *The Coronation of Marie de Medici*, 1622–5, Peter Paul Rubens.

FIGURE 4.7   Detail from *The Adoration of the Shepherds*, 1612–14, El Greco.

FIGURE 4.8   *Aurora*, 1621–23, Guercino.[7]

FIGURE 4.9   Bernini's *David* 1623–4[8] (compare with Michelangelo's *David*).

trumpet a vigorous new era. For the physical scientist, it was time to renounce the stultifying paradigm of Aristotle in search of a new intellectual adventure.

OBSERVATIONS ON AN INCLINE

With imagination and originality, Galileo inaugurated a new science. Through careful observations, quantitative measurement, controlled experiments, logical deductions, mathematical analysis, and creative insight, he made breakthroughs in understanding a variety of different types of motion: accelerated motion in free fall, down an inclined plane, uniform velocity motion in a horizontal line, and parabolic trajectories of projectiles. We can find the influence of Pythagoras (look for simple, underlying principles; describe the behaviour in mathematical terms), Aristotle (make close observations to find the supporting evidence), Parmenides (but do not be led astray by the evidence of the senses), Socrates (make logical deductions), Roger Bacon (make fresh observations), and Gilbert (devise experiments to manipulate the circumstances). A man of many parts, Galileo was the first to apply simultaneously many powerful approaches to fathom elusive secrets of nature.

Keeping his searchlight on *how* bodies move, without worrying too much about *why*, Galileo sought at first to describe and quantify two essential degrees of motion: *velocity and acceleration*. (Strictly speaking, velocity includes both speed and direction, but our focus here is on speed. Direction becomes important for circular motion in Chapter 10.) Ever since, the two crucial concepts have permeated physics. In early studies on free fall, he observed, as did Aristotle, that velocity grows with distance over which a body descends. Then there was a flash of insight: use *time* as the independent variable to examine motion. The notion of time had paid off handsomely in studying the swinging candelabrum. Time flow provided a simple understanding of velocity in terms of distance covered. In a fixed time interval, the larger the distance traversed, the higher the velocity. Velocity is the rate at which a moving object covers distance. More formally,

$$\text{velocity} = \text{distance travelled/time elapsed, or } v = d/t.$$

But an object in free fall drops too fast to measure easily and describe accurately. To slow down the fall, he deployed an inclined plane (Figure 4.10). Marking the positions of the ball at equal intervals of time, he measured progressively increasing distances. For example, if the incline was one degree, the intervals (in centimetres) were: 8.6, 25.7, 42.8, 60, ..., a series of increasing numbers showing no obvious pattern. But if he took the unit of distance to be the first interval (8.6 cm), and divided each number by 8.6, a familiar pattern immediately leaped to the eye:

$$8.6 \times 1, 8.6 \times 3, 8.6 \times 5, 8.6 \times 7, \dots, \text{ or } 8.6 \times (1, 3, 5, 7, \dots).$$

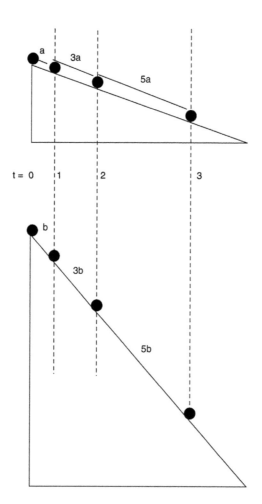

FIGURE 4.10   When a ball rolls down an inclined plane, the ratio of distance intervals covered in equal intervals of time increases as the odd numbers, while the total distance travelled increases as the square of the time. (Top) for one particular choice of the gradual incline, suppose that the distance covered in the first second is "*a*." In subsequent seconds, the distance intervals covered by the falling ball increase proportionally as: 3*a*, 5*a*, 7*a* ... (Bottom) for a steeper incline, the distance intervals are larger. Say the first interval is "*b*." Again, the ratios of the distance intervals follow the pattern of the successive odd numbers: 3*b*, 5*b*, 7*b* ... The essential point is that the ratios of the distance intervals always follow the odd numbers.

Increasing distances covered in equal intervals of time were in the ratio of the odd numbers. Clearly the velocity of the ball grew proportionately with time. In the first second, the distance travelled was one unit (i.e., 8.6 cm), in the next second the distance travelled was three units, and so on. The distances covered grew like the odd numbers. Symbolically we express "velocity grows proportionally with time," as: $v \propto t$ or $v = at$, where a is the constant of proportionality, called the "acceleration." The definition which emerges is: acceleration is the rate of change of velocity,

$$acceleration = change\ of\ speed/time\ elapsed$$

(Once again, acceleration also depends on direction, but here we will talk mostly about downward acceleration due to earth's attraction.)

Another remarkable relationship lurked in the pattern of odd numbers. Two thousand years earlier, Pythagoras discovered a mathematical fact of amazing simplicity: the sum of the first n odd numbers is n^2. For example, the sum of the first four integers is 4 squared: $1 + 3 + 5 + 7 = 4^2 = 16$ (see Figure 1.17). From this rule, it is mathematically clear that the total distance, d, which a falling object covers is proportional to the square of the time elapsed, symbolically expressed as:

$$d \propto t^2.$$

For Galileo it was a euphoric moment. He had arrived at the first quantitative laws of free-fall motion. Quantitative analysis shows that the precise relationship is:

$$d = (1/2)at^2,$$

where a is the acceleration, as before.

Galileo extracted simple, elegant, and quantitative laws for velocity and acceleration from careful observations and measurements of falling bodies:

(a) Velocity increases proportionately with time.
(b) The total distance travelled increases as the square of the time.

Any motion that obeys these particular laws is motion with constant acceleration. The relations hold true for objects moving down inclined planes with any angle of steepness. When the incline's slope increases, acceleration (a) also increases. By extrapolation, the inclined plane with an angle of 90 degrees should also have constant acceleration. But a 90-degree inclined plane is exactly equivalent to free fall. Therefore objects in free fall obey constant downward acceleration. It was an inescapable conclusion, one uniform rule for all objects on the face of the earth. Even a ball thrown straight *up* in the air moves with downward acceleration. Because of the downward acceleration, the body *loses speed* and comes to a stop. Not because it yearns to return to the centre of the earth, as Aristotle claimed, but because of the mathematical consequence of *downward* acceleration. The detailed paths of motion can be different depending on the initial velocity, but for all paths, the change of speed must be downward. For example, the trajectories can be parabolas or circles, topics for the future. In Newtonian terms (Chapter 11), the earth attracts all objects with the force of gravity, which causes acceleration. Labelled g for gravity, the value of acceleration in vertical free fall is 9.8 metres per sec, per sec (or m/sec^2), a very important constant for describing terrestrial motion.

Gone was Aristotle's long-winded rhetoric about the natural motion of earth and water to their natural resting places. Gone were the qualitative, confusing, and even contradictory descriptions of the past. Most remarkable was the ideal behaviour: free-fall motion is downward acceleration at a constant value, which is independent of the weight of the falling object. Why this law held true, Galileo did not know. For now, he only cared about how bodies moved. It was up to his successor, Newton, to make the next intellectual breakthroughs about gravity needed to understand why

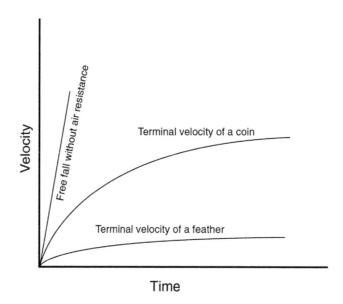

FIGURE 4.11   A comparison of the increase of velocity between free-fall motion and falling in presence of air resistance.

free-fall acceleration is the same for all bodies, and later for his successor Albert Einstein to decipher finally the full meaning for gravity's acceleration to be independent of mass.

Once Galileo exposed Aristotle's error, he could proceed to precise calculations about motion. But in order to establish conditions amenable to calculation he had to introduce ideals. In the absence of a medium, such as air or water, the velocity of a body, accelerating freely under the force of gravity, will increase in proportion to time, as Galileo calculated. But when bodies fall through resistive media, the velocity does not increase with time indefinitely; it approaches a maximum velocity, called the *terminal velocity*. After this, the body continues to fall with constant velocity, not constant acceleration (Figure 4.11). The reason that velocity ceases to increase is that air resistance (also a force) increases with the velocity of the falling body. Ultimately, the resistance becomes equal to the downward force of gravity, and acceleration becomes zero. Galileo did not derive the rest of the physics (involving air resistance forces) for calculating terminal velocity, but he did the crucial spadework to separate accelerated motion from the role of the medium.

Terminal velocity does indeed increase with weight, so a light feather reaches a lower final velocity than a heavy stone, as Aristotle observed. But, just like air resistance, terminal velocity also changes with the surface area of the body and with the density of the resisting medium. As a result of many such factors, terminal velocities can span a broad range. A penny falling through air will reach 9 m/sec, which is enormous compared to a floating feather's terminal speed of 0.5 m/sec.

MOTION WITHOUT CAUSE

Through his experiments with inclined planes, Galileo recognized another important asymmetry in Aristotle's physics (see Figure 4.3(c)): horizontal motion on a plane must come to a stop. Once again, evidence gathered from rudimentary perception yielded to Aristotle a clouded view of the world. Galileo used symmetry as a trusted guide to truth. A ball *gained* speed when it moved *down* an incline. When it moved *up* the incline, it *lost* speed because of downward acceleration due to gravity. So far the situation was symmetric. But why did an object moving along a horizontal plane *lose* speed when there was *no upward* incline? Since there was no slope up, nor any slope down, symmetry commanded to Galileo that there should be neither increase nor decrease in velocity.

Aristotle's "violent motion" idea was ready to fall under Galileo's next assault. *Ideally, an object moving with uniform velocity along a horizontal plane keeps going forward forever with that velocity rather than come to a stop.* It eventually comes to a stop not because of its nature, but because there is resistance to motion arising from "friction." Meaning "to rub" in Latin, friction is the resistance to motion as a sliding block rubs against the surface. A rolling ball or a wheel experiences less friction than a sliding block because its area of contact with the flat surface of the ground is greatly reduced.

Extrapolating from further experiments with inclined planes, Galileo exposed Aristotle's fallacy about horizontal motion through another light. Using two adjoining inclined planes, Galileo noticed that if a ball rolled down one incline, and immediately up a second identical incline, it reached a height that was very nearly equal to the height from which it started (Figure 4.12). He attributed the slight loss of height to the small friction. If there was no friction, the ball would ideally rise to exactly the starting height. Here was Galileo's breakthrough. Leaving the launch plane the same, if he progressively flattened the second inclined plane, the ball travelled farther along the second plane to reach the same height. Extrapolating this behaviour to its limit, Galileo made an intellectual leap: the ball would travel forever if the second plane became perfectly level.[9]

> Horizontal motion is eternal, for if it is uniform then it does not weaken for any reason, does not slow down, and is not extinguished.

It was a conclusion with profound and far-reaching implications, an astonishing triumph of human insight far surpassing knowledge gained from direct observation. Once again, Galileo stripped the dross of appearance to reveal the essential underlying law of horizontal motion, to find the hidden essence. Continued horizontal motion does not require a cause any more than rest requires cause. *Only change of motion requires a cause.* We adopt the word *inertia* for the natural property to maintain forever the state of uniform velocity motion. The word comes to us from the Latin word for

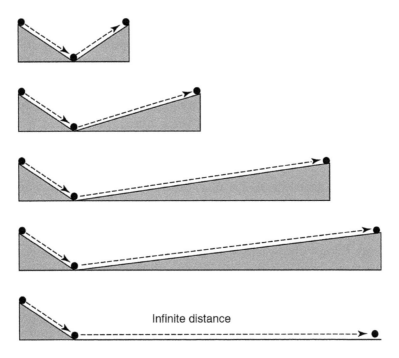

Infinite distance

FIGURE 4.12 As the slope of the inclined plane on the right decreases, the ball travels farther. On a horizontal plane, the ball will move forever!

"idleness" or "lazy." Just as an object at rest is inert, too "lazy" to move by itself, a uniformly moving body is also too "lazy" to change speed or direction all by itself. Motion with changing direction will be a fundamental topic for future discussions of circular motion. In the course of reaching the fundamental principle of inertia, Galileo used his powerful imagination once again to think away the resistance of friction and of air. As Einstein was later to phrase it so perceptively:[10] "Imagination is more important than knowledge."

Inertia can lead to some serious consequences. If you jump off a moving train, you fall down. Your feet come to an immediate stop on contact with the ground, but the inertia of your body continues to carry your body forward, as if it were still on the train. You can avoid falling if you anticipate the inertial principle, and start running as soon as your feet touch the ground.

Ideal horizontal motion is not a commonplace observation because of the presence of friction. Galileo pioneered the idea of a "frictionless surface" upon which the motion approaches the ideal. To visualize such a surface, he imagined a perfect plane, as smooth as a mirror, made of some hard material, such as bronze or steel. Upon this he pictured placing a ball which is perfectly spherical and also infinitely hard so that the ball touches the plane only at a single point. Such conditions approximated the ideal so that the ball travelled a great distance. A gliding ice-skater approaches the ideal frictionless motion sustained effortlessly across the length of an Olympic-size ice rink. Under pressure from the weight of the skater, the ice under the sharp blade melts into a water film for the skater to smoothly glide on.

GRACEFUL PARABOLAS

With conceptual and quantitative progress in vertical and horizontal motion, Galileo was well-equipped to tackle more complex movement. Yearning to find mathematical patterns in nature, he asked: "What is the mathematical curve which describes the complicated trajectory traced by a stone launched at an angle?" It was an immensely fertile question, with many lavish implications. Much as the swinging candelabrum stimulated his discovery of the pendulum principle, the graceful arches of water streams, shooting out from fountains decorating piazzas all over Italy, perhaps triggered his interest in the projectile problem. Not one to take nature for granted, Galileo proceeded to actively investigate, with careful observation and quantitative measurements, rather than hasty, overarching conclusions, to encompass the universe of motion. Similar to Renaissance artists, who brought linear perspective to art through detailed measurements, Galileo first measured the path of water shooting out from a hole in the bottom of a jar of water (Figure 4.13). It was a miniature laboratory fountain. With a ruler, he marked out equal distances along the *horizontal line*. At these marks, the vertical distances to the water track increased as the square of the horizontal distance. His thoroughness is reminiscent of Dürer's tutorial on linear perspective. With string, paper, and upright frame, the artist demonstrates (Figure 4.13) a technique for precise observation.

With the water track accurately marked, Galileo strove to identify the mathematical curve. His familiarity with classical mathematics revived from Alexandria proved crucial to recognizing the parabola. Why is the path a parabola? Galileo analysed the compound path by treating separately the *vertical* and *horizontal* motions of water drops, viewing each motion through passage of time. The vertical distance through which droplets fell increased with the square of the elapsed time, as expected for constant acceleration in free fall. But through those same time intervals, the drops *simultaneously* travelled equal steps in the horizontal direction. The horizontal velocity of droplets remained constant over flight as in the law of inertia. Once started in horizontal motion, water drops kept going. The droplets were *simultaneously* following the laws of vertical and horizonal motion.

Galileo introduced a novel idea: decompose two-dimensional motion into separate horizontal and vertical components for a methodical, step-by-step inquiry. Remarkably, the two components (vertical and horizontal) of motion proceed independently of each other. What goes on in one spatial dimension has no influence whatsoever over what goes on in other dimensions. If it were not for

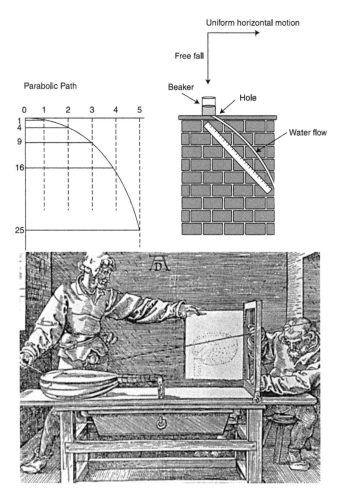

FIGURE 4.13 (Top) Galileo measures the path followed by a stream of water shooting out of a hole in the bottom of a jar. He had to keep the water level in the beaker constant (not shown). (Bottom) *Man Drawing a Lute*, Albrecht Dürer, 1525.[11] First place a piece of paper in an upright frame which can swivel aside so as to run a cord from the lute to the eye. Fix the point where the cord passes through the frame by crossing two movable cords attached to the corners of the frame. × marks the spot. Remove the long sighting cord, and swing the frame with the paper back into place. Mark the intersection point of the cords onto the paper with a dot. Continue the procedure from the same viewpoint to make a dense set of points on the lute. The dots on the resulting diagram project an accurate three-dimensional perspective on a flat canvas.

free-fall acceleration, the projectile would fly forever in a straight line, as its primal condition of inertia dictates.

Galileo pointed out a startling new result from the decomposition. Suppose from a cannon located on the top of a cliff (Figure 4.14) a gunner were to fire several balls horizontally with different explosive charges placed inside the cannon. Each ball would take *exactly* the same time to hit the ground, independent of whether it travelled horizontal distances of a few feet or 2,000 feet, as a result of the different propulsive firings. Each total flight time interval would be exactly the same, determined exactly by the time it took a ball to free fall from the height of the cliff to the bottom. Each ball behaved as if it were just dropped without ever being fired from the cannon. Even though the individual trajectories of each ball are different parabolas, depending on their initial velocities, the acceleration for all the balls is exactly the same, down to earth at a rate of $g = 9.8 \text{ m/sec}^2$.

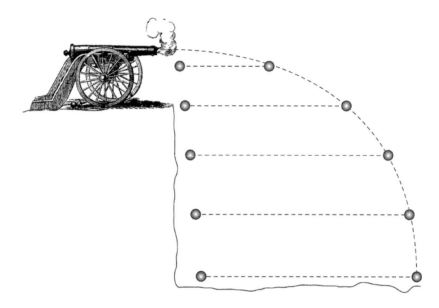

FIGURE 4.14   The time taken for a cannon-ball horizontally launched to hit the ground is the same even when the gunner changes the amount of charge in the cannon.

Breaking down compound motion, Galileo could explain another important aspect of projectile flight, well-known to artillery engineers. A shot travelled the *farthest distance* (called range) when fired at an angle of 45 degrees to the horizontal (Figure 4.15), a conclusion previously established by gunners on a purely empirical basis through centuries of military experience. As Galileo pointed out:[12]

> From the accounts given by gunners, I was already aware of the fact that in the use of cannon and mortars, the maximum range, that is the one in which the shot goes the farthest, is obtained when the elevation is 45° … *but to understand why this happens* far outweighs the mere information obtained by the testimony of others or even by repeated experiment.

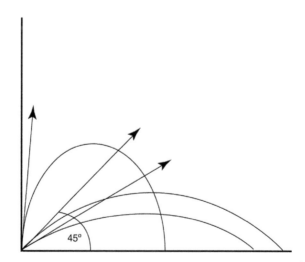

FIGURE 4.15   The range of projectiles launched at various angles. When the launch angle is 45 degrees the range is maximum.

Galileo understood the 45-degree angle. His reasoning may be cast along symmetry lines. If the launch angle is very shallow (near zero) the result is a low trajectory (Figure 4.15). The ball will stay in the air only for a short time. Even though the horizontal velocity component is large, the horizontal range will be short, because the ball will hit the ground in a very short time. At the other extreme, a launch angle near 90 degrees results in near vertical flight. Because the vertical velocity component is large (Figure 4.16), the ball will stay in the air for a long time. But it will not travel far along the ground at all because of the small horizontal component. Now the angle of 45 degrees symmetrically divides the two-dimensional velocity into *equal horizontal and vertical components*. Hence the ball will both stay in the air for a long time *and* it will have a high horizontal velocity to travel far. The total distance travelled is related to horizontal velocity × time in air. Highest values for both horizontal velocity and time spent in flight will yield the largest product. A 45-degree angle (half of 90 degrees) provides the optimal point between the two extremes of time and velocity to yield the maximum range.

Galileo finally understood the repeated experience of gunners. He followed the footsteps of Greek philosophers Thales and Pythagoras who struggled to understand abstract geometric principles underlying arbitrary rules and prescriptions they acquired from Egyptian and Babylonian cultures. As with Thales' similar triangles, success in understanding vertical and horizontal motion opened new applications. With the mathematics of the parabola, Galileo constructed a table for gunners to look up the range of projectiles launched at various angles and speeds. Enlightened by Galileo, the Venetian arsenal and artillery workshops maintained close contact with the wise and knowledgeable man.

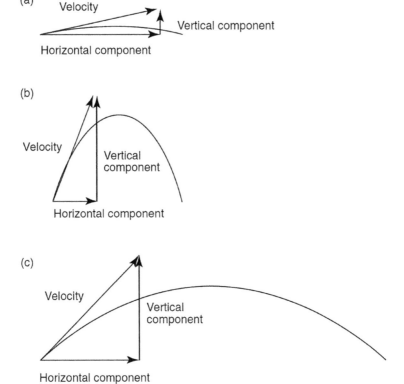

FIGURE 4.16 Vertical and horizontal components of the velocity of a projectile launched at an angle with respect to the horizontal.

PERFECT SYMMETRY

To continue the discussion of dynamic symmetries introduced in Chapter 1, important symmetries underlie the physics of free-fall motion, inertial motion, and parabolic trajectories. The spherical symmetry of earth (an idealization, of course) dictates that the value g for free-fall acceleration should be the same whether it is measured in Pisa, Rome, or Athens, anywhere on the face of earth. Deviations in g will of course be observed between mountain tops and sea level, which break the ideal perfect symmetry of the spherical earth.

Suppose we make a video recording of the motion of a bouncing tennis ball and play it backward in time (in the rewind mode) at regular speed. Thinking about time flowing backward is a helpful idealization for understanding symmetries. In reverse play, the motion appears normal, except the signs of the velocities are reversed. Upward motion takes the place of downward motion and vice-versa. But the sign of earth's *attractive* force and the corresponding sign of acceleration do not change with time reversal; g always points downward due to the attraction of the earth. A most important consequence from time reversal symmetry is that the reverse play case is also a valid motion in the real world.

A similar result prevails for horizontal motion. A movie of colliding marbles will look perfectly normal if run backward in time. Time reversal symmetry holds for the more complex to-and-fro motion of a pendulum. Ideally, the bob rises to the exact same height from which it is released. If it did not, it would become possible to tell which way time flows by recording a movie of the motion and playing it backward.

But time reversal applied to everyday events clearly yields weird results. Broken glass comes together, and spilled coffee returns to the cup. We never observe such events, so time reversal symmetry seems far-fetched. Yet in simple cases involving just a small number of objects, fundamental processes look quite normal upon time reversal. Only when the number of particles involved becomes very large, as for the trillion, trillion molecules of water in a cup, the probability of observing the same behaviour upon time reversal becomes incredibly small. That is why we never see spilt milk rise up miraculously to find its way back into the carton.

A deep reason that natural motion appears the same when time flows backward is that the laws of motion, indeed all laws of physics, are symmetric with respect to time reversal. (Weak nuclear force interactions – Chapter 2 – involving decay of nuclei do violate this symmetry, as will come up later.) Physics equations needed to describe motion under uniform acceleration are the same for a movie playing in reverse. The mathematically inclined can check this by substituting $(-t)$ for (t) in the appropriate equations. As long as they also change the signs of the velocities (as demanded by time reversal), the equations look exactly the same for (t) or $(-t)$. The sign of acceleration does not change with time reversal; falling bodies with downward acceleration due to gravity appear as rising bodies with slowing velocities, which is also downward acceleration. Distance travelled does not change sign since a distance interval has no direction.

Mirror symmetry is another universal symmetry for the laws of physics (Again, the weak nuclear force does not respect this symmetry.) Physical laws governing experiments in the real world are the same as in the mirror-world, even though any left–right asymmetry in the real-world will display as a right–left asymmetry in the mirror. A right hand appears as a left hand in the mirror. Even though the mirror image world may look different, the physical laws operative there are the same as in the real world. The crucial aspect about mirror symmetry is that mirror-image processes can also occur in the real world. Imagine hitting a squash ball toward a mirror instead of a wall (Figure 4.17). In the mirror, your image simultaneously appears to play the ball towards the mirror plane, so the sign of the ball's velocity in the direction of the mirror reverses. The essential point is that a mirror-image-player is also allowed by the laws of physics to be a real player.

A mirror reverses the sign of only one spatial coordinate. A more general transformation is a *parity transformation* which reverses the signs of all three spatial coordinates. Physical laws are invariant with respect to parity transformations, with exceptions involving the weak nuclear force.

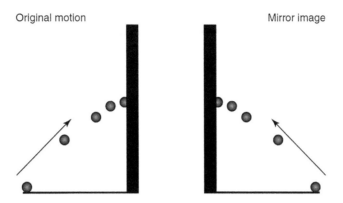

FIGURE 4.17   The mirror image reverses the sign of one spatial coordinate as well as the sign of the velocity along the same coordinate.

Physical laws show more general symmetries than time reversal, mirror, and parity symmetries. Galileo's law of inertia relates to a fundamental symmetry principle: laws of physics do not depend upon a choice location in space. Suppose three members of a movie crew stand 10 metres apart along a train track running through a featureless, open field (an idealization, of course.) Separately they record a passenger on a train passing them at constant velocity. There is no way to distinguish between the three video tapes of the observers standing in different places, even when the video is played frame by frame. In all displays, there is a uniform spacing between the train-passenger's location in successive frames. However, if the train were accelerating, the three results would appear quite different. When the train passed the first witness, the passenger would appear closely spaced between frames. The third observer's tape will show much larger separation between frames. In accelerated motion, inertial velocity is not invariant. Such motion is not symmetric with respect to spatial translation. Galileo's constancy of inertial velocity is equivalent to invariance of physical laws under spatial translation.

Consider another important symmetry. When describing the behaviour of physical systems, one moment in time is as good as the next to serve as the starting point. This is equivalent to the principle that laws of physics are invariant under time translation, just as laws are invariant under spatial translations. Consider a three-person movie crew filming an *ideal* oscillating pendulum that does not lose energy. Each member starts taping at an arbitrarily different instant of time. When examining the three tapes it is impossible to determine which camera started up first. However, if the pendulum loses energy through air resistance, then it is easy to tell first from last by the decaying amplitude of the swings. Invariance symmetries lead to powerful laws that help keep track of physical quantities, such as energy and momentum (covered later). The invariance of physical laws under spatial translations lead to the *Law of Conservation of Momentum*. Time translation invariance leads to the *Law of Conservation of Energy*.

Because of their deep origins in general symmetry, momentum and energy conservation hold even when motion is not ideal, for example, when pendulum swings decay, provided one takes into consideration the movement of parts external to the pendulum that are responsible for loss of momentum and energy. When it slows down, the kinetic energy of a swinging pendulum goes partly into the motion of the surrounding molecules which offer air resistance. Some goes into increasing the rapid internal motion of atoms of the supporting nail and suspending string through the friction between them.

From symmetry principles, physicists expect, and observations confirm, that all the fundamental laws of physics are symmetric with respect to time translation, space translation, and rotation through any angle. For *rotational symmetry*, every spatial *direction* is on an equal footing with another. It is equivalent to the statement that the universe must be isotropic in space. Rotational symmetry leads to another conservation law for angular momentum. A spinning skater uses this law to draw in his arms, and increase his rotation speed. The angular momentum from the arms decreases as they move toward the centre. To conserve angular movement the body spins faster.

A MOVING SYMMETRY

Galileo's decomposition of horizontal and vertical motion helps bring out a powerful consequence of another symmetry of physical laws. Consider a ball dropping along the mast of a ship, as viewed by a person on shore (Figure 4.18). If the ship is at rest relative to a shore observer, the ball executes free-fall motion, straight down to the deck for the observer on shore as well as for the one on the ship. But if the ship sails forward with uniform velocity parallel to the horizon, the person on the shore sees the ball moving in a parabolic path. The ship's horizontal motion combines with free-fall motion to result in a two-dimensional path of a parabola, just as for a projectile's combined horizontal and vertical motion. For different velocities of the ship the shore observer will see different parabolic paths.

But the passenger on the moving ship will witness a different path. She sees that the falling stone has *no horizontal velocity*. She *always* finds the ball in *vertical free-fall motion,* whether the ship is moving fast, slow, or is at rest. As long as the ship moves, the two observers disagree; the shoreman sees parabolic paths, while the passenger sees a straight-line path. Which is the correct trajectory, straight down or parabolic? Both are correct! The answer depends on whether the observer is stationary or moving. Two observers moving relative to each other have different perspectives from their vessels, and therefore different perceptions of the path. Each observation of motion is valid in its own "reference frame" or vessel. The path seen by the stationary observer can be calculated from

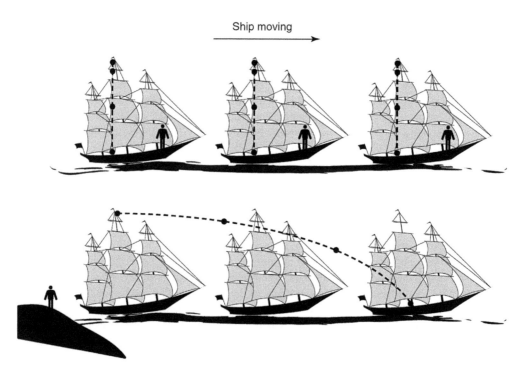

FIGURE 4.18 (Top) a passenger on the boat, moving with the boat, sees only downward accelerated motion of a stone from the top of the mast to the bottom. (Bottom) an observer on shore sees a stone following parabolic motion, compounded from free-fall accelerated motion due to gravity and the uniform horizontal motion of the ship.

the path seen by the ship observer by adding the horizontal velocity of the ship. In other words, the two paths can be reconciled by a *Galilean transformation* of adding velocities. Galilean transformation helps us switch between moving vessels.

Even though resulting paths are different, there is a crucial common denominator, a profound symmetry. Passengers on a moving vessel can use the same underlying laws of physics with identical equations of motion as persons on shore use. *The laws of motion on a ship do not depend on the velocity of the ship.* Just as equations of motion are invariant under time reversal, mirror reflection, spatial and time translation, laws of motion are also invariant of the speed of a moving ship, as long as the ship moves with constant velocity. You can bounce a basketball on an aeroplane flying at 500 mph and not tell any difference in the motion of the ball from the case when you bounce the ball in the plane parked at the runway. According to Galileo you cannot tell whether you are at rest or moving at some constant velocity by studying motion on the moving ship. As a universal consequence of Galileo's laws of motion, 45 degrees is always the correct angle to launch a projectile to cover the maximum range over deck, independent of the speed of a cruise ship. Later, Galileo was to use this idea to argue that we cannot tell, only by observing motion on earth, whether the earth is moving or at rest. Eloquently he explains:[13]

> I invite you to enter with me the cabin of a large ship. In it are gnats, ball players, and a bowl of goldfish into which drops of water are falling from above. Whether the ship lies at anchor or moves uniformly through the water, the gnats will fly with the same ease from wall to wall; the balls will hit with the same force in all directions; the fish will swim undisturbed; the drops will fall vertically upon the same spot. *No one can guess, by observing these processes, whether the vessel is at rest or moving.*

Galileo's assertion is a statement of dynamic symmetry. It connects our different descriptions of the same reality relative to each other. Known as *Galilean relativity*, the principle is a deep symmetry of nature, one with profound consequences for space and time that will emerge later with Einstein's theory of relativity.

THE FATHER OF SCIENCE

Soon after his marvellous insights about motion, Galileo resigned his position at Pisa. Not being paid enough to support his family, he accepted a mathematics professorship at Padua, the outstanding university of the period. Here Vesalius had carried out dissections at the *Theatre Anatomico*, and Copernicus had taken instruction in classics. Galileo's lectures were crowded with enthusiastic students, many of whom assisted in carrying out experiments. Besides the physics of motion, Galileo touched upon a range of subjects: bridge building, harbour planning, fortification, and artillery construction. Other professors grew envious of the 28-year-old newcomer who filled the Auditorium Maximus to capacity. Many students were noblemen from all over Europe, fascinated by natural phenomena and awed by his eloquence. They learned about new discoveries first-hand from lecture demonstrations. But to his professorial colleagues, who only knew how to theorize for or against Aristotle, experimental demonstration remained a form of witchcraft. In Chapter 9 we will turn to Galileo's discoveries in the heavens, his interpretations which ran counter to the Bible, and the reactions of the Pope and clergy.

Galileo was a Renaissance man *par excellence*. With enterprising spirit inimical to the merchants of the new middle class, he brewed trouble by questioning dictated dogma. Like Humanists before him, he returned to classical ideas. But he refused to cling to the past and stay bound by Aristotelian authority. Breaking Scholastic tradition, he emphasized the importance of independent thought. Like the artists around him, he awoke to the wonders of nature to observe motion in glorious detail. But with the eye of a painter he saw beyond the seen. While realists captured three-dimensional space on two-dimensional canvas, Galileo conquered time through the swinging pendulum. To project space in linear perspective, artists made quantitative measurements. Galileo measured trajectories to discover that motion can be described with mathematical accuracy. Pythagoras first introduced the idea that numbers govern nature. To Galileo[14], "the book of nature is ... written in mathematical characters."

As contemporary anatomists probed cadavers and navigators explored oceans, Galileo investigated nature with all the ingenuity he could muster. Going beyond the Renaissance, he transcended the revival of classical ideas. Rising with the tide of intellectual advance, he was part of the community of the Baroque artists who captured the vivacious sense of movement.

Galileo started science on a fresh path to new knowledge. As Albert Einstein said:[15]

> Pure logical thought cannot yield us any knowledge of the empirical world; all knowledge of reality starts from experience and ends in it ... Because Galileo saw this, and particularly because he drummed it into the scientific world, he is the father of modern physics – indeed, of modern science altogether.

MASS AND FORCE

With Galileo's ground-breaking efforts on *velocity* and *acceleration*, Newton, Descartes, and Leibniz laid the foundation for modern mechanics by imparting precision to the developing concepts of *force, mass, momentum,* and *energy*. These concepts came to a head with *conservation laws of momentum and energy*, reaching a climax with Einstein's discovery of the equivalence between mass and energy, immortalized by his equation $E = mc^2$. Our update section on relativity will include a simple derivation of that equation (albeit under special circumstances). The derivation will depend on the concepts of momentum *(= force × time interval for delivery of force)*, and energy *(= force × distance over which force acts)*.

A generation after Galileo, Isaac Newton embraced the principle of inertia, and advanced to the precise concept of *force*. His ideas and personality will dominate the synthesis of heaven and earth (Chapter 11). Force is the agent necessary to realize a change in the state of motion. Newton formulated his *First Law of Motion*: Without the action of an *external force*, an object will continue to move in a straight line with constant velocity forever. Force comes from the Latin word for "strength." Nature supplies a generous bounty of forces, and humans supplement these in good measure. Gravity makes objects fall to earth and projectiles describe graceful parabolas. Friction decelerates, so moving objects come to a stop. Archimedes of Alexandria applied muscle force to levers, pulleys, and the wheel-and-axle so as to single-handedly drag huge ships.

When a force (F) acts on a body it increases the acceleration (a) of the body, in direct proportion to the force:

$$F \propto a.$$

When the same force acts on different objects, the acceleration is different. What makes acceleration different is a unique property of the body which Newton identified and called *mass to signify* "quantity of matter." In his famous *Second Law of Motion*:

$$F = ma$$

Newton defined precisely and quantitatively the modern notions of *force* and *mass*. Like acceleration, *force* also has direction and is described by a vector. Direction becomes essential for understanding circular motion when the velocity, acceleration, and force all continuously change direction.

Newton's force law is the most useful physical law ever devised. One single law unifies the motion of all bodies under all types of forces. Knowing the force acting on a body with known mass, one can calculate, from $F = ma$, the motion of the body to predict its path. Conversely, knowing the paths of motion, one can infer the force(s) acting on a body.

Unification is the central theme of Newton's force law, the backbone of physics. With the concept of mass, Newton gave all matter a single nature. Mass is an intrinsic property of any object connected to its motion. It resists the change in the state of motion; the larger the mass, the higher the inertia. Mass governs how much acceleration there will be under a fixed force. Newton was the first to grasp and clarify the concept of mass, and to sharpen its distinction from the older and more familiar concept of weight. *Weight is a force* which earth exerts on a body. Weight is just one of the many manifestations of mass. The subtle meanings of mass will again crop up in Chapters 11 and 12 on synthesis.

MOMENTUM, IMPULSE, WORK, AND ENERGY

Galileo's successors Torricelli (Chapter 2 on vacuum and atmospheric pressure) and René Descartes made further progress connecting important concepts of motion: momentum, impulse (force × time), and energy (force × distance). These concepts will prove important in understanding the equivalence of mass and energy established by Einstein (in updates).

Building on the ideas of medieval scholar Buridan (Chapter 3) on "cargo of motion" (momentum) Torricelli clarified the concept of *impulse* to connect it with *momentum*. He compared the behaviour of a large galley ship to that of a little boat, each moored about 20 feet from the pier. A man can pull on the galley for a long time with all the force he can muster to give the massive ship some speed. When the galley hits the pier, the pier will shake violently due to the high *cargo of motion* (*momentum*) from its enormous mass.

$$\text{Momentum} = \text{mass} \times \text{velocity}$$

But he does not need as much force or as much time to build up the same speed for the small boat, and its effect is comparatively insignificant when it hits the pier. The boat with small mass does not acquire much momentum. Torricelli concluded that momentum builds up as a product of force × time applied.

$$\text{Momentum} = \text{force} \times \text{time} = \text{Impulse}$$

Torricelli probably referred to *mass* as *weight* at the time before Newton's clarifications.

Galileo's quest for ideal principles underlying motion had a strong influence on his younger contemporary, Frenchman René Descartes, who sought to describe all natural phenomena through their underlying *mechanisms*. According to Descartes, God created matter particles and endowed them with motion. After that, the universe evolved by laws of mechanics, moving perfectly like the gears of a giant cathedral clock. It was an appealing picture, easy to visualize, for such a clock was the centrepiece of every town square in Europe. As descending weights powered rotating hour hands, turning on axles, mechanical toy men rang bells, and cocks crowed regularly on the hour. One coherent mechanism generated a variety of operations, like the infinite phenomena of an infinitely complex machine we call nature. God was the clock-maker overseeing the complex workings of his orderly creation. Through flawless mechanical operations Descartes wished to prove the perfection of God's creation and so to vindicate God's rationality. If the universe was a machine that ran automatically, its operations were open to rational inquiry. The Cartesian universe took a firm grip on the minds of budding scientists, Newton and Huygens (whose progress in understanding circular motion we will address in Chapter 10).

Descartes seized Galileo's concept of inertial motion: a body set in motion continues to move forever in a straight line. Everlasting motion ensured the eternal operation of the clockwork universe. Once set running, the divine clockwork ran for ever, without needing winding or repairs, as God was the perfect clock-maker.

Turning his attention to agents responsible for change, Descartes submitted that particles influence each other through impact, interactions easy to visualize. Laws of impact governed the hidden workings of the universe. Descartes believed that in a universe fashioned by the perfect clock-maker, when individual particles alter their motion through interaction with other particles, the total "quantity of motion" during collisions must remain constant. God set the great mechanical system going at creation, giving it a set "quantity of motion," which never ever changes, even as matter goes through continual interactions and transitions. A religious man, Descartes framed the "principle of conservation" of motion on a divine foundation. Although vague and mystical, he glimpsed the first laws of impact. Descartes' conservation of "quantity of motion" later evolved into two separate laws: Conservation of Momentum and Conservation of Energy – dual linchpins of mechanics. Not surprisingly, Descartes erroneously merged concepts of momentum and energy. Both were in formative stages, to mature over the next two centuries. We will discuss Descartes some more in Chapter 10 for his breakthroughs on circular motion.

Following his predecessors, he accepted *weight* × *speed* as the "quantity of Motion" (momentum). One of the earliest forms of mechanical energy recognized was the concept of work done by a force. In modern terms,

$$\text{work} = \text{force} \times \text{distance}$$

As a result of work-done (by the earth's attraction for example), the body gains "*motive force*" (*energy*) equal to the work-done. Gottfried Leibniz (a contemporary of Newton) showed that the eminent Descartes was quite mistaken in equating motive force (energy) with quantity of motion (momentum). There is a fundamental difference between the two physical quantities. One cannot be obtained from the other. For example, if a body weighing 1 kg falls through 1 metre, it acquires one unit (*1 kg × 1 m*) of energy from the work done by the force of earth's attraction. Leibniz determined that the 1-kg body also acquires a speed of $\sqrt{2 \times 9.8}$ m/s according to Galileo's dynamics, and so its momentum (mass × velocity) is 1 kg × $\sqrt{20}$ m/s. The two quantities (and their units) are quite different.

In resolving Descartes' fallacy, Leibniz introduced a very new, and very valuable concept: living force, or vis-viva (modern kinetic energy). What mechanical quantity should one use to calculate the velocity from the work done by a force? What is the true measure of energy equal to work done? Velocity does not measure it, and momentum does not measure it. Leibniz showed that the product: *mass × velocity²* was the appropriate quantity to equate with motive force. For the first time the square of the velocity appeared in any mechanical quantity. He gave the quantity mv^2 the name of "vis-viva" or "living force." Collisions would have to satisfy the conservation of both momentum and energy transfer.

In modern language, the work done by a force equals the vis-viva (kinetic energy) gained. Leibniz's arguments equated vis-viva with work.

$$\text{Force} \times \text{distance} = \text{change in } 1/2 \times \text{vis viva}$$

$$F \cdot d = 1/2 \, mv^2$$

(The factor 1/2 came in when the above equivalence was derived from Newton's Laws of Motion.)

Leibniz argued with Newton about the relative merits of force and vis-viva as the appropriate mechanical quantities for analysing motion. Vis-viva did not carry Newton's approval. Leibniz's emphasis on vis-viva and his contradiction of Descartes sparked quarrels among scientists that continued for decades. For the English scientists who followed Newton, the laws of motion were settled once and for all with $F = ma$. New ideas, like *vis-viva* had no place in the physics of motion.

A man of many talents, Leibniz also invented the first calculating machine that could add, subtract, multiply, and divide. But he is best known through his contribution to mathematics with his invention of differential and integral calculus, independently from Newton (Chapter 11). With the new calculus, it became possible to precisely treat motions of great complexity, like elliptical planetary orbits.

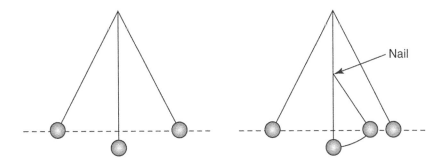

FIGURE 4.19   In the ideal motion of a swinging pendulum, the bob always rises to the same height during every swing. If a nail changes the pivot point by shortening the length of the string, the bob will again rise to the same height, but following a steeper arc. In the real motion, there is a slight loss of height after every swing due to air resistance.

KINETIC AND POTENTIAL ENERGY

During experiments on pendula and inclined planes, Galileo understood aspects of energy. Again, it was insightful interpretation following meticulous observation. If he caught the swinging string of the pendulum with a nail placed some distance below the point of suspension, the pendulum bob always climbed to the same height, even though the arc it climbed up was much steeper (Figure 4.19). Galileo realized that the velocity attained at the bottom of the descent was always great enough for an object to rise to its original height. When a falling ball reached the bottom of the first incline, it continued up the second incline to rise (ideally) to exactly the same height.

Today we identify the ideal possibility of rising to the starting height as the principle *conservation of energy*. When a pendulum bob "falls," it acquires, from the attractive force of the earth, energy of motion, called *kinetic energy*. At the bottom of the swing, the kinetic energy propels the bob back up to its original height. Here the bob comes to rest, so the kinetic energy is zero. But the energy is not lost. It is potentially available (*potential energy*), and released when the bob falls again. In modern terms, the pendulum's kinetic energy converts completely to potential energy at the top of the swing. The cycle of exchange between potential and kinetic energy continues for ever, as the pendulum swings back and forth. We think of potential energy as energy stored in the earth–body system. The greater the mass and height, the larger the potential energy. A massive boulder falling from a high cliff has the potential to cause severe destruction.

In another example, to raise a basketball from the floor to a hoop, a player must provide a certain amount of kinetic energy for propulsion. When the ball reaches the hoop, and comes to rest, all its kinetic energy is potential energy in the earth–ball system. When it falls from the hoop to the floor, the ball regains the original kinetic energy as it reaches the ground. During a perfectly elastic bounce off the floor, the ball comes to a momentary stop, absorbing one part of its kinetic energy by deforming its shape like a spring, and giving up the rest of its kinetic energy to the floor which depresses ever so slightly like a very stiff spring. The energy is now stored as potential energy components of the springy (elastic) ball and floor. As the elastic floor and ball return to their original shapes, they propel the ball up to return its full kinetic energy. With its kinetic energy completely restored, the ball rises again to the hoop.

UPDATES

A SPECIAL THEORY – EINSTEIN'S RELATIVITY OF SPACE AND TIME

According to Galileo's dynamics, the proper launch angle of a football (to use a contemporary example) for maximum range is always 45 degrees, whether the quarterback is in a football stadium in New York or whether he is on the deck of a speeding cruise ship to the Bahamas. Pouring drinks or even juggling oranges on an aircraft moving on a straight path at 500 mph makes no difference whether the jet flies smoothly at cruising altitude, or if the plane is parked at the gate. According to Galilean relativity, exactly the same laws of motion apply on a moving aeroplane as on the ground.

Instead of mechanics, consider a different example of physical behaviour that depends electricity, magnetism, or on Maxwell's electromagnetics (Chapter 3). Music streaming from a CD player on an aeroplane will sound exactly the same as when the plane is on the ground. A CD player's operating principles are based on Faraday's and Maxwell's physics of electricity and magnetism as well as on the physics of laser light inside the player that reads the CD. Just as for mechanics, laws of electromagnetism and light are exactly the same on the aeroplane as on the ground.

But there is a paradox here, first recognized by Albert Einstein in 1905. The paradox set off alarm bells for Einstein. Half a century earlier, Faraday and Maxwell had unified the physics of electricity and magnetism (Chapter 3). With a set of exquisite equations embracing the two fields, Maxwell made a monumental discovery. Light is an electromagnetic wave. The speed of light is an integral part of his equations that describe the intertwined behaviours of electricity and magnetism. The value of light speed comes out directly from two numbers, electrical and magnetic force constants measurable by laboratory instruments. The magnitude of light speed, $c = 3 \times 10^8$ m/sec = 186,000 mi/s = 670 million mi/hr, is a natural constant. As an electromagnetic physical quantity, *light speed should always be one and the same value*, whether it is measured by a passenger on a moving aeroplane or one standing still on the ground.

Here is Einstein's paradox in a modern form. In Galileo's dynamics, speed, as a physical quantity, depends on the motion of the observer who is measuring it. Suppose an aeroplane moves toward the sun at 500 mph (0.138 mi/s), to meet an arriving sunbeam. What would the pilot measure for the speed of the approaching light beam, if he needed to know this number for calibrating his navigational instruments based on electromagnetic science? According to Galilean relativity, the speed of the plane should be added to the speed of the approaching light beam since the beam is travelling toward the plane and light will arrive at the plane a bit sooner; the result will be 186,000 + 0.138 mi/s. And if the plane moves away from the sun, in the same direction as the light beam, light will take a bit longer to reach the plane, so that the plane's speed should be subtracted from the speed of light; the result would be 186,000 − 0.138 mi/s. In each case the answer for the speed of a sunbeam comes out *different* from the physical constant predicted by Maxwell's equations. And the measured speed of light depends on the speed of the plane. But this raises an immensely troubling question. Maxwell's equations predict a specific fixed number for light speed. What is that speed measured with reference to? According to Maxwell's fundamental equations for electricity and magnetism, the speed of light has to be 186,000 mi/s relative to the ground or to any vessel and to every vessel, regardless of the vessel's speed.

It would take very unusual thinking to see beyond the seen in order to resolve the paradox generated by Maxwell's laws and Galilean relativity. Einstein had that unusual temperament right from the start. He was born in 1879, the same year that Maxwell died, as if destiny intended to pass the torch. Growing up in a German-Jewish family that espoused liberal ideas, he showed talent for the Pythagorean pleasures of geometry and music (violin) at an early age. We saw Einstein's simple proof for Pythagoras' theorem in one of the update sections of Chapter 1. As a child, he withdrew from other children, preferring to erect complicated mechanical constructions on his own. With infinite patience, he constructed card houses with 14 floors. When 5-year-old Albert fell sick, his father gave him a compass magnet for amusement. Almost immediately, the curious boy wanted to know why the needle always turned to point in the same direction. When his father explained, in abstract terms, that earth has a magnetic field which influences the needle, Albert was astonished that something completely invisible could affect a real needle.[16] "Something deeply hidden had to be behind things," he recalled later. In high school, teachers complained he was always daydreaming at the back of the class. Such attitude destroyed student discipline, rated paramount in the German school system where regimentation and learning by rote was the gold standard. In college, rebel Einstein skipped lectures that strangled the "holy curiosity of inquiry"[17] and frequented cafes instead. But he enjoyed physics experiments, where he could experience nature in action, and he studied the latest physics texts on his own.

At the turn of the 20th century, physics had reached a pinnacle because of the new discoveries of Maxwell regarding electromagnetic radio waves, together with new inventions of electric light, telephone, and the phonograph, all based on electromagnetic technology. Newton held the last word in mechanics for two centuries; Maxwell now reigned supreme in electromagnetism. Prominent scientists declared physics near its ultimate pinnacle; only details needed to be worked out. "Everything that can be invented has been invented," declared the commissioner of the US Patent Office.[18] But he ignored the most prolific inventor of all, Thomas Edison, who cautioned:[19] "We don't know a millionth of one percent about anything." In the same arrogant belief, the Prussian patent office in Germany closed down.

Fortunately for young Einstein, just out of university, some jobs were still available at the Swiss Patent Office which remained open despite the malaise that beset European society at the turn of the century. As a youth, Einstein participated in the family enterprise of new electrical devices to develop a strong interest in electromagnetic gadgets. His father and uncle made generators in an electrical shop in the backyard. Later in life, Einstein patented several such devices himself.

The patent office job suited his style. With only eight hours of work required to pay the rent, he could spend the remaining time freely pondering nature. Little did anyone expect a junior patent clerk from Zurich to shake the bedrock of classical physics.

We can visualize Einstein riding on his bicycle to work at the patent office, pondering over the stark contradiction between Galilean dynamics, as for moving bikes, and electrodynamics, as for light beams (Figure 4.20). As an imagination exercise, Einstein moved ever faster trying to catch up with a light beam. "If I travel along with a beam of light can I come closer and closer to catching the beam in my hand if I speed ever closer to the speed of light?" According to mechanics, Einstein could keep pace with the light beam, making $c = 0$. But putting $c = 0$ in Maxwell's equations would

FIGURE 4.20   We can imagine young Einstein riding on his bike, wondering whether he could catch up with a beam of light.[20]

drastically alter all electromagnetic phenomena, playing havoc with electric motors and generators on moving vessels. Electromagnetics could tolerate no speed of light other than $c = 3 \times 10^8$ m/sec. Yet simple addition and subtraction of velocities work just fine for the mechanical world of moving bikes. Why should there be different ways for treating moving bikes and electromagnetic phenomena such as light beams?

In his seminal 1905 paper published at age 26, Einstein decided to boldly tackle the asymmetry between mechanics and electrodynamics by taking the invariance of physical laws as a fundamental postulate. Invariance dictates that the speed of light should also be the same for all observers. Einstein made it his first postulate. An observer on a moving ship will measure the exact same value, $c = 3 \times 10^8$ m/s, for the velocity of a light beam shining from shore to ship, independent of whether the ship is moving toward the shore (at a constant velocity) or whether the ship is moving away from the shore. Einstein was not aware of any overwhelming evidence about the constancy of light speed. He adopted invariance from a purely aesthetic point of view,[21] "grasping the essence of nature with pure thought, as the ancients had done."

Around the same period, but unknown to Einstein, American physicists Michelson and Morley showed by exquisitely sensitive measurements that the measured speed of light *does not* depend on the speed of the moving earth. Their method was sufficiently sensitive to detect the contribution, if any, of the 100,000 km/h (2.8×10^5 m/s) earth's speed in its orbit around the sun, even though this speed is only a minuscule fraction (0.0001) of c.

Einstein recognized that the paradox between electromagnetics and dynamics arises because the way we think about reality is seriously flawed in a fundamental way. The asymmetry is a sure sign that nature is trying to tell us something completely new and different. Ready to transcend inbred beliefs, Einstein realized:[22]

> Time cannot be absolutely defined, and there is an inseparable relation between time and velocity of light. With this concept, I could resolve all difficulties.

Astonishing consequences followed. Space and time must change for moving observers, not the speed of light. For the velocity of light to always come out a constant, clocks on moving vessels must give different readings than stationary clocks. Metre sticks on a moving vessel are shorter than the metre ruler for a stationary observer.

It is difficult to accept such consequences. Through everyday experience we have become wedded to the idea that time must flow uniformly the same for everyone. From experience at the limited velocities accessible to us, we have developed the hard notion that distances between locations will always be the same, independent of the speed of the vessel we are on. But Parmenides taught how common experience can be deceptive. Now Einstein was telling us: intuitive notions of absolute space and absolute time are false.

Still, a perplexed reader might wonder: whose watch is right, the moving or the stationary observer's? Whose ruler is accurate? Both are correct! The situation is analogous to Galilean relativity, where the straight, vertical path of the ball falling from the mast of the ship as seen by a person on the moving ship is just as correct as the parabolic path seen by the stationary observer on the shore. Two observers moving relative to each other have different perspectives and therefore different perceptions of nature. Each observer's perception is valid in his own reference frame. Galilean relativity provides transformation rules (add or subtract speeds) to determine the various parabolic trajectories for ships passing at different speeds. Einstein's special relativity provides transformation rules for how time and space measured by one person relate to time and space measured by another.

It is demanding to feel special relativity effects in our bones. The new theory plays havoc with our classical intuition of time and the absolute meaning it has assumed for all. Time no longer means the same thing for everyone, everywhere. If a group of travellers synchronize their watches at a common starting point and proceed on separate high-speed journeys through space, each participant's clock will read different times if their speeds are different. When they meet up again, their clocks will no longer read the same time. Each person carries their own time. Relative to an observer at rest, all time-dependent processes on a moving vessel slow down. Biological processes slow down. The heart beats slower. Pulse rate, thought processes, cancer growth rates, all proceed at a slower pace. To a stationary observer, the moving person appears to be living life in slow motion. However to the moving observer using his own clock, the pulse beats at the same rate (e.g., 80 beats/minute) as for the stationary observer using his stationary clock. It will take the moving observer the same time interval (e.g., two hours) to watch a typical movie. Distances between stars or between galaxies are not uniquely defined. Observers of galaxies moving at high speeds will have different catalogues of star distances.

In a powerful metaphor for the breakdown of the conventional notion of time, the surrealist painter Salvador Dali surprises the spectator by destabilizing him and so preparing him for enlightenment. Dali imposes a new vision of time. Instead of showing clocks as icons of stability, he depicts limp, malleable clocks suspended from dead trees wrapped around lifeless creatures, and hung over the sides of tables like wet napkins. One of Dali's most familiar works, the *Persistence of Memory* (Figure 4.21)[23] is a vast barren dreamscape without horizon, where time has come to a stop. Ants are eating a flipped-over dead clock. Looking at Dali's work in terms of how special relativity has altered our concept of time, we can appreciate how the artist's vision embraces not only the physically possible world but also the conceivable world.

Einstein delivered his first lecture explaining special relativity in the strangest of places, at the Carpenters' Union Hall. After rambling for an hour he asked the audience for the time, because he did not own a watch. Being the only one in the world to understand the true meaning of time, it was ironic that Einstein had lost track of time.

Treating velocity invariance of physics laws as a supreme principle, Einstein also uprooted our understanding of space. Space is not absolute, but relative. It depends on the speed of the observer. Imagine you are a space-age sports reporter, in a Goodyear super-blimp. Instead of hovering stationary over a circular sports field you whiz by earth at high velocity. At the instant you arrive over the stadium, you look down to find a normally circular field squashed into an oval shape, contracted

FIGURE 4.21   *Persistence of Memory*, Salvador Dali, Fundació Gala-Salvador Dalí, DACS 2019.

TABLE 4.1

Consequence of Relativistic Addition of Velocities

v_1	v_2	V_{total}
0.01c	0.01c	$0.019998c \approx 0.02c$
0.1c	0.1c	$0.198c \approx 0.2c$
0.5c	0.5c	0.8c
0.9c	0.9c	0.9945c

---

along your direction of motion. But there is no contraction perpendicular to the direction of motion. At near light velocity, the stadium contracts into a line. One conclusion you may be tempted to make: goals should be trivial to score! But you realize your error when you see each player in super slow-motion. Time has slowed down as well on the moving field.

From a moving spaceship the observed distance between stars depends on the speed of the spaceship. The faster the ship moves, the shorter the spacing. With space contraction and time dilation, Einstein gracefully removed the vexing inconsistency between how velocities add for Galilean relativity and for special relativity. By simply adding or subtracting velocities, measurement and intuition are in agreement for the mechanics of moving bikes. But measurements with bike velocities are not wrong, they are only *approximations*. When the speed of a bike is small compared to the speed of light, a simple addition of velocities holds imperceptibly close to true. Einstein determined the general (relativistic) law for the addition of velocities v_1 and v_2 that holds *over all velocities*. Instead of simple addition, $V_{\text{total}} = V_1 + V_2$, his treatment yields a new expression for velocity addition, equally valid for bikes and light beams.

$$V_{\text{total}} = \frac{V_1 + V_2}{\left(1 + \dfrac{V_1 V_2}{c^2}\right)}$$

Table 4.1 works out the consequences. At a speed of 0.01c in the first row, the exact treatment result hardly differs from simple addition. Slow-moving observers will never experience the effects of special relativity. But when velocities rise to 0.5c (row 3 in Table 4.1) the sum departs from simple addition by nearly 20 per cent. Einstein reconciled Galilean with special relativity.

If moving rulers shrink and moving clocks run slow relative to stationary rulers and clocks, does anything remain the same for all vessels, moving at arbitrary speeds? A certain mathematical combination (Δs) between distances and time intervals (between events) does remain the same for all observers on moving vessels. This is called the "distance" in spacetime between the events as specified by their location separation ($\Delta x, \Delta y, \Delta z$) and the time interval (Δt) between the events.

$$\Delta s^2 = c\Delta t^2 - \left(\Delta x^2 + \Delta y^2 + \Delta z^2\right).$$

SPACE TRAVEL

Now that space travel is possible, journeys near light speed are imaginable, though vastly impractical, and still in the distant future. And we will need Einstein's physics to travel to the stars. Future space explorers travelling fast enough could complete a distance-contracted trip to the nearest star within their time-dilated lifetime. According to clocks on earth, a spaceship travelling to the nearest star Alpha Centauri at 99 per cent of the speed of light will take 4.6 years, but according to the

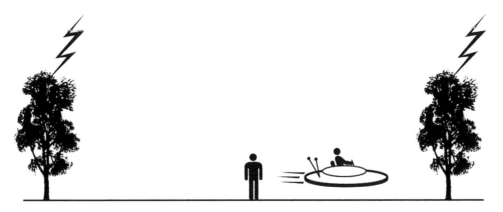

FIGURE 4.22   When events take place at different locations, the time-order depends on the observer's speed. Jack, who is exactly halfway between two trees, sees lightning strike both at the same time, since the light from both reaches Jack simultaneously. Jill, who is moving at high speed toward the right tree, sees lightning strike there before the left tree.

moving ship's clock only 0.64 years. Extending rocket speeds to the imaginary extreme reveals a startling consequence. At light speed, it takes no time at all to cross the zero (contracted) distance between stars. From the point of view of the fictitious traveller on a beam of light, all space reduces to single point, and there is no flow of time! The stars are in direct and immediate contact with the eye. But to reach light speed is physically impossible, a consequence of special relativity that will come later. Even for a 10 kg satellite to achieve a speed of $0.9c$ will take more than ten times the total energy consumed by US gasoline usage per day. It will take a tremendous advance in technology.

Special relativity also violates another common place notion: simultaneity. Events at two separate locations that are simultaneous for a stationary observer are not seen simultaneously by a moving observer. Suppose lightning strikes two trees located 10 km apart (Figure 4.22). For Jack, positioned exactly midway between the trees, the strikes appear simultaneous because the light signals from each tree reach his midway location at exactly the same time. But Jill, on a rocket-ship travelling at near light speed *toward* the tree on the right, will see lightning strike the right tree *before* the left tree. Since Jill moves toward the right tree, light has to travel a shorter distance to reach her. Moving *away* from the left tree, light must travel a longer distance to reach Jill. Light from the right tree reaches Jill earlier than light from the left tree. In disagreement with Jack, Jill thinks that lightning struck the right tree first. The shocking conclusion is that people who move with respect to each other will disagree about the time-order of events when these events take place at different places. There is no disagreement, however, for the events which occur in the same place. Relativity preserves the relationship between cause and effect at two different locations because information can travel no faster than the speed of light.

BLOCK UNIVERSE

Here are some other mind-bending consequence of special relativity. An event can be in the future from one observer's perspective from one vessel, but that same event can be in the past for another observer in a different vessel moving. The two events just exist in space and time (spacetime) at different co-ordinates for different observers in motion; they have no communication between each other, and do not violate cause and effect.

In familiar space and time, I can choose where I want to be in space, in Boston, or in New York City, and I can move through and explore 3D space freely. All of space is always "out-there." But I cannot choose *when* I want to be. I can only yearn for the past or look forward to the future. There exists only the present. All the places of the past are still there, but the events of the past are gone. Or are they?

We live inside a four-dimensional *block universe*, where the "block" is all of space *and* all of time, i.e., spacetime. If a fictional "Luke Skywalker" could step out of the spacetime block into a fifth dimension, he would be able to access all of space *and* all of time. For Luke, not only does all of space exist for him, but all of time exists as well, from outside the block of spacetime. Luke sees all the past and all the future simultaneously. He can see dinosaurs roaming the earth 60 million years ago. Your birth is out there in spacetime, and your death is out there too! Luke can see what will happen in your future, which may make you wonder about whether your future is determined or whether your free will will lead to an uncertain future by chance, and therefore does not yet exist for Luke to visit.

We don't know if the future advances will reveal the existence of a fifth dimension or if deeper knowledge of the nature of spacetime, or some unknown connections between spacetime and quantum physics (Chapter 12) will invalidate the block universe idea to reveal that there is indeed something special about the time dimension that makes it flow only in one direction.

TIME TRAVEL (RESTRICTED)

Suppose in an exceptional technology era, astronaut Jill travels to Alpha Centauri at $0.99c$, leaving behind her twin brother Jack. When Jill *returns to earth*, an interesting conversation might ensue, if the twins are unaware of special relativity effects:

> "What happened to you dear brother? Your hair is grey."
> "Well, what do you expect after 35 years?" replies Jack.
> "35 years!" exclaims the stupefied Jill, "Why, it's been only
> 5 years, Jack? Have I travelled into your future?"

At which point, Einstein's ghost steps in to clear up the mystery. It is no surprise to him that space traveller Jill returns 30 years younger than Jack because Jill's moving clock slowed down.

Now an apparent paradox arises. According to Jill, she saw brother Jack along with the entire earth recede from her spaceship at $0.99c$. According to Einstein's theory, Jill should find Jack 30 years younger. So who is younger when they meet on earth? The symmetry between the twins gives contradictory results. They quiz the ghost.

With a kindly smile, symmetry expert Einstein reveals that the situation is not really symmetric between the twins. They are not always moving at a *constant velocity* relative to each other. When Jill takes off from earth, she has to accelerate. As rockets blast off, she feels the formidable tug of acceleration to light speed, an unmistakable signal that she is the one who is in motion relative to Jack. But there is more. When she turns around at the star, she has to decelerate to stop, then re-accelerate toward earth. Acceleration (and de-acceleration) breaks the symmetry between twins and makes it possible to distinguish who is moving. Einstein gently explains that his general theory of relativity (Chapter 11), which came a decade after his special theory, clears up the mystery. Without a doubt, Jill will take the time leap into Jack's future ending up 30 years younger than Jack when the twins re-unite at earth. The special theory does not apply to Jill's observations of Jack, because, without doubt, accelerating Jill is the moving twin.

Although time dilation slows down clocks, one must not equate the effect with the long-sought elixir of perpetual youth. The slowing of time equally affects all events in the moving vessel. If the heartbeat slows down along with the ticks of the clock, a human will live the same number of ticks. You cannot see more movies in your rocket-ship lifetime, because the rate at which a movie plays also slows down.

What about travel to the past, suggested by the limerick?:[24]

> There once was a lady named Bright
> Who could travel much faster than light.
> She went out one day,

In a relative way
And returned on the previous night.

No. It is impossible to go back in time. It would be necessary to move at greater than light speed, which is denied by energy considerations to be discussed. Travel to the past also leads to serious logical contradictions stemming from the reversal of cause and effect. If it were possible to go back in time, you could change the past. Such an action could affect the present. You could eliminate your existence, and therefore your ability to travel back in time.

If a fifth dimension other than the familiar spacetime is accessible, can Luke travel to the past? Yes, but he cannot change the past. The past is there. If Luke travels to the past, it is an event that has already taken place, and is part of the past. He cannot change the future because the future is already out there according to the properties of the Block Universe.

PRACTICAL APPLICATIONS OF RELATIVITY

Even though we stand at the threshold of space travel, relativistic effects, such as time dilation and length contraction, are still negligble at the speeds accessible to present technology. Yet there are important practical applications of special relativity (and general relativity). Satellite-based, global positioning systems (GPS) are devices of modern living, commonly found in cars for everyday navigation. To properly resolve distances over the surface of the earth of the order of 10 metres, a GPS must keep time to an accuracy of 10 metres divided by c, which is equal to 33 nanoseconds (billionths of a second). A satellite revolves around the earth at a radius of 40,000 km, an enormous circle compared to earth's radius of 6,400 km. There are about 30 GPS satellites operating in space, with four needed at any given time to determine accurate location. If the satellite always stays above the same spot on earth (as for a geosynchronous orbit) it orbits at a speed of 4,000 m/s, much faster than any car equipped with GPS, rotating with the earth at 465 m/sec (930 mph). Relativistic effects between earth and satellite clocks give different readings with the high-precision, caesium atomic clocks on the satellite, by nearly 7,000 billionths of a second per day. Without relativity corrections, the GPS would fail in one hour; boats would be wrecked and cars would be lost. There is another – larger – correction from general relativity which will be discussed in Chapter 11. The combined correction is 38,000 billionths of a second. Today GPS systems are in routine use and installed in our smart phones. We walk around with relativity literally in the palm of our hand.

UNIFYING MASS AND ENERGY

One of the most famous consequences of special relativity is the equivalence *mass* and *energy*. Mass can change into energy and vice-versa, like dollars can be changed into cents. The conversion takes place according to the celebrated equation that has become an icon for the genius of Einstein:

$$E = mc^2$$

It applies to all forms of energy as a deep and general result. Through mass–energy equivalence, the *law of energy conservation* extends to *conservation of mass and energy*. Relativity embraces not only space and time, but also matter. Matter is frozen energy. The conversion factor of mass to energy results in spectacular consequences. All the energy Americans use in one day, if converted to mass, would amount to a mere 2 kilograms. Mass to energy conversion powers the sun and stars (Chapter 8). In the depths of fiery nuclear furnaces, when hydrogen atoms fuse into helium, 1 per cent of their mass converts to pure energy. Just 10 grams of uranium in the atomic bomb unleashed the energy of massive destruction that fell on Hiroshima. That single blast released a billion times more energy than liberated by 1 ton of water tumbling down Niagara Falls from a height of 100 metres.

Einstein established the mass–energy equivalence principle, but his formula gave no clue as to *how* to convert mass into energy. Contrary to the popular misconception that sullies his name, he played no part in the project which developed the nuclear physics and engineering of the deadly device that can do the job.

Mass–energy equivalence shows that all forms of energy have mass. When an object speeds up it acquires energy of motion, *kinetic energy* discussed earlier. Mass–energy equivalence implies that the faster moving object is also more massive. It has more inertia. As mass increases with speed, so does the innate resistance to acceleration, built in to the fundamental meaning of mass. When mass continues to grow with increasing velocity, it becomes increasingly difficult to raise the velocity further because of the excessive amount of energy needed. At 90 per cent of light velocity, the mass of a moving object is larger than its rest mass by a factor of 2.29. A mass of 10 kg will become 22.9 kg. This may not seem like much, but you have to think about the energy needed to supply a mass of 12.9 kg. From $E = mc^2$, that energy is ten times the energy used by US consumption of gasoline. Or you can consider that the atomic bomb dropped on Hiroshima converted just 1 kg of (uranium) mass into pure destructive energy. Moving at 99.99 per cent c, the mass increases by a factor of 70!

The general derivation of $E = mc^2$ is somewhat complex, involving the momentum and energy of light beams. We provide a simpler derivation using the same concepts. Recall from the historical evolution discussed in earlier sections of this chapter that

(1) A force applied over a certain time provides impulse which results in momentum increase.
(2) A force applied over a certain distance provides work which increases energy.

Now suppose a body is moving at a speed close to c. A uniform force (F) applied to the body will pump both momentum and energy into the body. The force will increase the speed but not by much, since the speed is already close to c. Imagine F is applied for one unit of time. Since the speed is nearly constant at c, the distance over which the force will last for one unit of time will be c metres (*distance = speed × time*). So the *energy gained E* will be *force × distance*, as discussed earlier.

$$E = F \times c, \quad E = Fc \tag{4.1}$$

The increase in energy (E) will of course be the increase (m) in mass, since there is no gain in speed.

F will also add momentum to the body. Since the body is always moving a constant speed c all the momentum gain comes from the mass gain; there is no gain in speed. The momentum gain is equal to the mass gained (m) times the speed, which is always c. But the momentum gain (mc) can also be computed from *force × time*

$$\text{Momentum gained} = m \times c = F \times 1, \text{ or } F = mc \tag{4.2}$$

Going back to Equation 4.1 and using $F = mc$ for force from Equation 4.2 gives

$$E = Fc = mc \times c$$

$$E = mc^2$$

which is the mass–energy equivalence we set out to establish.

An object which has any mass whatsoever when at rest can never reach light speed because, as the mass becomes infinitely large, it takes infinitely more energy to get it there. Relativity enforces a speed limit through increasing inertia. How then can particles of light reach light speed? Because all the mass of the light beam is pure kinetic energy. If we think of light as particles – photons – they have zero rest mass and can move at exactly light speed. For the same reasons, light must also obey a minimum speed limit; a photon moving in vacuum cannot slow down below c.

SYMMETRIES RESOLVE PARADOXES

During the miracle year of 1905, Einstein rose from obscurity to the pinnacle of his profession with breakthrough papers on atomic motion (Brownian motion – Chapter 1), the quantum nature of light (Chapter 12), and special relativity – landmark works written in the space of four prolific months. Without academic credentials and no direct contact with physicists, he put forth ideas that demanded the most unusual thinking.

It was unnerving to deny the everyday perception of simultaneity. Even more discomforting was the idea that space and time are not absolute, but relative. It is no surprise that such radical concepts were difficult to accept. When Einstein received the Nobel Prize in 1921, it was not for relativity, but for his other two papers of 1905. We will return to Einstein's general theory of relativity in Chapter 12. This theory addresses the deeper meaning of gravity.

Galileo resolved paradoxes in vertical and horizontal motion left behind by Aristotle's common-place notions on the nature of motion. Symmetry principles guided Galileo to the radical concepts of eternal horizontal inertial motion with constant velocity, and uniform acceleration under gravity, independent of mass. Through idealization, he could see beyond the seen to transcend inbred beliefs. Galileo truly grasped the importance of time in understanding velocity and acceleration in motion.

Einstein resolved a vexing asymmetry that cropped up between electromagnetism and dynamics to overturn our commonplace notions about space and time. The constancy of light speed tangled with the nature of time and space. Both clocks and rulers change so that the speed of light always comes out to be a universal constant. With special relativity, he extended his grasp of physical reality well beyond the restricted realm attainable by common-sense perceptions.

Part 2

The Heavens

5 Celestial Motion: A Heavenly Romance with the Solar System

The spacious firmament on high,
With all the blue ethereal sky,
And spangled Heavens, a shining frame,
Their great Original proclaim.

Th' unweary'd Sun from day to day
Does his Creator's power display;
And publishes to every land,
The work of an immortal hand.

Soon as the evening shades prevail,
The Moon takes up the wondrous tale;
And nightly, to the listening Earth,
Repeats the story of her birth:

Whilst all the stars that round her burn,
And all the planets, in their turn,
Confirm the tidings as they roll,
And spread the truth from pole to pole.
What though, in solemn silence, all
Move around the dark terrestrial ball;
What though no real voice, nor sound
Amidst their radiant orbs be found:

In reason's ear they all rejoice,
And utter forth a glorious voice;
Forever singing as they shine:
'The hand that made us is divine.'

"An Ode," Joseph Addison[1]

REGULAR MOTIONS, CAPRICIOUS MOTIONS

Starting with the ancient civilizations' recognition of the regular movements of the sun and moon, chains of observations bind our conceptions of the heavens. It was comforting to connect cyclic changes in the position of the sun on the horizon with recurring seasons. The regular waxing and waning of the moon provided a convenient time interval for calendar design. With the help of familiar star patterns, called constellations, ancient travellers oriented themselves at night and reckoned the passage of hours without the benefit of consulting the sun. By the regular appearance and disappearance of well-known constellations, they tracked the seasons. The faithful path of the sun through the constellations of the zodiac, a circle of animal-shaped star patterns, provided a reliable, annual calendar visible in the sky.

Thus ancient civilizations depended more on the heavens than we ever do. Rhythms of heavenly motion regulated lives, and influenced culture and religion. By detecting cycles, ancients recognized symmetry in heavenly motion. Meticulous astronomic observations spanned over 4,000 years of human civilization. The quirks and customs of our calendar bear the marks of their intense scrutiny of the heavens. Our day is divided into 24 hours, our week into seven days. Because of ancient calendar reckoning, 13 became an unlucky number, and 7 signalled magic.

But science languished. There was no rational understanding of the cyclic changes in the position of the sun, the recurring seasons, or the length and direction of shadows. The waxing and waning of the moon only gave rise to colourful myths. There was no explanation for the periodic appearance and disappearance of constellations with the seasons. At this stage, the vast catalogue of observations seemed totally unconnected. The transition from gathering facts to gaining genuine understanding was slow, painstakingly so.

While the symmetry of cyclic motions provided soothing reassurance to civilizations across the face of the globe, the sky often presented many perplexing features: lunar and solar eclipses, the Milky Way, and unpredictable comets, with dagger-shaped tails threatening disaster (literally meaning "bad star"). Upsetting the symmetry of the heavens, strange happenings gave rise to bizarre superstitions. At first, anomalous events led to explanation through colourful mythological accounts. Eventually heavenly irregularities exposed new aspects of the cosmos far more fascinating than the regular motion of heavenly bodies.

Astute observers of the sky detected five bright celestial lights, later called planets, which distinguished themselves from the multitude of stars. They wandered slightly among the zodiac constellations, some from night to night, others from month to month. Some planets moved fast, others slow, sometimes even going backward. Ancient civilizations invented planetary gods with special attributes to account for such puzzling irregularities in heavenly motion. Planetary traits stirred the imaginations of poets and artists. Their creations intertwined with fantasy and enchanting myths which sought to connect humanity with the universe, reflecting an intensely human reaction to the drama taking place in the heavens.

Planetary quirks inspired temple-priests. They came to believe in deep connections between the movement of heavenly bodies and the unfolding of human experience. In its regular motions, the sun assured day and night as well the seasons. Clouds and rain from the heavens decided the fate of crops. It seemed plausible that planets played crucial roles in guiding human destiny. Because heavenly bodies grew so important to culture, sky-watchers grew intensely aware of minuscule details.

Our education teaches us that the earth rotates around its axis once every 24 hours and that it revolves around the sun once every year. But how did we come to these remarkable realizations? We cannot "sense" the earth spinning. By all appearances, it would seem that the sun goes around the earth. Our everyday language reflects this simple observation, when we speak so casually of "sunrise," "sunset," and the "coming" of seasons. What eventually forced a drastic change in viewpoint was the struggle to understand *minute irregularities* in the motion of heavenly bodies. Asymmetries in nature's behaviour provided essential clues to the solar system design. Proper interpretation of capricious planetary motions finally revealed the moving earth. But it took 3,000 years of careful observation and profound thought.

CELEBRATING THE BIRTH OF THE SUN

Practical necessities motivated ancient Egyptian astronomy as early as 2500 BC. The need to mark the seasons, the desire to pinpoint ideal times to plough the land and sow the seed, all drove the evolution of an accurate sun-based calendar. Priests found a calendar essential to mark dates of festivals and reigns of rulers. A calendar held people together. They could make common plans, such as when to deliver goods, or when to assemble the market. It helped anticipate future events.

Egyptians drew up their calendar by recording the sun's horizon position as well its altitude over the course of a year. Accustomed to the daily rising of the sun in the east and setting in the west, they attached special significance to east–west directions. Pharaoh Tutankhamen's (ca. 1350 BC) tomb had four chambers, each with a ritual purpose. Following the sun's motion, the chamber of birth faced east. West was the chamber of departure. Pyramids stand on the west side of the Nile to shelter departed kings for eternity.

Through centuries of painstaking observations, ancient cultures recognized that the arc which the sun traces through the sky is not the same from day to day (Figure 5.1). Sunrise and sunset points on the horizon change daily, moving southward from summer to winter by about two sun-diameters every day. The sun also changes its zenith position from high in summer to low in winter. Its cyclic, north–south swing spans a period of 365 days, the first estimate for the length of a year.

Egyptians noted important correlations between the north–south position of the sun and the length of each day. When the sun's arc is at its northernmost location, the summer day is the longest day of the year (June 21, on our calendar). We call this day *summer solstice*, meaning that the "sun stands still" in the swing of its annual path. The long twilight of the summer solstice held a strong appeal to many cultures. In *A Midsummer Night's Dream*, Shakespeare dwells on its enchantment.

As the arc begins its southerly drift, the midday position of the sun descends in the sky. When the arc reaches its southernmost location, the sun's altitude is ominously low; that winter day is the shortest. We call it *winter solstice* (December 21). On the next day the path moves northward again, and the annual cycle repeats.

There are two other special times of the year, called *equinoxes*, from *nox*, which means "night" in Latin. Occurring when night and day are exactly 12 hours in length, equinoxes mark a special symmetry of the sun's location. One of these days, called *vernal equinox*, occurs in spring on March 21. The other, called *autumnal equinox*, occurs in fall, on September 23. *Vernal* comes from the Latin word for green, as in verdant spring. There is a very special feature about the path of the sun on days of the equinox. The sun rises in the *exact east* and sets in the *exact west* during both

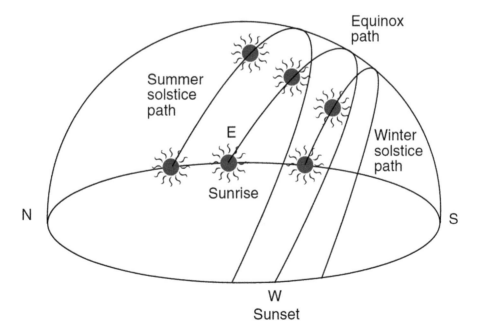

FIGURE 5.1   Daily and seasonal motion of the sun as viewed from a northern latitude. At the extremes of its north–south swing are the *summer solstice* (June 21) and the *winter solstice* (December 21). In the middle of the cycle are the *vernal and autumnal equinoxes*.

equinoxes. The number of days separating two spring equinoxes defines the length of the year as 365 days, which Egyptians discovered as early as 4200 BC.

Solstices and equinoxes held special significance for various civilizations. Techniques emerged to determine the special days, and customs evolved to commemorate them. To indicate the precise time of the summer solstice, Egyptians designed a special feature in the solar temple, Amun-Ra at Karnak, a village on the Nile. On that auspicious midsummer day, the rising sun cast a long shaft of light along the main corridor so that a narrow beam of sunlight fell upon the wall of a darkened sanctuary. Architect priests designed temple complexes at Karnak to integrate Egypt with the cosmos. When the sun rose on the morning of the summer solstice to renew the universe, it illuminated a straight path along a massive colonnade of pillars. Thus the sun served both as calendar-keeper and powerful god. When they built the Great Pyramid at Giza, Egyptians oriented it so that, on the day of the equinox, the sun's rays simultaneously touched both the east and west faces.

Mayas, Aztecs, Celts, and Chinese cultures were similarly interested in places on the horizon where the sun rises and sets. Some marked solstice positions with mounds; others used giant stones (megaliths). Still others aligned the horizon location of the sun on the special days with conspicuous mountain peaks. Around 2500 BC the Celts constructed the giant pillars of Stonehenge on Salisbury Plain to mark the solstice sunrise. In its awesome arrangement of mammoth stones, a central avenue points to the northernmost rising of the sun, accurately delineated by a distinguishing stone, now called the "Heel Stone." On the day of the summer solstice, there is a spectacular sunrise precisely over the Heel Stone (Figure 5.2). Descendants added massive stones which define a circle. Each stone weighs close to 60 tons, while the Heel Stone weighs 35 tons. Engineering skills of these prehistoric civilizations still baffle technology historians. Other notable sight lines point toward extreme horizon positions of the moon's celestial path. Aubrey Burl, interpreter of many mysterious features of this prehistoric site, poetically depicts its commanding presence: "The ravaged colossus rests like a cage of sand-scoured ribs on the shores of eternity."[2]

FIGURE 5.2   On the morning of the summer solstice, the sun can be seen rising just above the Heel Stone of the prehistoric Stonehenge structure. There are 900 other structures like Stonehenge throughout the British Isles.[3]

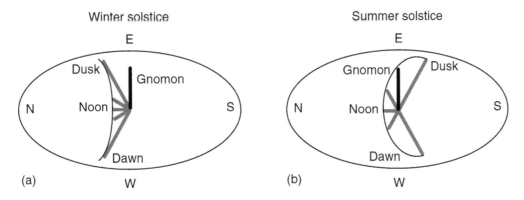

FIGURE 5.3   (a) Progression of shadows through the day of winter solstice, when the sun is low in the sky and (b) the same for summer solstice. In both cases, the noonday shadow points to the north. The shadows described here are for the location in the Northern Hemisphere at mid-latitude.

On days following the winter solstice, agricultural communities of ancient cultures held grand festivities to celebrate the end of the sun's fearful descent in the sky. As it turned around to rise again, they were joyful that the cycle of the sun's motion would continue for another year, as would the cycle of life. Succeeding civilizations continued winter solstice festivals. Romans celebrated the rebirth of the sun by holding magnificent feasts from December 17 to December 24. As the popular custom continued through the rise of Christianity, the Church grew concerned that pagan revelry, bordering on sun worship, distracted people from pure Christian practices. Rather than abolish the popular custom and face serious discontent, they decided to absorb the auspicious day by giving it a very Christian significance. Around AD 300, they declared December 25 to be the day when Jesus was born, since the historically precise birth-date of Jesus was unknown. Thus ancient astronomy continues to hold sway over Western culture.

Astronomical events link to other facets of modern religions. For example, there were 12 disciples of Christ because there are 12 months in a year. John the Baptist had 29 apostles to correspond to the number of days in the lunar cycle. Christians celebrate Easter on the first Sunday after full moon, after the spring equinox. Jews fix Passover by the full-moon phase.

SHADOWS

For countless centuries, the sun's shadow served as a universal measure of time and direction for many cultures. The earliest method to tell the time of day was based on measuring the lengths of shadows cast by a vertical stick (Figure 5.3) called the *gnomon*, from the Greek word "to know." Astronomy's foundation literally stems from the gnomon, or "knowledge" stick. Shadow tracking eventually developed into the ubiquitous sundial. However, early morning and late evening shadows are too long to be useful for telling time. Shadow clocks become practical only when the sun is higher in the sky. At midday (local noon), when the shadow is shortest, it always points exactly north, for locations well in the Northern Hemisphere. The shortest shadow thereby serves as a crucial reference point. The size and direction of the midday shadow helped keep track of the seasons (Figure 5.3).

MOONSTRUCK

With its charming shapes, the silvery moon presents a most impressive sight in the night sky. With regular and comforting nightly progression, the shape changes from a bright disc to a thin luminous crescent. At the end of the cycle it is totally dark. On the next night, a thin sliver of a new crescent appears at twilight with "horns" pointed away from the setting sun.

Since the shape of the moon changes regularly and perceptibly from one night to the next, it provides a convenient basis for a calendar. A lunar calendar was natural for nomadic tribes who did not cultivate the land, and had no need to keep track of seasons. From the moon's slowly changing shape, hunters could anticipate bright nights to venture out and dark nights to stay safe in their caves. Archaeological digs have uncovered prehistoric bone fragments cut with moon-shaped marks that kept track of lunar cycles.

As another illuminating example of ancient cultures' attention to lunar details, many tales attempt to interpret the light-grey, irregular patches on the face of the moon, easily visible on a clear, full-moon night (Figure 5.4). A Jewish account claims that Moses banished "the man in the moon" for collecting sticks on the Sabbath. In *The Tempest*, Shakespeare, who never missed an opportunity for astronomical reference, refers to the charming folklore:[4]

Caliban: Hast thou not dropped from heaven?
Stephano: Out o' the moon, I do assure thee, I was the man in the moon, when time was.
Caliban: I have seen thee in her, and I do adore thee, my mistress showed me thee, and thy dog and bush.

The moon's changing appearance has fascinated all cultures. Some were comforted by the faithful regularity of its gentle waxing and waning. Others were fearful of the awesome power of full-moon light. Still others were troubled by its fickle appearance. In Shakespeare's most popular romance, Romeo's pledge distresses Juliet:[5]

Romeo: Lady, by yonder blessed moon I swear
That tips with silver all these fruit-tree tops –
Juliet: O, swear not by the moon, the inconstant moon,
That monthly changes in her circled orb,
Lest that thy love prove likewise variable.

FIGURE 5.4   A high-resolution modern photograph of the full moon showing dark patches.[6]

It is particularly intriguing how so many myths had already anticipated the crucial link between moon phases and the sun, as the later Greek rational explanation clearly established. The Sani tribe of Southern Africa (a living tribe, often referred to as Bushmen) believe the sun in its wrath hacks the moon down to smaller and smaller size. When the moon begs for mercy, the sun relents. But, as soon as the moon grows threatening again, the fiery orb cuts it down again. In a Hindu myth, dead souls ascend to the moon which swells with the light of their breath in the first half of the month, till it is full to the brim. As it wanes, it transfers the light of departed souls to the sun. In another legend, the boat-shaped moon sails across the sky carrying a load of dead souls, which it passes gradually to the sun. In a delightful twist, this fable even relies on the relative positions of the sun and the moon. When the moon-boat is farthest from the sun, it grows full with soul-light. No transfer of souls is possible. As nights pass, the moon approaches the sun and shrinks as souls transfer out. Just as it passes directly near the sun, the moon-boat becomes empty and nearly dark. Such creative explanations could only follow perceptive observations of the relative locations of the sun during the various phases of the moon.

Babylonians identified important correlations between phases and relative sun–moon positions (Figure 5.5). Picturing the horizon as a circle, they recognized how the full moon rose diametrically opposite to the setting sun, a situation called *opposition*. On the other hand, the crescent moon was always near to the rising or setting sun, a juxtaposition called *conjunction. Later in the development of astronomy, opposition and conjunction of planets played an important role overturning the earth-centric system (Chapter 7).* Babylonians also incorporated these connections into myths where the Creator commands the moon to go through its cycle of fascinating shapes.[7]

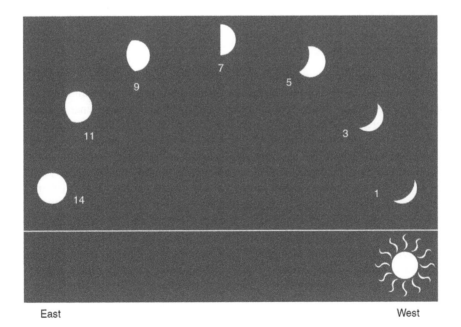

East   West

FIGURE 5.5   Phases and positions of the moon after sunset during the displayed number of days after the new moon. Note how early in the lunar month (day number 1) the crescent moon is close to the sun and sets soon after it. Seven days later, the half-moon is at its high altitude at the time of sunset. After that, the shape turns *gibbous*, borrowed from the Latin word for "hunchback." Another week later (day number 14), the full moon is in opposition to the setting sun at dusk.

At the beginning of the month,
Thou shalt shine with horns to make known six days;
On the seventh day with half a tiara
At the full moon thou shalt stand in opposition to the sun, in
the middle of each month.
When the sun has overtaken thee on the foundation of heaven,
Decrease the tiara of full light and form it backward.
At period of invisibility draw near to the way of the sun.

Although a myth, the account reads partly like a scientific record of observations. Making such correlations, Babylonians were well on their way to understanding moon phases. Centuries later, the geometrically minded Greeks arrived at a complete explanation for the cyclic changes in the shape of the moon from the relation of its position with respect to the sun.

Our pearly white companion inspired various beliefs about the influence of the moon's cycles on human affairs. With links between moonlight and departed souls, the crescent became a symbol of immortality in some cultures, frequently adorning funeral headstones. As the moon became the pre-ferred method to fix days of religious ceremonies, it assumed a supernatural significance. Sacred practices in the Muslim religion are determined by phases of the moon. During the holy month of Ramadan, when Mohammed received holy revelations from God, the series of ritual fasting is not to start (or stop) until heralded by the appearance of the first crescent sliver, even if it is delayed by clouds. The new moon plays an important role in marking the festivals of Easter and Passover in Christian and Jewish religions.

Moon-based fantasies persist in modern culture. The word *croissant* for French pastry bread arises from the characteristic shape. Our word "lunatic" warns how the brilliant rays of the full moon can strike with dementia, whereas the "moonstruck" are smitten with love. Illegally brewed liquor earned the name "moonshine" as if it was distilled surreptitiously by the light of the silvery moon. The accidental connection between lunar and human fertility cycles gave rise to words such as "menstrual" and "menopause." Fertility connections endowed the moon with life-giving aspects.

CALENDARS – UNLUCKY 13

Compared to the long year, the month provided a conveniently short interval to mark time passage. But moon phases bore no correlations to changing seasons, important for agricultural communities to track carefully. Accordingly, there were many serious efforts to reconcile lunar and solar calendars, some leading to creative arithmetic schemes as well as persistent superstitions that arose from the incongruity of lunar and solar cycles. Expert at fractions, Babylonians were among the first to tackle the challenge (ca. 700 BC). We have inherited vestiges of both their ordering plan as well as their superstitions. For simple calendar-reckoning, a month is roughly 30 days long. The sun takes *approximately* 12 lunar cycles to complete its circuit of the heavens. This was satisfactory agreement with a very early (but inaccurate) reckoning of the length of the year as $12 \times 30 = 360$ days. The numerology was the origin of the special significance which the Babylonians attached to the number 360, leading to their signature division of the angle around the centre of a circle into 360 equal parts we call degrees. It came as quite a shock to early Babylonians when they recognized, through the precise number of days between two vernal equinoxes, that the true length of the year is 365 days. Five days of rituals marked the extra days of the year.

Tying the lunar calendar to the 365-day solar calendar proved quite difficult. As one catch-up solution, some years were just decreed to have a leap month. During the intercalary, 13th month, all regular activities were suspended. Eventually the 13th month became unlucky. To this day, the number 13 continues to carry the negative connotation stemming from the trials and tribulations of adjusting the two calendar systems.

ANIMALS IN THE SKY

In the clear and unpolluted night sky over ancient Egypt, with none of today's background glow of bright cities to dim the brilliant points of light, ancient Egyptians marvelled at the changing pageant of the heavens. Why did stars only come out at night? Many cultures believed that stars took flight from the sky on the appearance of the sun. Some imagined that stars were campfires, burning out at dawn and rekindled every night.

Civilizations kept track of stars by arranging the multitudes into patterns that looked familiar, called *constellations*, the first attempt at a classification of the vast multitude of heavenly lights. Ancient constellation charts dating back to 2000 BC appear on the lids of pharaohs' coffins. Constellations presented distinct and unchanging patterns whenever and wherever seen, generation after generation. That is why they were called *fixed stars*, and the night skies the firmament. Along with the faithful course of the sun and the regular cycle of moon phases, the fixed stars guaranteed law and order in an otherwise unruly universe. They helped keep track of time and seasons, oriented travellers, and told them their location on earth.

Ancients identified clusters of stars with shapes of animals, such as a dog, a bear, a bull, a lion, a bird, a fish, or a snake. Sumerian, Egyptian, Greek, and Native American cultures identified the northern constellation Draco as a dragon, serpent, or crocodile. Frequently, people identified patterns with mythological gods. Around the 16th century BC, Egyptians identified a group of stars with their principal god, Amun-Ra. We call this constellation Aries, the ram. Sometimes ancient civilizations identified groups as familiar objects of the farm or household, such as a plough or a ladle, sometimes a musical instrument, such as a lyre. Making up stories to remember assigned names, they taught children the constellations, much as we teach capitals and countries.

Unchanging patterns of constellations inspired poets and artists, ancient and modern. Homer told a charming story about Aquila, the eagle. Jove sent his sacred eagle to earth to fetch the most handsome young man in Troy to be his wine-bearer. Kidnapping the young Ganymede as he tended his flock of sheep on the slopes of Mount Ida, Aquila brought the attractive boy to Mt Olympus, home of the gods. Delighted by the boy's beauty, Jove rewarded the eagle by placing it forever among the stars.

POINTER STAR

Much as the positions of the sun and shadows defined east and west, the location and movement of stars provided a way to define orientations at night. Through a single night, the path followed by stars appears quite different depending on whether one faces east, west, north, or south. Stars travelling across the sky appear to rise from the east and set in the west (Figure 5.6). Unlike the sun, however, stars always maintain the same rising and setting points on the horizon.

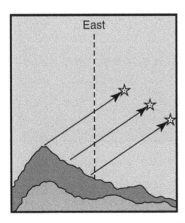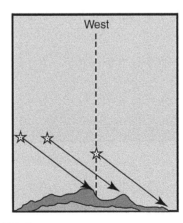

Latitude about 50 degrees

FIGURE 5.6 (Left) turning east, stars climb the heavens as night progresses. (Right) turning west, stars depart from the heavens.

Turning toward the north, stars do not appear to rise or set with respect to the horizon. (This is true for those who live in the upper, northern latitudes.) Rather, they move in circles around one motionless, bright star, called Polaris. It is part of the constellation Little Bear, also known as the Little Dipper or Wagon. The North Star appears not to take part in the general motion, but remains stationary through the night (Figure 5.7). The concept of north as a cardinal direction grew out of these observations.

To the Egyptian pharaohs, the North Star held a special significance, the entrance to heaven. The 5,000-year-old, Great Pyramid of Cheops has a steep ramp leading from the burial vault in the crypt below to the single entrance on the ground above. The builders aligned the steep shaft so that it points straight to the Pole Star. It was the royal ramp for the buried Pharaoh, the son of heaven, to ascend to eternal rest. Six of the nine remaining pyramids at Giza share the same feature.

The steady North Star stirred the imagination of poets over all time. Always full of astronomical references, Shakespeare compares Julius Caesar's steadfastness to Polaris:[8]

> I am constant as the northern star,
> Of whose true fixed and resting quality
> There is no fellow in the firmament.
> The skies are painted with unnumb'rd sparks,
> They are all fire and everyone doth shine;
> But there's but one in all doth hold his place.

William Bryant responds to the regular rising of the stars from the east:[9]

> The sad and solemn night
> Hath yet her multitude of cheerful fires;
> The glorious host of light
> Walk the dark hemisphere till she retires;
> All through her silent watches, gliding slow,
> Her constellations come, and climb the heavens, and go.

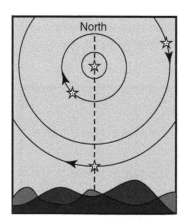 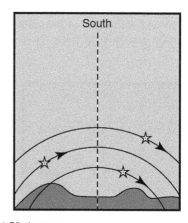

Latitude about 50 degrees

FIGURE 5.7   (Left) facing north, stars move in circles around a particular point in the sky that *defines* the North direction. The North Star, Polaris, is located near that fixed point. Stars farther from the pole also travel in circles, partly hidden below the horizon. Representative paths are for a location at middle latitude of 50 degrees North. At this latitude, stars which lie within 50 degrees of the Pole Star never fall below the horizon. As the poets saw it, these stars never take a bath in the ocean. Visible through the entire night, they are called *circumpolar stars*. (Right) facing south.

And in his "Hymn to the North Star," Bryant captures the everlasting significance of the stationary centre of celestial motion, the star that never sets for voyagers in the Northern Hemisphere:[10]

> And thou dost see them rise,
> Star of the Pole! and thou dost see them set.
> Alone, in thy cold skies,
> Thou keep'st thy old unmoving station yet,
> Nor join'st the dances of that glittering train,
> Nor dipp'st thy virgin orb in blue western main.

A CLOCK IN THE NIGHT SKY

As constellations progress across the sky through the course of a single night, they change orientation, but not their defining patterns, or their relative spacings. Pisces will always be a fish and remain at the same distance from Aquarius. Nineteenth-century poet and philosopher Ralph Waldo Emerson lauds their permanence:[11]

> Teach me your mood, O patient stars!
> Who climb each night the ancient sky,
> Leaving no space, no shade, no scars,
> No trace of age, no fear to die.

Inhabitants in the Northern Hemisphere can read the Dipper orientation to tell the time, much as we read a clock. The segments of Figure 5.8 show how the northern constellations circle Polaris through one night. As night progresses, the Dipper changes its orientation, but always looks like a ladle, sometimes pouring, sometimes upright. At 6 p.m., it is hanging down from the handle. At 8 p.m. it rotates eastward so that by 1 a.m. it is nearly horizontal.

WINE MAKING UNDER THE CONSTELLATIONS

Ancients could anticipate the seasons by keeping track of the changing selection of visible constellations. Canis Major (the large Dog), and its brilliant star Sirius, were not visible in Egypt from mid-spring to midsummer, but they were quite conspicuous through the rest of the year. Well before the arrival of the first philosophers around 640 BC, early Greeks were astute observers of the night sky and learned how to put astronomy to work for them. Homer named many constellations, connecting them with the seasons, and also referred to bright Sirius:[12]

> Among men is called the dog of Orion
> Rising ever in autumn, radiantly glowing
> Brighter than other stars, at the dusky hour of milking.

Both Egyptians and Greeks exalted Sirius' splendour, but more often they dreaded it by association with the blistering hot days of summer. In the words of King Priam from the walls of Troy, Homer compared the dreaded advance of Achilles to Sirius:[13]

> blazing as the star that cometh forth at Harvest-time,
> shining forth amid the host of stars in the darkness of the night,
> the star whose name men call Orion's Dog.
> Brightest of all is he, yet for an evil sign is he set,
> and bringeth much fever upon hapless men.

(a)

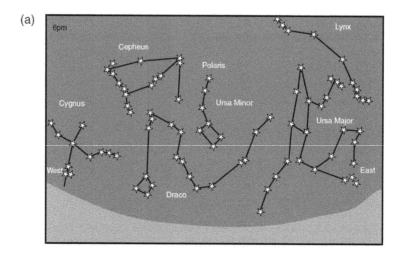

(b)

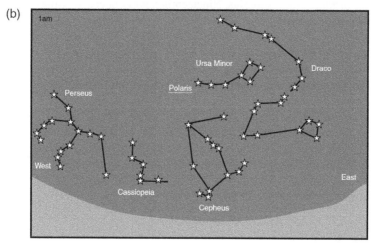

FIGURE 5.8   Constellation patterns formed by fixed stars help chart the movements during the course of a night. Looking in a northerly direction from a location on earth at 40 degrees latitude, stars appear to revolve around the North Star, Polaris. Note the change in the location of stars in the constellation Ursa Minor (Little Bear, or the ladle) as the night progresses from (a) 6 p.m. when the ladle is hanging down to (b) 1 a.m. The ladle is now nearly horizontal. The relative positions of stars in the constellations show that star patterns do not change; stars remain in the same positions relative to each other. The appearance of the sky is on February 10, 1997, from latitude 40 degrees, 45 minutes North, longitude 74 degrees West, the coordinates of New York City. The overall field of view shown here spans an abnormally large angle of 126 degrees, which is why the horizon is represented as slightly curved, rather than straight.

The association of Sirius with the hot, sultry days also gave rise to our phrase "the dog-days of summer."

Another Greek poet, Hesiod (ca. 8th century BC), drew inspiration from correlations between changing constellations and varying seasons, to give excellent advice to farmers:[14]

> When the great Orion rises, set your slaves to winnowing ... holy grain ...
> But when Orion and the Dog star move into mid-sky ... then pluck the clustered
>   grapes, and bring your harvest home ...
> When the great Orion sinks, the time has come to plough;
> and fittingly the old year dies.

In another example of homespun astronomy, Hesiod instructed his brother Perses, the farmer, that the best time to pick grapes from the vine was when Sirius crossed the north– south line overhead and reached its highest elevation at dawn, also marked by the annual reappearance of another bright star, Arcturus. The time of the year was about the middle of September:[15]

> When Orion and Sirius are come to the middle of the sky, and the rosy fingered
>   Dawn confronts Arcturus, then Perses, cut off all your grapes and bring them home
>   with you.
> Show your grapes to the sun for ten days and for ten nights.
> cover them with the shade for five, and on the sixth day press out the gifts of bountiful
>   Dionysus into jars.

JUST WATCH THE BEAR

The stars had a crucial impact on the life of seafaring people. In the *Odyssey*, Homer discussed how his epic hero, who personified the rugged nation of sailors, used stars for navigation:[16]

> Odysseus sat by the stern-oar steering like a seaman. No sleep fell on his eyes; but he watched the Pleiades and the late-setting Waggoner, and the Bear … which wheels round and round where it is, watching Orion, and alone of them all never takes a bath in the ocean. Calypso had warned him to keep the Bear on his left hand as he sailed over the sea.

In this beautiful piece, Homer revealed how the northern constellations circle the Pole Star and how the Great Bear never dips below the ocean when viewed from any of the Greek islands. During long-distance journeys, travellers kept track of how far north (or south) they ventured by judging the location of familiar stars and constellations relative to the horizon. When they travelled to higher latitudes, the altitude of the northern constellations increased. They found the Pole Star farther away from the horizon. Bryant reminds us how modern travellers continue to rely upon Polaris for defining north:[17]

> On thy unaltering blaze
> The half-wrecked mariner, his compass lost,
> Fixes his steady gaze,
> And steers undoubting, to the friendly coast;
> And they who stray in perilous wastes by night
> Are glad when thou dost shine to guide their footsteps
> right.

The modern charts of Figure 5.9 show changes in the appearance of the night sky between latitudes 20 degrees N and 30 degrees N. Notice how Polaris, the Dipper, and Draco appear much higher in the sky – farther away from the horizon – when examined from the higher northern latitudes.

SACRIFICES FOR THE SUN AND MOON

Day after day, night after night, month after month, the sun and the constellations faithfully appeared the same, while the moon reliably repeated its phases. On rare occasions, however, frightening events occurred. Understanding such anomalies ultimately opened new windows to the operations of the universe. Unexpectedly, in the middle of a bright day, the earth would slowly plunge into a ghostly darkness. Flowers closed, and an eerie silence fell as birds ceased chirping. Such anomalous events provoked crises. Some cultures thought a venomous sky-serpent was maliciously

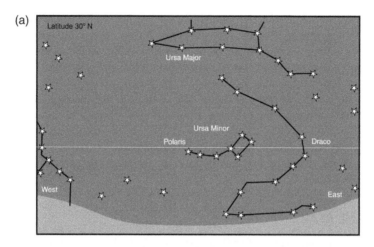

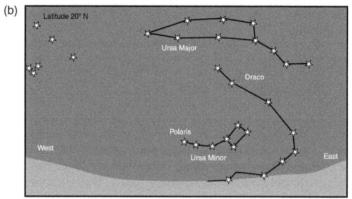

FIGURE 5.9   Ancient travellers used star charts to determine location. Compare (a) the appearance of the sky at midnight in January 1999 BC from latitude 30 degrees N (above the equator), with the sky as it appears in (b) at midnight from latitude 20 degrees N. Note that almost all of Draco is visible from the higher latitude but only a part of Draco is visible from the lower latitude. Note also that Polaris and Little Bear are closer to the horizon when viewed from the south. The field of view is about 100 degrees.

chewing up the sun. In the most dramatic manifestation, the entire sun was blotted out (Figure 5.10). A completely unpredictable and awesome occurrence, the *total solar eclipse* caused considerable consternation among ancient peoples. Some act was necessary to save the sun from the grip of evil.

The Chinese shot arrows toward the heavens to scare off dragons, while Peruvian natives thrashed dogs so that their howling would keep evil spirits away. Romans beat pots and pans or held torches high in the air. Some cultures demanded a human sacrifice. In India, beggars ran through the streets

FIGURE 5.10   (Left and middle) partial solar eclipses, and (right) a total solar eclipse.[18]

shouting for alms so that the stranglehold of evil would release the sun. In *Popular Religion and Folk-Lore*, Crooke writes:[19]

> A high-caste Hindu will eat no food that has remained in the house during an eclipse, and all earthen vessels which are in the house at the time must be broken. During an eclipse all household business is suspended, and eating and drinking is prohibited, even sleep at such times is forbidden, for it is then that the demons and devils are most active. The most effectual way of scaring the demon … and releasing the afflicted planets is to bathe in some sacred stream.

Of course, whatever ceremony they performed, it always succeeded in reversing the eclipse!

Eclipses similarly horrified the ancient Greeks. According to historian Herodotus (ca. 5th century BC), a solar eclipse occurred while the Lydians of Persia and the Medians of Asia Minor were preparing for battle. At war for nearly five years, the two countries prepared for bloody combat once more. Suddenly a ghostly veil of darkness subdued the day. Terrified by the disappearance of the sun, both parties laid down their arms and made peace. This particular solar eclipse was well remembered for saving thousands of lives.

If the sun could suddenly disappear from the sky, what other catastrophe was nature capable of bringing upon the world? The mystery of the unknown charged humanity with fear. Greek poet Archilochus (ca. 648 BC) wondered about other seeming impossibilities. If day could turn into night, could sheep live in the oceans? Could dolphins move to dry land?:[20]

> Nothing is strange, nothing impossible,
> Nor marvellous, since Zeus the father of gods
> Brought night to midday when he hid the light
> Of the shining sun. Grim fear has smitten us,
> And anything can happen to mankind.
> Let no man marvel if he sees the flocks
> Yield up their grassy pasture to the dolphins
> And seek the salty billows of the deep,
> Grown dearer to them than their native meadows,
> While to the fishes sweeter seem the mountains.

Around 450 BC, lyric poet Pindar bemoans the disappearance of the sun as a challenge to understanding:[21]

> Beam of the sun!
> What wilt thou be about, far-seeing one,
> O mother of mine eyes, O star supreme,
> In time of day
> Reft from us? Why, O why has thou perplexed
> The might of man
> And wisdom's way,
> Rushing forth on a darksome track?

Total solar eclipses also gave an important clue about stars to ancient observers. Upon spotting a few shining stars during the middle of the day on the rare occasion of a perfect and total solar eclipse, they recognized that stars do not disappear during the day. They are always present in the skies.

A few times every year the shape of the full moon would also change very quickly, within the space of a few hours, and then return, just as promptly, to its original, fully lit shape. The sudden and unpredictable disappearance of the moon was unnerving. Many myths arose to account for the erratic and eerie disappearance. Some cultures believed that abrupt changes predicted disaster, averted only by making the appropriate sacrifice to appease the offending evil.

Babylonians (after 700 BC) made the first breakthrough toward understanding lunar eclipses. By faithfully tabulating the occurrence of lunar eclipses over centuries, they discovered interesting numerical cycles, with short periods, long periods, and a super-period of 18 years. Recognizing cycles reduced the fear of erratic eclipses. With the compilation of tables, Babylonian priests increased their prestige. They gained the magical power to forecast eclipses correctly. However, Babylonians had no explanation for lunar eclipse cycles; that came much later, after the Greeks developed a geometrical model for the heavens.

THE CELESTIAL NILE

Standing outdoors on a crisp, clear, and starry spring night, perceptive observers could identify a faintly glowing band of hazy white light, best observed on a moonless night after waiting a few minutes for the eyes to adapt to the reduced light intensity. Facing north, the band stretched along the sky from high in the west and down to the horizon in the north. It was not a wisp of cloud nor a passing mist, because the night was crystal-clear. Night after night they could find the same hazy patch weaving through the same constellations: Gemini, Auriga, Perseus, and Cassiopeia (Figure 5.11).

Throughout the centuries, various cultures attempted explanations of this faint band of light. To Egyptians, who lived in a river-dominated society, the hazy band represented a heavenly river. Gods travelled at night along a celestial Nile. The hunter-gatherer tribe of Namibia in Africa saw it as the ashes of a heavenly fire. Fishermen from China saw it as a school of divine fish frightened off by the hook-shaped crescent moon. Ancient Jews thought of the celestial strip as a long bandage around the heavens. Another culture described it as a crack or a seam where the two halves of heaven joined imperfectly, giving mortals a faint glimpse of the glorious light of heaven beyond the darkness of the night. A French legend attributes the hazy strip to the glimmer of lights held by angels guiding dead souls into heaven. Following their obsession to lay down roads to make their empire accessible, Romans called the heavenly band "via lacta," for milky road. To the pilgrims of the Middle Ages,

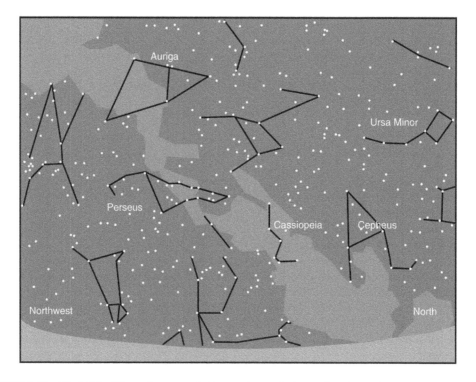

FIGURE 5.11   The Milky Way as it appears facing northwest during spring from mid-northern latitudes.

FIGURE 5.12   *The Origin of the Milky Way,* 1575-80, Jacopo Tintoretto.

it showed the trail to Rome. In a Scandinavian legend, two lovers, after 1,000 years of separation, build a bridge of starry light. Meeting at the completion of the heavenly arch, they dissolve into each other's arms to form the brightest star, Sirius.

A most colourful Greek myth explained the hazy band as a spurt of milk from the breast of the goddess Hera. In *The Origin of the Milky Way* (Figure 5.12)[22] Tintoretto shows Jealous Apollo pulled the baby Herakles from Hera's breast, spilling her milk across the heavens to form a milky circle. Souls of the dead collected to draw nourishment. Borrowing from that Greek myth, astronomers called the pale glow the "Milky Way'," also adopting the Greek word *gala* for "milk" when naming our *galaxy.*

THE HEAVENLY SWORD

Sudden apparitions of fuzzy roving stars with hairy tails occasionally alarmed devoted sky-watchers (Figure 5.13). We call these *comets*, from the Greek word *kometes*, meaning long-haired. When comets appeared, they crossed the sky slowly as their tails elongated. Sometimes they took a month or two to complete their trip. Much like the eclipse, the random appearance of a comet caused much distress. Many cultures believed comets to be a warning sign of impending evil, disease, pestilence, or famine. Sumerians spoke of comets as vultures, or birds of death. Members of other cultures thought of a comet as a heavenly dagger hanging over the world, threatening disaster. In *Paradise Lost*, the 17th-century poet Milton invoked a similar image as he recreated the wrath of God sweeping out the sinners from Paradise with his mighty sword:[23]

High in front advanced,
The brandished sword of God before them blazed
Fierce as a comet.

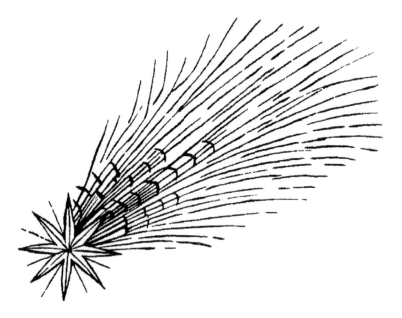

FIGURE 5.13   A very early sketch of a comet that appeared over 7th-century Europe. The artist treated the body of the comet as a star, probably unaware of Aristotle's ideas about comets as atmospheric phenomena.[24]

A few months after the assassination of Julius Caesar (44 BC), a comet passed over Rome, taken by the populace as a sign of the great man's ghost passing over his beloved city. To commemorate the occasion, the Republic issued silver coins with the stamp of the comet. Only the Chinese attached a positive meaning, describing comets as heavenly brooms that swept evil away.

THE RISING OF SIRIUS, THE FLOODING OF THE NILE

Tracking the sun's location with reference to stars rather than with respect to artificial megaliths and mountain peaks changed the course of early astronomy. One such crucial turn took place in Egypt (ca. 4000 BC).

A key component of Egyptian agriculture was the annual flooding of the Nile river, central artery of Egyptian life. Every year, with unwavering regularity, the waters of the world's longest river gushed to inundate the valley. With the arrival of summer, melting snow from mountains in the south swelled the Nile as it flowed toward the Mediterranean. Depositing layers of rich silt over arid land, the annual flood provided a country-size strip of fertile soil amid a vast, lifeless desert. Egyptians planted their wheat and barley crops in the alluvial mud left behind after floods subsided. With the bounty of the Nile, the soil was so rich and soft that there was no need for manure or ploughs. Luxuriant crops grew in regions that never received a single drop of rain for years. Without the gift of the Nile, Egypt would have joined the lifeless desert.

It is not surprising therefore that the magnificent Nile shaped the life, thought, religion, and cosmology of Egyptians. At cosmic creation, the primordial male and female gods – Shu and Tefnut – procreated Geb and Nut, separating them into distinct earth and sky. The striking parallel with the biblical account of the separation of heaven and earth should give cause for reflection. Like a mighty river (Figure 5.14), the male god (Geb) stretched out his arms and legs extending like tributaries. Patterned with leaves, he represented life-bearing earth. High above, star-spangled goddess Nut represented the heavens. Protecting Geb, she stretched over him in an elongated arc, much like a celestial river. The sun and the moon traversed the celestial river in barges that resembled barques daily plying through the Nile.

FIGURE 5.14   The overwhelming presence of the Nile river shaped the Egyptian's view of the cosmos.[25]

One early scorching-hot July morning, while offering thanks for the arrival of the long-awaited Nile flood, an Egyptian priest noticed that bright Sirius was rising almost exactly at the same time as the sun was about to peer out over the horizon. It must have been a breath-taking coincidence of earthly and heavenly events. What was particularly surprising to that knowledgeable priest was that exceptionally bright Sirius, source of many a colourful legend, had been invisible for a few months prior – between early May and early July. It now re-appeared above the horizon at sunrise, on about the same day that the Nile river started to show its first sign of flooding.

Here was a reliable way to predict when the Nile river would start to swell and make a sea of the lowlands. Now temple priests could warn herders to move their flocks up the dikes to safety. Ever since, the remarkable concurrence between Sirius-rise, sunrise, and the Nile flood marked the beginning of the cycle of seasons for Egyptian culture. It was to become the most auspicious time of the year, observed with great ceremony. From that time forward, Sirius became the Nile star.

For astronomy, temple priests had initiated a key advance: the practice of observing heavenly bodies with respect to the firmament of fixed stars. Those who kept close track of solstices, equinoxes, and the heliacal rising of Sirius served as the mediators between heaven and earth, between humans and the supernatural world.

A CORRECTIVE LEAP

Keeping regular watch on the sun's position with respect to Sirius, astronomer-priests made another surprising discovery. The relative positions of the sun and Sirius changed slightly from year to year. Precisely 365 days after observation of one coincident rising, they were disappointed to find that Sirius rose slightly ahead of the sun on New Year's dawn. After a long period of observation, they realized it takes four 365-day years plus one full day to keep the sun in exact coincidence with

Sirius. The Sirius cycle was $365\frac{1}{4}$ days long.

A correction scheme became necessary to keep track of the heliacal rising. Neglecting the extra $\frac{1}{4}$ day would bring Nile floods more and more out of step with the sunrise appearance of Sirius, making it impossible to predict the flood. If priests had chosen to add an extra day every four years, the sun and Sirius would remain in the same position relative to each other over many years, even over many centuries. It would have been a major advance in calendar precision, and would keep the solar and stellar cycles synchronized. But the arbitrary intercalation of days would be an embarrassment, upsetting the familiar regularity of the old calendar and religious festivals already based upon it. Priests tried to keep the new discovery a secret. Whenever they inaugurated a new Pharaoh, they made him take an oath at his coronation that he would not alter the long-established 365-day calendar.

Eventually, they overcame their compunctions. In 45 BC, an Alexandrian astronomer, Sosigenes, helped Julius Caesar introduce what is now called the Julian year of $365\frac{1}{4}$ days. The legacy of the Egyptian-based Julian calendar persisted well into the Middle Ages, until the time of Copernicus – a tribute to the accuracy of the temple-astronomers. We continue to maintain their correction scheme: three years of 365 days followed by a leap year of 366 days. Additional minor corrections are also necessary, as we will see.

A CIRCLE OF ANIMALS

Great strides in describing celestial motion emerged from neighbouring Mesopotamia, where Sumerians flourished under the Babylonian dynasty (ca. 1800 BC). One of their crucial contributions detailed the path which the moon takes through the stars. As the month progresses, the moon travels through one special band of constellations, a star-studded corridor circling the sky. In this heavenly strip, the Babylonians identified 12 constellations, and named these after animals, such as fish, lion, and bull. They called these the constellations of the *zodiac*, which means *the circle of animals*. Our names are: Gemini, Cancer, Leo, Virgo, Libra, Scorpio, Sagittarius, Capricorn, Aquarius, Pisces, Aries, and Taurus. Figure 5.15 shows the patterns of stars in some of the zodiacal constellations. Every night, the animal constellations move from east to west. Over the course of one month, the

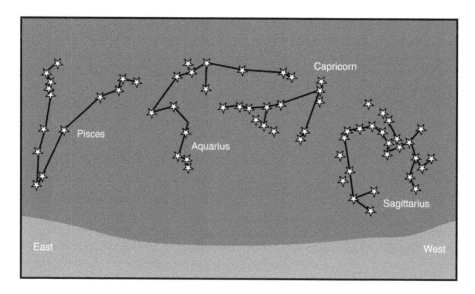

FIGURE 5.15   Some constellations of the zodiac. Only the brightest stars are displayed.

moon threads its way from west to east. Because there are 12 prominent constellations lying along the path of the moon, the Babylonians divided night-time into 12 hours, and symmetrically daytime into 12 hours, giving rise to the 24-hour day.

Watching the sky just before sunrise, ancient astronomers noticed that the sun also rose among the constellations of the zodiac. For example, during the first month of spring, the sun rose in Aries. That same evening, the sun also set in Aries. Through one particular day, the sun maintained its position within one constellation (see Figure 5.16). But the next day, the sun slightly changed its position toward a neighbouring zodiac constellation.

Over a period of two months, the sun crawled all the way through two zodiac constellations (Figure 5.17). As the year progressed, the location of the sun slid through all 12 constellations of the zodiac, making one grand circuit. For the Babylonians, the sequence of the sun's passage through the zodiac constellations provided a most convenient solar calendar in the twilight sky. Summer solstice occurred when the sun was in Cancer, winter solstice when the sun was in Capricorn. During the spring equinox, the sun was in Aries, shifting to Libra by the time of the autumnal equinox.

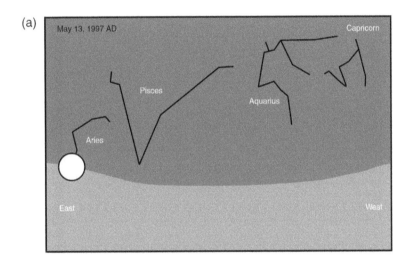

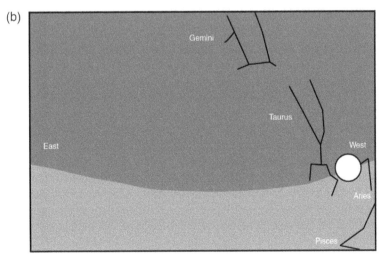

FIGURE 5.16   The position of the sun relative to the constellations at sunrise and sunset. (a) The sun *rises* in constellation Aries on May 13. (b) Same day, 7 p.m.: the sun *sets* in the *same* constellation Aries.

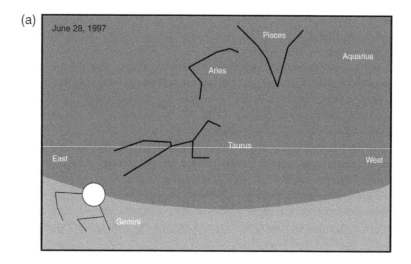

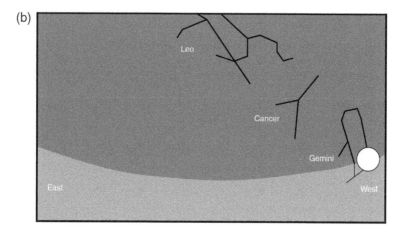

FIGURE 5.17 Every day the sun slightly changes its location as it progresses along the zodiac constellations. Nearly two months later, on June 28, (a) the sun rises in the constellation Gemini. (b) On the same day the sun sets in the *same* constellation Gemini.

While the constellations move *east to west* through one night, the sun's position slips from *west to east* through one year. The sun appears to drift backward (relative to the nightly motion of the constellations) along the zodiac about 1 degree per day. It follows a special path (Figure 5.18) which the Babylonians called the *way of Anu*, the sky god. We call the *Anu* path the *ecliptic*. (Its connection to eclipses will become clear later.) Even though both the sun and moon move within the zodiac belt looped around the firmament (Figure 5.19), their paths are not exactly identical. The moon's path is inclined by about 5 degrees with respect to the ecliptic. Although slight, the angle plays a major role in our ability to see a full moon every month, and in the occurrence of eclipses.

The changing monthly residence of the sun among the zodiac constellations inspired artists through the ages. In some of the most picturesque portrayals, medieval artists illustrated calendars and prayer books in the belief that the sun's zodiac position controlled seasonal activities. Reviving mythical themes from the classics, Renaissance artists painted imaginative connections. On the eastern wall of the *Room of the Months* at the Palazzo Schifanoia, Francesco del Cossa painted 12, three-tiered allegorical frescoes. The middle tiers show the sun rising in the appropriate constellation, Taurus, the Bull, for the *Allegory of April* (Figure 5.20[26]). One of the oldest, most commonly recognized constellations, it symbolizes strength and fertility in many civilizations. The maidens

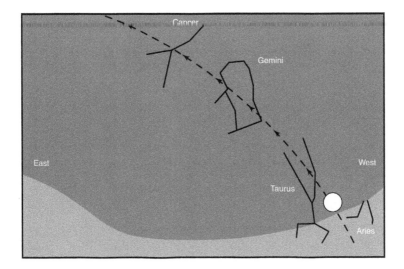

FIGURE 5.18   Aries to Gemini path of the sun through the zodiac constellations between May and June. *Ecliptic* is the name for the sun's path (the dashed line). The zodiacal constellations emerged as an excellent reference system to keep track of the path of heavenly bodies, such as the sun and moon. Whereas the sun moves from east to west in its daily path, it drifts *from west to east* along the ecliptic in its annual path.

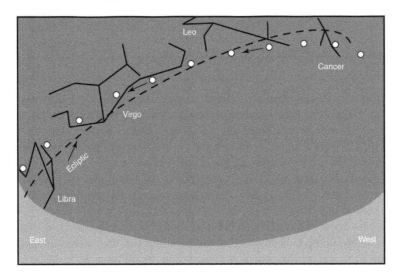

FIGURE 5.19   The path of the moon (circles) and the sun (dashed line) through the band of zodiac constellations. The moon's path is near to but not the same as the ecliptic. It is slightly inclined with respect to the sun's path. These properties are important for lunar eclipse cycle timings.

Hyades and Pleiades accompany Taurus. These groups of bright stars were to play important roles in the future of astronomy. The upper tiers depict activities appropriate to the chosen month. We see, on the right, youths, maidens, and musicians surrounding two young lovers kneeling in embrace (Figure 5.20). Clearly, love is in the April spring air. Even rabbits huddle together.

EVENING STAR, MORNING STAR

Besides the obvious moon, ancients noticed five prominent celestial bodies of unusual brilliance. Even when haze blocked out the stars, some of the special night-lights were still visible. Remarkably,

FIGURE 5.20   *Allegory of April,* 1476-84, Francesco del Cossa.

the five bright lights also move within the band of zodiac constellations, the heavenly highway for the sun and moon. Because of their meandering motion through the fixed constellations, Greeks later called them *planets* – from their word for "wanderers." Our names are Mercury, Venus, Mars, Jupiter, and Saturn. The long struggle to understand the bewildering motions of planets finally led to an understanding of the solar system design and earth's place in it.

Venus appears as the brightest planet. It is so bright you can sometimes see it in the sky during a clear day if you know exactly where to look. For many months it appears as an *evening star*, glowing soon after sunset, as if destined to lead out its companions (Figure 5.21). A few hours later, after most of the stars have come out, it will sink below the horizon, its appointed task complete. As an evening star, it is visible in the west at twilight for a month or two, and then it disappears from the sky. A few months later, it reappears, but now as a *morning star* (Figure 5.21). Venus is the last star visible at dawn, tarrying to herald the arrival of the sun. It rises from the horizon a few hours before the sun. On months that you can see the evening star you cannot see the morning star. Venus only appears in conjunction with the sun (on the same side of the horizon as the sun), never in opposition.

In the *Iliad*, Homer referred colourfully to the evening and morning stars:[27]

Evening star (Greek Hesperos, later also called Vesper for evening):
The great spear, which gleamed like the finest of all the stars of heaven, Hesperos brilliant in the dark night.
Morning star:
But when Eosphorus came, proclaiming light on earth, and after him the saffron mantle of Dawn spread o'er the sea, at that hour did the flame of the funeral pyre die down.

Pythagoras assimilated many common features about the evening and morning stars, reaching a brilliant deduction – the two are one and the same (ca. 530 BC) planet.

(a)

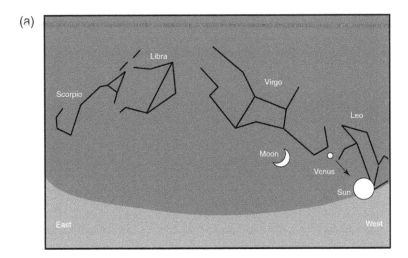

(b)

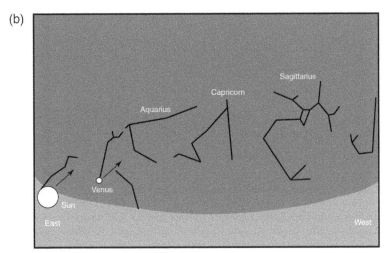

FIGURE 5.21 (a) Venus sets soon after the sun in late July, 650 BC. (b) Venus rises a little before the sun in late February, 650 BC. Venus appears to follow the sun. The arrows indicate the direction of motion for a planet within a single night, and for the sun during a single day. The daily (nightly) motion is otherwise known as diurnal motion.

A GALLERY OF GODS

At first, ancients referred planetary positions and movements with respect to the sun. Venus and Mercury always stay close to the sun, like the crescent moon, a juxtaposition called *conjunction*. Mars, Jupiter, and Saturn show different behaviour. Sometimes they appear close to the sun (in conjunction), but often they stray far away from where the sun rises or sets over the horizon. When they are about 180 degrees across the sky from the sun, they are said to be in *opposition*.

Intrigued by the same vagabond stars, Babylonians (around 700 BC) began to keep close track of their path among the 12 zodiacal constellations. It was the beginning of a long odyssey to understand heavenly motion. The circle of animals provided a most convenient reference system to study the prominent lights. Like the sun and moon, planets move against the sense of their nightly motion. While constellations move from west to east through one night, planets slip east to west from night to night, similarly to the motion of the sun through the year. See Figure 5.22 for the paths of Mercury and Venus over roughly one month. Through regular scrutiny of the conspicuous lights,

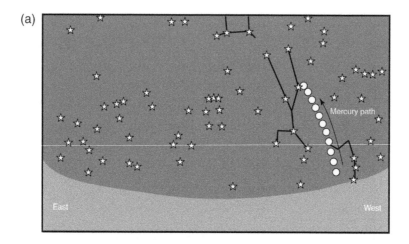

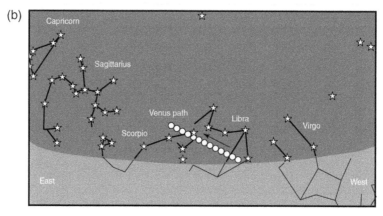

FIGURE 5.22   (a) The path of Mercury through the constellations from March 21 to April 9 in the year 640 BC at 7 p.m. (b) The path of Venus from September 10 to September 27, 650 BC. Notice how planetary paths change with respect to the pattern of fixed stars. Mercury moves faster through the zodiac than Venus.

Babylonians came upon some rather curious facts. While moving among the zodiac band, the moon and errant stars do not follow the exact path of the sun god, Anu, but dip above and below it, like sheep who like to escape from the fold.

Based on their speed of motion relative to fixed stars, and their position relative to the rising and setting sun, ancients assigned special attributes to the five bright celestial lights. Over time, fascinating folklore developed about these properties. More important, planetary qualities which Babylonians recorded meticulously over centuries of sharp observations became the basis for geometric models of the heavens subsequently developed by the Greeks. For example, Babylonians discovered that each heavenly body has its own period for completing the zodiacal circuit. The moon takes the shortest time – one month. Next come Mercury (three months), Venus (seven months), sun (12 months), Mars (23 months), Jupiter (12 years), and finally Saturn (29 years). This order provided a rationale for Greeks to locate heavenly bodies relative to the earth, the fastest ones the nearest. Ultimately the arrangement was destined to change, when the sun-centric system replaced the earth-centric, but the period of planetary cycles around the zodiac continued to play a key role in the theory of gravitation.

To Babylonians, Mercury was a swift and fleet-footed god, because he moved quickest. He became the patron of messengers and fast-moving thieves. With its reddish glint, Mars reminded them of blood; hence they named it after the god of war. His symbol became a shield and sword (the male symbol). About every two years the gory god of war would return to the sky as if from a distant campaign. At times he appeared dim, barely discernible, as if depressed from defeat. At other times he returned glowing victoriously, even brighter than Jupiter. Saturn earned the label of old man because he moved slowest through the constellations. For the Greeks, he became god of time (Saturn's Greek name was Chronus) because he took 29 years to cross the zodiac. Jupiter moved neither fast or slowly. His moderate speed reflected temperate character, full of good judgment. Unlike the variable brilliance of Venus and Mars, Jupiter's steady glow reaffirmed his restrained character, likened to a wise king among the cast of planetary gods of wilder character. Showing great restraint, Jupiter never strayed more than 1.5 degrees from the sun's path, while others wandered about by several degrees.

Our name for Jupiter is the Roman one for Zeus Pater, or Father Zeus. In *Paradiso*, the 14th century poet Dante captures the qualities of Jupiter when his hero encounters this most sensible planet in his imaginary journey through the heavens:[28]

> Turning, I perceived
> The whiteness round me of the temperate star …
> And in that torch of Jove, I was aware
> Of sparkles from the love within it warm …
> O sweet star, with how many and rare a gem
> Didst thou prove that the justice we obey
> Proceedeth from the heaven thou dost begem!

PLANET OF LOVE

Venus, brightest among planetary gods, commanded special attention from many cultures. To the Babylonians, she was a dangerous female constantly pursuing the sun god, never straying far from the object of her affection. But she was fickle. At certain times of the year, she came very close to the sun, at other times she ran away shyly, as if fleeing from her lover. When first to appear at dusk soon after the setting sun, she drifted eastward every night, shying away from the sun, and remaining longer in the night sky. Then, surprisingly, she turned around and headed towards the sun, as if she longed to cross his fiery disc. A Babylonian tablet gives an enchanting description about Venus (who they called Ishtar):[29]

> Ishtar is clothed with pleasure and love
> She is laden with vitality, charm and voluptuousness.
> In lips she is sweet; life is in her mouth.
> At her appearance rejoicing becomes full.
> She is glorious; veils are thrown over her head.
> Her figure is beautiful; her eyes are brilliant.

The best time for romance was when Venus was highest in the sky! The Babylonian association of Ishtar with love continued with Greek poets who named the planet Aphrodite, goddess of love. For Greek sailors, a glistening Aphrodite rose from the sea to grace their resplendent sunsets and sunrises over the ocean. In an early Greek artistic representation (Figure 5.23), Aphrodite is simple and perfect, a Greek ideal.

Through the centuries, Venus continued to inspire poets to tell the story of love. The 19th-century poet Tennyson uses florid language to compose a homage in *Maud*:[30]

> For a breeze of morning moves,
> And the planet of love is on high,
> Beginning to faint in the light that she loves

FIGURE 5.23   *Birth of Aphrodite: Graecia Ludovisi Throne*, ca. 470–460 BC.[31]

On a bed of daffodil sky,
To faint in the light of the sun she loves,
To faint in his light, and to die.

Our culture has hardly forsaken the amorous associations of Venus. Its symbol has now become a symbol for the female sex. Her sign resembles a hand-held mirror, befitting the goddess of beauty. One need only recall the danger of "venereal" disease.

As much as she beguiled poets, painters, and sculptors in all cultures and at all times, Venus had a strong impact on astronomers. She unveiled new truths about heavenly motion. The attachment of Venus to the sun later gave Copernicus a major clue that Venus and all the planets must revolve around the sun, not around the earth. Half a century later, when Galileo saw a new beauty of Venus through his penetrating telescope, he found renewed faith in the Copernican system.

Moving capriciously among the zodiac constellations, the five planets with their mythical attributes inspired artists. Greek sculptors represented Zeus with long, curling hair and dignified beard, clad in flowing drapery and seated on a mighty throne holding a thunderbolt in one hand, ready to hurl it if seriously provoked. Reviving themes from the classics, Renaissance artists painted fanciful features for planetary gods. In a detail from *Parnassus* (Figure 5.24[32]), Andrea Mantegna shows the warrior Mars and the gorgeous Venus looking down at the activities of their fellow gods. At the fountain in Villa Medici, an arabesque statue of *Mercury* (Figure 5.25) balances precariously on a column of air issuing from the mouth of the west wind, Zephyr. As water flows over his lips, it gives the illusion of a swiftly floating Mercury, rising upward. By contrast, the slow-moving Saturn's age clearly shows through his flowing beard. As god of the time-consuming activity of agriculture he carries a scythe (Figure 5.25).

LUCKY SEVEN AND THE ORIGINS OF ASTROLOGY

With the sun, moon, and five planets, ancients counted seven heavenly bodies, clearly set apart from the vast multitude of stars. Over time, the number seven assumed magical significance. Some cultures adopted seven as a sacred sign of divine planning. God created the world in seven days. Jews,

FIGURE 5.24   Detail from *Parnassus,* 1497, Andrea Mantegna.

FIGURE 5.25   (Left) sprightly young *Mercury,* 1580, Giovanni Bologna. (Right) old and hunched over *Saturn,* ca. 1451–1500, Agostino di Duccio. The movement of planets reflects traits ascribed to each planetary god.[33]

ancient Egyptians, and Sumerian civilizations adopted the division of the week into seven days, a convenient subdivision of the lunar cycle of 28 days into four parts. Our names for days of the week originated from the seven heavenly lights. In French we say: Lundi (Monday) for the moon, Mardi for Mars, Mercredi for Mercury, Jeudi for Jupiter, Vendredi for Venus, Samedi for Saturn, and of course, in English, Sunday.

Later Babylonians imagined connections between the celestial movement of planetary gods and terrestrial activities of humans. It was obvious that the sun, moon, and constellations influenced events on earth. It is hot when the sun is in Cancer, and cold when in Capricorn. They noticed how the varying height of tides followed lunar phases. With remarkable regularity, tides are especially high during a full moon and new moon. Eventually tidal cycles played an important role in Newton's theory of gravitation. Given striking correlations between heaven and earth, we can understand why Babylonians came to believe that heavenly bodies must also influence human destiny. Thus began the activities of astrology.

Priests became obsessed with lunar phases and eclipses as well as with the exact rising and setting times of planets. Astrologers made prolific sets of tables, mostly for the purposes of divination. Their predictions grew especially important to the ruling elite who believed themselves to be direct descendants of gods. Whatever the reasons, the close scrutiny which Babylonian priests gave to the irregular motions of the wandering stars turned out to be crucial to the development of astronomy. Hundreds of Babylonian stone tablets carry presumed connections between the shape of the moon's horns and events on earth, such as famine, floods, locusts, sickness, and death. But buried amidst vivid astrological lore and predictions are precious gems of astronomy. Here is an inscription from a very early tablet which describes the comings and goings of Venus as the evening and morning star:[34]

> When Ishtar (Venus – as the morning star) is *visible in the east* on the sixth of Abu (month), there will be rain and destruction. Until the tenth of Nisan (month) it remains easterly; on the eleventh it vanishes and remains invisible for three months. If, then, it *glows in the west* (evening star) again on the eleventh of Du'uzu (month), there will be fighting in the land, but the fruits of the fields will prosper.

Over centuries of close planetary observations and accompanying astrological predictions there appeared some rather disconcerting facts. Planetary paths were not only *irregular*, but they were sometimes *unpredictable*. This meant serious trouble for astrologers obliged to come up with predictions for weather and the outcome of wars, as well as horoscopes for kings and their offspring:[35]

> When the planet Ishtar enters the Scorpion, floods take place, as is well known. This time, however, there were none, for as soon as Ishtar came up to the breast of the Scorpion, *she was snatched away.* Although she touched the Scorpion, she did not penetrate it.
> As to Marduk (Jupiter), I recently reported. My prediction was based on his position in the Anu path (the ecliptic). I now report he has been delayed. *He has truly run backward;* therefore my interpretation was mistaken. But it would not have been mistaken if Marduk had remained on the Anu path. May my Lord King understand this.

What a shock to find out that planets were slowing down in their tracks, stopping, and at times even turning around, as if to retrieve something they had lost along their way. The astrologer was pleading for the king to spare his life. The erratic motions of Venus and Jupiter had frustrated his ability to make a correct prediction. Such subtle variations in planetary movement only became apparent to patient observers, who were inspired to extract some crucial meaning from capricious motions.

While solar and lunar astronomical lore arose from the needs for time keeping, calendar reckoning, orientation, and navigation, planetary astronomy arose mostly out of a desire to connect the activities and fate of humans to the universal whole. Astrology and astronomy were born as twin offspring. For centuries they thrived side by side. While the significance of each arena was drastically different, their practices and results maintained a filial relationship. For a long time, astronomy flowered for the benefit of astrology. Later, astronomers drew freely upon the wealth of astrologers' data. Elaborate tables of planetary positions led to computational techniques for predicting planetary locations. Astrology continued to play an important role in the lives of future astronomers, Tycho Brahe and Johannes Kepler (Chapter 8) in particular, who drew upon its popularity for financial support from royal patrons.

Starting in Mesopotamia with Sumerians and Babylonians and continuing with the Chaldeans (600–500 BC), astrology migrated to Greece, Alexandria, India, the Roman Empire, the Mediterranean, and ultimately to all Europe. The practice continues to hold a significant place in the modern world. There are tens of thousands of astrologers in the United States. Many people know their signs and read their horoscopes, if only just for fun. If you are born in August, your sign is Leo, because, during August, the sun used to rise in the constellation Leo, according to Babylonian records. Your sign supposedly determines your character. According to some astrologers, Leo is forceful, with a logical mind. An Aries is vivacious and adventurous, but lacking in self-control and perseverance. Taurus is stable and balanced, but is prone to occasional bursts of rage. Gemini is anxious and unstable, while ingenious and clever. Cancer is sensitive and shy. Virgo is simple and well-ordered. Libra is moderate and harmonious. Scorpio is violent and passionate. Sagittarius is thoughtful and reasonable, while more philosophical than practical. Capricorns are reserved and patient. Aquarius has a light and airy nature. And impressionable, indecisive Pisces tends to fantasize. Astrologers may differ from one another in the assigned characteristics.

Ancient astrologers divided up the sky into 12 sections, or houses. To each house they assigned a special influence over the affairs of human beings. Thus the first house in the eastern horizon became the house of life, the next, the houses of riches, children, health, marriage, death, and so on. During the course of the year, the constellations appear to move sequentially through the houses. Moving at different speeds, planets traversed houses, taking up different positions among the constellations. A special astrological quality signified each planet; for example, Jupiter can mean male, white, and gold. To make a forecast for a new-born child, an astrologer started out with the exact day and time of birth. If, at that time, Jupiter was in the second house (riches), the male baby was to become successful, fair-complexioned, and rich. To complicate matters further, astrologers also factored in the position of the moon. For example, when the moon was in Aries, it brought enthusiasm, energy, and creativity, but it could also be responsible for restlessness and a quick temper.

Planetary influences became even more important when two or more came together during what we now call a *conjunction*. For those preoccupied with celestial omens, the simultaneous appearance of planets in the same constellation of the zodiac held deep cosmic significance. In medieval times especially, planetary conjunctions became dreaded as harbingers of devastating floods and other cataclysms.

As a set of capricious rules grafted onto disconnected observations, astrology's distinguishing feature is its thorough inconsistency. Nevertheless, astrology remains strong in contemporary culture, as evidenced by the horoscope section carried in almost every modern newspaper. The fact that two newspapers can forecast different predictions for the same person on the same day should cause any thoughtful reader to wonder about the claim of astrology to reveal the future.

UPDATES

THREE NEW PLANETS

In the contemporary account of the planetary system (Figure 5.26), the sun is at the centre, surrounded by Mercury, Venus, Earth, Jupiter, and Saturn. Earth is indeed a planet, as Copernicus (Chapter 7) dared to propose in his revolutionary view. But Saturn is not the outermost planet. Galileo opened the heavens with his telescope, a special story for Chapter 9. Telescope advances over the next three centuries after Galileo revealed three new planets, Uranus, Neptune, and finally

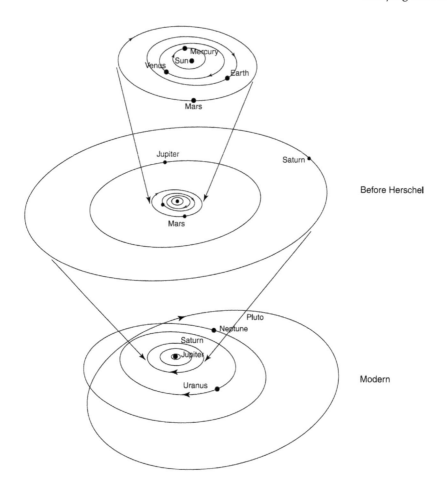

FIGURE 5.26   (Top and middle) the solar system of the ancients. (Bottom) the solar system as we know it today.

Pluto, to complete the solar system. Pluto is now banished from our planetary system, with some new candidates under consideration.

When a music teacher named William Herschel (1738–1822) developed an amateur interest in astronomy, he began to spend his hard-earned money on celestial star charts and astronomy books instead of musical scores. On clear nights he would gaze at stars through telescopes rather than teach music lessons. But telescopes available for rent were of a disappointingly poor quality. And he could not afford to buy good ones on his meagre teacher's salary. Herschel set out to make his own telescopes with help from his sister Caroline. After putting in a full day's work as a musician, he would sit up all night, recording star positions, with only five-minute breaks every few hours to warm himself, even on nights that were so cold that his inkwell would ice over. The frigid nights of winter most clearly revealed the most wondrous sights. In the course of one meticulous search, he came upon a startling discovery in 1781 – a heavenly body that appeared to drift ever so slowly from month to month in relation to the fixed stars. Ancients had discovered motion relative to the stars, one of the key features distinguishing planets from stars. Could it possibly be a new planet? Herschel dared to wonder. It would take 84 years to complete its zodiac orbit, far longer than old man Saturn's 30 years. In one fateful night, the Herschels revised the 2,000-year-old history of the solar system.

News of a brand-new planet spread like wildfire. A musician had discovered a new planet with a home-made telescope, not from the Royal Observatory of England, but from his own backyard.

There was a vast universe out there, waiting to be explored. There were resounding accolades. Astronomers wanted to name the new planet after Herschel, who modestly wanted to name it Georgium Sidus (George's star), in honour of King George III. Amidst all the turmoil of England's defeat in the American War of Independence, the king of England was glad to receive a bit of good news. Besides, Herschel was trying to flatter the king into increasing support for scientific research. Finally they agreed to name the new planet *Uranus*, father of Saturn, and grandfather of Jupiter, keeping the tradition of naming planets after Greek gods. In his poetic trilogy about the achievements of scientific pioneers, Alfred Noyes contemplated the discovery of Uranus:[36]

> Then, as I turned on Gemini,
> And the deep stillness of those constant lights,
> Castor and Pollux, lucid pilot-stars,
> Began to calm the fever of my blood,
> I saw, O first of all mankind, I saw
> The disc of my new planet gliding there
> Beyond our tumults, in that realm of peace.

Soon after the phenomenal discovery of a new planet, astronomers, now trained to look for the smallest possible deviations, noticed that Uranus appeared to wander off its perfect orbit, calculated by Newton's laws of gravitation (Chapter 11). Could known laws of physics break down at the outskirts of the solar system? Those with utmost faith thought a comet must have struck Uranus. Others conjectured there was some "extraordinary and unknown influence," perhaps another planet, acting on Uranus. Once again, Alfred Noyes captured the tantalizing mystery:[37]

> To know why Uranus, uttermost planet known,
> Moved in a rhythm delicately astray
> From all the golden harmonies ordained
> By those known measures of its sister-worlds.
> Was there an unknown planet, far beyond,
> Sailing through unimaginable deeps
> And drawing it from its path?

While browsing in a bookstore, a young undergraduate student at Cambridge, John Adams (1819–92), picked up a book by Astronomer Royal George Airy discussing the deviations of Uranus' orbit. Adams' faith in known laws of mechanics was rock solid. The problem could never be with physical laws of planetary motion. The right explanation was to look for an unknown planet which caused deviations of Uranus through an extra gravitational attraction. After four years of intense mathematical calculations, trial and error, everything fit. He submitted his prediction together with the accurate location of a new planet to the Royal Observatory at Greenwich, asking them to conduct a search. But Astronomer Royal Airy paid no attention to the wild request from an upstart theoretical astronomer. Mathematics was not a tool which could tell humans where to point their telescopes, thought the misguided head.

Only a few months later, reputed French astronomer Jean Leverrier (1811–77) independently arrived at the same prediction. Frustrated that none of his French astronomer colleagues would take his predictions seriously either, he sent his calculations to the head of the observatory at Berlin:[38]

> Direct your telescope to the point on the ecliptic in the constellation Aquarius, in longitude 326 degrees, and you will find within a degree of that place a new planet.

Recognizing the monumental importance of such a find, unlikely though it seemed, the director of the observatory decided to search for the planet on the very same evening that Leverrier's letter arrived. It did not take him long to find the new planet at the predicted location. Now the number of planets grew from the ancient five to seven. Capturing the wonder of the feat in his *Pioneers of Science*, physicist Sir Oliver Lodge wrote:[39]

> To predict in the solitude of the study, with no weapons other than pen, ink and paper, an unknown and enormously distant world, to calculate its orbit … and to be able to say to a practical astronomer, 'point your telescope … and you will see a planet hitherto unknown to man' – this must always appeal to the imagination with dramatic intensity.

Again, there was a strong sentiment to name the new planet after the astronomer who predicted it, but Leverrier himself suggested the name "Neptune," god of the ocean, presumably because of the slight greenish hue of the planet.

More discoveries lay in store for the watchful eye poised at the base of the most powerful telescope. Scientists closely studied Neptune. In time, it became clear that deviations of Neptune and those of Uranus were again larger than expected on the basis of the influence of known planets. Could there be yet another planet? Quite appropriately, Planet X was the provisional name. Among hundreds of starry photographs taken over many careful sweeps of the skies, one could have revealed a new planet. Unfortunately, the image of the planet fell directly onto a small flaw in the photographic plate, as discovered later. Faithful astronomers continued the search for 25 years after prediction, but in vain.

American astronomer Percival Lowell (1855–1916) refined calculations. To increase the reliability of observations, he launched a brand-new observatory in Arizona, where the air is most stable. At one time, Lowell thought he observed canals in Mars, an illusion. Despite his heroic computational and observational efforts, he could not find Planet X. His successor Clyde Tombaugh meticulously continued the search at the Lowell observatory. Finally, in 1930, he had a new idea: to photograph the same part of the sky on two different nights, one week apart. Superimposing the two photographs, the new planet finally showed up in the expected region. Between successive nights it had moved a minuscule amount relative to the fixed stars. Farthest from the sun, it earned the name "Pluto" for god of darkness, as suggested by an 11-year-old school-girl from Oxford who was fascinated by both mythology and astronomy. Pluto's symbol is formed from the letters PL, the same as Percival Lowell's initials, a coincidental, though fitting, tribute to the persistent scientist.

Uranus, Neptune, and Pluto increased the number of planets from antiquity's five to nine (if we include the earth as a planet). The new planets quadrupled the extent of the solar system. Figure 5.26 compares orbit sizes of planets known since antiquity to the orbit sizes of the new planets. Even though Jupiter and Saturn are much farther away from earth than Mercury, Venus, and Mars, the ancients could see them clearly because these remote planets are much larger in diameter. Figure 5.27[40] shows modern views of all nine planetary worlds of the solar system from powerful telescopes and space telescopes.

PLUTO BANISHED

Pluto has five known moons. The largest, Charon, is nearly as large Pluto, so the Pluto and Charon are really twin "planets." Which raises one of the many questions about calling Pluto a planet. If all the bodies orbiting the sun are to be called planets, what about the status of asteroids like Ceres, along with Vesta, Juno, Pallas, and a million other large rocks? Since asteroids populate a belt between Mars and Jupiter, asteroids can be excluded from planet classification.

In the region of Pluto, other small bodies – Eris, MakeMake, Haumea, Sedna, Orcus, Quakoar, and Varuna – came into the picture. Eris is actually bigger than Pluto. If Pluto is classified as a planet then all these new bodies are also eligible, increasing the total number of planets to more than 20, and still counting, a cumbersome situation.

FIGURE 5.27   *Modern Views Of Nine Planetary Worlds.*

After much controversy and debate the International Astronomic Union finally adopted three standard criteria for planet definition. Even though Pluto is in orbit around the sun and has a nearly spherical shape, it fails the last of the new criteria established for qualifying a planet; Pluto has not cleared its orbital zone of other bodies, and shares its orbital region with other bodies of similar size. It even crosses the orbit of Neptune. In 2006, Pluto was therefore re-classified as a dwarf planet along with Eris, Sedna, and others, as planet spotters began to demand that new bodies be adopted as planets. Many are still unhappy about losing Pluto since they learnt early in school about the planetary system.

6 Spherical Models: Scaling the Cosmos

All the world over and at all times there have been practical men absorbed in 'irreducible and stubborn facts': all the world over and at all times there have been men of philosophic temperament who have been absorbed in the weaving of general principles. It is this union of passionate interest in the detailed facts with equal devotion to abstract generalization which forms the novelty in our present society.

Alfred North Whitehead[1]

OUT OF THE SHADOWS

As with the study of motion, as with the composition of matter, Greeks were the first to postulate the existence of rational organizing principles behind the celestial drama taking place in the night sky. They searched for a system in the paths of the sun, moon, planets, and stars. They asked questions about celestial motion similar to their queries about the substance of the universe and the reasons for its diversity. Is there some unifying principle to the various motions of heavenly bodies? Is there a grand order that organizes the universe? In seeking answers, they were guided by fundamental aesthetic principles. Nature is beautiful; explanations for the motions of heavenly bodies must be elegant. They searched for order and symmetry in nature's operations. Speculation, observation, reason, and insight all played crucial roles in the advance from gathering knowledge to genuine understanding. Living in a culture that accentuated geometry in art and architecture, they used geometry to synthesize astronomical lore gained from Egypt and Babylon. Their models brought order into prevailing confusion and uncertainty. With the emergence of a unifying picture, scattered lists of antique observations slowly began to turn into a coherent whole. Unrelated phenomena dissolved into geometric constructions, with increasing complexity. Further still, it became possible to quantify the universe for the first time in history. Ever since their pioneering initiatives, mathematical models have dominated the evolution of physics.

Greeks drew upon centuries of observations and knowledge accumulated in older civilizations. Thales taught sailors how to steer their ships at night using the constellation Little Bear, and compiled a detailed astronomy manual for mariners. But philosophers wished to go much further. Dismissing folklore widespread in Egypt as well as at home in Greece, they wished to understand celestial phenomena using natural explanations. Rather than think of the sun and moon as gods transported in chariots, they viewed them as material bodies moving on regular courses around earth. One recognized that moonlight is merely a natural reflection of glorious sunlight. Another viewed the night sky as a dome-shaped vault studded with stars. Stars were fixed to a rotating canopy, so they stay in exactly the same position relative to each other, forming never-changing patterns. He explained the nightly motion of constellations by postulating that the vast hemispherical canopy spins clockwise from east to west, returning to the same position every 24 hours. Regular changes in the moon's appearance, together with the faithful progression of the sun, and the nightly repetition of stellar motion inspired Greeks to seek a rational design governing the universe. They also tried to extend natural explanations to bewildering phenomena, such as eclipses, comets, and the Milky Way.

MUSIC OF THE SPHERES

A major scientific advance came with the first attempt to formulate an overarching model for heavenly motions. The Greeks called their models *simulacra*. We use the word "simulation" with the same meaning. Advancing beyond early thinkers, Pythagoras formulated the first mental picture of the earth in relation to the heavens. Why should the sky be only half a spinning sphere? It was far more beautiful to think of the heavens as a complete sphere. Symmetry commanded the imagination. As abundantly obvious from their art and architecture, Greek culture valued symmetry. It was appealing to imagine a spherical earth freely poised within a spherical cosmos without any gods, or water, to support it. With earth at its immobile centre, the whirling, stellar sphere became the outer boundary of the Greek universe (Figure 6.1). Over time, the spinning, celestial sphere grew into a powerful model embracing many previously disconnected facts into one coherent picture. For instance, the celestial sphere must spin about an axis which passes through a point near Polaris. That is why this one star, the North Star, remains motionless, while the others move around it, slowly sweeping celestial arcs during the night. Facing east, stars appear to rise; facing west, they appear to set with the rotation of the stellar sphere (Figure 6.2). Characteristic features developed for the sphere, such as a celestial equator parallel to the earth's equator, as well as north and south celestial poles aligned with earth's poles. Plausible and convincing, the celestial sphere dominated cosmology for 2,000 years.

In a symmetrical system, heavenly bodies must follow perfectly circular paths. Therefore the sun and moon make complete circles around the earth. But the strange, irregular motion of the seven heavenly bodies intrigued Pythagoras. If the celestial sphere rotates *clockwise* (so stars move from east to west) through a single night, why do the sun, moon, and planets move *counter-clockwise* through the constellations of the zodiac over many nights? Even more puzzling, planets make occasional bizarre turns. After first appearing at dusk as the evening star, Aphrodite (Venus) moves away from the sun from night to night (Figure 6.3(a)). After a few weeks, she reverses direction and heads for the sun. She shows the same pattern of behaviour when she reappears as a morning star.

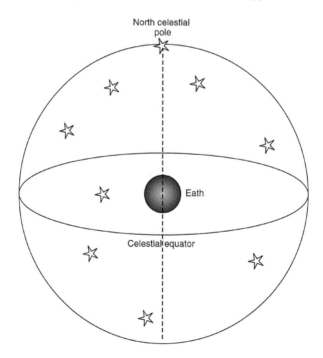

FIGURE 6.1   The celestial sphere is oriented so that the celestial equator is parallel to the earth's equator. It rotates from east to west, one turn every 24 hours, around an axis which points toward the North Star, Polaris.

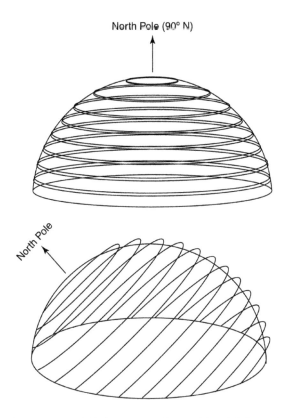

North Pole (90° N)

North Pole

FIGURE 6.2 In the rotating sphere model it is easy to understand the nightly motion of stars. (Top) from the North Pole on earth, it appears that all stars make circles around a stationary point. (Bottom) from latitude 45 N, the Pole Star is at an elevation of 45 degrees. Stars rising from the east set in the west, and make circles around a stationary point in the north.

Aphrodite is not alone in wayward behaviour. Examine the path of Ares (Mars) in Figure 6.3(b). Through most nights of its journey across the zodiac he follows the general counter-clockwise heavenly traffic flow of the sun, moon, and other planets. But then he goes through a short period of regression, travelling against heavenly traffic. Why do planets occasionally stop and turn around in their paths?

If the cosmos is to be simple, perfect, and ordered, there cannot be such blatant chaos. Indeed, the Greek word *chaos* is the antonym of *cosmos*. To banish chaos, Pythagoras submitted that a clockwise-rotating celestial sphere holds the constellations but there must be a separate counter-rotating celestial sphere for each of the seven heavenly bodies. Each sphere must rotate independently from the star-studded celestial orb (Figure 6.4).

Because of his love for simple number ratios, Pythagoras guessed there must also be a fundamental numerical relationship between the sizes of spheres holding the seven celestial lights (planets). Again, he was inspired by music. When the lengths of two strings were in a simple numerical ratio, such as 2:1, or 3:2, the combined chord was harmonious; the interval between the notes was an octave, or a fifth. From music, he expected it should somehow be possible to couch the relative sizes of circular planetary orbits in simple numerical ratios. Harmonic ratios must also apply to the time taken by various celestial bodies to complete their orbits of the earth so that the speeds of revolution of the seven spheres would bear harmonious proportions to each other.

To the mystical Pythagoras, the movement of the heavens was "the music of the spheres." Although he never succeeded in finding rational number ratios for the sizes of the spheres, he believed future astronomers would ultimately be able to discover harmonies of the universe by

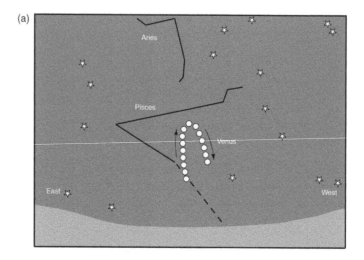

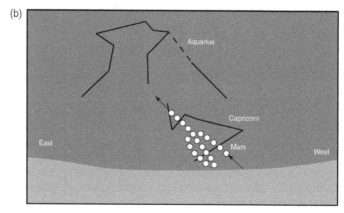

FIGURE 6.3   (a) Retrograde motion of Venus observed between January and March, 649 BC. (b) Retrograde motion of Mars showing the loop-the-loop path. Normally, Mars moves eastward from night to night, through the band of zodiac constellations. When it retrogresses, it moves westward. At this time, it also appears brighter than usual.

FIGURE 6.4   Pythagoras proposed seven independently rotating celestial orbs carrying seven heavenly bodies, in addition to the outermost celestial sphere. Earth is at the centre, then come the moon, Mercury, Venus, the sun, Jupiter, and finally Saturn.

contemplating the regular motion of the heavens. By applying mathematics to the cosmos, they would capture the cyclic rhythms as nature's underlying harmony.

Pythagoras' love for harmony persists through our language when we describe a beautiful symphony as "music of the spheres." Whimsical though it was, his desire to encompass natural phenomena into an elegant and harmonious mathematical framework continues to permeate scientific endeavour. His predilection for spherical symmetry and circular paths pervaded astronomical thinking for millennia to come. Even Copernicus, who overthrew the earth-centric scheme with a heliocentric model, continued to think of planetary motion in terms of perfect circles around a new central point, the sun. In *Arcades*, the blind poet Milton echoes the scientist's deep longing to "hear" celestial harmonies:[2]

> But else in deep of night when drowsiness
> Hath lockt up mortal sense, then listen I
> To the celestial Sirens harmony …
> Such sweet compulsion doth in music ly,
> To lull the daughters of necessity,
> And keep unsteady Nature to her law,
> And the low world in measur'd motion draw
> After the heavenly tune, which none can hear
> Of human mould with grosse unpurged ear.

Because they loved to challenge each other in intellectual debate, some of Pythagoras' followers, Philolaus (5th century BC) and later Heraclides of Pontus (ca. 390–310 BC), proposed that earth rotates one turn per day, not the stellar sphere, to give *the appearance* of stars whirling around the North Star. But the radical idea violated the naturally obvious fixity of earth. How absurd to imagine earth in motion! It has no reason to move. Being at the centre of the cosmos, it has no place to go. A moving earth only goes against the dictates of common sense. As another challenge, Heraclides proposed that Venus and Mercury orbit the sun, which orbits the earth. Subsequent Greek thinkers quickly abandoned such wild notions. But both ideas were destined to reappear. Most essential was the courage early Greeks showed for independent thought. Followers of Thales reflected the same trend by not immediately accepting the master's claim that water is the fundamental substance of the universe. Healthy competition between philosophers prevented one dominant mode of thought, a welcome contrast to the Egyptian temple-priests who, by closely guarding their methods and knowledge, pre-empted independent thinking. Free flow of ideas and debate in Greek culture created a wholesome climate for intellectual growth.

HORNS, TIARAS, AND HUNCHBACKS

Although records have been obliterated, making it impossible to trace properly the evolution of ideas, it is clear that by the 4th century BC one of the collaborative Greek successes was a rational explanation for the nightly changes in the moon's shape. The key is the relative location of the moon with respect to the sun as the moon makes its circuit around the earth in 28 days (Figure 6.5). Since the sun and moon traverse almost the same path through the zodiac, their circles lie in nearly the same plane. During a full moon, they arrive on diametrically opposite parts of the horizon. As Babylonians observed, the moon stands *in opposition* to the sun. So positioned, the sun can fully illuminate the moon. At dusk, the full moon rises while the sun sets on the opposite part of the sky. About two weeks later, the setting moon comes near to the setting sun. As it passes directly in front, the sun can only illuminate the spherical moon on the side that is far from earth. Since that lighted hemisphere is invisible to us, the moon appears dark. On the next night, the moon overtakes the sun. Being slightly ahead of the sun, we glimpse the first sliver of a crescent moon. Since the moon appears near the sun, the crescent moon sets soon after the sun at dusk. A day or two just before new

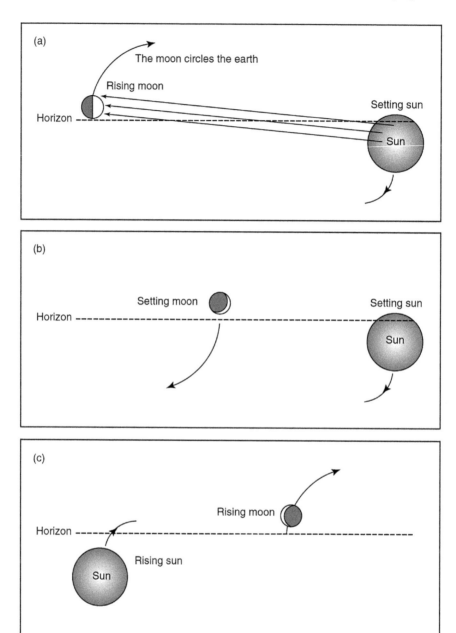

FIGURE 6.5 Correlating the phase of the moon with reflected light from the sun. (a) Full moon rises as the sun sets. (b) About two weeks later, at the beginning of the first quarter cycle, the moon appears crescent and sets soon after the sun. (c) About two weeks earlier, at the end of the last quarter of its cycle, a crescent moon rises before the rising sun at dawn.

moon, the moon also appears close to the sun; but it lags behind the sun. As a result we see a slice of the moon at dawn, just before sunrise. But the lighted sliver of the rising moon is on the east side as compared to the westerly crescent of the new moon. The juxtapositions of Figure 5.5 cover the first and second quarters of the lunar cycle. During the third and fourth quarters of the lunar cycle (not shown), the moon rises in daylight. A few hours before sunset, a *gibbous* moon (from Latin for hunchback) appears in late afternoon towards the east.

As the moon circles earth, it travels through the zodiac constellations. Moonrise progresses every day by $\frac{1}{28}$ of a day (equivalent to 50 minutes). Moonset is also nearly one hour later each day. Those who kept track of tidal rhythms saw important correlations between moon phase and high tides, as well as the fact that every day the high tide would shift by about one hour. Such correlations gave rise to speculations that the moon is somehow responsible for tides. Newton would be the first to figure out exactly how, through gravitation (Chapter 11).

NIGHT AT MIDDAY

Geometric explanations took hold among the educated populace of Greece. When his soldiers refused to fight during a lunar eclipse, Pericles (495–429 BC) tried to calm their fears by offering reason. He even demonstrated the principle of an eclipse by slowly bringing his cloak in front of the sun to show how a part seemed to disappear, then reappear, as the cloak passed. Pericles learned much from his philosopher-teachers Zeno and Anaxagoras, who turned him into a staunch rationalist. In their quest for a rational order, Greek thinkers understood there was nothing dire about eclipses. It was simple geometry. During a solar eclipse, as the moon passed in front of the sun, it obscured the sun, temporarily casting a shadow upon earth. Similarly, the earth cast its shadow upon the moon during a lunar eclipse.

Careful observers from antiquity had already noticed correlations between eclipses and lunar phases. But they had no rational explanation. The Greek geometrical model for the motion of the sun, moon, and earth successfully brought together seemingly unrelated facts about phases and eclipses. Eclipses can only take place during certain lunar phases. From the relative positions of the sun and moon in Figure 6.6, it is clear that a lunar eclipse will take place only during a full moon, whereas a total solar eclipse can occur only during a new moon. Another puzzling phenomenon: even though earth comes between the sun and moon every month, the earth does not cast an eclipse over every full moon, because the paths of the sun and moon around the earth are not exactly identical. We will return to this point for a fuller explanation. During a solar eclipse, the extent of the moon's shadow upon the earth covers only a very small area. Therefore, a total solar eclipse is only visible by people living in a small area of the earth's surface. On the other hand (see Figure 6.6), a lunar eclipse can be observed simultaneously from many places on the earth.

Rational explanations dispelled anxiety, but for the romantic minds they removed some of the mystery and wonder from natural processes. Thomas Campbell, a modern poet, regrets how scientific understanding deprives the creative imagination from dreaming up quixotic explanations:[3]

> When Science from Creation's face
> Enchantment's veil withdraws
> What lovely visions yield their place
> To cold Material Laws!

But our deeper sense of wonder at nature's operations hardly diminishes with deeper understanding! It continues to drive us to ponder and discover nature's secrets. Cosmic mysteries stir up scientific imagination and open astounding new perspectives, new symmetries, and fascinating mathematical patterns.

HEAVENLY TRUTH

Victorious in battle over the Persians, Athens became the leading cultural centre of Greece, spawning great philosophers. Herodotus described with optimism the triumph of Athens as the victory of freedom. Every field of intellect and artistic expression flourished in a Golden Age that followed. With Socrates, Plato, and Aristotle, Greek thought reached a pinnacle.

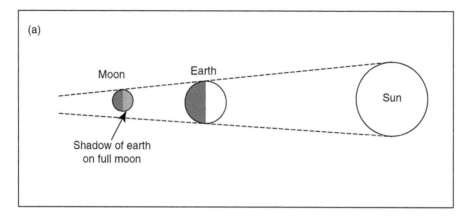

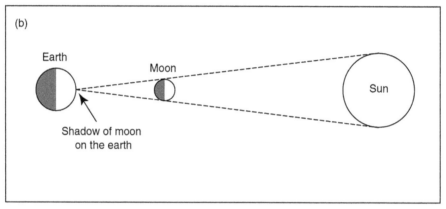

FIGURE 6.6   (a) A lunar eclipse takes place when the earth comes between the moon and the sun so that the shadow of the earth falls on the moon. The eclipse will occur only when the path of the moon crosses the ecliptic, i.e., when the moon crosses in the annual path of the sun. (b) A solar eclipse takes place when the moon comes between the sun and the earth. The moon blocks out the sun.

Plato's idealism and love for mathematics extended to astronomy. He subscribed to Pythagoras' belief that distances to heavenly bodies must be related in a simple numerical way. All that remained for future astronomers to complete the heavenly model was to find a scheme to cast the irregular motion of the planets, and their inexplicable twirls, in terms of the individual rotation of eight spheres. Planets are not really wanderers; they move on well-defined paths according to the natural elegance of the cosmos. They only *appear* to wander and retrogress. Plato wrote eloquently about the heavens:[4]

> Had we never seen the stars, and the sun, and the heaven, none of the words which we have spoken about the universe would ever have been uttered. But now the sight of day and night, the month and the revolutions of the years, have created number, and have given us a conception of time, and the power of inquiring about the nature of the universe; and from this source we have derived philosophy, that which no greater good ever was or will be given by the gods to mortal man.

But he never ascribed much importance to gazing at stars. He worried that limited sensory faculties gave misleading information about the changeable world. He believed that only pure reason was capable of penetrating the ideal principles of nature's operations, that the heavens were perfect, and heavenly bodies moved in perfect circles. A simple and elegant principle embodied the truth about the motion of the sun, moon, stars, and planets. Plato posited that our senses trick us with illusions, such as irregular motion and varying brightness. A proper geometric model would ultimately transcend appearances.

A NEW INCLINATION

Eudoxus was Plato's devoted disciple at the famous Academy in Athens. But he was not a member of the elite. Poor and unable to afford housing in Athens, he walked five miles every day from the seaport of Piraeus to the Academy. After he established his reputation as a competent geometer, he migrated to Egypt to build an astronomical observatory on the banks of the Nile, where he made new observations. When he returned for a visit to the Academy, Plato arranged an elaborate celebration in his honour.

Not satisfied with the sweeping generalizations of Pythagoras and Plato, Eudoxus wanted to know details about the heavens. What causes the seasons? Why do we have eclipse cycles? For the sun, it was clear that the existence of one celestial sphere would be insufficient to explain both the daily and annual motions. Instead of viewing the sun's tilted arcs relative to the local horizon (Figure 6.7), and the Pole Star at the local angle of elevation (e.g., 38 degrees in Athens), it was more revealing to view the sun's path inside the celestial sphere surrounding the spherical earth, with the celestial pole directly above the north pole of the earth (Figure 6.8).

A potent geometric scheme, this model of celestial spheres unifies several disparate observations. Every 24 hours, the sun goes around the earth from east to west along a circle parallel to the equator to give day and night on earth. Since the celestial sphere, which carries the constellations, also rotates in the same direction and once every 24 hours, the sun remains synchronized in the same zodiac

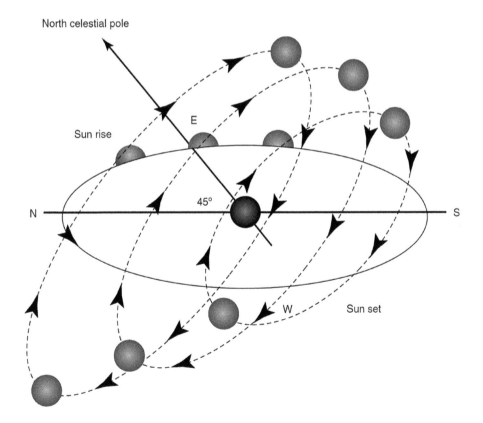

FIGURE 6.7   The daily path of the sun at different times of the year as viewed from earth at a location at latitude 35 degrees N. Daily, the sun moves from east to west. A re-orientation in the next figure results in a simpler picture.

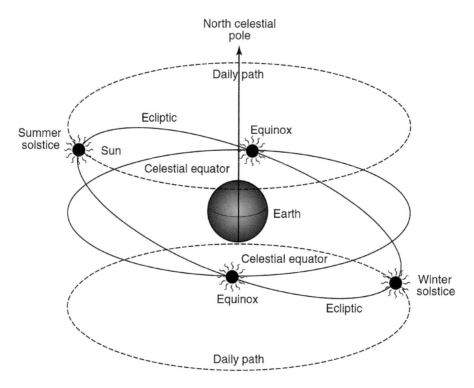

FIGURE 6.8 Explanation of the seasons from the inclination of the sun's annual path. Daily, the sun moves clockwise (east to west) and parallel to the equators. Through the course of the year, however, the daily circles shift uniformly, remaining parallel to the equators, as the sun follows the ecliptic path. The ecliptic is inclined with respect to the equator by an angle of about 23.5 degrees. Intersection points of the ecliptic and celestial equator are called equinoxes.

constellation between sunrise and sunset. Besides daily motion, the sun also has an annual motion, along the "ecliptic" path, which lies in the zodiac band. As the year progresses, the sun therefore appears to migrate, month to month, from one zodiac constellation to the next, passing through all 12.

Since the ecliptic path appears tilted relative to the earth's equator, it is also tilted relative to the celestial equator, the two equators being parallel. (The tilt angle is 23 degrees and 27 minutes as succeeding astronomers determined.) The inclination is crucial; it explains the seasons. Our word "climate" comes from the Greek word *clima*, meaning "inclination," referring to the inclination of the sun's path.

When the sun reaches the celestial equator, it shines symmetrically over northern and southern hemispheres. Days and nights are equal in length (equinoxes), and the climate is temperate in both hemispheres. There are two such positions, one corresponding to spring, and the other to autumn – hence two equinoxes. When the sun reaches the extreme solstice position of its annual path, its proximity to the northern hemisphere causes summer there, and winter in the south. When the sun moves to the other extreme, the winter solstice position, it is winter in the north, and summer in the southern half. Of course, the weather does not turn hot exactly on the day of the summer solstice. It takes about one month for the earth and its atmosphere to warm up. Even though the northern summer solstice is in June, July and August are the hottest months in the Northern Hemisphere

CELESTIAL SPHERES MULTIPLY

With the scheme of multiple spheres for each heavenly body, Eudoxus modelled the nightly and monthly motions of the moon with three spheres to make the moon's path through the zodiac tilted by an angle of 5 degrees with respect to the sun's path. The deviation of the lunar circuit explained why ancient "lunstice" observers found the extreme positions of the moon on the horizon to migrate 5 degrees farther north and south compared to extreme positions (solstices) of the sun.

Details are important for the progress of science! The 5-degree tilt explains why we see a fully lit moon, even though the earth comes directly between the sun and moon once a month. If the moon moved precisely on the ecliptic, sharing the exact same path with the sun, the earth would eclipse the moon every month at the full-moon position. The slight inclination between solar and lunar paths also explains eclipse cycles. An eclipse occurs when the path of the moon intersects the path of the sun, which is why we call the path of the sun the *ecliptic*. The moon's path cuts the ecliptic at two opposite points in the sky, called *nodes*. A lunar eclipse occurs only when the moon is full *and* when it is exactly on the ecliptic circle, i.e., when the moon is at a node. Similarly, a total solar eclipse occurs only when the moon is new *and* at one of the nodes.

But the moon's orbit is even more complicated, which is why Eudoxus needed three spheres. The angle of tilt slowly varies between the extremes of plus 5 degrees and minus 5 degrees about the ecliptic, over a period of 18 years. Hence the nodes move around, and eclipses occur in a complicated pattern with a super-period of 18 years, as Babylonians determined from careful records of eclipse tables (Figure 6.9).

Eudoxus' approach reflected an emerging cultural trend: realism through emphasis on detailed observation. Ideal principles aside, what are the real movements of the heavens? Appealing to an audience wider than just the elite, Praxiteles, a dominant artist of the time, created sculptures of planetary gods in settings more typical for ordinary people than for deities. In preparing for a bath, Aphrodite drapes her clothes over a water vase (Figure 6.10). Breathing life into marble, the artist is interested in hair curls, anatomical features, and articulation of muscles. The goddess stands in a relaxed S-shaped pose, a spontaneous stance that contrasts with the rigid classical pose. This more realistic pose shows that the human body is flexible and moves by continuously shifting weight from one supporting leg to another. Praxiteles was a penetrating observer. Legend has it that his nude Aphrodite was so realistic that the goddess herself exclaimed upon seeing the statue:[5] "Where could Praxiteles have seen me naked?"

The more Greek astronomers studied the skies, the more details they learned about the real motion of the seven heavenly bodies. They discovered regularities in the brightness variations. Mars usually appeared brightest when it was in opposition to the sun. At those times, its brightness rivalled even that of Jupiter. When it appeared in conjunction with the sun, it paled in comparison. Planetary regressions were also periodic. Mars, Jupiter, and Saturn only retrogressed when located in opposition to the sun. They also appeared brightest at these times. Such real features of planetary

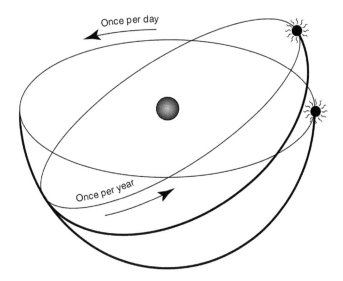

FIGURE 6.9 Eudoxus' two spheres to account for the motion of the sun.

FIGURE 6.10   *Aphrodite*, Praxiteles, copy of 4th-century BC original.[6]

motion demanded serious refinements to the spherical model of eight perfect spheres. Eudoxus was able to construct a loop-the-loop motion to simulate periodic retrogression (Figure 6.11). With multiple spheres for each planet, Eudoxus assembled an elaborate system of 27 concentric spheres. As each rotated at its own tilt and speed, heavenly bodies appeared to move in bewildering fashions.

Eudoxus' planetary model showed improved agreement for the motions of the sun, the moon, Jupiter, and Saturn, but failed to account satisfactorily for the known motions of Mars and Venus. There was a devil in the details, or was it the hand of God? The struggle to penetrate the secrets of nature had intensified. Reflecting upon the time, Milton wondered in *Paradise Lost* whether humans would ever be up to the daunting task:[7]

> From man or angel the great Architect
> Did wisely to conceal, and not divulge,
> His secret to be scanned by them who ought
> Rather admire; or, if they list to try
> Conjecture, he his fabric of the Heavens
> Hath left to their disputes, perhaps to move
> His laughter at the quaint opinions wide.

CRYSTALLINE SPHERES

Greek sailors plied their ships between what we now call latitude 45 degrees N to 22 degrees N, a region large enough to observe changes in visible constellations. For example, as they travelled north, the Bear refused to take a bath in the ocean, as Homer described. The angle between Polaris

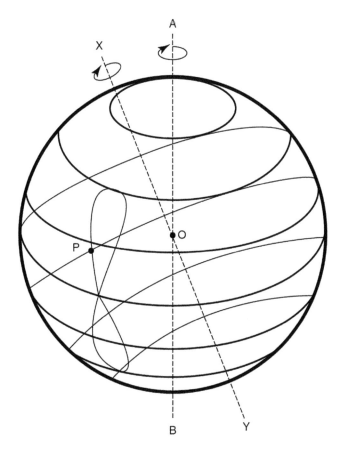

FIGURE 6.11   Simulation of retrograde motion using the simultaneous rotation of two concentric spheres, one tilted. In a simple version of his scheme, there are two concentric rotating spheres, but the rotation axis of one sphere (delineated by light lines) is tilted relative to the axis of the other sphere (heavy lines). As the latter sphere rotates, it carries the axis of the former sphere with it. A point P on the equator of the thin sphere traces a figure-eight curve, called a *hippopede*, from the word for horse-fetter.

and the horizon increased (Figure 6.12). Travellers to the south spotted new stars. Bright Canopus, invisible in northern Greece, was clearly visible in southern islands and climbed higher in Egypt. Aristotle understood that such changes in the firmament were only possible on a spherical earth. A simple application of spherical geometry shows that the angle to the Pole Star increases from 0 degrees at the equator to 90 degrees at the Pole. The altitude of the Pole Star is identical to the latitude of that location on earth.

Dissatisfied with a purely mathematical description of heavenly motion by combinations of perfect spheres, Aristotle introduced a new concept into Eudoxus' model. He wished to view the universe as a connected, interrelated system, akin to a functioning organism. After all, biology was Aristotle's strong suit. If the sky rotates, it must be a solid sphere. Spheres that hold planets are not pure geometrical constructs, but concrete. They rotate because they are solid, fashioned from heavenly ether, a weightless, transparent, absolutely pure crystalline material. Nothing could pass between them. By linking heavenly spheres together with additional counter-rotating spheres, Aristotle doubled the number of spheres proposed by Eudoxus from 27 to 54. From Pythagoras' eight to Eudoxus' 27 to Aristotle's 54, the complexity of the heavens steadily increased in ungainly proportions.

For more difficult features, such as the Milky Way and randomly appearing comets, Aristotle devised rational explanations. The Milky Way arises from the eternal motion of the stellar sphere.

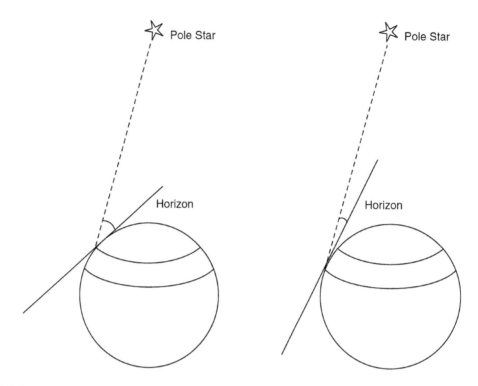

FIGURE 6.12   Because the earth is round, the angle between the Pole Star and the horizon decreases as one travels south toward the equator.

A pale glow appears because small accumulations of ether caused by the motion of the celestial sphere ignite within the band, giving the rarefied, wispy quality. Hairy-tailed objects that sweep across the night sky on rare occasions cannot possibly belong to the heavens, claimed Aristotle after examining the evidence. Any true celestial event, such as a lunar eclipse, always occurs regularly with predictability. But comets appear at random. Heavenly bodies never change their appearance. But comet tails change in length and brightness as they cross the sky. If comets belong to the heavens they would either move along the heavenly highway, or be permanently in the same place as the firmament of constellations. But cometary paths often lie *outside* the region of the zodiac, unlike the planets, which always keep to the celestial highway as do the sun and moon. Like clouds, rainbows, thunder, and lightning, Aristotle classified comets as phenomena occurring in the upper air and dealt with them in his work *Meteorologica*. He even proposed rational explanations for comet formation and the catastrophes they inflict upon the earth. The fear that comets bring evil was commonplace since antiquity. In the upper parts of the atmosphere, where the shell of air and the shell of fire coincide, air and fire can mix, thought Aristotle. A comet appears when the mixture spontaneously ignites. Because comets are mixtures of air and fire, bad weather and pestilence usually follow when comets gasp hot, dry air, making conditions ripe for disease and epidemics.

Although many of the presumptions and models of Aristotle and his predecessors ultimately fell out of favour, their lasting contribution was their approach of using rational methods which led to an overarching conceptual scheme to explain both regular and irregular motions. Astronomers who followed attempted to fit new observations into Aristotle's grand framework, or to account for deviations by introducing new levels of refinement into his comprehensive scheme. A crucial missing component was scale, an estimate for distances to heavenly bodies. Here Alexandrian astronomers initiated a major advance. The first was Eratosthenes, who put a scale to the size of earth.

SIMILAR TRIANGLES IN THE HEAVENS

With its declining political and economic power, the centre of Greek culture shifted from Athens to Alexandria, where the Ptolemies maintained Greek civilization at a peak. The Museum emerged as a centre of vigorous research through this immensely fertile epoch. Archimedes, the mechanical genius, built an impressive planetarium using an elaborate system of gears to display the motion of the sun, moon, and five planets. The technical simulation was quite an engineering feat. Aristylles and Timocharis recorded the position of the brightest stars. Aristarchus, Apollonius, and Hipparchus were innovative thinkers who tried to fathom cosmic mysteries, generating new patterns in scientific thought. Capping the pinnacle of Alexandrian astronomy, Ptolemy synthesized the prolific efforts of the pioneer astronomers.

With the conquest of Babylon by Alexander, Greeks gained direct access to a vast store of Babylonian and Chaldean astronomical knowledge. No doubt they imbibed the growing astrological lore. Comparison with Babylonian records revealed many inadequacies in Eudoxus' spherical model. Around 260 BC (see timeline, Figure 6.13), a radical scholar from the island of Samos visited the research institute at Alexandria after studying at the Lyceum in Athens. Recall that Pythagoras was also from this intellectually fecund region standing at the crossroads between Asia and Europe. Now Samos was ready to release another great mind to the world of science.

Aristarchus (310–230 BC) interpreted the irregular motion of planets with a much simpler and graceful design than Eudoxus' 27 concentric spheres, or Aristotle's cumbersome scheme of 54 interlinked spheres. He proposed that all planets revolve around the sun rather than circling the earth (Figure 6.14). Even the earth revolves around the sun, like a planet. Only the moon revolves around the earth, judging from the regular occurrences of solar and lunar eclipses.

Reviving early ideas of Philolalus, Aristarchus claimed that the earth also rotates about its own axis that runs through Polaris, thus accounting for the sun's apparent daily motion as well as the

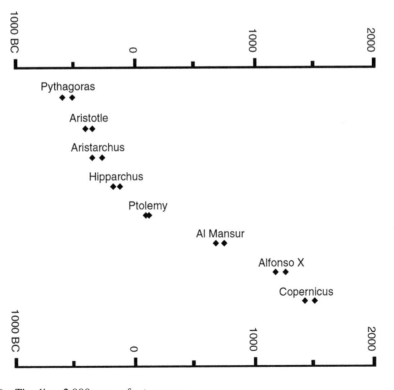

FIGURE 6.13   Timeline: 2,000 years of astronomy.

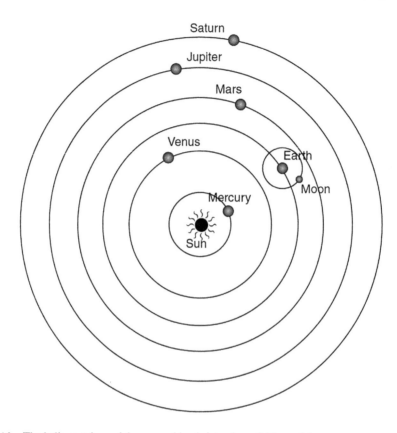

FIGURE 6.14   The heliocentric model proposed by Aristarchus of Alexandria.

nightly progression of stars. The sun is stationary, and the constellations remain in fixed positions. Relative to prevailing wisdom, Aristarchus' ideas were daring contradictions. They defied the sages who educated him in Athens. What led Aristarchus to such radical suggestions? Aristarchus' works are lost to us forever, probably consumed in book-burnings at Alexandria. We only know his ideas about the heliocentric theory through accounts from his younger contemporary, Archimedes. As a result, we are not sure how he developed the heliocentric model nor how far carried it. But we do know from one surviving treatise, *On the Sizes and Distances of the Sun and the Moon*, that he was a great geometer who made some surprising discoveries. Applying similar triangles to the heavens revealed that the sun is 19 times larger in size than the moon. But the sun is also 19 times farther from earth than the moon was. Our senses trick us into thinking that the sun and moon are of comparable diameter. More astounding was the discovery that the sun was about seven times larger than earth. In the light of these startling discoveries, it probably appeared unlikely to Aristarchus that a giant sphere like the sun should be circling a paltry earth. In the grand design of the universe, why would a Titan pay homage to a dwarf? It was more likely to be the other way around.

   With seminal application of geometry to astronomical observations, Aristarchus went beyond comparisons between the sizes of the sun, moon, and earth; he also estimated the relative distances between them. As a first step, he argued that the finite distance to the sun must lead to a slight asymmetry in the cycle of the lunar phases (Figure 6.15). From the asymmetry he estimated the distance to the sun. It was a remarkable feat, involving careful observation of astronomical events, insightful reasoning, and mathematical analysis, all the makings of a modern scientific venture. It merits close examination.

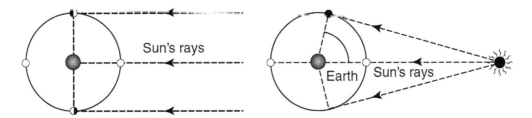

FIGURE 6.15   (Left) if the sun were infinitely far away, its rays to the moon and to the earth would be parallel to each other. Both the sun–moon–earth angle and the sun–earth–moon angle would be 90 degrees. In such a symmetric arrangement, new moon, first-quarter moon, full moon, and last-quarter moons would all fall on equally spaced locations around the moon's circular orbit. (Right) but the sun is a *finite* distance away. Therefore, the time between new moon and first quarter is slightly shorter than the time between first-quarter moon and full moon. Once again, a slight asymmetry opened a new door. The actual time difference is about 40 minutes out of 28 days. Without an accurate clock, Aristarchus estimated 11 hours out of 28 days which led him to a sun–earth–moon angle of 87 degrees.

On the first quarter of the moon's cycle beginning with a new moon, the sun illuminates exactly half the moon. Therefore, the sun's rays must arrive perpendicular to the lunar diameter, forming a precise right angle to the earth–moon line (Figure 6.16(a)). If he could determine another angle of the right triangle formed by the sun, moon, and earth, Aristarchus could solve all the interesting distances by similar triangle analysis. Aristarchus estimated 87 degrees for the sun–earth–moon angle. In a right triangle, where one angle is 87 degrees, the ratio of the short adjacent side to long hypotenuse is 1:19. Thus Aristarchus concluded that the sun is 19 times farther away from earth than the moon.

Advancing to a greater challenge, Aristarchus determined the size of the sun relative to the moon by applying geometrical analysis to the situation of a total solar eclipse, when the disc of the moon almost covers the disc of the sun (Figure 6.16(b)). By similar triangle analysis, the ratio of the sizes is equal to the ratio of the distances. Therefore the sun is 19 times larger than the moon. It appears to be nearly the same size as the moon because it is also that much farther away from us than the moon is. Our moon–sun combination creates a spectacular coincidence of size and relative distance to cause a total eclipse. Like Eratosthenes and Archimedes, Aristarchus was paying close attention to shadow details and scales.

Proceeding further and with greater geometrical dexterity, he used additional observations made during a lunar eclipse to arrive finally at the size of the sun relative to earth. When the earth comes between the sun and the moon, the moon enters the earth's shadow, taking a certain time to become eclipsed (Figure 6.16(c)). Then it stays in the earth's shadow for nearly the same time as the first time interval. Aristarchus concluded that the width of the earth's shadow must be twice the diameter of the moon. By two consecutive applications of similar triangle geometry (not shown), he concluded that the diameter of the sun is about seven times that of the earth.

Modern measurements show that the angle Aristarchus measured is not 87 degrees, but 89 degrees and 50 minutes. It is so close to 90 degrees that only precise instruments can measure the slight difference from 90 degrees. Therefore we know that the sun is 395 times farther away than the moon. Applying the same methods, modern measurements also show that the sun's diameter is 109 times larger than that of the earth.

Errors in Aristarchus' results need not concern us. Inaccuracies are common at the cutting edge of science. Even though his measurements and final results were off, his geometric approach was logically sound. Breaking rich, new ground, Aristarchus took geometrical methodology one notch further than Eratosthenes, providing for the first time a relative distance scale for the heavens. Following Eratosthenes' breakthrough, Aristarchus applied careful observation and precise

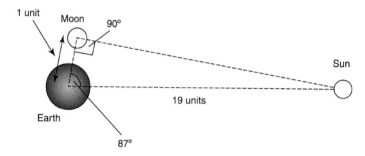

1 unit

Moon

90°

Sun

19 units

Earth

87°

(a) Earth-moon and earth-sun distances

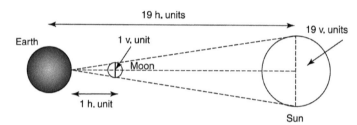

19 h. units

19 v. units

Earth

1 v. unit

Moon

1 h. unit

Sun

(b) Eclipse of the sun

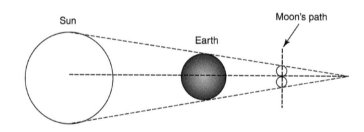

Sun

Moon's path

Earth

(c) Eclipse of the moon

FIGURE 6.16   Aristarchus applies triangle geometry to the heavens (a) during the night, when exactly half of the moon is illuminated, (b) during the total eclipse of the sun, and (c) during the eclipse of the moon.

measurement to the heavens. Instead of shadows of sticks on earth, Aristarchus used shadows of the earth on the moon. His aim and methods fully warrant 19th-century poet Browning's exhortation:[8]

> Ah, but a man's reach should exceed his grasp,
> Or what is heaven for?

Aristarchus' contemporaries rejected the heliocentric view as irreverent speculation, and indicted him on a charge of impiety for reducing earth to the lowly status of another wanderer among stars. How could earth be insignificant compared to the sun? Centuries earlier, Athenians chastised Anaxagoras for claiming the sun to be a flaming rock, far larger than a Greek island. More disturbing was the idea that the earth is not at the centre of the heavenly drama. A moving earth contradicted common sense and everyday observations. If earth moves, Olympic jumpers would land in the spectator stands. Star patterns would change over the course of half a year due to parallax, an important concept that played a central role through the development of astronomy

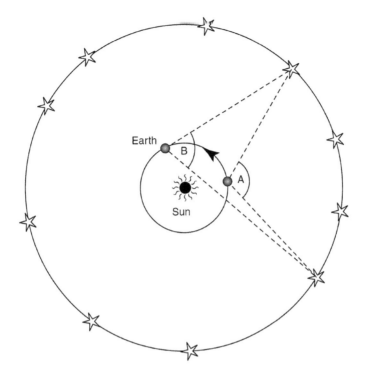

FIGURE 6.17   Annual parallax of stars due to the motion of the earth around the sun. The angles (A and B) between two stars on the celestial sphere should appear different between summer and winter, if earth orbits the sun.

(see update section for this chapter). The word comes from *parallaxis*, meaning to take the value of an angle. Any new theory had to withstand cogent geometric arguments in a culture that valued geometry. During summer, when the earth is on one side of a solar orbit, the angle between two stars on the celestial sphere can be larger than the angle between those same stars when earth moves to the opposite side during winter (Figure 6.17). Science historian George Sarton[9] suggests that Aristarchus answered the challenge by proposing the celestial sphere to be infinitely large compared to the extent of the earth's orbit to account for the fact that there is no such observable parallax.

Both egocentric prejudice and technical objections from his opponents nipped Aristarchus' premature vision right in the bud. Perhaps he would have made a more compelling case had he used the heliocentric model to explain key irregularities in planetary motion, as for example, their varying brightness or retrograde motion. Eudoxus' scheme with the earth at the centre could never account for varying planetary brightness, since all planets always maintain the same distance from earth in the concentric arrangement. Aristarchus' bold proposition failed in technical aspects. The heliocentric view retired to obscurity, to be revived 1,500 years later by Copernicus.

THE PIROUETTE OF THE PLANETS

The quest for a scheme more satisfactory than Aristotle's 54 spheres did not disappear. Nor did the creativity of planetary astronomers. Apollonius of Perga (262–190 BC) invented clever variations to the earth-centric scheme to make more accurate predictions for planetary positions. To explain the change in planetary brightness, he displaced the centre of planetary circles from the centre of earth. He too was ready to give up the idea that the earth was at the exact centre of the universe. But this slight shift was hardly as unsettling as Aristarchus' radical overhaul of the earth from its privileged central position.

What about retrograde motion? Apollonius added a novel device. Each planet describes small circles, like a pirouetting ballet dancer, around a moving centre that traces a larger circle around the earth (Figure 6.18).

Epicycles did the trick. When compounded with the deferent, it could account for planetary retrograde motion in a simple way. In Figure 6.19, just follow the path of a planet along the epicycle, as the epicycle revolves around a large circle. Apollonius reduced the number of circles to 14, two for each of seven heavenly bodies, a major simplification over Eudoxus' scheme of 27 spheres, which later burgeoned to 54 under Aristotle. With the help of such devices, numerical predictions

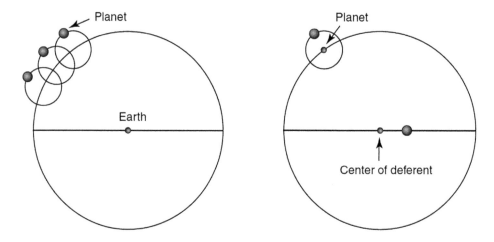

FIGURE 6.18 Instead of nested spheres, Apollonius and Hipparchus came up with epicycles and deferents to explain the motion of planets. (Left) the small circle is called the *epicycle* and the large, carrying circle the *deferent*. (Right) when the centre of the deferent is not the earth, the deferent circle is called an *eccentric*.

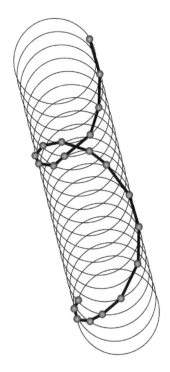

FIGURE 6.19 How the scheme of epicycles explained retrograde motion.

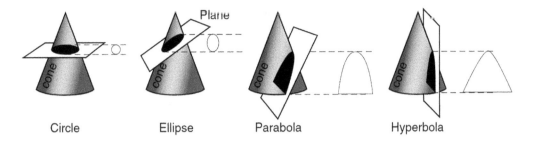

Circle   Ellipse   Parabola   Hyperbola

FIGURE 6.20   The conic sections of Apollonius extended Archimedes' parabola. If a plane cuts through a cone parallel to the base, one obtains a circle. A cut across the cone, but not parallel to either side, yields an ellipse. A parabola results from a cut made parallel to one side of the cone. A vertical cut gives a hyperbola.

of planetary position could better find their mark. Aristotle's crystalline spheres disappeared, but only temporarily. Scholastics of the Middle Ages, following in the footsteps of the Arabs, ardently revived the interlocking spherical architecture of the heavens.

Apollonius is best remembered not for his epicycles but for his breakthrough mathematical discovery, the conic sections. This is a special family of curves which emerges when a plane intersects a cone at various angles (Figure 6.20). The curves are the *circle, ellipse, parabola,* and *hyperbola.* Just as Euclid thoroughly investigated triangle and circle geometry, Apollonius explored mathematical properties of conic sections. What is truly astonishing is that while Apollonius invented the epicycle to preserve circular motion, he also provided the mathematical key to his successor of a thousand years. Johannes Kepler (Chapter 8) finally deduced that the true planetary path is an ellipse. Even more remarkable was the subsequent development by Newton, who rigorously proved that the correct path of bodies orbiting the sun could be *any one* of the conic sections. Apollonius' physics of heavenly motion is dead, but his mathematics lives on in the heavens.

Time and again, developments of mathematical ideas long precede the recognition of corresponding mathematical patterns in nature. Pythagoras recognized the sum of n odd numbers is n^2, which played a pivotal role in Galileo's analysis of vertical free-fall motion. Archimedes' parabola also influenced Galileo's analysis of projectile motion (Chapter 4).

THE AGE OF AQUARIUS

The search for an accurate scheme continued with renewed vigour under Hipparchus (190–120 BC). Making good use of Apollonius' devices, he further reconciled prevailing discrepancies between observed and predicted planetary motion. More importantly, Hipparchus spent a good part of his life gathering accurate data for heavenly motions to become the most astute sky-watcher since the beginning of positional astronomy. To improve observations, he devised new instruments. To interpret volumes of data, he adopted and advanced trigonometry, which he learned from Babylonians and Hindus. It was a systematic discipline for computing angles and distances in general triangles.

As the history of science repeatedly shows, those engaged in making careful observations are likely to reap rewards of important discoveries. Hipparchus' lifelong efforts are truly exemplar. Alert Hipparchus was early to earn his prize. During one of his careful night sky surveys, he thought he observed a bright new star in Scorpio. How could the already perfect and unchanging heavens exhibit a new star? Had ancient sky watchers missed some? Star records at the library of Alexandria, where librarians diligently copied and compiled books from all over the world, showed no such bright star in that position. Perhaps it was a matter of poor records. To prevent such an uncertain situation from recurring, Hipparchus decided to create his own detailed and accurate star map. By establishing a proper star catalogue he wished to provide all posterity with a complete record of the heavens. Future astronomers could keep track of whether new stars appeared, or moved in their constellations over long periods, or whether any stars ever faded out. (Later astronomical discoveries showed that all of such changes do indeed occur, as we shall see.) The principal idea of a star map was not new; Babylonians had recorded zodiac constellations with great care to keep track of the seven heavenly bodies. A century before Hipparchus, two Alexandrian astronomers, Aristylles and Timocharis, recorded the positions of the brightest stars in the firmament. But Hipparchus had in mind a grander venture; to list *all* stars visible to the naked eye, not with respect to arbitrary constellation patterns but with a consistent scheme of latitude and longitude coordinates on the celestial sphere. And all the stars are not the same; they show different brightness. From a special observatory he arranged to be erected on the Greek island of Rhodes, he classified the stars according to faintness, putting them into six categories, a classification system still in use today. In this scheme, fainter stars earn larger numbers. (When called the brightness scale, as is often the case, the numbering scheme becomes counter-intuitive, of course.) For example, bright Vega is magnitude 0 and dimmer Polaris is +2. The faintest star Hipparchus could discern had a magnitude of +6. In the modern use of his scheme, a very bright object gets a negative rating; Sirius, the brightest star earns –1. Dominating the night sky, the moon has brightness of –13, and the sun is a blinding –28. Each magnitude change means the apparent luminosity changes by a factor of 2.5 (strictly the fifth root of 100). A magnitude 1 star is 100 times brighter than a magnitude 6 star.

It must have been a solemn and dreary project to spend night after night over a good portion of his life, carefully mapping the position of 850 stars. But Hipparchus wished to observe and describe the heavens in live, infinitesimal detail. Some day in the future, the excruciating details would become important to the big picture.

The ambitious sky-mapping project stimulated new advances in positional astronomy. Instruments and methods Hipparchus devised were to serve astronomy for posterity. He used a *quadrant* (Figure 6.21) to find elevation and azimuth (angle on the horizon), a *sextant* for measuring the angular distance between stars, a gnomon and clepsydra (water clock) to keep track of time, and a 15-foot diameter (armillary) sphere to note down the positions of heavenly bodies. Through detailed and precise observations made possible by these devices, he defined star positions to an

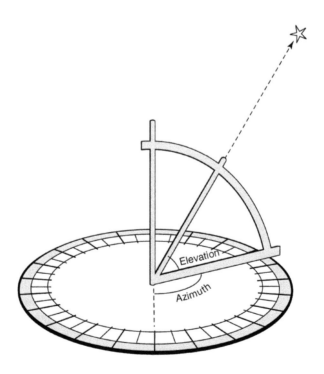

FIGURE 6.21   A quadrant determines the elevation and azimuth of a star.

accuracy of *10 arc-minute angles*, setting a new standard for accurate measurements. To appreciate that level of precision, try viewing a finger-breadth from 15 feet distance.

As always through the development of science, precision opened a new door. Armed with the best techniques to spot the smallest of discrepancies, Hipparchus discovered a brand-new feature of heavenly motion. The ecliptic, that unwavering reference path of the sun, is not really stable. It drifts, albeit exceedingly slowly. Comparing his star map to records made centuries ago by Timocharis and Aristylles, as well as with earlier observations made in Babylon, Hipparchus found a small shift in the position of the rising sun on the auspicious morning of the spring equinox observed and celebrated by many cultures for marking the arrival of spring. Around 1000 BC, ancient observers reported the spring sun to rise near the constellation Aries (Figure 6.22). But Hipparchus found the spring sun located in Pisces (Figure 6.22(b)). The regular path of the sun through the zodiac constellations was changing slowly from century to century. To exaggerate the shifts, Figure 6.22(c) shows the locations of the sun even earlier, in 2000 BC, when the sun was well inside Aries. Around 4000 BC, Sumerians, who rigorously followed the path of the sun through the constellations, made this record on stone tablets, "The bull marks the beginning of spring." On the first day of spring, the sun rose in Taurus. That was *two constellations* away from the spring sunrise at the time of Hipparchus. Going back even further in time, to about 6000 BC, Egyptians retained an ancient tradition in which their paintings of the zodiac put Gemini at the centre of the picture presumably because they found sunrise in Gemini at the beginning of spring. Gemini is *three constellations* away from Pisces. Over time, this drift of the sun came to be called *precession of the equinoxes. The drift is about 10 arc-minutes over 12 years, large enough to be detectable over that time period with Hipparchus' precision.*

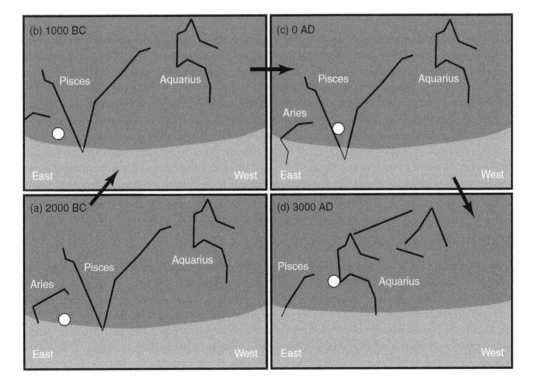

FIGURE 6.22 Changing location of the sunrise on spring equinox, clockwise from lower left: (a) and (b) sun rises in Aries around the year 2000 to 1000 BC. (c) Sun rises in the constellation Pisces during Hipparchus' time, about AD 0. (d) The spring sunrise in the year AD 3000 will move to Aquarius.

In present times, the sun appears in Pisces during spring equinox. Since Hipparchus' time, the spring equinox has moved but not enough for the sun to completely leave Pisces. In a few centuries, it will escape from Pisces and move toward Aquarius. Figure 6.22(d) shows the location of the sun well within Aquarius in the year AD 3000. Hence comes our popular song from the 1960s musical, *Hair*, representing the spirit of the new generation. It heralds the arrival of a new epoch with "the dawning of the age of Aquarius." Those still interested in astrology should take note that even though precession has moved the constellations, astrologers still fix the "sign" of a person according to the constellation that locates the sun on that person's birthday, but using the sun's location among the zodiac constellations as recorded by the Babylonians. Those upset by this revelation may take heart in the dubious consolation that all the signs will come back to realignment with the Babylonian scheme in 24,000 years!

Hipparchus' discovery of equinox precession was a remarkable payoff for the skilful precision of observations. But the mere ability to detect a discrepancy does not naturally lead to an appreciation of its significance. It took special insight to interpret volumes of data within the framework of the powerful model of the celestial sphere and the ecliptic path of the sun. Why does the sun not always return to exactly the same patch of stars in the zodiac? Another group of astronomical observations relates to the same slow shift. As we see it, the axis of the celestial sphere goes through the North Celestial Pole, near Polaris. But around 3000 BC, the rotation axis pointed toward a different star, Thurban, the brightest star in the tail of Draco. Thurban, not Polaris, was the North Star for ancient Egyptians. They constructed the oldest pyramid to point toward Thurban as the royal entrance to heaven. Just as the vernal equinox migrates among the constellations of the zodiac, the celestial pole migrates among a circle of constellations that girdle the northern sky: Draco, Ursa Minor, Cepheus, Cygnus, and others. The pole shifts about one degree over 72 years. It will take 26,000 years for the celestial axis to complete its sluggish circuit through the northern constellations. The celestial sphere slowly changes its axis of rotation, contrary to the original concept of an eternally permanent celestial sphere (Figure 6.23).

Why does the rotation axis of the celestial sphere keep changing? Precession continued to baffle astronomers for 2,000 years. The real cause is not the wobble of the celestial sphere axis, but the precession of the earth's axis of rotation. Copernicus, who replaced the whirling celestial sphere by the spinning earth, was the first to recognize an additional component of earth's motion besides daily rotation and annual revolution (Chapter 7). The earth's rotation axis makes one complete gyration every 26,000 years. Newton finally understood the cause of precession as a consequence of the earth's shape, in particular its deviation from spherical symmetry (Chapter 11). Earth's rotation axis precesses due to an imbalance in the gravitational attraction of the sun and the moon on earth's protuberant equator, causing the earth to wobble like a spinning top, but ever so slowly. It was one of the many successes of Newton's synthesis of celestial and terrestrial motion.

A MAJESTIC ENCYCLOPAEDIA

With the pragmatism of Rome and the spirituality of Christians in ascendancy, Alexandrian astronomy began to languish after Hipparchus, staying nearly stagnant for 300 years. Hellenistic culture came under attack from Roman legions and Christian bishops. Scholars retreated from new observations and original thought to passive study and commentaries on great books of the past. Awe and respect for the great masters led them to believe it was impossible to rival their powerful work. They compiled encyclopaedias in response to demands from Roman military captains. The pedantic attitude prevailed until the arrival of Claudius Ptolemy (no relation to the Ptolemaic kings). In his passion for astronomy, Ptolemy proclaims:[10]

> In studying the convoluted orbits of the stars, my feet do not touch the earth … seated at the table of Zeus, I am nurtured with celestial ambrosia.

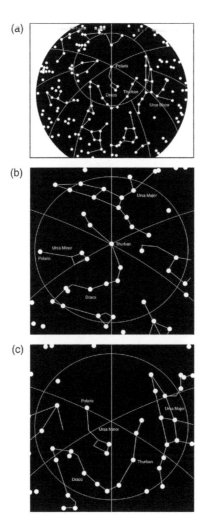

FIGURE 6.23   (a) The celestial sphere in AD 2000 with rotation axis passing through Polaris. (b) In 3000 BC, the rotation axis of the celestial sphere passed through Thurban, a star in Draco. (c) By AD 1000, the precession progressed so that the celestial pole was between Thurban and Polaris, the North Star of present times.

With fresh observations from an observatory at Canopus, a city named after the bright star, Ptolemy discovered new inconsistencies between observed and predicted planetary motions. To rectify these, he added more circles and epicycles. But the agreement remained unsatisfactory. For better precision, he introduced another complexity (Figure 6.24). The velocity of a planet was uniform about a new equant point. But the equant was not the centre of the circle which a planet described, nor was it the centre of earth. It was a clever idea that helped many errors go away, but it was contorted.

At first encouraging, the simplicity of Apollonius' epicycles evolved into a hopelessly corrupt system. With every fresh fact discovered, Ptolemy figured out a new ingredient to weave into the heavenly tapestry. He ended up with a *separate* system for each heavenly body without any connections between individual constructions. It was the product of a tortured era. Gone were the simple Greek ideals and aesthetic principles. By the time he was done, Ptolemy introduced 80 circles and epicycles to account for the paths of the seven heavenly bodies. Even though his calculations were extremely cumbersome, Ptolemy's results were *sufficiently accurate* to fulfil the needs of calendric and astrologic computations. For 1,500 years astronomers used Ptolemaic constructs of phantom

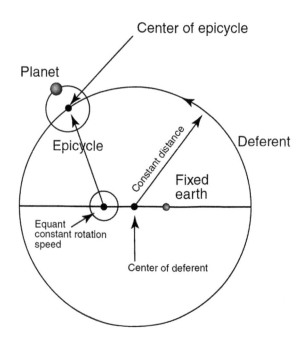

FIGURE 6.24 Ptolemy further complicates Hipparchus' scheme by adding the equant.

circles to predict planetary positions. In *Paradise Lost*, the 17th-century poet John Milton captures the monumental effort to save the spherical rule of the cosmos:[11]

> When they come to model Heaven
> And calculate the stars, how they will wield
> The mighty frame, how build, unbuild, contrive
> To save appearances, how gird the sphere
> With centric and eccentric scribbled o'er
> Cycle and epicycle, orb in orb.

It is the hallmark of science to create models to explain observations. It is equally important to make accurate observations with increasing precision, and carefully consider whether new findings fit within the existing framework. Optimistically, the ultimate goal is a simple and quantitative design that joins together scattered observations into a concise and comprehensive form. If the observations do not fit, the first tendency is to improve the model, provided there is sufficient faith in the accuracy of the new data. New discoveries thus lead to refinements in the prevailing conceptual structure. But successive mutations soon become cumbersome. Increasing complexity screams of some underlying deficiency, demanding an overhaul to root out preconceived notions, and weave a brand-new fabric of thought.

Graceless though it was, Ptolemy's apparatus provided a scheme to find a scale for the extent of the universe (Figure 6.25). With the help of elaborate constructions, Ptolemy devised estimates for distances to all the seven heavenly bodies, a new regime for predictions. Facing the enormity of the universe, he abandoned the practice of quoting distances in miles (or stadia). He found the moon to be 64 earth radii away, impressively close to the modern determination of 65.5. Under the idea that no heavenly body trespassed on the territory of another, and that there could be no wasted space between their territories, he obtained the greatest distance of Saturn to be nearly 20,000 earth radii. If we use Eratosthenes' estimate of the radius of the earth to be about 4,000 miles, Ptolemy's celestial sphere was at most 80 million miles in radius (8×10^7 in scientific notation). With Saturn,

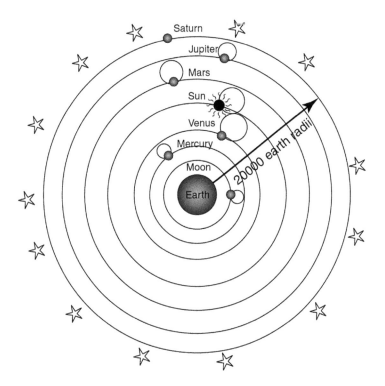

FIGURE 6.25   Ptolemy's scheme for deriving the size of the cosmos.

Ptolemy assumed he approached the boundary of the heavens, the edge of the celestial sphere, or so he thought. For the first time in history humans had attached a scale to the entire cosmos, although it was hopelessly off the mark. But it was based on a rational model, with serious calculations. Astronomers have long since abandoned his devices, but not his purpose.

Gathering together the wisdom of philosophers who preceded him, Ptolemy assembled a grand collection of astronomical knowledge. He called it *Megale Syntaxis*, meaning the great (mathematical) collection. Later, Greeks named it *Megistee*, meaning "greatest," and Arabs renamed it *Almagest*, meaning "the greatest" of all books. In 13 volumes, he laid out his elaborate systems of circles, epicycles, deferents, eccentrics, and equants. Ptolemy made every attempt to gather, comment upon, and improve predecessors' contributions. For example, he included Hipparchus' catalogue of stars, expanded to more than 1,022 stars in 48 constellations, each listed according to celestial longitude, latitude, and brightness. Faithfully he jotted down seven faint cloud-like apparitions in the heavens, later (Chapter 9) identified as *nebulae* (from Latin for cloud). One of these was later found to be our neighbour galaxy Andromeda. Reconsidering Aristarchus' radical heliocentric system, he rejected it on the grounds that observational evidence contradicted the idea of a moving earth:[12]

> Certain thinkers … have concocted a scheme … that the heaven is motionless, but that the earth rotates about one and the same axis from west to east, completing one revolution approximately every day … such an hypothesis must be seen to be quite ridiculous … everything not actually standing on the earth must have seemed to make one and the same movement always in the contrary sense to the earth, and clouds and any of the things that fly or can be thrown could never be seen travelling toward the east.

Ptolemy believed a moving earth would show significant annual parallax between summer and winter. His own estimates for the distance from earth to the sun, and from earth to the stars, would yield a ratio of $1,210/20,000 = 0.06$, equivalent to a parallax angle of 0.06 radians. Since an angle of π radians is equal to 180 degrees, 0.06 radians convert to 3.6 degrees which is seven times larger than the moon's apparent diameter. That would be an easily observable parallax shift for any acute observer of stars on the celestial sphere. The angle is so big that the shape of the constellations would be noticeably different between summer and winter. No wonder Ptolemy had to reject Aristarchus.

With trigonometric tables, diagrams, formulae, long computations, and lists of numerous observations, the *Megale Syntaxis* served as the prime textbook of astronomy, remaining the astronomer's best guide until the Renaissance, much as Euclid's elements served as the primary text for geometry. Ptolemy was equally famous for another tome, this one on astrology, called the *Tetrabiblios*, a work in four parts. In his preface, he provided a defence of astrology, thereby assuring its long and dark place in history.

ISLAMIC HEAVENS

Among the few works to survive the senseless destruction of the great library at Alexandria were the *Almagest* and the *Tetrabiblios*. There is a fascinating tale about how Arabs first came upon the lost *Megale Syntaxis*. In the newly founded city of Baghdad (around AD 780), beloved Caliph Al-Mansur took severely ill to the stomach, and hovered at death's door. His loyal subjects' only hope was to bring down a Christian monk, named Yishu, from the monastery in mountains 150 miles away, where he ran a hospital. Yishu came down to Baghdad and cured the terminally ill Caliph after consulting with the stars and invoking astrological charts. Monks who were driven out from Constantinople had settled in the mountains outside Baghdad, bringing with them a copy of Ptolemy's *Syntaxis* translated into Syriac language. No doubt they hoarded an equally rich chest of astrological lore.

Arabs of Baghdad were not ignorant of astronomy before medical-astrologer Yishu. They maintained an avid interest to satisfy navigational needs through vast tracts of featureless desert, as well as for timekeeping, calendric, and orientation necessities. Whether at home or in foreign lands, the Muslim had to submit to Allah by prostrating himself at the correct hour, in the direction of Mecca, as prescribed by Islam. Astronomy guided the muezzin to climb the minaret five times a day at the precise hours of daily prayers. Mohammed dictated that the life of a Muslim should be regulated by lunar cycles beginning with the blessed appearance of the crescent. As a symbol of submission to heaven, the crescent moon dominates the flags of Muslim nations.

With the discovery of Ptolemy, Al-Mansur recovered a treasure chest of powerful astronomical knowledge. As part of their fascination with astrology, Arabs continued to advance positional astronomy. In new observatories scattered through the Arab world, Muslims verified ancient texts, enriching Ptolemy's collections with their own observations. Caliph Sharaf-al-Daulah installed advanced instruments such as a wall quadrant 25 feet in radius at the Baghdad observatory, and a sextant 65 feet in radius. Later European astronomers (Chapter 8) were to emulate the grand scale of Islamic instruments to improve the accuracy of small angle measurements.

With creative energy, Arabian scholars advanced the *astrolabe* (Figure 6.26), meaning from the Greek "to take a star." This scientific instrument, originally invented by Apollonius, accurately modelled the geometric architecture of the heavens. From a basic device to aid in navigation, the astrolabe evolved under the Arabs into a complex instrument with many levels of refinement, a primary tool for future astronomers. For travellers, it helped determine location by comparing the position of the stars with previously recorded star charts. Miniature astrolabes decorated with astronomical lore became the pocket watch and portable navigator of the educated Arab.

Arab astronomy was not just a blind repetition of the classics. Arab star catalogues were not mindless copies of Ptolemy's lists. In 905, Al-Sufi produced a catalogue of stars that improved on Ptolemy's, adding Arabic names when Greek ones were missing. Thus we inherit: Aldebaran, Mizar, Altair, Denebola, Achernar, and Betelgeuse among others. Remarkably, he recorded a faint cloud in the heavens, about four times the width of the moon, later identified as the Andromeda Nebula (magnitude 3.7), the only galaxy visible to the naked eye on a crisp clear night. (Andromeda

FIGURE 6.26   Two sides of an astrolabe.[13]

will play a major role in modern astronomy – Chapter 9.) In flat contradiction to the Greeks' insistence on circular orbits, an astute Arab astronomer mentioned (in 1081) that the planet Mercury must describe an oval-shaped orbit. Bold enough to depart from the symmetrical circle, he vaguely anticipated Kepler who came four centuries later. Arab astronomers corrected errors found in the *Almagest* and prepared new tables. Calculators added more epicycles to Ptolemy's. Al-Farghani reworked the dimensions of the spheres of the moon, the planets, and the celestial sphere. As poets sang praises of the heavens, and artists decorated astrolabes, astronomers continued to scrutinize the positions of heavenly bodies. The heavens were still wide open.

UPDATES

HIPPARCHUS' LEGACIES: COUNTING STARS AND GALAXIES

From a launch pad in remote French Guyana, the European Space Agency fired Ariane rockets to launch a High Precision Parallax Collecting Satellite, Hipparcos, in 1989. It was more than 2,000 years after Hipparchus looked at the heavens to launch the field of "astrometry." Emulating its namesake of 2,100 years ago, Hipparcos' mission was to acquire data for a modern, three-dimensional catalogue of 1 million stars in the Milky Way galaxy. It was a painful birth for the capsule, loaded with expensive, delicate instruments, designed and assembled over many years of painstaking effort. When booster rockets failed, Hipparcos missed its geostationary orbit where it would have stayed above one point over earth to send continuous streams of precious data. Instead, the lopsided orbit threaded through Van Allen belts above earth populated by high-energy cosmic ray particles swirling in the earth's magnetic field. Intense radiation could have crippled its life-sustaining, solar energy panels. Fortunately, the panels stood up to the challenge. So did the mission staff who set up three stations over earth in an heroic salvage operation. Assisted by amateur astronomers, they successfully collected two-thirds of the hoped-for data on distances, luminosities, masses, sizes, and ages of countless stars.

Scanning a 2.5-metre diameter mirror continuously over the skies for five years, the ill-fated space telescope harvested more than 120,000 stars of brightness down to magnitude 12.5 with an angular accuracy of 0.001 arc-second, the limit of the technology. The milli-arc-sec parallax is sufficient to pin down distances of stars out to about 3,300 light years. Over the same period, Hipparcos accumulated data for a more extensive compilation of 2.5 million stars down to luminosities of 11.5, and accuracy of 0.01 arc-second. With sharp eyes alone, the ancient astronomer Hipparchus had assembled the very first catalogue of 850 stars down to a faintness magnitude 6, and a position accuracy of 10 arc-minutes. Stars captured by modern Hipparcos are more than 100 times dimmer. And the gain by a factor of a million in acuity is even more spectacular. If two lines extend from a point in the eye to touch the sides of a half-inch thick finger located 15 feet across a room, the angle between the lines (called the *subtended angle*) is 10 arc-minutes, but that same finger would have to be 2.5 miles away to subtend an angle of 1 arc-second, and 2,500 miles away (across the continental United States) for 0.001 arc-second. If the human eye ever gained such acuity, it would be possible to spot a virus flitting about, or a person's hair growing, from across a room. Hundreds of astronomers, engineers, and data analysers took four years to process massive quantities of data from Hipparcos by 1997.

The Global Astrometric Interferometer for Astrophysics (GAIA) satellite will continue the long legacy of star charting to catalogue a billion stars 200 times more accurately than the Hipparcos mission. The satellite launched in 2013 will build a 3D map of the Milky Way. As the earth rotates and revolves around the sun, GAIA will maintain its position to always point away from the sun, and never pass through the earth's shadow to stabilize instruments. GAIA's mission is to determine the position, parallax, and annual motion of 1 billion stars of brightness magnitude 15 with an accuracy of about 20 microarcseconds (μas) which is equivalent to measuring a star-distance of 150,000 light years. Named after the Greek goddess, Gaia, the mother of all life, the secondary objectives are to find up to 10,000 planets beyond our solar system. We will continue the quest to map the heavens in Chapter 9 with galaxies.

A COSMIC DISTANCE LADDER

Astronomical distances in the observable universe are so immense that one must abandon the expression of them in miles and kilometres. Already cognizant of this challenge, Ptolemy resorted to earth-radii, estimating the edge of the universe to be 20,000 times larger than earth's radius. After Copernicus moved the centre of the universe to the sun, astronomers advanced to the basic unit of the earth–sun distance, calling it one astronomical unit (AU) roughly equal to 24,000 earth radii, larger than Ptolemy's entire universe. At the edge of the ancient universe, Saturn is 9.5 AU away. The farthest planet we know today, Pluto, is 36 AU distant.

In a stark escalation of magnitudes, the appropriate scale to use today is the distance which light travels in seconds, minutes, and most often years, appropriately called the light year (LY). One LY is equivalent to 9.5 trillion km, or 63,000 AU. The moon is a couple of light-seconds away, and the sun is 8.2 light-minutes from us. Saturn is just 77 light-minutes away, and the farthest planet of our solar system, Pluto, is about 5 light-hours distant from the sun.

Moving out of the solar system, our sun's neighbour star, Proxima-Centauri, is 4.2 LY away. Stars in the solar neighbourhood span a 20 LY region carrying well-known bright members such as Sirius, Procyon, Altair, among others. Moving into the 75,000 LY Milky Way galactic realm, populated by 100 billion stars, our sun is 30,000 LY away from the galactic centre. We can wonder with poet Elizabeth Carter about the possibilities lurking in such incredible numbers:[14]

> Throughout the Galaxy's extended line
> Unnumbered orbs in gay confusion shine,
> Where ev'ry star that gilds the gloom of night
> With the faint trembling of a distant light
> Perhaps illumes some system of its own
> With the strong influence of a radiant sun.

Distances to foreign galaxies are so stupendously large that it is convenient to express them in mega-light years (MLY). A quarter MLY from earth are the satellite galaxies to our Milky Way, the Magellanic Clouds, first spotted by intrepid explorer Magellan in the southern skies during his pioneer expedition to circumnavigate the globe. Andromeda, spotted by Al Sufi, is a large, "nearby galaxy" at 2.2 MLY. When we look at a faint and fuzzy patch of light arriving from the phenomenal distance of Andromeda, we are also looking back in time, at the state of the galaxy more than 2 million years ago, when humans barely evolved from their ancestor species. A local archipelago of galaxies sprawls out over a 5 MLY region. But this is just one among many galaxy clusters inhabiting a supercluster region extending over 150 MLY.

SCALING THE COSMIC LADDER

How did we come to reckon stupendous distances of light years, millions of light years, and even billions of light years? German astronomer and mathematician Friedrich Bessel (1784–1846) made the first measurements that broke out of the solar system. When the German government charged him to construct a giant, new observatory to compete with others in Europe, Bessel fulfilled his mission and used the facility to establish exact locations of 50,000 stars with angular resolution exceeding the state-of-the-art. Armed with a precise map, he was in an excellent position to detect any possible movement among stars. Were the stars in the firmament really fixed? To his delightful surprise, he discovered that star 61 Cygni in constellation Cygnus (Swan) appeared to be making very slight oscillations with respect to the background of distant stars, and with a period of *exactly one year*. The magical period could only stem from the earth's motion around the sun. Vigilant Bessel had stumbled upon the annual parallax of the stars (Figure 6.27) to vindicate finally Aristarchus' brave proposition and Copernicus' revolutionary heliocentric system (Chapter 7).

Imagine a triangle where the base is the diameter of the earth's orbit around the sun, and the apex is the nearest star (Figure 6.27). One can say that the base *subtends* the apex angle. The earth's

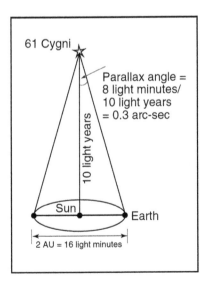
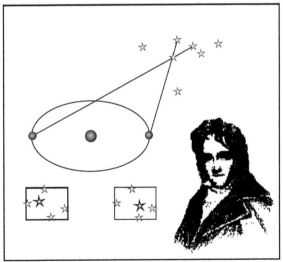

FIGURE 6.27 The angle of parallax to the nearest star is exceedingly small and difficult to measure, until telescopes of 300 (or higher) magnification became available. (Left) at a distance of 10 LY from the sun, the parallax angle of a star (due to the motion of the earth around the sun) is given by simple geometry as 0.3 arc-seconds. (Right) at last, Bessel, with his sharp telescope, finds the parallax of star 61 Cygni, as small shifts compared to the fixed background of distant stars.

orbital diameter (2 AU = 300 million km) is the longest baseline accessible to earth-bound astronomers. From simple trigonometry, the ratio of base-distance to star-distance is a measure of the apex
angle in radians. By definition, the *parallax angle* is half the angle at the apex. The farther away an
object, the smaller the subtended angle, and the smaller the parallax. Using this simple surveyors'
triangulation technique, Bessel extended Aristarchus' triangle geometry from the sun and moon out
to distant stars. Estimating a parallax angle of 0.3 arc-second, Bessel fixed star 61 Cygni to be 10
LY away. Light takes ten years to reach earth from that star. It was the first determination of a stellar
distance, setting a new scale for the universe. Later, a nearer star, Alpha-Centauri in the Southern
Hemisphere, was measured at 4.3 LY. (When a star's parallax angle is 1 arc-second, professional
astronomers call the distance 1 *parsec* (pc), equivalent to 3.3 LY.) With the help of photographs
taken over time through improved telescopes, trigonometric parallax techniques improved to detect
angles as small as 0.02 arc-second, extending distances out to 150 LY, and reaching several thousands of stars.

Aristarchus was on the right track to think that stars are "infinitely" far away. With Hipparchus'
best naked-eye angular resolution of 10 arc-minutes, it was impossible to see sub-arc-second level
parallax expected between summer and winter for even the nearest stars. Observations with the very
best ground-based telescopes are accurate to a few tenths of an arc-second. Out in space where telescopes do not suffer from atmospheric blurring, satellites can measure hundredths of arc-seconds
corresponding to distances to several thousand stars. Satellite Hipparcos' milli-arcsecond resolution improved upon Bessel's accuracy by a factor of 300. (A milli-arcsecond is one-thousandth of an
arc-second.) Whereas it confirmed previous distance measurements on nearby stars, such as Sirius
(8.6 LY), Hipparcos greatly improved the distance values for remote stars such as Polaris at 430 LY,
and Deneb at 3,200 LY. Since both stars had served as heavenly guideposts over millennia, their
reassessed distances forced serious revisions of astronomical standards.

WEAVING COLD HARD FACTS INTO GENERAL PRINCIPLES

As Whitehead describes eloquently in the quote at the beginning of this chapter, science needs
those with a temperament to pursue the cold hard facts as did Eudoxus and Hipparchus. Modern
astronomer-groups, who launched the Hipparcos, GAIA, and Hubble space telescopes (Chapter 9),
continue the quest for stubborn details of nature on ever-widening scales. Their efforts bear strong
testimony that fresh empirical content is essential to the progress of science.

By contrast, Pythagoras, Aristotle, Apollonius, Aristarchus, and Ptolemy were men of a more
philosophic temperament, aiming to weave together general principles. Scientific progress relies
heavily on creative inspiration. The union of the two passionate interests led to overarching schemes
for the cosmos.

At each stage, fresh evidence stimulated imaginative new directions of thought. With continuous
adjustments forced upon early spherical models by increasing accuracy of observations, Greek geometrical constructs grew ungainly at every step. Only a radically new model could break free from
the growing complexities. A new overall philosophy had to emerge to re-direct scientific imagination. Looking ahead, Copernicus (Chapter 7) was another thinker absorbed with abstract generalizations. With a radical re-orientation he established a new order, restoring elegance to the cosmos.
Destiny coupled Tycho Brahe and Johannes Kepler into a fruitful pair (Chapter 8), one relentlessly
pursuing stubborn facts and the other seeking general principles to weave into the crucial laws of
planetary motion. Newton's crowning achievement was to unify the physics of heaven and earth
(Chapter 11) through the force of gravitation by drawing on valuable information gathered by Brahe
and organized by Kepler. Einstein penetrated the deep meaning underlying Newton's universal
gravitation to reveal together with Edwin Hubble an expanding universe (Chapters 9 and 12).

7 Reformation and Revolution: Changing Perspectives

As the sky is to the light of the sun, so is the mind to the light of truth and wisdom. Neither the sky nor the intellect ever receive rays of light when they are clouded, but once they are pure and clear they both receive them immediately … the divine cannot be spoken or learned as other things are. However, from continued application and a matching of one's life to the divine, suddenly, as if from a leaping spark, a light is kindled in the mind and thereafter nourishes itself.

Marsilio Ficino[1]

As we saw in Chapter 3, science retreated after the dissolution of Alexandria. A Dark Age descended upon Europe. Far from breaking out into a brand-new direction, Greek knowledge headed for oblivion. The mental image of our earth degenerated from round to flat. With newfound emphasis on the soul, the Christian intellect withdrew from exploring the cosmos. Geometric architecture of the Greek universe evaporated into spiritual forces.

When Europe awakened slowly from the stupor of the Dark Ages, the wisdom of the classical age began to trickle in. Astronomy came back in service. Priests stressed the need to keep track of heavenly bodies to maintain the Christian calendar. Along with Ptolemy's *Almagest*, ancient astrological lore flowed in with his four-part compendium on astrology, the *Tetrabiblios*. Eager to learn about heaven's intentions, persistent astrologers found new freedom to compile copious horoscopes. Star patterns became guiding lights, first for sailors in the Mediterranean, then for explorers venturing out into the open ocean. Navigators consulted hefty astronomical tables and almanacs on long voyages.

Reliability remained a concern. Calculations from Ptolemy's tables no longer agreed with observed positions. Over intervening centuries, errors propagated into serious shifts in the paths of the seven heavenly bodies. New tables had to be compiled, and new calculations carried out using the age-old system. But the original quest for understanding the heavens remained at a full stop. Instead, Christian theological dogma usurped the Greek spherical model of the universe. Christians believed that from His place in the starry heavens, God put humans on earth at the stable centre of the universe. He controlled the motion of the heavens.

With cultural transformations sweeping over a reawakened Europe came serious challenges to established doctrines, as well as to entrenched patterns of thought. A few bold souls were prepared to question the unquestioned, to think the unthinkable. Martin Luther challenged the authority of the Church, declaring that the path to salvation need not pass through the Holy Office of the Pope. He launched the Reformation of religion. Copernicus challenged the centrality and immovability of the earth. He launched a revolution in scientific thought. Both ultimately changed human destiny.

Faced with new ideas that violated cherished beliefs, die-hards clung tenaciously to established notions, vigorously opposing new patterns of thought. It was important not to move God's place in the universe. There would be grave consequences for those who tried to change that structure.

THE TEACHERS' RETURN

Following the destruction of the Roman Empire, and the obliteration of the classic Greek heritage, Christianity remained the single thread of continuity through the fabric of Western culture. But Christian thought withdrew further and further from science. Rejecting the trauma of the world unfolding before them, Christian thinkers became preoccupied with the afterlife. Writings of Church Fathers rejected the relevance of any connections through astronomy between humankind and the cosmos. One wrote:[2]

> The magical relationships in the heavens existed only up to the appearance of Christ. Through Christ man was raised from the servant of the stars to the master of the stars.

A famous orator, Bishop of Gabala (AD 400), claimed that the heavens could not be spherical because, in Isaiah, the scriptures say that God[3] "stretcheth the heavens as a *curtain* and spreadeth them out as a *tent* to dwell in."

Priests fought against assigning astronomical events any religious significance. They ignored the necessity of keeping track of heavenly bodies for timely observance of religious festivals. To study planetary paths was to give undue reverence to pagan ideals. There was too great a risk that the intellectual freedom at the root of Greek ideas might once again win over human minds. Christianity was based on authority, and Christian doctrine was to be the ultimate source of that authority.

Lost in the intellectual desert, the European mind returned to a flat earth. Angels provided the push for wandering planets through the heavens. Detailed movement of heavenly bodies was irrelevant to spiritual life and to matters of salvation. Heaven was no longer the celestial sphere studded with stars, but the place for the soul to go after death, if the spirit earned salvation. It was more important to study the scriptures than the stars. In his *Confessions*, Father Augustine (later made a saint) wrote:[4]

> There is another form of temptation even more fraught with danger. This is the disease of curiosity ... It is this which drives us on to try to discover the secrets of nature, those secrets which are beyond our understanding, which can avail us nothing, and which men should not wish to learn ... In this immense forest, full of pitfalls and perils, I have drawn myself back, and pulled myself away from these thorns. In the midst of all these things which float unceasingly around me in my everyday life, I am never surprised at any of them, and never captivated by my genuine desire to study them ... I no longer dream of the stars.

What a tragedy! Negative attitudes toward science left astronomical knowledge in shambles. According to traditional prescription, Easter was celebrated on the first Sunday after full moon, after spring equinox. If no one kept track of equinoxes, solstices, the solar calendar, or even the phases of the moon, how could Christians prepare to observe the correct day to rejoice the resurrection of Jesus? Eventually, calendar uncertainties grew so intolerable that the Pope decided he had to send an emissary to Spain to find out from Arabs the day for the next spring equinox.

With equal vehemence, the Church denounced astrology. If human character depends on planetary positions, how could people be held responsible for their deeds? What would be the meaning of free will, the freedom to choose good over evil? Surely, Almighty God could not be subject to the course of the stars. People should put their trust in God, not the constellations. Even though Church Fathers fought hard to curb astrology, they were unable to completely stamp out the popularity of ancient astrological beliefs and predictions. Superstitions survived. Thirteen remained an unlucky number. Seven continued to hold its magical charm and divine significance.

After the Holy Wars (starting in the 11th century), Arabic knowledge began to dribble into Western Europe, which was slowly re-awakening from centuries of stupor. Moors spread the cultural heritage of Islam across the heartland of Spain. Stemming the tide of Islam's expansion, Christian Crusaders recovered Spain. Monks who followed them discovered Ptolemy's majestic encyclopaedia of astronomy. Eager to share the rich contents with fellow monks, they translated the *Almagest* from Arabic to Latin. With the classical world of Greece and Rome on the verge of re-discovery, the thick veil of cultural darkness began to lift. In the new light of classical wisdom, abbeys and monasteries transformed from remote corners of rigid Christian thought to active centres of learning. Like a flood through an open dyke, the accumulated experience of the ages began to pour upon medieval Europe. Plato, Aristotle, and Ptolemy re-appeared on the scene in a lively re-enactment of the Greek intellectual drama. In the east, when Roman Christian Constantinople fell before advancing Muslim Turks, droves of eastern scholars fled back to Italy, bringing with them important lost books along with the abandoned tradition of Greek scholarship.

A WORD OF ADVICE FOR GOD

Rediscovered and accepted, the writings of Aristotle and Ptolemy stimulated a re-birth of classical studies and astronomy. Now that Christianity was securely rooted in the mind of Europe, the Church's intellectual and spiritual authority was complete and dominant; Greek learning no longer posed a threat. Priests could be true to feasts by observing them on the correct day of the year. With market activities revolving around days of the Christian festivals, a proper calendar became essential to the commercial life of growing towns.

Astrology was back in service to dominate all aspects of life. Working astrologers-cum-astronomers produced nautical almanacs for casting horoscopes. Political treaties had to be signed at the right hour, and battles waged only when stars were propitiously positioned. Astrologers marched with the armies, giving the signal for when the time was right for cavalry to mount or for infantry to charge. Daily life was subject to the stars. Only certain days were right for christening and marriage. For royalty, astrologers had to be present during labour to record the exact hour of birth. Medicinal herbs had to be gathered on the auspicious day and administered on certain nights for the right effects.

But first astrologers had to re-chart the movement of stars and planets for proper predictions. As knowledge expanded about planetary positions and paths, the goals of matching observations with predictions became increasingly difficult. Over 13 centuries since Ptolemy, planets had shifted considerably from bearings determined using the antique system. Besides, computations demanded by prescriptions in the *Almagest* were complex and tedious. Alfonso X of Castile (1221–84), Spain (see timeline, Figure 6.13), appointed a commission of Arab and Jewish astronomers to compile a new set of astronomical tables to update those of Ptolemy and the Arabs. He personally presided over scholars, revised their work, and wrote an introduction to their publication. The *Alfonsine Tables* provided a new standard for 300 years. After struggling through laborious calculations needed under the Ptolemaic system, he declared:[5] "If God had asked my advice, I would have suggested a simpler design for the Universe." Although the sarcasm of the far-thinking man was really directed against Ptolemy's preposterous system, his son eventually used the "blasphemous words" to force his abdication, ending the father's climb toward the crown of the Holy Roman Empire.

Within decades, inaccuracies began to crop up once again. Over a century, large discrepancies emerged for notable celestial events, such as equinoxes, lunar eclipses, and even the anticipated night of the full moon. It became necessary to compile new tables at frequent intervals, a sophisticated, time-consuming, and demanding task for the priests, calendar makers, navigators, and astrologers.

Well-versed in astronomy and astrology, medieval poet Geoffrey Chaucer (ca. 1340–1400) wrote a treatise on the astrolabe and used the *Alfonsine Tables* to predict the eclipses and conjunctions which filled his tales with allegories. In *Nun's Priest Tale*, for example, he casts Chanticleer, the cock, and his seven wives as the sun and seven stars of the constellation Pleiades. Many of the fortunes and misfortunes that befall them through the story, Chaucer calculated from heavenly conjunctions.

THE DIVINE COMEDY

Dark clouds were gathering over the horizon. Some aspects of Aristotle's rediscovered cosmology were repugnant to the Holy Fathers. Most objectionable was the Greek concept of an eternal universe. No primary act of creation brought the universe into being. No divine act of destruction could end it. As Aristotle taught,[6] "Everything that comes into being must arise from what is." Greek thought excluded a supreme article of Christian faith. God created the world. Aristotle had to be repudiated.

Averting a repeat of the intellectual catastrophe that destroyed Alexandria, Thomas Aquinas deftly tackled Aristotle's most offensive premise. The Greek master's claim for the eternity of the world lacked formal proof, and should therefore be rejected. Matters such as creation of the universe and salvation of the soul had to rest on grounds of faith alone. After Aquinas divorced faith from reason (Chapter 3), medieval Christendom was largely free to accept the rest of Greek cosmology, with few, but crucial, exceptions. Indeed, Aristotle's separation of heaven and earth into distinct realms fit the creation account of *Genesis*. The earth is the lowliest place in the universe where change is rampant and nothing eternal. Humans are mired in corruption, decay, and sinful pleasures. If we achieve redemption, our soul can rise from the changeable and degenerate world up to heaven, the eternal world of pure spirit, the home of God.

With growing interest in Greek thought, scholars strove for a logical interpretation of the scriptures. Aristotle's overpowering view of the cosmos gradually gained acceptance, eventually to become reigning dogma, even sacrosanct. Whenever universities installed a new professor, part of the inaugural ceremony was to take an oath agreeing with the teachings of Aristotle, which gradually gained stature nearly as sacred as articles of Christian faith. The geocentric universe satisfied an important theological position. It fit the concept of the Vatican as the centre of the universe, conveniently endorsing the seat of the Church's power.

The spherical architecture of the rediscovered cosmos exerted a profound influence on theology as expressed through art and literature. Figure 7.1 portrays how God created the spherical heavens

FIGURE 7.1   *Creation of the World*, ca. 1534, anonymous, from Martin Luther's Bible of 1534.[7]

and earth expressly for humanity. We are His special creations, thriving under His embrace. He watches over us and cares for us. In his *Divine Comedy* (Figure 7.2), 14th-century poet Dante Alighieri casts the geocentric system into eloquent verse, paralleling Aristotle's hierarchical arrangement with a Christian cosmos. A mesmerizing combination of astronomical and theological aspects, Dante's epic artfully blends Greek cosmology with Christian theology, Aristotle's elemental spheres with Heaven and Hell. To reach the roots of the classical universe, the poet starts his fanciful journey from the navel of the spherical earth, as from the navel of an orange, and traverses through its miserable bowels to enter the domain of the Devil incarnate. Hell is a stepped conical pit with successive circles for various classes of sinners. Returning to the earth's surface at a point diametrically opposite the navel, the Christian voyager begins his graceful ascent to the home of God. Rising upwards, he encounters Mount Purgatory, the peak of which protrudes

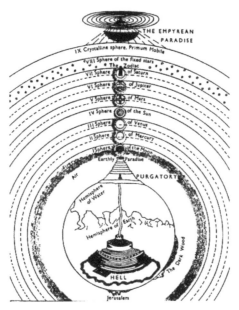

FIGURE 7.2   (Top) allegorical Portrait of Dante, looking over Mount Purgatory, 1530, unknown Italian painter.[8] (Bottom) medieval cosmology as described by Dante in the *Divine Comedy*.

through the atmosphere (see Figure 7.2). Souls have to suffer in Purgatory for their earthly sins before they can gain final access to Heaven. Paralleling Hell, Purgatory rises in successive circular ledges that hold various classes of penitent sinners. After floating through spheres of air and fire, Dante reaches the glorious Heavens above, nearing ultimate ecstasy. It is a starry Heaven layered with crystalline planetary spheres:[9]

> From thence we mount aloft into the sky,
> And look unto the crystal firmament,
> There we behold the heaven's great hierarchy,
> The stars' pure light, the spheres' swift movement.

Trekking through each divinely arranged celestial sphere, the epic voyager converses with angelic spirits residing among planets. His lovely heroine, Beatrice, explains astronomical technicalities such as eclipses, epicycles, and retrograde motion. With astrological flair, Dante paints Venus' position amidst the constellations:[10]

> The radiant planet, love's own comforter
> was setting the eastern sky a-smile
> veiling the fishes that escorted her.
> Translation: Venus was rising in the east with the stars of the constellation Pisces at
>   dawn.

Past Saturn, beyond the celestial sphere, he reaches his final destination where he contemplates God's Throne in the last Empyrean Sphere. Dante loved the stars; each of the three books of the *Divine Comedy* ends with the Latin word, *stelle*, for stars.

Thanks to Aquinas and Dante, Greek planets once again threaded their errant course through zodiac constellations. But angels pushed heavenly bodies along epicycles, lovingly guiding them through ethereal spheres. Planetary motion and its mysteries recaptured the European intellect. In the mind of artists, Christ in Heaven was the original mover, as in the *Creation of the Heavens Mosaic* (Figure 7.3). He arranged the motion of the celestial spheres for the benefit of humanity.

ON THE SPICE TRAIL

The Holy Wars between Christians and Muslims catalysed the European discovery of the eastern world. Asia became the legendary infinite source of valuable commodities: silk, cotton, tapestries, jewels, ivory, sugar, and above all, rare spices. People used pepper, cinnamon, cloves, ginger, and nutmeg for curative drugs and preservation of meat. Arab traders moved their precious wares by the shortest sea route between the Red Sea and India, and thence by a caravan of camels over land to the markets of the eastern Mediterranean in Alexandria and Constantinople. Southern Arabia had long been a trading centre for frankincense and myrrh, legendary gifts that the three wise men brought to the baby Jesus. For centuries, Arab traders kept the true source of the spices (such as the islands of Malaya) a carefully guarded secret. To protect their markets, they spread intimidating stories about how cinnamon grew only in deep valleys infested by poisonous snakes.

Arabs traded wares of the exotic East with the merchants of Venice, who prospered greatly by selling precious items in dreary European markets. Bold and enterprising Venetians took caravans across the old "Silk Road" through Iran and Afghanistan to the cities of China. Among them was the pioneer Marco Polo (1254–1324), who travelled to Persia, China, and Malaya. On his return, he reported Hangchow to be[11] "the greatest city in the world where so many pleasures may be found that one fancies himself to be in Paradise." His fascinating accounts of strange cultures widened medieval Europeans' horizons, as well as their concept of geography. With captivating stories, he

FIGURE 7.3   Creation of the Heavens Mosaic, 12th century.[12]

portrayed the Indies as a land of limitless wealth, encouraging Italian merchants to seek out these areas, and break the monopoly of overland Arab traders.

As the Islamic empire expanded westward towards Constantinople, it upset the flow of eastern commerce through the old silk road to China, pushing Europeans to explore Mediterranean waters to reach directly the vibrant commercial centres of Alexandria and Constantinople. Sea trade began to flourish. Wealthy from traffic of the plundering crusades, Italian city-states established commercial flotillas trading in wool, fur, wine, cloth, and olive oil for the spices of the East. Venice grew immensely prosperous as it enjoyed a monopoly on active trade with the East. Italians sold spices to northern and western Europe at exorbitant prices. From their huge profits, they further expanded trade and prosperity. Venice built warships to protect its merchant fleet from greedy pirates. Through its commanding position, the prosperous Italian peninsula evolved into the most cultured region of Europe, future breeding ground for artists and scientists, such as Michelangelo and Galileo.

With improvements in shipbuilding and sail rigging, together with the advent of the compass from China and the astrolabe from Islam, ships could sail at any time of the year, day or night, in clear weather or foul. The pace and number of voyages increased, as did the volume of trade, the size of the ships, and the number of investors. Sailing on the open ocean loomed as an inviting challenge. Was the vast, unbounded ocean indeed navigable? Geographers were eager to learn the location of spice-rich islands. But the impenetrable ocean always formed a watery prison wall. Wild stories circulated about the mysteries of vast oceans and far-away lands. The open sea was replete with slimy serpents and monsters. Who would dare to venture south of the tropics where oceans boiled, and excessive heat turned men black? And woe to the mariner who dared to breach the mythical underwater Pillars of Hercules that blocked the ocean at the conjunction of Africa and Spain (now Gibraltar). These were the mythical limits of the navigable sea.

The intellectual heritage of Greece and Alexandria had a major impact on seafarers. Through their voracious hunger for classical texts, they learned how to read the constellations. Fascinating patterns of stars in the night sky served once again as guiding lights. Translation and dissemination of Ptolemy's *Geography* caused quite a stir. Included in that classical treatise was an atlas of the known world, spreading from the Canary Islands in the west to China in the east, and from Iceland in the north to Madagascar in the south.

The open oceans carried no landmarks for explorers. Location of latitude and longitude was absolutely necessary. Ptolemy described Eratosthenes' methods of measuring the earth, marking charts with a systematic system of latitude and longitude. Powerful techniques finally fell into the hands of eager navigators.

EXPLORERS

Portugal was farthest from the glittering East, and paid the highest price to prosperous Italian merchants for coveted silk and spices, monopoly of the Venetian traders. Having beaten back the Moors into Africa, Prince Henry of Portugal dedicated himself to reaching the legendary land of fabulous wealth by circumnavigating the continent of Africa, *terra incognita* up to this time. From Sagres (meaning sacred point), a bit of Portugal jutting out onto the ocean, Henry organized and dispatched hundreds of expeditions that inched their way around the coast of Africa.

Although he personally never ventured out into the open ocean, Henry installed a navigational institute and staffed it with German mathematicians, Italian map-makers, and Hebrew and Muslim scholars, directing them all to map the earth. He taught sailors how to use the compass and the latest navigational techniques for long-distance sailing. Captains had to keep accurate logbooks and charts so that knowledge about the seas and new lands could accumulate, while maps grew more accurate by pooling information. Succeeding travellers could benefit from the experiences of their predecessors. Spreading the knowledge of navigation and cartography became a key factor in the rapid advances made by explorers. Ship after ship sailed out from Portugal, each expedition masterminded by the inquisitive Prince. After attaining the Azores, Madeira, and Canary Islands, the adventurous sailors inched their way forward along the rugged coast. When they finally reached the peak of the western bulge of Africa, their patron prince died, a life-long ambition unfulfilled.

Henry's successors continued past the bulge. As ships sailed farther south, determination of latitude started to became more challenging. The trusted Pole Star was no longer visible in the southern latitudes. Sailors grew nervous about their navigation capabilities when they saw familiar constellations dip below the horizon.

Italian explorer Christopher Columbus read about wonders in the Indies from Venetian adventurer and traveller Marco Polo. Vivid tales incited him to seek an expeditious route to the East. Marco Polo's accounts led him to the false conclusion that Asia extended much farther east than European scholars believed. If he sailed directly in a westerly direction, he estimated the trip from Europe to Asia to be a mere 3,000 miles, not the actual 12,000 miles of unknown, treacherous ocean. To estimate the earth's circumference, Columbus used an erroneous number which had reached him through recent translations of Ptolemy's *Almagest. It was not the 40,000 km estimate of Eratosthenes.*

As an explorer for Portuguese ventures, Columbus submitted his ambitious plan for a westward route to the Indies, an aggressive alternative to the tortuous circumnavigation of the African continent. The Portuguese king's geographers rightly rejected the maverick sailor, ridiculing the short distance he claimed. They were aware of larger estimates for earth's circumference. To sail to the coast of Asia would require a voyage of three years. By any scholar's reckoning, it was a foolhardy proposal. It would be impossible to survive such a long stint at sea. They recommended that the ongoing circumvention of Africa was still the best sea route to the Indies. After all, the Cape of Storms at the southern tip of Africa had recently been discovered. It would not be long before Africa would be circumnavigated. In anticipation, they renamed the Cape of Storms the Cape of Good Hope. Portuguese explorers successfully rounded Africa, but only after another 15 years of treacherous voyages and shipwrecks.

Failing in Portugal, Columbus tried to interest royal patrons in his native Genoa as well as England. No one would support his preposterous quest despite ten years of persistence. In 1492, he turned to Queen Isabella, whose recent marriage to Ferdinand united the kingdoms of Castile and Aragon into the monarchy of Spain. In the afterglow of victory at Granada over the last remnants of Muslim rule in Spain, the exuberant monarchs agreed to support the renegade sailor. It was a stroke of luck, but it came to one who doggedly pursued his ambition, eager to try out a new approach.

Columbus set sail on August 3, 1492 with three rickety old ships, carrying a crew of 120 men, including many prisoners, released for the express purpose of making up a full crew. He followed a path parallel to the equator, never deviating from that single line of latitude. Sailing the open ocean for two long months without any sign of land, his crew was about to mutiny. On the same day, an ugly storm was brewing. As skies blackened, a miraculous sight appeared. A glow of purple streamers filled the air from the tips of the masts, shooting up to heaven. It was lightning, pure and simple. Before an awe-stricken crew, Columbus seized the propitious moment proclaiming that St Elmo, patron saint of sailors, sent his holy fire as a blessing for their voyage. God commanded the search for land to proceed.

At 2 a.m. on October 12, after seven anxious weeks on open seas, when nearly ready to abandon the lunatic project, they sighted land. The two-month long voyage changed the world forever. For the Portuguese, who rejected Columbus' idea, Vasco da Gama finally fulfilled the age-old dream of a route around the continent of Africa to the fabulous riches of Asia. Every expedition east or west brought news of new territories, new products, and new people. Europe launched colonization and started to build empires around the world.

EASTER IN DECEMBER

Ocean navigation intensified demand for accurate astronomical information, a need only to be filled by more careful observations and more reliable predictions. Techniques of celestial observation advanced. Renewed interest in astronomy encouraged improvements in design and construction of the compass and astrolabe. Navigators and inventors turned to rediscovered Alexandrian astronomy to design new tools, and Greek geometry to solve new problems.

Johannes Mueller (1436–76) erected an observatory to make his own star charts, and equipped it with a printing press. From there he sent out volumes of newly compiled astronomical tables and calendars. Like the Humanists, he forged a new identity by Latinizing his name to Regiomontanus, for "royal mountain," after his birthplace.

Regiomontanus was first among many to warn the Church about a serious problem cropping up with the Julian calendar, a slippage in the date of Easter. Put into effect by the administrative talents of Julius Caesar, the Julian calendar replaced the arcane Roman calendar which had acquired a variable number of days from year to year. Adding to the confusion, months no longer synchronized with the seasons as Roman calendar-makers ignored the advances of ancient civilizations. To rectify these deficiencies, Julius Caesar consulted with Egyptian temple priests. Merging the Sirius cycle with the solar cycle, Egyptian temple-astronomer Sosigenes (Chapter 5) recommended adoption of the 365.25-day year. A leap day added every four years would keep March 21 on the vernal equinox. Caesar decreed the defining year, 46 BC, to be a 445-day year, making up the lag between the old and new calendars. A few centuries later, in AD 325, the Church declared Easter as the first Sunday after the first full moon, on or after the vernal equinox, a reliable prescription for celebrating resurrection in spring.

Now here was the problem. The Julian calendar was based on the 365.25-day year, as the Egyptians determined from that dazzling event of the heliacal rising of Sirius (Chapter 5). But more corrections became necessary over centuries. The length of a year is actually 365.2422 days, not 365.25 days, as recognized by the Islamic astronomer Omar Khayyam. The minuscule difference, equivalent to 11 minutes and 14 seconds per year, seems like a small interval compared to the half-a-million minutes in a year. But to ignore it year after year amounts to one day in 128 years, or three days in 400 years, causing the Julian calendar to fall seriously out of step with the seasons by the Middle Ages. In AD 1200 the outspoken scholar Roger Bacon found an error of more than one week. Vernal equinox was falling on March 11. Eventually Easter would slide from March into December, and end up on the same day as Christmas! Corrections had to be applied. The Pope called Regiomontanus to Rome to reform the Julian calendar, but he died prematurely. The problem was not corrected until after Copernicus.

After Hipparchus' 850 stars, there were many grand attempts to map the universe. As European navigators of the Middle Ages set out to explore oceans, stellar maps proliferated to fill explorers' needs in a world with expanding frontiers. Mariners carried hefty astronomical tables on voyages. Columbus' favourite were prepared by young Johannes Muller (1436–76), a child prodigy, who published his first astronomical yearbook at age 12. In 1603, a German lawyer, Johann Bayer, produced *Uranometria*, a milestone work of art and science. A total of 51 plates codified each of the 48 Ptolemaic constellations, plus one map for the newly discovered southern skies. Exquisite drawings in lively, Baroque style imaged famous constellations, accurately mapped stars, and portrayed apparent brightness by proportional discs. Later astronomers, John Flamsteed, Johannes Hevelius, and Johann Bode, followed suit with elaborate atlases based on their own observations. (By sheer coincidence, men named John compiled many of these catalogues!) Figure 7.4 shows picturesque constellations from Bayer and Hevelius. As positional astronomy continued to advance in accuracy, every generation of stargazers assembled landmark compilations. With the advent of photography, half a century of mapping from observatories all over the world fattened catalogues to 6 million entries from over 100 million stars imaged. Even today, astronomers often refer to bright stars by their Bayer or Flamsteed designations.

FIGURE 7.4   Constellations Cassiopeia by Bayer and Taurus by Hevelius. Note the artistic rendering and the astronomical accuracy.[13]

A MADMAN MAKES THE MOON DISAPPEAR

Columbus' favourite tables were those prepared by Regiomontanus. On one of his later voyages to the New World, Columbus was stranded in Jamaica, his ship leaking, and the native population angry with European invaders for plundering their villages. No longer in awe of white aliens, they refused to obey routine demands for food and supplies. As the weary explorer retired to his cabin, nervous over his predicament, he noticed, while looking over the Regiomontanus tables, a prediction for an upcoming total lunar eclipse on 28 February. And there were additional notes for the exact duration of the eclipse. Carefully, he applied a correction for the longitude difference between Germany (Regiomontanus' home) and Jamaica to figure out the exact local time. Even though he was not quite sure of the accuracy of his calculations, he decided it would be worth the risk to carry out his devious plan. On the day before the published eclipse, he haughtily summoned the native Jamaicans to warn them with arrogant confidence. If they resisted cooperation, the moon would disappear from the sky the following night, right before their eyes. But the chiefs only laughed at the madman.

Columbus was incredibly lucky. Neither the tables nor his reckoning were sufficiently accurate. Errors of a day or more were quite common for eclipse predictions. At the appointed hour, the moon disappeared slowly, spreading panic among the natives. Terrified tribe leaders immediately relented, pleading with Columbus to restore their moon-god to the sky. Returning to his ship, Columbus waited for the expected duration, and emerged just before the moon would start to reappear. God had answered his prayers, he announced; He would promptly restore the moon to full brightness. The next day, grateful chiefs reverently returned with abundant quantities of food and supplies. Once again apprehensive of the awesome power held by the white alien, they continued to cooperate with Columbus over the two years it took for his rescue party to arrive. With similar intentions, Jesuit priests who followed explorers to become missionaries in Asia cultivated a strong interest in astronomy to make eclipse predictions that impressed and intimidated the people they aimed to colonize.

REFORMATION AND INQUISITION

From the late Middle Ages to the Renaissance, the age of artists, writers, and explorers (Chapter 3) was also the age of religious reformers. Exceptional scientists who emerged during this turbulent period, Nicolaus Copernicus, Tycho Brahe, Johannes Kepler, and Galileo Galilei, were swept up by the tide of events stemming from the battles between the Reformation and Counter-Reformation. Conflicts between the powers of the Church and ascending states engulfed their lives. The rampant frenzy of religious ideology brought brutal judgements against their bold ideas.

With cultural transformations sweeping Europe came serious challenges to established doctrines as well as to entrenched patterns of thought. The Church was becoming materialistic and wildly extravagant. A way of life devoted to pleasure became the norm for bishops, cardinals, and popes. Like princes, they lived in opulent splendour. Along with grand papal palaces spread over Italy, the Vatican launched an extravagant new project to rebuild the Church of St Peter (St Peter's Basilica), commissioning renowned architects and artists. It took more than a century and cost the equivalent of more than 5 billion dollars. With daunting power over the daily lives and eternal souls of common people, corrupt Church organizations exacted heavy tolls through mass sales of papal indulgences to fund their grand projects.

Hailing the beginning of the Protestant Reformation, a German monk named Martin Luther nailed a sheet of paper to the door of a church in Wittenberg in 1517. On it he presented 95 theses that defied the doctrines of the Church and condemned abhorrent practices:[14]

Number 27. There is no divine authority for preaching that the soul flies out of purgatory as soon as the money clinks in the bottom of the chest.

Number 28. Christians should be taught that, if the Pope knew the exactions of the indulgence-preach-ers, he would rather the church of St Peter were reduced to ashes than be built with the skin, flesh and bones of his sheep.

Luther challenged the corruption of powerful clergy, spreading the idea that the fate of a person's soul depends on the quality of faith alone. Every man was directly answerable to God alone. Defying papal authority, Luther flooded Germany with pamphlets printed with the newly invented press. His vernacular translation made the Bible accessible to common folk. Besides the printing press, the vernacular became a powerful instrument for the democratization of ideas. Other rebels would resort to both. Luther set the standard for the German language. Naturally, the reigning Pope Leo X excommunicated the rabble-rouser as an agent of the Devil.

In the age of emerging nation-states, the religious disputes triggered a break-up of broad clerical authority. Petty princes of Germany seized the opportunity to stem the power of the Church which was draining the wealth from their realms. It was in their interest to protect Luther, to support his challenge, and to facilitate the spread of Protestantism. In reaction to the Italian Church's sweeping power, many German princes enthusiastically received Luther's eloquent writings. Others remained faithful to the traditional Holy Order. Luther's insurgency pitted town against town. Erupting from Germany and France (through Calvin), the Protestant revolt spread across Europe.

The age of creators, explorers, and reformers was also the age of inquisitors. Responding to Protestant challenges, the Catholic Church initiated a purge of abuses. Under the leadership of St Ignatius Loyola and the order of Jesuits he founded, they launched the Counter-Reformation. Popes of the Counter-Reformation used diplomacy, persuasion, and force to crush the spread of heresy. St Dominic formed the Inquisition, an age-old practice used to weed out witchcraft and sorcery. As dogs of the lord, they succeeded in stamping out Protestant influence in Italy, Spain, Bavaria, Austria, Poland, and Belgium. In Spain, King Ferdinand and Queen Isabella took on the role of protectors of the Catholic Church and set up the brutal Spanish Inquisition to eradicate both heresy and opposition to their iron rule.

PTOLEMY'S FRANKENSTEIN

Into this turbulent period of the early 1500s came Nicolaus Copernicus to turn the universe inside out. From his native Poland at the remote outpost of Christendom, Copernicus travelled to Italy, where universities excelled all others in Europe. Supported generously by a rich uncle, Copernicus could afford to be a student to the ripe age of 30. Under Scholastics at the University of Padua, the most famous university in Europe, he studied Latin, Greek, law, theology, mathematics, astronomy, and medicine, a broad humanities education for the archetypal humanist. It was in Padua's anatomi-cal theatre that Vesalius and other bold physicians carried out secret dissections of human cadavers.

While Copernicus was still a new student, Columbus discovered America, and Vasco da Gama found a seaway to India. New books were coming off presses every month. When he returned to Poland after taking Holy Orders in Rome, Frauenberg's Cathedral installed him as canon. Clerical life attracted young men of the time, even those who were not religious, because it provided security, a decent living, respect, and education. As a canon, Copernicus oversaw church estates, headed a group to reform local currency, and even organized a free clinic. His medical abilities endeared him to the bishops. Humanists like Erasmus and Regiomontanus Latinized their names in a symbolic proclamation of their new awareness, a kind of second baptism. Forging a new identity, Copernicus Latinized his name from Koppernik. The family had long since progressed beyond the wholesale copper business, from which the name derived.

Columbus' discoveries made a deep impression upon the young Copernicus' intellect. Flouting established practice, the ambitious navigator's courageous turn in a new direction had paid off handsomely. Copernicus set off on his own voyage of discovery, searching for a simpler system for the heavens. Simplicity bore the sign of truth. To say that earth was at the centre of the universe, and that all heavenly bodies moved in combinations of perfect circles, appeared to be deceptively simple at first glance. But the Ptolemaic system acquired close to 80 simultaneous motions for just seven celestial bodies. What an immense number of wheels to run the planetary spheres! When Copernicus read Ptolemy directly, he was disappointed to find that the master had set down *different rules of motion* for each planet. The complexity of the whole system was horrifying. Burdened by laborious calculations, astronomy was getting more and more impossible:[15]

> It is as though an artist were to gather the hands, feet, head and other members for his images from diverse models, each part excellently drawn, but not related to a single body, and since they in no way match each other, the result would be a monster rather than man.

The Ptolemaic approach had produced a Frankenstein, to use an anachronistic turn of phrase. Researching ancient books, he hoped to find better ideas. Between Ptolemy, Islamic astronomers, and recent European revisions, various sets of tables now circulated all over Europe, claiming to best predict the future positions of planets. Each had major errors. Purchasing 800 stones and a barrel of lime from his Church's workshops, Copernicus arranged construction of a small tower observatory. From there he planned to make new charts. He measured the obliquity of the ecliptic, but his result was not as accurate as the value obtained by Arabs. European astronomers were still far behind in technique as well as quality of instruments. It was not his observations, however, but his theoretical propositions that were destined for revolution. It was not the tower he built, but the system he forged that penetrated the heavens.

Ptolemy's elaborate devices were particularly bothersome to Copernicus. There was no simple way to provide both uniform and circular motion about one single point. One had to think of the motion of a planet as uniform in velocity about the *equant* (Chapter 6), and circular about a different point. And *neither of these points was the centre of earth*, in stark contradiction to the crucial feature one would demand of any *earth-centric* system.

Copernicus had to get rid of the equant. Although Ptolemy's artifice was ugly, it was technically successful. Copernicus found that without the equant he had to introduce even more epicycles into the dance of the planets. Looking for a radically new approach, Copernicus turned to the views of Aristarchus, whose ideas of a spinning earth and planets orbiting a stationary sun had entered Europe as a discredited scientific opinion.

As a first step, Copernicus allowed the earth to rotate about its axis so he could greatly simplify the motion of the sun. Between the daily arc and the annual seasonal motion, the sun described 183 separate circles in the sky (Figure 7.5). By discarding the deeply rooted belief in an immobile earth,

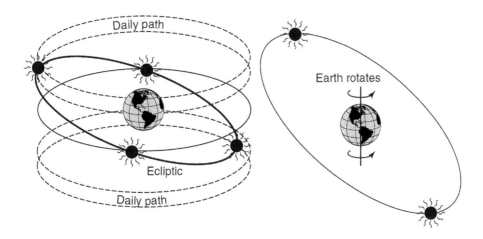

FIGURE 7.5   In the first stage of his revolution, Copernicus replaced the daily motion of the sun with the rotation of the earth, simplifying the 183 circumlocutions of the sun into *two circular motions*, one for the sun's annual path, and the other for earth's daily rotation.

Copernicus could account for the motion of the sun with *just one tilted circle* around a spinning earth. In a single step, the tangled solar spiral dissolved into one circular path of amazing simplicity.

THE LAMP AT THE TEMPLE'S CENTRE

Contemplating the regular paths of Venus and Mercury among the band of zodiacal constellations, a remarkable connection struck Copernicus. The two fastest moving planets always rise and set with the sun. When it is a morning star, Venus rises just before the sun. As an evening star, Venus sets just after the sun. So also for Mercury. Copernicus was on the threshold of a stunning geometric revelation about the bright celestial that so thoroughly captivated the imagination of artists, poets, and astronomers through the ages. Venus' love affair with the sun stimulated Copernicus to rethink the sun's position on the celestial stage.

Although Venus and Mercury always appear in conjunction with the sun, Mars, Jupiter, and Saturn travel freely through the entire zodiac, quite independently from the sun. Sometimes they appear in conjunction, sometimes in opposition to the sun. Why is there such a stark difference between the regions over which planets move? Copernicus wondered whether Venus moves with the sun because it is attached to the majestic lamp. Perhaps Venus does not circle the earth. Rather, it circles the sun (Figure 7.6). The same is true for Mercury. Now, if Venus and Mercury circle the sun, why should other planets circle earth? Why should only two planets be orbiting the sun? Perhaps they all revolve around the sun. Even the earth revolves around the sun!

> Who indeed, in this most beautiful temple would place the light-giver in any other part than whence it can illumine all other parts … Indeed the sun, reposing as it were on a royal throne controls the family of planets which surrounds him.

thought Copernicus.[16]

It was a momentous step on the historic path to the scientific revolution! Copernicus was not alone in his attraction to the sun. Through liberal education in Italy, the fresh ideas of 15th-century humanist Marsilio Ficino strongly influenced his thinking. From the Platonic Academy of Florence, Ficino translated Greek classics into Latin and wrote eloquent commentaries. (See his quote at the beginning of this chapter.) What he wrote about the sun rang true to the young astronomer.

Copernicus was poised for a breakthrough in understanding why Venus and Mercury never stray from the sun, while Mars, Jupiter, and Saturn rove freely through the zodiac to sometimes appear

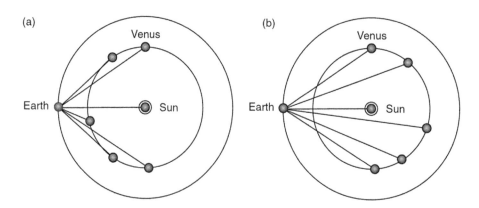

FIGURE 7.6   Since Venus' orbit lies inside earth's orbit, it is easy to understand why Venus always appears in conjunction with the sun, and never in opposition. This is true whether Venus is (a) close to earth or (b) far from earth.

in opposition to the sun and other times in conjunction. It was a geometric insight, pure and simple. Venus and Mercury are *interior* planets; their circles lie inside the earth's orbit. As observed from earth, the angle between Venus and the sun can vary only between 0 and 46 degrees. Being closer to the sun, the angle between Mercury and the sun can vary only between 0 and 22 degrees.

Mars, Jupiter, and Saturn are *exterior* planets. Their orbits lie outside earth's orbit. They will appear in conjunction when they are far from earth, and in opposition when they are near to earth (Figure 7.7). In fact, exterior planets can appear at any aspect with respect to the sun.

Copernicus found a unified geometrical explanation for a host of planetary irregularities accumulated over millennia of observations. Venus appears as an evening star for some months and as a morning star for other months. It is totally invisible in between. All distinct appearances are direct consequences of one heliocentric geometry (Figure 7.8). When Venus is to the *west of the sun*, it appears as an evening star, setting in the west after the sun. When it is to the *east of the sun*, it will appear as a morning star, rising in the east before the sun. When Venus is directly in line with the sun, it will be invisible in the light of the sun. There will be two such in-line positions (conjunctions) during which Venus will die in the light of the sun (Figure 7.8). All the wonderful myths and romantic notions about Venus' fickle motions, her love affair with the sun, yield to the simple sun-centric geometry.

There was much more to come. With this profound geometric re-orientation, Copernicus could elegantly explain retrograde motion (Figure 7.9), an irregularity which baffled astronomers through the ages. In fact, it is merely the consequence of the planets' speed relative to earth speed. As they orbit the sun, *outer planets* (Mars, Jupiter and Saturn) move *slower* than earth. When the swifter earth *overtakes*, slow-moving planets appear to "move backwards." An observer on a fast-moving boat passing a slow-moving boat sees the sluggish vessel receding, a familiar trickery of the senses. All superior planets appear to regress when faster-moving earth overtakes them. On the other hand, interior planets appear to regress when they overtake the earth. Indeed, the senses had fooled astronomers for centuries, stuck on the centrality of the earth.

The brightness variations of planets also fell into line. Mars appears 25 times brighter when it is in opposition to the sun than when it is in conjunction. As an exterior planet, when Mars is near earth, and bright, it is also in opposition to the sun. Mars appears dim when it is far away from the earth. In this position, it appears in conjunction with the sun. When Mars is close to earth, faster-moving earth overtakes it, and Mars appears to go backwards. Regression, opposition, and brightness variations all stem from pure geometry.

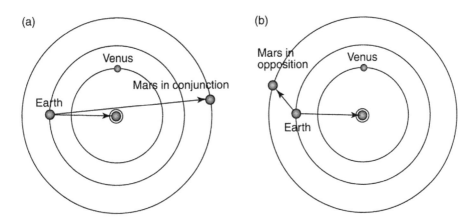

FIGURE 7.7   With the orbit of Mars lying outside the orbit of the earth, Mars appears (a) sometimes in conjunction with the sun, and (b) sometimes in opposition. When Mars is near to earth, and appears *in opposition*, it will also appear brighter than when it is far from earth and in conjunction.

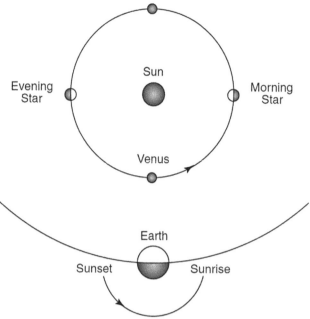

FIGURE 7.8   When Venus is in line with the sun and earth, it is lost in the glare of the sun and therefore invisible. When the sun is to the east of Venus, it rises and sets before the evening star. When the sun is to the west of Venus, it rises after Venus, visible as the morning star.

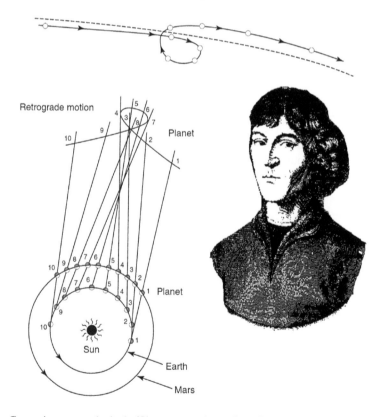

FIGURE 7.9   Copernicus unravels the baffling retrograde motion of planets with his revolutionary sun-centric system. Moving in a smaller orbit than Mars, the earth moves faster and therefore overtakes Mars. When earth overtakes its outer neighbour, as it travels through points 3, 4, 5, 6, 7, Mars *appears to move backward*. Copernicus showed that retrograde motion is a simple consequence of relative motion.[17]

Reflecting on the power of the heliocentric model to provide a single and elegant geometric explanation of many apparently unrelated phenomena, Copernicus confidently declares:[18]

> I think it is easier to believe this [the heliocentric model] than to confuse the issue by assuming a vast number of Spheres, which those who keep the Earth at the centre must do. We thus rather follow Nature, who producing nothing in vain or superfluous often prefers to endow one cause with many effects.

Aesthetics guided Copernicus. Nature avoids superfluous operations. The simpler explanation, involving fewer mechanisms, is likely to be closer to the truth. At first glance, nature may appear replete with complex and unrelated phenomena. But, when viewed through the proper perspective, disparate effects unify to spring from a single cause. Copernicus changed our perspective by stepping outside the planetary system. To understand nature we must be prepared to change our outlook and accept what we are not used to. Motivated by aesthetics, Pythagoras had grasped a most essential aspect of heavenly motion. With proper geometry, it should be possible to re-compose the apparent meandering of heavenly bodies in terms of simple motions. When Copernicus put the sun at the centre, and allowed earth to join the planets as a swift wanderer among stars, abnormalities of planetary motion evaporated as illusory aspects of earth's motion:[19]

> So we find underlying this ordination an admirable symmetry in the Universe, and a clear bond of harmony in the motion and magnitude of the Spheres.

Symmetry and harmony once guided Pythagoras to deep insights about nature's baffling operations. Clearly, Copernicus had taken the sage's lessons to heart. But he had gone much further. Overcoming the timeless obsession to keep humanity at the centre of the drama, he had the courage of his conviction to upset the traditional order. Ever since Copernicus put the earth in "revolution" around the sun, the word has become synonymous with overturning established order.

But the new system placed a disturbing demand on the human psyche. Instead of being the stable centre of the cosmos, earth now became subservient to the sun, from headliner to extra! Copernicus defied the sacred alliance of theology and cosmology to raise troubling questions. What becomes of humanity's place in the ancient cosmic arrangement? Are we still important? Are we still special in God's mind?

Reviving classical ideas, Renaissance artists of Copernicus' time had built a bridge to the Golden Age of Greece. Mingling images of God with those of his creations, they resurrected the classical belief of humanity's position in the universe, showing a new intimacy between God and humanity. On the ceiling of the Sistine Chapel, Michelangelo expressed humanity's special relationship with God by painting a lightning arc between God's finger and man's outstretched hand. Now, in one quick stroke, Copernicus knocked humanity off its special pedestal.

But Copernicus was not alone. Parallel threads were emerging in the fabric of a culture under transformation. Revolutionary artists' prescient imaginations often transcend common sense about order and priority. To help change our thinking, artists try to trouble us, by disturbing our thought habits. Instead of always putting the clear and meaningful subject at the foreground, front and centre, artists of the late Renaissance and Baroque periods strained to break free from classical traditions. Consider the classical, Renaissance treatments of *The Virgin and Child* (Figure 7.10)[20] and *Presentation of the Virgin* (Figure 7.11).[21] Subjects are always in the foreground, invariably at the centre. Compared to central characters, the landscape is remote and far less important. However, in later works, Tintoretto feels no such compulsion. Joseph and Mary are no longer at the physical centre (Figure 7.12).[22] Yet they are clearly the focus of the painter's theme. An orange sky background holds attention as strongly as the halo over Mary. Lush trees are no longer remote and inessential details; they envelop the entire landscape. In a unique interpretation of *Rest of the Flight into Egypt* (Figure 7.13),[23] Elsheimer's star-studded night-sky overwhelms the scene. Were it not for a divine light that softly illuminates the centre, one could hardly find Mary, Joseph, and baby Jesus.

FIGURE 7.10   *The Virgin and Child with St John the Baptist, La Belle Jardiniere,* 1507, Raffaello Santi.

FIGURE 7.11   *Presentation of the Virgin,* Calvaert, Denys (1540–1619).

FIGURE 7.12   *Flight into Egypt, 1583–7*, Tintoretto, Jacopo.

FIGURE 7.13   *Die Flucht nach Agypten*, (1609), Elsheimer, Adam.

They are fused into an harmonious unity with their landscape. Elsheimer was one of the first artists to represent the constellations accurately. And he did not forget to paint in the meandering Milky Way. Strong departures from core centrality and dominance are equally evident in *Presentation of the Virgin* (Figure 7.14[24]) by Tintoretto. At first glance, one wonders who is the main subject. Careful study reveals how the Virgin, though far in the background of the spiralling staircase, is indeed at the mental centre of attention, for all eyes are fixed upon her ascending grace. Rather than placing the Virgin at the physical centre of a rectangular frame, Tintoretto puts her at the tip of a spiral, creating a visual tension that heightens the impact of his theme.

A TILTED WOBBLY HOME

For the new system to be credible, it was imperative to provide satisfactory explanations for previously well-understood behaviour. If the sun does not move in a tilted orbit around earth, how to explain seasonal temperature variations? Copernicus had a simple answer. The sun's orbit is not tilted; the earth's axis of rotation slants relative to the plane of its solar orbit. Pointing toward Polaris, the axis makes an angle of about 23.5 degrees with respect to the earth's orbital plane (Figure 7.15). As a result, the Northern Hemisphere leans closest to the sun during summer solstice. At the same location, the Southern Hemisphere is farthest from the sun and it is winter in the south. As earth circles the sun, its axis always points to the same direction in space, explaining the fixity of the North Star.

If heavenly bodies revolve around the sun, how do we know the moon remains an exception and orbits earth? Our silvery companion has cunningly misled us since time immemorial to generalize that all heavenly bodies must circle earth. Lunar and solar eclipses continue to testify that the moon orbits the earth (Figure 7.16). As earth revolves around the sun, the moon tags along, but continues its monthly circuit around the parent.

Copernicus' model of the spinning earth with tilted axis provides the basis of a clear mechanism for Hipparchus' subtle effect, precession of the equinoxes. There is a novel aspect about the earth's

FIGURE 7.14   Presentation of the Virgin in the Temple, Tintoretto, Jacopo (1518–94).

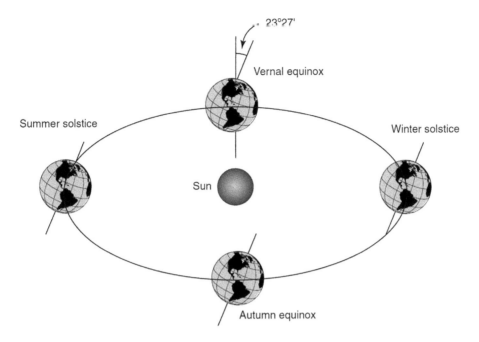

FIGURE 7.15 Copernicus put earth in orbit around the sun and tilted the earth's rotation axis to explain the apparent inclination of the sun's path.

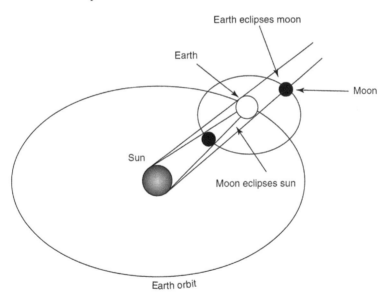

FIGURE 7.16 Solar and lunar eclipses viewed in the heliocentric system.

rotation axis. It gyrates slowly over 26,000 years to trace out a cone. Thus Copernicus gave the earth a three-fold motion: rotation, revolution, and precession. Conical motion of the earth's axis explains two coupled effects: precession of the equinox and the circuit of the North Star (Figure 7.17).

PREDICTIVE POWER

The strength of a novel theory lies not just in its ability to embrace what we already know. It must be capable of making new predictions and projecting a new horizon. When predictions come true,

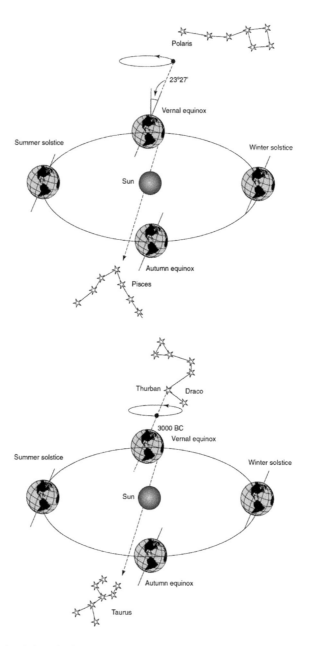

FIGURE 7.17 The axis of the spinning earth does not always point in the same direction. It drifts in a slow circle that takes 26,000 years to complete. (Top) today, the earth's rotation axis points to Polaris. When we face north at night, stars appear to circle the stationary North Star. Looking east at dawn on an early spring day, the sun rises in Pisces. (Bottom) in 3000 BC, the axis pointed toward a different star, Thurban, in the constellation Draco. The spring sun rose in Taurus.

they boost confidence in the theory. Having placed the sun at the centre of the heavenly drama, Copernicus came to another profound realization: *planets are all dark bodies* that reflect the glorious light of the sun. It was well-known since antiquity that the moon does not shine on its own. It merely reflects the light of the sun. Phases occur due to its juxtaposition with the sun. Now Copernicus made a new prediction. If planets merely reflect sunlight, they must also show phases like the moon. If only our vision could somehow be made sharper, the phases of planets would be visible:[25]

If we agree ... that planets are themselves dark bodies that do but reflect light from the Sun, it must follow, that if nearer than the sun, on account of their proximity to him they would appear as half or partial circles; for they would generally reflect such light as they receive upwards, that is toward the Sun as with the waxing or waning Moon.

Venus should show phases like the moon. There was another surprising forecast. Venus should show variations in size, and appear larger in size when near earth. In today's scientific climate, such predictions would immediately launch a flurry of ideas to develop techniques to verify effects, and put theory to test. But Copernicus' contemporaries viewed the same predictions as weaknesses. To the naked eye, Venus showed no phases and no changes in size. His forecast made the new system even more incredible. Everyone could see clearly that planets are shapeless, celestial lights. Although Copernicus' system was conceptually simpler, as well as computationally less cumbersome, it was not decisive. It did not provide immediate and incontrovertible proof to definitively rule out the Ptolemaic view.

Half a century later, Galileo fulfilled both Copernicus' dream and his predictions. Enhancing vision with the telescope, he discovered that Venus does indeed show phases similar to the moon, as well as expected variations in size. Science is replete with examples of predictions fulfilled posthumously.

DISCORDANT NOTES

Putting the sun at the centre, Copernicus untangled the Ptolemaic complexities of epicycles, eccentrics, and equants, with contrary motions and multiple centres. Six planets orbited the sun in six circles. Of all the heavenly bodies, only the moon remained a loyal satellite of earth. The planetary irregularities of the past dissolved into one *unifying theory*. Planets now moved coherently in *one direction*, counter-clockwise around the sun. Their speeds diminished uniformly with increasing distance from the sun. It was an act of beauty. A geometric harmony prevailed from the hub of the cosmos out to the farthest planet.

A nagging problem remained. Despite the simplicity, elegance, and harmony, the accuracy of the seven-circle system was nowhere near the 10 arc-minute precision of the 80-circle Ptolemaic system. Copernicus' simplest model failed the demanding test of experience. In retrospect, we know his monumental blunder. He stuck to the tradition of circles as the perfect figure of motion for heavenly bodies. To make the new theory work, he had to resort once again to Ptolemaic devices, and re-introduce the complexity of epicycles and eccentrics into the sun-centred cosmos. To account for the small, but well-known irregularities, he used 34 circles and epicycles in all. Later he added more. At this point in the story, the complete Copernican system turned out to be no more accurate than the Ptolemaic system in its ability to predict planetary positions. Even though the sun-centric system was conceptually appealing in its unified geometry, even though its 34 circles were less intricate than Ptolemy's 80, it was far from totally satisfying. Ptolemy's ghost continued to haunt the universe. What a bitter disappointment!

With hundreds of calculations, Copernicus painstakingly computed all the orbits. Table 7.1 gives his ratios for the average orbital radii of planets in comparison to the radius of the earth's orbit. It established the proportions of the solar system to within 5 per cent of modern values, a tribute to the accuracy of his analysis.

SAFE COMMENTARIES

Copernicus hesitated for 36 years before daring to publish his radical ideas:[26]

The thought of the scorn which I had to fear on account of the novelty and incongruity of my theory well-nigh induced me to abandon my project.

TABLE 7.1

[27]**The Size of the Planetary Orbits Relative to the Size of the Earth's Orbit around the Sun in the Heliocentric System**

Planet	Modern	Copernicus
Mercury	0.387	0.376
Venus	0.723	0.719
Earth	1.000	1.000
Mars	1.523	1.520
Jupiter	5.202	5.219
Saturn	9.554	9.174

The magnitude of the radius of the earth's orbit around the sun is now referred to as one Astronomical Unit (AU). The modern value is 0.15 trillion metres (1.5×10^{11} metres)

Perhaps the Inquisition and heresy trials conducted in Catholic countries also intimidated him. In the hostile climate of Counter-Reformation, the Church persecuted radical ideas. At home, the King of Poland remained a repressive Catholic who burned heretics at the stake. The new bishop of his diocese decreed that anyone who read or possessed heretical books would be punished by death. Jews were systematically harassed; 240 were burned at the stake in Vienna on a single day in 1421. In France, the Church's dreaded agents burned Calvinists. In Paris, they used the wheel to break people who dared to challenge established authority. In Spain, they burned Moors. In Germany, they burned witches and magicians. In Italy, they tortured physicians who dared to perform dissections to study human anatomy. Could free thinkers about heaven and earth hope to escape the dreaded flames?

Timidly, Copernicus prepared a list of "Short Comments" to circulate among friends, carefully communicating dangerous ideas only through handwritten manuscripts. There were immediate and strong objections. Direct perception of the senses spoke against the stock absurdity of a moving earth. It was difficult to break free from the iron grip of Aristotle's cosmology and its captivating fusion with theology under Dante. How could earth possibly be a planet, sailing among stars? Clearly, earth did not shine like a planet. The earth could never be part of the heavens. If earth was a celestial body, residing among the stars, where was heaven, and where was hell? Astronomers should desist from changing the plan of the universe and fiddling with the permanent hierarchy of the cosmos.

Two thousand years of astronomy repeatedly and forcefully contradicted Copernicus' claims. The authority of the Bible dictated against the stationary sun. Even Luther, who dared to reform religion, bitterly opposed reformation of cosmology. He laughed at Copernicus' foolish theories. Did not the prophet Joshua lengthen the day by commanding the moving sun to stand still until the nation of Israel took vengeance upon its enemies? Luther put his faith solidly in the Bible as the sole source of truth. And it was not just one obscure Bible passage. There were many:[28]

> The Lord is king, in splendour robed; robed is the Lord and girt about with strength; And he has made the world firm, not to be moved. [Psalm 92:1]

Who would dare to contradict the written word of God?

The 19th-century poet, playwright, and philosopher Wolfgang von Goethe (1749–1832) put his finger on the heart of the problem:[29]

> Humanity has perhaps never faced a greater challenge; for by admission, how much else did not collapse in dust and smoke; a second paradise, a world of innocence, poetry and piety, the witness of the

senses, the conviction of a religious and poetic faith … no wonder than men had no stomach for all this, that they ranged themselves in everyway against such a doctrine.

To the scriptural objections of Catholics and Protestants alike, Copernicus lashed back:[30]

It may fall out, too, that idle babblers, ignorant of mathematics, may claim a right to pronounce a judgement on my work, by reason of a certain passage of Scripture basely twisted to suit their purpose … I make no account of them … Mathematics are for mathematicians.

Copernicus took refuge in the position that only qualified mathematicians best understand astronomy. After seeing the power and import of the new computations, competent mathematicians would finally be able to cast off old anchors of dogma and appreciate the need for radical re-orientation.

There were rational objections, raised once before when Aristarchus proposed a moving earth. What about the terrific wind that would constantly blow if earth rotated? Birds would never be able to fly fast enough to keep up with the spinning earth. How could there ever be a force large enough to move our giant earth? The dizzying idea of trees and houses, cities and countries, mountains, oceans, and vast continents whirling with the spinning earth as it rushed around the sun simply staggered the mind.

Copernicus skilfully answered the age-old objections. There is no gushing wind on the spinning earth because the air moves along with the earth, just as the coat of a man remains with him when he walks around. A man in a moving boat may think for a moment that trees outside are in motion; but he quickly realizes his error, knowing that trees are firmly rooted to the ground. So also, from a moving earth we may think that it is the sun which moves. Like trees to earth, the sun is rooted to the centre of the cosmos. It was a paradoxical yet compelling deduction.

Another rational objection stemmed from analogy with a potter's wheel. Would not the rotating earth break apart if it spins as rapidly as one turn per day, speeding around the sun at the same time with breath-taking speed? Copernicus pointed out that the Ptolemaic system required orbital revolutions of the sun, a body many times larger than earth. Its enormous distance from the earth implied that the sun moved at far greater speed and accompanying risk, if such risk were indeed any real factor. Would it not be much less of a strain on the mind to think of the earth as spinning than to accept immense heavenly spheres revolving at phenomenal speeds?

For scholars educated in traditional cosmology and Aristotelian physics, a planetary earth defied the traditional notion of gravity. All objects fall toward the earth. If the sun is the true centre of the universe, why do objects not fall toward the centre of the sun? To this philosophical objection, Copernicus had no answer. It had to wait for the genius of Newton's universal gravitation. Earth and all its attachments do indeed "fall" toward the sun, as they whirl in in their orbits.

It is remarkable how Copernicus' early contemplations on gravity hold the seeds of very modern ideas:[31]

Gravity is a natural inclination, bestowed on the parts of bodies by the Creator, in order to combine the parts in the form of a sphere and thus contribute to their unity and integrity. And we may believe this property present even in the sun, moon and the planets.

Copernicus was looking through a thin veil, dimly sensing gravity's power to unite.

Critical thinkers revived an important geometric argument against an orbiting earth. Why is there no *parallax*? As one sways back and forth on a swing, objects appear to veer left and right, nearby ones more so than far away objects. If earth swings around the sun, stars should show a parallactic shift between summer and winter (Figure 6.16). When earth approaches a certain group of stars they should appear to open up. Conversely, angular separation between stars should shrink when earth recedes. To these objections, Copernicus' explanation was that *all* the stars are very far away from us:[32]

Yet so great is the Universe that though the distance of the Earth from the Sun is not insignificant compared with the size of any other planetary path … it is insignificant compared with the distances of the Sphere of the Fixed Stars.

We now know that the distance to the nearest star is 275,000 times the radius of the earth's orbit around the sun. No wonder the parallax effect was too small to detect at the time of Copernicus, and for centuries beyond. It was not until 1832, when sufficiently accurate telescopes became available, that Bessel (Chapter 6) was finally able to measure parallax of the stars arising from the annual motion of the earth around the sun. By this time, scientists already accepted the heliocentric view.

It was bad enough to displace earth from the hub of the celestial drama, an anathema to so many. Now Copernicus was asserting that earth was insignificant, and the abode of stars was infinitely far away, thrusting the heavens beyond reach of human imagination. If stars are so far away from earth, is there any celestial sphere? If there is no celestial sphere, where is heaven? According to hybrid Christian-Aristotelian cosmology, vividly cast in the epic of Dante's *Divine Comedy*, heaven was located just beyond the celestial sphere. The tectonic shift delivered devastating blows to traditional beliefs!

REFORMATION OF THE CALENDAR

The mathematical simplicity of Copernicus' system eased the burden of essential calendric calculations. Problems with the Julian calendar continued to be worrisome. The rapid expansion of commerce and accompanying growth in communications cried out for its reform. In 1514, Pope Leo X, hearing about the accomplished mathematician and astronomer, asked Copernicus for assistance. Still busy computing motions, the astronomer politely declined the honour of an invitation to Rome. But he could not resist offering the suggestion that his heliocentric system of 34 circles could prove an effective tool. Calendar reformation would first demand a reformation of astronomy. When Pope Gregory XIII finally introduced the superior Gregorian calendar in 1582, Church scholars indeed carried out the needed computations using the simpler Copernican system. But they emphatically denied the idea that earth moves around the sun. They only used the new system because it was a more convenient tool for the demanding calculations.

To correct for the accumulated slippage, calendar reckoners advised Pope Gregory to ordain that October 4, 1582 should be followed the next day by October 15, a ten-day jump. After such a correction, the following March 21 would re-set to the spring equinox of the original Julian calendar, bringing Easter into synchronization with the seasonal calendar. Many protested against the missing ten days. Would people still have to pay a full month's rent for 20 days? Servants demanded pay for extra days. Employers refused to pay wages for missed days. Some felt cheated out of their precious time: "Give us back our ten days!" became a popular slogan. Protestant countries, such as England and British colonies in America, refused to follow the calendric dictates of a Roman Catholic prelate. As a result, the old Julian Calendar persisted in England and America for nearly two centuries. In 1752, England and the colonies finally inserted the extra days, and one more to correct for additional slippage. Today, we celebrate George Washington's birthday on 22 February, even though records say he was born on 11 February.

The precise reason for slippage is that the solar year (equinox to equinox) is not 365.25 days long – but more precisely, 365.2422 days. Therefore the scheme of adding one day every four years overcorrected by roughly one day per century. The ten-day correction rectified the overcompensation. But how to prevent a recurrence of the annoying effect? To compensate on a regular basis, the Gregorian scheme has additional features. It skips the leap year on years that end in 100, even though a century year is divisible by four, and therefore obeys the commonly used rule for assigning leap years. For example, even though 1800 and 1900 are both divisible by four, the years 1800 and 1900 were *not* leap years. An easy calculation will show that the prescription of skipping century years still leaves a small discrepancy of about one-quarter of a day every 100 years. The next obvious correction is to add in a leap year every 400 years, i.e., if the century year is divisible by the number 400. This means that the year 2000 was a leap year, even though it was a century year. Following the Gregorian plan, it will take another 3,000 years for our calendar to be out of step by one day, a comfortably long period. For those who remain uncomfortable, there is the modified

Gregorian system in which years 4000, 8000, and 12,000 will be non-leap years to correspond to the more accurate year.

PUBLISHED UNDER FALSE PRETENCES

Copernicus continued to spread his ideas cautiously, only by word of mouth, and only to those few properly initiated in astronomy and mathematics. Fortunately, there appeared upon the scene the eager young Joachim Rheticus (1514–74), a Lutheran professor of astronomy and mathematics at the new University of Wittenburg. From this centre of Protestantism, Luther's right-hand man, Philipp Melanchton, hand-picked Professor Rheticus to visit the Polish cleric to investigate the new-fangled ideas and their attendant dangers. Rheticus already knew how to charm his quarry. He arrived with gifts of the first printed editions of Euclid and Ptolemy in original Greek. But it was Rheticus who was charmed. After studying the secret manuscripts, Rheticus felt compelled to broadly publicize the new system and its virtues. Since Copernicus wrote his manuscript "for mathematicians," Rheticus undertook the challenge of making it generally accessible by preparing an introductory narrative explanation. His action was remarkably courageous, considering Luther's position on the Bible and on Copernicus. Sticking his neck out once more, Rheticus prevailed upon the master to publish the entire manuscript and so commit the radical theory and calculations to posterity.

By the time publication started, Copernicus was already an old man. So long had he hesitated to publish that he barely got the chance to see his finished work. The story is often told that printers delivered an advance copy of the book to his deathbed, as he lingered half-conscious from a stroke.

On the Revolutions of the Heavenly Spheres came out in 1543. It was the same year as the first printing of another rebel manuscript, *On the Fabric of the Human Body*, by 28-year-old Andreas Vesalius. In his own preface, Copernicus warns:[33]

> The theory of the earth's motion is admittedly difficult to comprehend, for it runs counter to appearances and all of tradition. But if God wills, I shall in this book make it clearer than the sun, at least for mathematicians.

Knowing full well the radical demands, Copernicus accurately anticipated overwhelming reactions.

As the first comprehensive text that could rival the *Almagest* in precision, it soon became the standard reference for astronomers. Sad to say, the only place where he gave credit to Aristarchus, the heliocentric proponent of antiquity, is a note in the margin of his manuscript. Yet, upon examining Copernicus' sweeping accomplishments and detailed computations, we are forced to admit that the ancient Greek's insight was indeed premature lightning.

It is equally sad to see that Copernicus totally ignores Rheticus' role in making the theory both available and understandable. Perhaps he was trying to avoid the controversy that would have arisen because Rheticus was a Protestant. Not more than a few years had passed since the Catholic Church excommunicated Martin Luther, the reformist. To fend off trouble that the new ideas might generate, Copernicus dedicated the work to Pope Paul III.

After turning over the supervision of the printing to Rheticus, Copernicus had little control over the final product. Besides, printing took place in Nürnberg, far away from Fauenberg, where the author lived. But Rheticus soon ran into political trouble. Forced to flee Nürnberg after being involved in a scandal, Rheticus left a Lutheran minister, named Andreas Osiander, in charge of the final product. As a theologian and opinionated preacher, Osiander took it upon himself to include yet another preface, in which he labelled the new system just a mathematical scheme, convenient for astronomical computations:[34]

> These hypotheses need not be true or even probable; if they provide a calculus consistent with the observations, that alone is sufficient.

Osiander's aim may have been to prevent the work from being immediately declared as heretical and therefore unpublishable. But any good intentions he had in crafting the clever preface were probably harmful to the long-term advancement of science. Some opponents, who thought that Copernicus wrote the disclaimer, conveniently dismissed the author as someone who did not believe in his own theories. Fifty years later, Johannes Kepler (Chapter 8) pointed out that Osiander, not Copernicus, had written the notorious preface condemning the new physics. There was no contradictory self-denial by Copernicus.

BURNED AT THE STAKE

Fifty years after Copernicus, Italian writer and philosopher Giordano Bruno ignored the Church's heightened sensitivity to its doctrines concerning earth's privileged position in the cosmos. He blatantly spread the heliocentric view along with fanaticism about sun-worship. Extrapolating Copernicus' ideas, Bruno raised provocative questions. If earth rotates, there is no longer any reason to postulate a stellar sphere. If the distance to the stars is immeasurably large, maybe the stars spread out over an infinite universe, the first appearance of the bold concept. With poetic words, Bruno's vivid imagination burst the starry celestial sphere:[35]

> Henceforth I spread confident wings to space;
> I fear no barrier of crystal glass;
> I cleave the heavens and soar to the infinite.
> And while I rise from my own globe to others
> And penetrate ever further through the eternal field,
> That which others saw from afar, I leave far behind me.

Bruno was neither an astronomer nor an astrologer, merely an audacious thinker. Everywhere he went he stirred up youth with eloquent, animated sermons. The sun, moon, and planets are not our slavish attendants. Earth is not the centre of the universe; even the sun is not the centre, just one among millions of stars. The cosmos has no centre. Why would God in his infinite power create just one solitary world?

With the clear hindsight of 400 years, we realize that Bruno was a lucky prophet with a luxuriant imagination. In the form of dialogues, he wrote several books containing astronomical speculations and stimulating conjectures, many borne out by posterity. It was not the firebrand's astronomical views that condemned him. These mostly earned him ridicule. Bruno questioned the foundation of the Church's teachings in a reckless fashion, writing that the outlook of the Bible was no longer relevant to his time. Such ideas flagrantly violated doctrinal sources of Church authority, causing considerable consternation among religious powers. The Dominican Order expelled Bruno and excommunicated him. This hardly stopped him. With a biting combination of poetry and satire, he continued to ridicule antagonists, stinging them with Humanists' barbs. God is revealed not through Holy Scriptures but through His Creation of nature. The laws of nature are the same everywhere, on earth and in the heavens.

Bruno's reckless challenge to human chauvinism strengthened Church opposition to the Copernican view. Brought before the Inquisition for glorifying the sun, worshipping the sun, and other heresies, prosecutors threatened him with death. But he refused to recant. In 1600, they burned him at the stake. As a fierce warning to those who contemplate blasphemy, executioners pierced his tongue and palate with iron spikes. His garish execution profoundly intimidated thinkers in Catholic states. But the stakes that pinned his tongue could hardly silence his eloquent cry to scientists and poets:[36]

> Open wide the door for us, so that we may look out into the immeasurable starry universe; show us that other worlds like ours occupy the ethereal realms.

UPDATES

Exoplanets, Is Anyone out There?

Italian writer and philosopher Bruno was burned at the stake in 1600. But poets who echoed Bruno's wild flights of thought about other suns and other worlds could take heart through vindication from modern discoveries in the last two decades.

In 1995, Mayor and Queloz of the University of Geneva announced their pioneering discovery of the first exoplanet, 51 Pegasi b orbiting the star Pegasi, nearly 50 light years from earth. They launched the exciting hunt for exoplanets in our galaxy. They discovered the planet by detecting slight but regular velocity changes in the star as it wobbled slightly in synchronization with the orbit of its large nearby planet. The star's motion caused slight doppler shifts in the wavelengths of its light spectrum. The doppler effect is the same principle police use by deploying radio waves to measure your car's speed. Since the first find, the doppler shift technique has discovered about a hundred exoplanets.

A more sensitive method for the exoplanet search is to detect planetary transits across a star by analysing continuously and very accurately the brightness of a star. If the brightness dips (example, by about 0.01 per cent) on a regular basis, it indicates an extrasolar planet crossing the star in its regular orbit around the star. The transit method also helps determine the radius of the orbiting planet by the depth of the dip. The planet's orbital size can be calculated via Kepler's Law (Chapter 8) from how long it takes the planet to complete its orbit around the star.

In 2009 NASA launched satellite Kepler to hunt for planets by continuously monitoring the brightness of 150,000 stars in a small field of view (about 10 degrees) over our galaxy. Unlike previous searches, Kepler is sensitive to planets as small as earth, and with orbital periods as long as a year. Over nine years of its life, Kepler confirmed nearly 4,000 planets in 3,000 planetary systems. Kepler observed at least three transits across a star to confirm the dimming by a planet. If a planet orbits too close to a star it is too hot; if too far from the star, too cold. In its field of view, Kepler discovered that about 20 per cent of sun-like stars have an "Earth-sized" planet in the "just right" Goldilocks (habitable) zone.

As good examples, Kepler discovered planet Kepler-22b at a distance of 600 light years from earth. It is about two and a half times larger than earth and orbits a sun-like star in the habitable zone once every 290 days. With an estimated balmy surface temperature of about 70 degrees Fahrenheit, the planet could possibly be our cosmic sibling. The temperature is suitable for liquid water to exist on the surface, and therefore there is a decent probability of life. Other good candidates residing in habitable zones are Kepler-62f, Kepler-186f, and Kepler-442b, all at a distance of about 1,000 light years.

Assuming there are 200 billion stars in the Milky Way, we can hypothesize from the small Kepler data sample that there are about several 100 billion exoplanets in the Milky Way, about one or two per star on average. Most planets are so near their star that they take only a few hours to orbit, and others so far that they take thousands of years to orbit. The near type is likely to be too hot and the far type is too cold to support life. Of the more than 100 billion exoplanets, about 40 billion are earth-like planets of which 10 billion planets orbit their sun-like stars in Goldilocks, habitable zones. That is indeed a huge number for our galaxy alone.

The nearest exoplanet in a habitable zone is Proxima Centauri b, located 4.2 light years from Earth and orbiting Proxima Centauri, the closest star to the sun. But Proxima Centauri is not a sun-like star. It is a faint red star, also called a "red dwarf." Rocky planets not much bigger than earth are common in the habitable zones around red dwarfs. With about 60 billion such red dwarfs in the Milky Way the exoplanet hunt suggests there are an additional 10 billion rocky-type planets in the habitable zone around red dwarfs. Since the red-dwarf star is much cooler than a sun-like star, the habitable zone is much closer to the star than the earth is to the sun. Altogether, between sun-like stars and red-dwarfs there are about 20 billion earth-like planets in habitable zones in our galaxy.

With the demise of the Kepler mission, Space X (a private company run by billionaire Elon Musk) launched TESS (Transit Exoplanet Survey Satellite) in April 2018 on a Falcon rocket. The ambition of TESS is to survey the whole sky for planets around bright nearby stars, numbering about 200,000. Together with studies from other telescopes it will be possible to study the composition and atmosphere of some of the planets by analysing the spectra of light (Chapter 13), giving us another tool to evaluate the hospitality of a planet to life.

The discovery of the vast number (20 billion) of exoplanets with some earth-like features has intensified interest in the search for extra-terrestrial life, with special interest in planets that orbit a star's habitable zone. It is possible for liquid water, a prerequisite for life on earth, to exist on the surface when the temperatures are moderate. The study of planetary habitability also considers a wide range of other factors in determining the suitability of a planet for hosting life.

Are conditions favourable for life anywhere else in our own planetary system or in any of the exoplanetary systems? What is the probability of finding life, or even intelligent life in our galaxy?

In our own solar system, Mars is the only other planet barely in a habitable zone. Surface temperatures at the Mars equator may reach 70 degrees F in some areas. The Phoenix Probe analysed the soil of Mars for water to find clear favourable evidence. Long ridges on the surface of Mars suggest water flow billions of years ago in streams and rivers, carving out canyons. In addition Mars has polar ice-caps visible with a telescope. With water, conditions on Mars were probably conducive to life at some point in time. Much of the water has since been lost to space. In 1996 NASA released possible evidence of life on Mars. About 15 million years ago a meteorite was ejected into space by an asteroid impact on Mars. The meteor orbited the sun until it impacted earth in Antarctica about 13,000 years ago. The composition of the planetary rock is consistent with the surface of Mars. But the most encouraging feature is the evidence that the rock may contain fossilized micro-organisms.

With 200 billion stars populating a galaxy, and 100 billion or more galaxies spinning all over the visible universe, can earth bask in the privilege of being the only habitable planet with intelligent life? In his universal prayer, Alexander Pope pleads:[37]

> Father of all! in every age,
> In every clime adored …
> Yet not to earth's contracted span
> Thy goodness let me bound,
> Or think Thee, Lord, alone of man,
> When thousand worlds are round.

What is the probability of life and even intelligent life in our galaxy? How many exoplanets could develop life, and how many could develop intelligent species with communication capabilities?

We present very crude estimates based on the evolution of life forms on our earth, our solitary example. (There are other ways to make such estimates.) If the appearance of life on a given planet was very unlikely, one might expect it to take a long time for life to appear. On the other hand, the first forms of life on earth appeared quite fast. After earth formed in the solar system about 5 billion years ago, fossil evidence indicates that it took just less than 0.5 billion years to create the first bacteria. Accordingly, the probability of evolving elementary life-forms is quite high, optimistically 10 per cent. The further process of biological evolution on earth was very slow. It took 2.5 billion years, to evolve from the earliest cells to multi-cell animals, and another billion years to evolve through fish and reptiles, to mammals. Altogether it took 4 billion years to form 10 million species of large creatures. Humans developed only 0.2 million years ago. If a planet exists for 5 billion years, but "human"-level life only forms in the last 0.2 million years of the planet's life, the probabilities are very small. Again, a very crude estimate for planets with human-like creatures would be 0.2 million/5 billion = 0.00004. Finally, humans developed sufficient intelligence, science, and technology to communicate using radar only about 200 years ago, which gives the miniscule probability of advanced-technology "humans" with basic communication capability to be 200/5B = 0,00000004. All the above estimates of course presume earth's example to be typical, a hugely optimistic model.

We can turn these optimistic probability estimates into the probable number of exoplanets with life, intelligent life, and life with communication skills. Estimates based on the Kepler satellite discoveries are that there exist about 20 billion earth-like planets residing in habitable zones around sun-like and red-dwarf stars. Of these, 10 per cent or 2 billion will have some early form of life, 800,000 (0.00004 × 20 billion) will have large animals and even human-like creatures, but there can be only 800 (0.00000004 × 20 billion) exoplanets with advanced civilizations capable of communication. Hence crude but optimistic estimates, based on relative evolution times, suggest that several hundred advanced civilizations in our galaxy have cropped up at some time over a planet's existence. Some of these may have thrived and died out due to one of many possible causes. All these civilizations may not achieve a sufficiently advanced state in the "same" time frame to successfully cross-communicate. The "same time frame" for two civilizations to convey messages has to be within a span of 50,000 years for light-signals to reach one another, since the Milky Way is about 100,000 light years across.

Even if we are a factor of 100 too optimistic, there are/have been probably about ten advanced civilizations in the Milky Way galaxy. If we are a factor of 1,000 too optimistic, then we humans are surely alone in our galaxy.

The numbers of advanced civilizations become much more prolific when we go beyond our galaxy and consider the observable universe. As we will see in Chapter 9 there are 100 billion galaxies in the visible universe, based on observations from the Hubble telescope. If each galaxy has a few to a few hundred advanced species, *there are/have been/will be trillions of planets with advanced life-forms* in the visible universe, some perhaps more advanced than our own. But all these planets are immensely distant. Even our neighbouring Andromeda galaxy is 2 million light years away, so that signals originating from any planet in Andromeda would have to be launched more than 2 million years in the past to reach us now. The chance of inter-galactic communication is close to zero.

With more than 100 billion stars in each of 100 billion galaxies, the probability of intelligent, communicating life forms existing at some time is substantial. But we do not have the definite answer to: Are we truly alone? As we wait for the answer, Buckminster Fuller's reverie provokes us:[38]

Sometimes I think we're alone. Sometimes I think we're not. In either case, the thought is staggering.

EARTH MOVES

It took more than three centuries after Copernicus for solid evidence of the earth's motion to emerge. The most striking proof for earth's rotation came from Galileo's pendulum (Chapter 4). Jean Foucault arranged a spectacular experiment in 1851. Inspiration came from an unusual discovery at a lathe in his metal shop. While centring a thin metal rod mounted in the lathe chuck, he accidentally twanged the rod into up–down oscillations. When he rotated the chuck he was surprised to find that the up–down plane of oscillations did not change with the orientation of the chuck. Paying close attention to details others may have considered frivolous, he spun the lathe, to find that the plane of oscillations remained the same. The swinging pendulum ignored the rotation of the attachment point. Pursuing the effect further, he suspended a long pendulum vertically in a rotating chuck to find the same behaviour. In a single startling jump, Foucault recognized that if he replaced the chuck with the rotating earth, the pendulum would also maintain oscillations in the same plane while earth twisted under that plane (Figure 7.18). The effect is simplest to visualize at the pole (Figure 7.19).

Ironically, a church provided the large space needed for the first visible demonstration with a very long pendulum. Foucault carried out his experiment from the ceiling of the dome in the Pantheon of

FIGURE 7.18   The first demonstration of Foucault's pendulum under the dome of the Pantheon Church in Paris.[39]

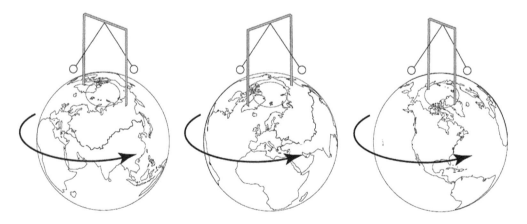

FIGURE 7.19   In the Northern Hemisphere the plane of the Foucault pendulum swing appears to turn clockwise, because earth rotates anti-clockwise at 15 degrees per hour, and less at lower latitudes. At the equator there is zero twist.

Paris. Under the dome, he suspended a 28-kilogram iron ball, 60 centimetres in diameter, by a steel wire more than 60 metres long. The pendulum ended in a spike that just cleared the floor but made marks in a layer of sand sprinkled on the floor. After spectators gathered, he dramatically set the pendulum in motion by setting fire to the cord that tethered the ball. It was also important to avoid the slightest wobble when the pendulum's motion started, so there would be no scepticism about the slow change in direction of oscillations. As the pendulum swayed to and fro over hours, the marks in the sand clearly showed the pendulum plane rotating clockwise. As the world turned counter-clockwise, it left the swinging pendulum behind. For the first time, human eyes could directly witness earth's rotation.

Absence of measurable parallax remained one of the strongest objections to the heliocentric theory until Bessel found annual parallax (Chapter 6). But a century before Bessel, English astronomer James Bradley (1693–1762) discovered another effect due to the orbital motion of earth. It was a new line of attack. Expecting that the angular position of overhead stars would shift slowly as his perspective changed with the earth's motion around the sun (Figure 7.20), he pointed a precision telescope straight up at the zenith and kept faithful track of stars that passed directly overhead. At this angle, distortions of starlight from atmospheric refraction were also minimal. A guiding plumb line, naturally vertical, told him how much the aim of the telescope had to be altered through the year to bring the star back into the centre of view. Bradley was delighted to detect a small angular shift over a few months of earth travel, but it was rather large compared to expectations from failed parallax attempts. Other stars near the zenith showed the same behaviour. It was a paradox. And once again, paradox opened the door to brand-new revelations.

One day, when Bradley went sailing along the River Thames, he noticed that every time the course of the boat changed, the vane at the masthead changed slightly in direction. (Like a flying-pennant, the weather-vane tells which way the wind blows.) Bradley remarked to the sailors that it

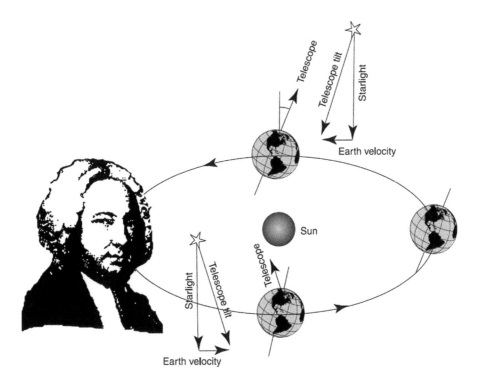

FIGURE 7.20   James Bradley discovered the aberration of starlight to demonstrate the revolution of earth around the sun, and determine the velocity of light.

was too much of a coincidence for winds to shift each time the ship changed course. Sailor friends assured him that such vane shifts were quite familiar. But there was no corresponding change in wind direction. Wrestling with the puzzle of shifting vanes for days, Bradley finally realized that on a moving boat the vane did not point purely in the wind direction, as on a stationary boat. The vane's direction was a combined result of the boat's velocity and the wind velocity. He had to account for both direction and speed. Each time the boat changed course, the vane swerved with the boat under the influence of the boat's new heading.

Bradley reasoned that light from a star approaching the moving earth is like a gust of wind. If earth moves, the direction from which light arrives is different from when earth is at rest. He had to tilt his telescope to catch the light in the tube, just as a man running through a vertical rainfall would have to tilt his umbrella into the oncoming rain to avoid getting wet. The ratio of earth's orbital velocity to the velocity of light gave the tilt angle.

But the tilt is a minuscule angle. Since earth moves at 3×10^4 m/sec and light comes down the tube at 3×10^8 m/sec, the ratio, and therefore the angle, is equal to 0.00001 radians, which is the same as 20 arc-seconds. Such a small angle would require the acuity of spotting a thumb at a distance of 100 metres. Bradley's techniques were up to the challenge. When earth velocity changes direction on the opposite side of its solar orbit, the tilt changes by 40 arc-seconds over a period of six months. Called "aberration of starlight," the shift is much larger than the parallax angle to the nearest star (about 1 arc-second), determined a century later by Bessel (Chapter 6). Bradley had stumbled upon an outstanding effect to demonstrate earth's orbital motion around the sun.

Unlike parallax, the aberration of starlight depends on how fast earth travels, not on how far it moves over its orbit around the sun. The effect is largest when the line of sight to the stars is perpendicular to the direction of the earth's motion. There is zero effect for a star which lies in a direction parallel to earth's orbital motion. In one stroke, Bradley demonstrated that earth moves, and determined the velocity of light to be 3×10^8 m/sec. By his time, however, the earth-centric view had already retired to obscurity, and a first estimate had already been made for the speed of light, as we will see later (Chapter 9).

8 Laws of Motion in the Heavens: Opening New Doors through Precision

The roads that lead man to knowledge are as wondrous as the knowledge itself.

Johannes Kepler[1]

Just as the ears were made for sound and the eyes for colour, so the mind is made for quantity and precision. It is lost in darkness when it leaves the realm of quantitative thought.

Johannes Kepler[2]

The cosmic Revolution had only just begun with Copernicus. Tycho Brahe (1546–1601) and Johannes Kepler (1571–1630), the dual heroes of this chapter, took the Reformation of astronomy to a new level, ushering in an age of precision and rigour. They were binary astronomers with richly contrasting characters. One had a passion for cold, hard facts collected with elaborate instruments and tireless energy. The other was possessed by a spiritual longing to penetrate the design of the universe to reach its long-sought harmonies. Brahe pursued the ant's view of precise details, Kepler the overarching bird's view. Their union was an epic confluence of minds, approaches, and discoveries, leading directly to Newton's breakthrough concept that gravity is the force which rules the solar system, and the law that runs the universe. Following the trail-blazers, precise measurements of natural phenomena and exacting comparisons with existing theoretical structures have become the twin forces which have guided progress in understanding nature.

SPARKED BY AN EXPLODING STAR

Born three years after Copernicus died, Danish astronomer Tycho Brahe was destined to become the greatest observer of stars since Hipparchus. With diligent attention to the heavens, Brahe's Alexandrian/Greek mentor discovered the curious phenomenon of equinox precession. Heaven rewarded Brahe even more handsomely. His unwavering devotion to details in the night sky met with stunning new revelations.

Tycho's fascination with astronomy started as a young boy when he saw his first solar eclipse. Struck with wonder that the eclipse actually took place on the day it was predicted, he later described his amazement:[3] "To know of such a thing in advance makes man almost a god."

Ever since that fateful eclipse, Tycho was star-struck, obsessed with how he could understand the motion of heavenly bodies well enough to forecast their relative positions. He purchased Ptolemy's *Almagest*, and spent all his spare money to buy the latest astronomical instruments. An affluent uncle, who raised him, preferred Tycho devote his energies to studying law. The Brahes came from a leading family of Danish nobility, whose forefathers held a proud record of service to the crown. To keep the boy on the tried and true path, the uncle hired a tutor to monitor Tycho's activities. When Tycho wanted to study mathematics, his tutor dutifully told him stories about mathematicians who went mad because they thought too much about impossible problems. Pursuit of astronomy was aimless and unsuitable for a man of noble birth, whose presumed destiny was to attend court.

To young Tycho, the prospect of a nobleman's luxurious existence was thoroughly boring. Endless preoccupation with horses, hunting dogs, and drinking bouts was futile. Tycho preferred to look around God's beautiful world. While his watchdog tutor slept, Tycho sneaked off to make measurements in the night sky. Eventually the tutor realized his young charge had turned his mind toward the heavens, far from the base pleasures of a nobleman's estate. He even helped Tycho to obtain books and instruments.

Faithfully keeping a systematic record of observations and checking them against published tables, Tycho was surprised to find that actual planetary positions differed substantially from calculated positions, even if he judged by the best available works. At one time, there was a spectacular conjunction when Jupiter and Saturn came so close together as to be almost indistinguishable. Ptolemy predicted the event to take place a month later, and Copernicus a few days later. Bothered by such large discrepancies, Tycho pestered his professors with questions about astronomy. Obviously almanacs and tables were in poor shape. Here was an area where he would make important contributions.

At the same time, Tycho began to dabble in astrology and poetry. These subjects remained important preoccupations for the intelligentsia, including astronomers. A recent lunar eclipse stimulated Tycho to write a poem predicting the death of the Turkish Sultan, feared and hated all over Europe as the detested conqueror of the once glorious capital of the Eastern Roman Empire, Constantinople. When Suleman the Magnificent actually died near the appointed time (1568), Tycho became famous to take on the mantle of leading astrologer. Whenever a royal child was born, the monarch would ask Tycho to send out a horoscope. Tycho usually obliged with several hundred pages, bound in green velvet. For this he was paid handsomely.

Poets of the time dipped into astronomical and astrological themes with equal versatility. Shakespeare describes Romeo and Juliet as a pair of star-crossed lovers. In *Henry VI*, Bedford beseeches the spirits:[4]

> Henry the Fifth, thy ghost I invocate:
> Prosper this realm, keep it from civil broils,
> Combat with adverse planets in the heavens!
> A far more glorious star thy soul will make
> Than Julius Caesar or bright.

But in *Julius Caesar*, the wise poet clearly shows scepticism about the power of stars over the fate of mortals:[5]

> The fault dear Brutus lies not in the stars
> But in ourselves.

Upon the death of his uncle, Tycho's financial position improved. Finally he could put aside the cheap draughtsman's compass that barely filled his need and begin a serious career in astronomy. From skilled German artisans, he ordered elaborate instruments of superior quality: a huge quadrant for measuring the altitude of stars, a sextant for measuring the angular distance between stars, and a globe, 5 feet in diameter, for mapping the universe. It took 20 men to carry the giant quadrant to its destination and erect it for observations.

A heavenly gift rewarded the diligent observer. In 1572, Tycho spotted a stunning sight, a brightening star in the constellation of Cassiopeia:[6]

> Amazed … astonished and stupefied, I stood still, gazing … with my eyes fixed intently upon it and noticing that star placed close to the stars which antiquity attributed to Cassiopeia. When I had satisfied myself that no star of that kind had ever shone forth before, I was led into such perplexity by the unbelievability of the thing that I began to doubt the faith of my own eyes.

By now he was quite familiar with the stars populating each constellation to realize that this strange light was a newcomer to the heavens. Of course, other astronomers also noticed the brilliant guest star. The most natural explanation they jumped to was another comet. Tycho dismissed them. Heightened perception from years of meticulous observation endowed him with a deeper insight about happenings in the sky. The new star could not be a comet because it remained in the same place relative to fixed stars, night after night, over a period of one and a half years, as if the brand-new star was nailed to the celestial sphere. Past experience taught that a comet clearly changed its position relative to fixed stars over a few days, moving through the skies, even faster than a planet does. Rather than showing a tail as most comets do, the unexpected visitor changed intensity, growing more brilliant than Venus. He could even spot it in daylight. Over months it faded and became suffused with a reddish tinge. But it remained visible for a year and a half before finally dying out. Tycho named the new star *stella nova*. (See Figure 7.4 for artist Bayer's rendering of the stunning appearance.)

How could such a changing body possibly belong to the heavens? Greek astronomers had made it abundantly clear that the stellar sphere never changed. The Bible taught that God ceased the work of creation on the seventh day. Were not the heavens unchangeable for all eternity ever since? And yet, this startling change in the heavens was bright enough for all to see.

For Brahe, there was a decisive way to find out if the nova belonged to the heavens: parallax. With his recently acquired five-and-a-half-foot sextant, more accurate than any of his contemporary astronomers' instruments, Brahe showed that the nova was much farther away than the moon. Night after night, Brahe found the visitor to remain in exactly the same place among stars of Cassiopeia. It was a sensational discovery.

When amateur astronomers and astrologers flooded the market with treatises claiming that the new star was a comet, Brahe berated them as[7] "blind watchers of the skies." Some of Brahe's contemporary stargazers predicted a dire outcome; others hailed a glorious new era. Some believed that the celestial arrival proclaimed the Second Coming of Christ. Others forecast doom, the end of the world. Indeed Brahe's nova was to herald a change. But it was not to be a flood of biblical proportions. Rather it was an intellectual cleansing, the first trickle of a deluge destined to wash out the Aristotelian cosmos.

Even Brahe could not resist interpreting the astrological significance of the event. By now he was adept at providing juicy forecasts for his eager audiences. It was a brilliant display of heavenly fireworks. At first, since the nova was bright and beautiful like Venus, he forecast that its effect would be pleasant. As it grew dim like Mars, a period of wars and destruction of cities would come. Finally, as it faded like Saturn and eventually disappeared completely, the nova warned of a period of famine, pestilence, and death.

A CASTLE FOR ASTRONOMY

After the nova report, Brahe's fame continued to spread. From his remarkable reading of the new sign from heaven, his work gained respect abroad. Making many friends in Germany, he seriously contemplated migrating there. Alarmed by the possibility, King Frederick II of Denmark prevailed upon Brahe to stay in Denmark. A good court astrologer would be essential to avert evil from his kingdom. And Brahe could continue his astronomical discoveries to enhance the prestige of Denmark. Determined not to lose the new celebrity, the Danish king offered Brahe the island of Hven near Elsinore to set up his dream observatory in 1574. It was a luxurious 2,000 acres of land only 14 miles from Copenhagen, the active centre of Denmark. Here was Brahe's opportunity to set up the best observatory of all time. Hven was the first large-scale enterprise in the history of physics. Sparing no expense, he assembled the finest instruments possible. Capping it all was a heavenly castle, a red brick palace with pointed turrets, which he called Uraniborg, after Urania the muse of astronomy. Brahe's palace cost the king more than a ton of gold, a major financial commitment for a small country. And Brahe invested a large sum from his own funds.

From Ptolemy's description of Hipparchus' clever astronomical tools, Brahe reproduced quadrants, sextants, octants, and armillaries. Some were fabricated with wood, others with expensive brass. Some were modular so they could be disassembled and carried away to a new location if necessary. The instruments were larger, more stable, and better calibrated than any before (see Figure 8.1 for the size of one of his quadrants). His mural quadrant had a radius of 6 metres, took 40 men to install, and had a portrait of the egotistic man carved inside. On the huge arcs that fit within the gigantic apparatus, he marked smaller fractions of a degree than ever before used to map the skies. A careful eye could accurately read off angles as small as a single minute of arc. (Compare 1 minute of arc to the 30 arc-minutes for the angular size of the moon's face.) He came up with new ways to make oversize devices rotatable in the vertical and horizontal planes. Alternatively, he could lock the instruments in position so they could track the same star, night after night. Using mercury in clepsydras (old Egyptian water clocks), Brahe and his staff kept accurate track of time intervals. With duplicate sets of equipment, his staff could carry out simultaneous and independent observations of notable celestial events.

The King of Denmark assigned a large staff of skilled workers and students to Brahe's command. They built workshops to repair instruments, a windmill to grind food and pump water, fish ponds, flower gardens, and a printing press supplied by its own paper mill to record and disseminate volumes of precious observations. In an extensive library, he stocked astronomical manuscripts collected from all over Europe. Among the large entourage of assistants were crews of human calculators. Operations like long divisions and fraction manipulation were still very tedious. Trigonometry, the most advanced level of mathematics, entailed hours of computational drudgery, rather than the instantaneous result from the push of a button on a modern electronic calculator. Teams of

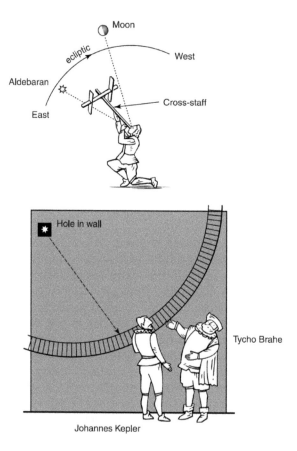

FIGURE 8.1 Compare the size and quality of Brahe's quadrant (bottom) with the crude staff of Copernicus (top). Brahe installed a giant quadrant that could mark off angles as small as 1 minute of arc.

number crunchers were essential for any serious project aimed to predict occurrences of eclipses and conjunctions.

At every expected lunar eclipse, all of Brahe's assistants spent the night plotting the course of the moon. Extending the customary practice of making observations only during notable celestial events, such as planetary conjunctions or eclipses, Brahe organized continuous sky surveys with the goal of compiling a new star catalogue. His was no haphazard survey. With a degree of scrutiny unprecedented in history, he established accurate data on 777 fixed stars over 33 years. It was the first comprehensive star catalogue since Hipparchus and Ptolemy. Every few years, he repeated observations, comparing them against older sets.

Brahe inaugurated a new era in the development of science, the era of rigour. With measurements accurate to 1 minute of arc, he surpassed Ptolemy's and Hipparchus' observations by a factor of ten. In another vital innovation, Brahe made error estimates for observations, giving successors a high level of confidence in his numbers. Brahe's achievement was all the more spectacular considering that the invention of the telescope and the regulated clock were still a century in the future.

What drove the indefatigable Brahe through years of repeated observations, painstaking calibration, recalibration of elaborate instruments, and detailed cataloguing of volumes of measurements? Like artists of the Renaissance, he felt compelled to describe nature in live infinitesimal detail, exemplified by the dazzling realism of Albrecht Dürer's *Hare* and *The Large Turf* (Figure 3.7). In the *Craftsman Handbook*, a Renaissance art teacher, Cennino Cennini, instructs would-be artists on how to learn from nature. It could just as well serve as a lesson to scientists in Brahe's time:[8]

Take pains and pleasure in constantly copying the best things which you can find done by the hand of the great masters … Mind you, the most perfect steersman that you can have, and the best helm, lie in … copying from nature. And this outdoes all other models; and always rely on this with a stout heart.

In this same spirit, Brahe meticulously observed and catalogued the stars, striving for exactitude without precedent. Precision was the new frontier.

Another force drove Brahe to the precision frontier: a passion to figure out the true scheme of the universe. Captured by the centrality of earth, Copernicus' bold theories did not captivate him. Unable to measure parallax of the stars due to the orbital span of the earth around the sun, Brahe was ready to reject the heliocentric model. Little did he know that his superb, arc-minute precision came nowhere close to the arc-second accuracy needed to observe parallax for even the nearest star (Chapter 6). Apart from the absence of parallax, Brahe found it difficult to accept that an object as large as the earth would have the ability to move, although the moving sun did not seem to bother him too much. Copernicus would no doubt have enlightened Brahe about such a logical fallacy. Brahe's idea was that Copernicus was only partly right. Planets revolved around the sun. But the sun revolved around the earth. There was a certain degree of spiritual comfort in making such a compromise to keep earth at the stationary centre. It did not bother him that in such a scheme the orbit of the sun would intersect the orbits of Mars, Mercury, and Venus. Brahe painted a picture representing his cosmology on the dome of his grand observatory. On the walls of his castle he commissioned portraits of great astronomers of the past from Timocharis to Copernicus. To these he proudly added a mural of himself, and one of his unborn descendant. Ironically, even though he rejected Copernicus, Brahe's precise measurements ultimately tilted the balance in favour of the heliocentric theory.

A COMET SHATTERS THE CRYSTALLINE SPHERES

In 1577, at the peak of Brahe's career, the heavens smiled once again upon the great Dane. A spectacular comet appeared. Bright as Venus and with an enormous tail, the length spanned an angle of 22 degrees across the sky. It was an apparition terrifying to the superstitious. As customary for new comet appearances, astrologers, including Brahe, immediately started preparing dire forecasts.

Always, the unpredictable comets spelt disaster (from Latin dis-astra – against the stars), such as the fall of a mighty ruler, or a deluge of Biblical proportions. Terrified eyewitnesses let their imaginations run wild, reporting visitors as bloody-red and multi-headed mythological beasts, equipped with savage talons. Comets were blamed routinely for bad weather and pestilence. One Pope even excommunicated a comet which was causing great panic in Europe.

Shakespeare makes liberal use of cometary warnings, sometimes with scepticism. In *Henry VI*, Charles responds to news of a signal light:[9]

> Bastard of Orleans:
> See, noble Charles, the beacon of our friend;
> The burning torch in yonder turret stands.
> Charles:
> Now shine it like a comet of revenge,
> A prophet to the fall of all our foes!

But in *Julius Caesar* the bard complains how illogical it is that:[10]

> When beggars die, there are no comets seen
> The heavens themselves blaze forth the death of princes.

Even today, many cults cite comets as messengers of doom. As recently as 1997, the arrival of Hale-Bopp triggered a mass suicide among the California Heaven's Gate cult who believed that death would hasten their encounter with an alien spaceship parked behind the comet, as their leader foolishly proclaimed.

Brahe studied the comet of 1577 with the eye of an astronomer. As with the nova he scrutinized five years earlier, he used parallax to determine (Figure 8.2) whether the comet belonged to the nearby atmosphere, as Aristotle claimed, or whether it roved in the heavens beyond the moon.

If the comet was as far as the moon, its lunar parallax displacement over six hours of the earth's rotation would have been very nearly a whole degree. But this comet showed a substantially smaller

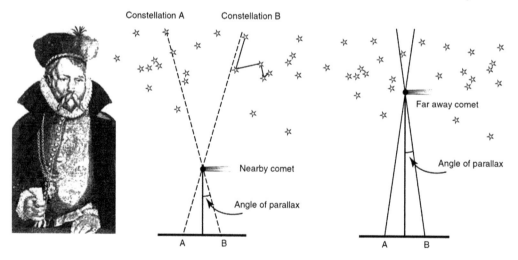

FIGURE 8.2 Tycho[11] (left) placed comets beyond the moon by parallax measurements. The farther away a comet, the smaller its angle of parallax as measured from two places on the earth, A and B. (Centre) if the comet is close to earth, observers at A and B will find the comet to be in different constellations. (Right) if the comet is distant, observers at A and B will claim it resides in the same constellation. Brahe did not employ a second observer. He just waited six hours until earth rotation carried him far enough to detect a true lunar parallax of about 1 degree. In the same period the comet shifted less than 10 arc-minutes (one-sixth of a degree).

displacement. It was at least six times farther away than the moon. Brahe was ready to stick his neck out again. Like the nova, the comet belonged to the heavens. It never came anywhere close to our atmosphere. Over the next two decades, Brahe used the same technique to place four other comets well beyond the moon. Comets are truly celestial objects. Brahe's assertion raised a number of troubling questions. If comets actually move through celestial spheres, they would pierce right through them! How could Aristotelian spheres have any physical reality? Atmospheric comets and crystalline spheres were headed for the dustbin of history.

THE EVICTION OF THE GREAT DANE

Brahe was flamboyant, full of contradictions. He ran his observatory tyrannically, reprimanding staff for wasting precious time on pleasures of hunting, feasting, lusting, and constant duelling, even though he frequently partook of the same revelry and debauchery. His personal game preserves attest to his love for hunting. He took the daughter of one of the peasants as his mistress. At grand feasts held in a regal banquet room, lavishly decorated with huge cages filled with rare birds, he would impress royal guests. Always present at banquets was his favourite clairvoyant, whose sayings he treated as oracles. Whenever the gypsy spoke, Brahe ordered everyone to quiet down so the prophecy could be duly heard and recorded. Like a despot, he ruled over the farmers of the island, incarcerating in his private prison any who defied his rules. Yet he would administer medicine to sick peasants at no charge, arousing the animosity of professional doctors, who labelled him a quack.

Even as a young man, Brahe displayed a quick temper. Once he challenged a drinking partner to a foolish duel, not over love, but about who knew more mathematics. Fortunately, all he lost was a slice of his nose. For the rest of his life he wore a false nose of gold and silver, scaring fishermen and peasants of Hven, who feared him for a sorcerer. Brahe was very sensitive about his metal nose. According to stories, he carried around a little box of glue to dab on whenever the appendage grew wobbly. One of his colleagues joked behind his back that Brahe used his metal nose as a trusted guide to set the sights on his instruments. Sadly, his gold nose attracted more interest among Europeans than his astronomical findings.

When Brahe's royal benefactor died, the successor, Christian IV, did not share his predecessor's broad cultural outlook. Besides, the Danish treasury was running dry from military adventures. The new king could not tolerate the rude, arrogant, and extravagant Dane. Claiming that Brahe's astronomy was a trivial hobby, he appointed a commission to inquire into the value of his astronomical work. The advisors obliged with a report the boy-king wanted, declaring that Brahe's work was not only useless, but dangerous. As rumours spread about undesirable activities, Brahe grew unpopular. Blaming Brahe for mistreating peasants, the young monarch withdrew the crown's financial support, grabbed the island of Hven and made it a gift to his favourite mistress. It was a sad and abrupt ending to Brahe's 21-year reign (1597).

An incensed Brahe finally abandoned Denmark for Austria, claiming that wherever his fate took him the same stars would shine for him to study. In Austria, an enlightened emperor was most interested in the services of the world-famous astronomer, especially his forecasts on the destiny of his kingdom. Brahe took his precious instruments to Prague. His foresight in making the essential ones portable now paid off. More important, he carried all the valuable data accumulated over two decades of painstaking observations. Brahe received a generous allowance and another castle, this one at Benatky, 22 miles from the capital at Prague. A few years after his departure, ruin fell upon Uraniborg. All that remains today is a ghostly outline of its once glorious foundations. Everything went up in smoke, except for one great brass globe.

A BUMBLING MYSTIC LOOKS FOR CELESTIAL HARMONIES

Besides shattering the crystalline spheres, the great comet of 1577 triggered another landmark event in the history of physics. It aroused young Johannes Kepler's interest in astronomy. When he was a child, his mother showed him the sensational comet with its spectacularly long tail. Another night, his father took him out to see a lunar eclipse. Unlike Brahe's noble lineage, Kepler's father was a poor German mercenary soldier. Criminally inclined and quarrelsome, he narrowly escaped a hanging. Eventually he abandoned his wife and child. The mother tended to her father's inn where young Johannes served as tavern potboy. Only a miracle could turn the family fate. It came in the form of the provincial Duke's charity school programme. From an impoverished family of hard-working and God-fearing Christians, one wretched youth would receive a scholarship. That youth was to be Kepler. Completing his education, Kepler landed a post as a teacher of mathematics and astrology in the small Austrian city of Graz, temporarily under Protestant rule.

Cataclysms of the Reformation and Counter-Reformation wracked the religious situation in Germany. Secular princes, embroiled in religious disputes, expropriated Church property, inviting the wrath of Catholic prelates. Citizens of each German state had to conform to the religion of their state's prince. As a Lutheran minority in a Catholic community, Kepler suffered from the antagonism of opposing parties. While Graz remained under Protestant rule, Kepler taught mathematics and astrology. But Catholic victors soon uprooted him.

Deeply engrossed in his own thoughts, Kepler made a poor teacher. His students were usually bored, except when the near-sighted teacher, with his sickly, frail constitution, tripped over himself, drawing hearty jeers. As an astrologer he fared much better. One of his duties was to publish annual forecasts and calendars giving dates of eclipses and prospects for weather and harvests. In his first calendar, he made a lucky guess, correctly predicting a spell of bitter cold weather and a Turkish invasion. On another occasion, he warned of a severe thunderstorm two weeks in advance. People were duly impressed when the storm arrived punctually and struck with gale force. It only took a few successful predictions for people to forget and ignore the many dismal failures of astrology. Respect as a competent astrologer made Kepler's position at Graz secure.

At heart, Kepler was a Pythagorean mystic who fervently believed that God constructed the universe according to elegant geometric principles. As a young university student, he ardently defended Copernicus in a public debate. The sun enthralled Kepler as captivatingly as it had enticed Ficino, Bruno, and Copernicus. The geometrical simplicity of Copernicus' heliocentric model beguiled him. What an astonishing triumph of reason! Like Pythagoras, Kepler imagined that the clue to God's mind lay in simple geometrical order, numerical relations, and harmony. Surely the Architect of the universe was neat, orderly, and precise. Why were there six, and only six, planets, he wondered? To Kepler, the number six held a profound significance stemming from an elementary number property that $1 + 2 + 3 = 1 \times 2 \times 3 = 6$. Surely geometry ordained the number six for the planets. What was the mathematical relationship between the diameters of planetary orbits? Copernicus had skilfully worked out the values, but the master was completely silent about the tantalizing question of any *relationships* among the table of numbers. Kepler was on the lookout for numerical and geometrical links between orbit sizes.

One day, while totally self-engrossed in delivering an astronomy-astrology lecture, he drew two circles, a large one to represent the orbit of Saturn and a smaller one for the orbit of Jupiter (Figure 8.3) The ratio between the diameter of the orbit of Saturn and the orbit of Jupiter was about 1.8, according to Copernicus, as shown in Table 7.1. Examining the space between the two circles, he made a startling find. He could almost fit an equilateral triangle symmetrically between the two circles so that the outer circle touched the triangle vertices while the inner circle just touched the three sides.

Kepler hastened to insert the next perfect figure, a square. Could a circle drawn symmetrically inside the square represent the orbit of Mars? The ratio of the diameters did not agree too well. After the square, the pentagon and hexagon followed as third and fourth regular figures. He continued the sequence of regular figures to draw a pentagon inside the circle of Mars, a circle inside the pentagon, and so on (Figure 8.4). Could he have stumbled upon God's geometric secret for the arrangement of the planets? To his bitter disappointment, the ratios of the resulting circle diameters did not quite follow the progression of planetary orbit sizes Copernicus had so painstakingly determined (Table 7.1).

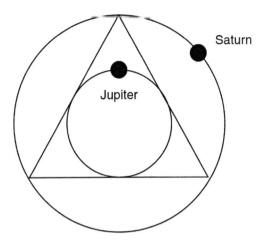

FIGURE 8.3   In the first stage of his lifelong pursuit of a geometric relationship between planetary orbits, Kepler thought he hit upon the key. Circles inside and around an equilateral triangle very nearly represented the orbits of Jupiter and Saturn.

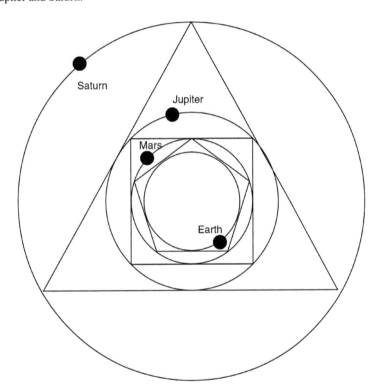

FIGURE 8.4   Kepler thought he found a geometrical pattern for the arrangement of all the planetary orbits.

Kepler was not ready to give up so early. Space is three-dimensional, not planar. There is no mathematical imperative for a series of planar figures to end abruptly in the octagon, as would be necessary for such a sequence to represent a planetary series that ends with Mercury. To Kepler, the universe had to be proportioned after the *five* perfect solids, not after the *infinite* number of regular plane figures. Through perfect solids, Kepler hoped to reach the harmony which eluded astronomers ever since Pythagoras initiated the quest for a geometrical structure underpinning the universe.

For Kepler, the most interesting property of Platonic solids was that he could situate each symmetrically inside spheres. All corners of a solid touched the surface of the sphere. Equally fascinating was the property that a smaller sphere fit symmetrically inside each solid so that the inner sphere touched every solid face at the exact centre. Making models for solids out of sticks, he suspended each with strings to nest them inside each other. The space between the five solids was just large enough to accommodate six spheres of nearly the correct diameters of planetary orbits (Figure 8.5). Could this be the invisible skeleton for the solar system, he wondered?:[12]

> Within a few days everything fell into its place. I saw one symmetrical solid after the other fit in so precisely between the appropriate orbits that if a peasant were to ask you on what kind of hooks the heavens are fastened so that they don't fall down, it will be easy for thee to answer him.

At age 25, Kepler published his first work, *The Secret of the Universe*. At the very beginning of his treatise, Kepler endorsed Copernicus, calling *De revolutionibus orbium coelestium*[13] "an inexhaustible treasure of truly divine insight into the wonderful order of the world and all bodies therein." Constructing the universe according to the model of perfect solids, God made the cosmos into one exquisite geometrical work of art, thought Kepler. Having read the mind of God, he could "hear" the tunes of the celestial music box, celestial harmonies that mystics and poets through the ages had dreamed about. English poet John Milton (whose life overlapped Kepler's) expressed the deep intellectual longing to hear the celestial harmonies:[14]

> Ring out, ye crystal spheres!
> Once bless our human ears
> (If ye have power to charm our senses so;)
> And let your silver chime
> Move in melodious time,
> (And let the base of heaven's deep organ blow;)
> And with your ninefold harmony
> Make up full concert with angelic symphony.

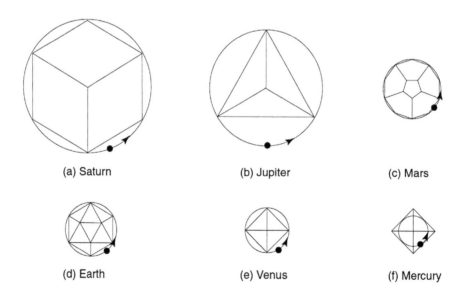

(a) Saturn   (b) Jupiter   (c) Mars

(d) Earth   (e) Venus   (f) Mercury

FIGURE 8.5   Kepler's planetary model based on five perfect solids and spheres. A sphere containing the orbit of the outermost planet (Saturn) could exactly envelop a cube. The cube circumscribed the sphere of Jupiter's orbit. Then a tetrahedron fitted inside Jupiter's sphere and so on.

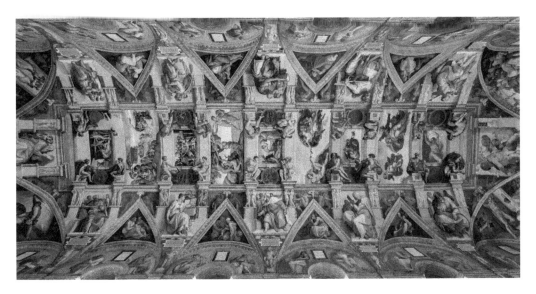

FIGURE 8.6 *Sistine Chapel Ceiling, Vatican Palace*, Vatican State. Michelangelo (1475–1564).

Kepler's geometric model for the solar system was as much a product of the Renaissance as Michelangelo's masterpiece of the Sistine Chapel ceiling (Figure 8.6).[15] Revival of the classics created a yearning for an orderly world expressed through the clarity and precision of geometry. On the 70-foot-high curved ceiling of the Vatican chapel, Michelangelo arranged his frescoes in a geometrical pattern of still rectangular, triangular, and circular frames that imposed a central spine into the complexity of human figures, organizing the intricate iconography narrated by 300 figures. Through the tangled tales of the Testament, Michelangelo delivered the biblical message: Humanity will be exalted by the coming of the Redeemer. Balance and symmetry dominate the entire ceiling, as well as individual frescoes within each frame. For example, Christ's ancestors, Eleazar and Mathan, balance each other while fitting perfectly inside spherical shells that define the lunette (Figure 8.7).[16] Order, harmony, and geometric patterns captivated the minds of the Renaissance astronomer and the artist.

"LET ME NOT HAVE LIVED IN VAIN"

Even though the scheme of perfect solids was enchanting, Kepler was acutely aware that the fit between data and model was far from satisfactory. The stubborn facts just did not conform to the captivating theory. In his bones he felt that better astronomical data would ultimately resolve the remaining discrepancies and vindicate his elegant construction. It was too beautiful to be wrong! Who better to supply such data than that famous Imperial Astronomer, Brahe? Surely he would find final corroboration for his theory of nested perfect solids:[17]

> Let all keep silence and hark to Brahe, who has devoted thirty-five years to his observations … he shall explain to me the order and arrangement of the orbits.

The religious situation in Austria continued to grow more precarious by the month. Fanaticism spread with equal fervour throughout Catholic and Protestant regions. Persecution was on the rise. Anyone with non-conformist ideas was at great risk. In 1600, the same year that Bruno burned at the stake in Italy, an official decree proclaimed that all Lutheran preachers and schoolteachers must leave Graz within eight days, or face punishment of death. Kepler fled. It was time to go to Prague, where he could join the famous Brahe.

FIGURE 8.7   *Eleazar and Mathan. Lunette*, Sistine Chapel, Vatican Palace, Vatican State, Michelangelo (1475–1564).

Word of Kepler's claim to fathom cosmic mysteries had already reached Brahe. The seasoned master was quite impressed by young Kepler's astronomical knowledge, in particular, by his mathematical ability to come up with such an ingenious model. But Brahe's encounter with the comet of 1577 had already shattered his belief in crystalline spheres, so he could hardly give much credence to Kepler's nested spheres and solids. Yet Brahe was thrilled at the prospect of finding a competent astronomer-assistant. Perhaps it would take a mathematician of Kepler's calibre to finally make sense out of Brahe's wealth of observations. Brahe invited Kepler to join him. Now 53 years old, Brahe knew he lacked the time and the prowess to fit the volumes of exquisite data into his favourite hybrid model. He hoped Kepler would agree to prove his conjecture. It was a fateful day when Kepler set out on his journey from Prague to meet Brahe on January 1, 1600. It was to be an eventful century of fascinating discoveries.

Brahe and Kepler tried to work together for nearly two years. They quarrelled with great frequency. Fortunately for science, they reconciled with equal regularity. Brahe was always reluctant to share his data openly, afraid that the sharp mathematician would beat him to the secret of the universe. In one letter, Kepler noted bitterly that Brahe gave him[18]

no opportunity to share in his experiences. He would only, in the course of a meal and in between conversations about other matters, mention, in passing, today the figure for the apogee of one planet, tomorrow the nodes of another.

In retaliation, Kepler often threatened to quit. But there really was no hope for Kepler to return home, where religious-political uprisings aimed to wipe out Protestantism. Kepler waited patiently for his opportunity to get at the treasure:[19]

> My opinion of Brahe is this: he is superlatively rich, but he knows not how to make proper use of it, as is the case with most rich people ... Any single instrument of his cost more than my and my whole family's fortune put together ... Therefore, one must try to wrest his riches from him.

After many bitter quarrels, Brahe grudgingly agreed to hand over all the data on Mars, just one planet. Brahe knew that Mars would be an impossible challenge for any astronomer. Both the Ptolemaic and Copernican models failed miserably for the wayward planet. All previous attempts to compute its orbit had failed to the extent of a few degrees, a huge error. Overjoyed at finally receiving a healthy set of data, the confident Kepler boasted he would determine the true orbit of Mars within eight days. It kept him befuddled for eight years. At times, it nearly drove him insane. The battle to imprison the god of war into a calculable orbit turned out to be a fertile quest. It finally led Kepler to the ellipse.

By stages, Kepler gained the master's total respect. Ultimately Brahe, nearing death, was ready to bequeath all his observations, records, and instruments to Kepler. But he hoped Kepler understood that any new tables of planetary motions should be drawn up according to Brahe's model. Brahe suffered an unfortunate and premature medical calamity. He died of a burst bladder after drinking too much at a royal banquet. He could not, by custom, insult his royal guest by leaving in the middle of the feast to relieve himself. On his deathbed, he beseeched Kepler over and over, "let me not have lived in vain."[20] The greatest astronomer since Hipparchus was buried with pomp and ceremony befitting a royal prince.

REFORMATION OF ASTRONOMY

Kepler remained convinced that the true orbits of planets would ultimately succumb to the aspirations of Pythagoras and find expression in simple and elegant mathematical relationships. Undaunted by the extent of its aberration from the circular, he continued to focus efforts on the recalcitrant planet, Mars. On what curve did Mars move over 20 years of Brahe's meticulous observations? To put Mars into a calculable orbit demanded the solution of both Mars' and earth's orbits. After about 70 trials, using circles upon circles, placing the sun in various positions, slightly off-centre by different amounts, Kepler finally produced an intricate combination that appeared to agree quite well with one extensive volume of Brahe's meticulous observations. Kepler felt sure he had defeated the god of war. Then, to his utter dismay, he found that the same curve, when continued beyond the data he had just matched, disagreed with the rest of Brahe's observations. The discrepancy was only a slight amount, just 8 minutes of arc. Copernicus himself would have been delighted with an accuracy of even 10 minutes of arc. But Kepler knew that Brahe's observations were far more accurate, with a margin of error of less than 2 minutes of arc:[21]

> Since divine goodness has granted us a most diligent observer, Brahe, from whose observations the error in this calculation of eight minutes in Mars is revealed, it is fitting that we recognize and make use of this good gift of God with a graceful mind ... if I had believed that we could ignore these eight minutes, I would have patched up my hypothesis accordingly. But since it was not permissible to ignore them, those eight minutes point the road to a complete reformation of astronomy.

Kepler was putting theoretical construct to the most severe test, demanding that it fit the trustworthy data, or be called worthless. A scrupulous scientist must never force facts to fit the theory. Kepler was willing to go where his observations led him, ready to bow to obstinate facts. With this attitude, he rejected two laborious years of hard computations. He even wrote 39 lengthy chapters to describe his failure. But Kepler realized where he had gone wrong. Perfect circles were doomed:[22]

> This was our punishment for having followed some plausible, but in reality false, axioms of the great men of the past.

THE ELLIPSE, AN HONEST MAIDEN OF NATURE

Faith in elegance and simplicity gave Copernicus the courage to demolish Ptolemy's intricate apparatus. Kepler was ready to go one step further. If planets are sometimes closer to the sun, and sometimes they are farther away from the sun, then the *simplest*, most logical conclusion is that their *paths cannot be circles*. Some other mathematical figure would better describe planetary orbits. As he started to explore various oval orbits for Mars he saw a glimmer of success. As a consummate mathematician, Kepler was aware of the conic sections of Apollonius of Alexandria, still a pure mathematical curiosity. When Kepler tried the ellipse, it fit perfectly (see Figure 8.8 for the definition of an ellipse). He no longer needed multiple circles to account for the changing distances between Mars and sun. The ellipse gracefully provided the needed variations.

Encouraged, he spent a long time scrutinizing the orbits of the rest of the planets. To his utter amazement, he found that ellipses worked for *all* the planets. Kepler's *First Law of Motion* for planets was born (1609):

> All planets describe elliptical orbits with the sun as one of the foci.

Kepler[23] "gazed with astonishment on a new light." Nature does conform to mathematical patterns, and he had found the right one. In one decisive stroke, he could finally sweep away the classical rubble of deferents, eccentrics, epicycles, and equants, banishing at last and forever the ghost of Ptolemy. With six elliptical orbits for six planets, Kepler erected a new order. Rather than compounding multiple circles, the fresh ellipse was far more elegant. Although mathematically more demanding than a circle, the ellipse was the first curve to fit all planetary orbits as accurately as observations demanded.

But Kepler's new design imposed a serious demand; it displaced the sun from the centre of the heavenly drama to the focal point of the ellipse. Alas, another symmetry had to be abandoned. A faithful Copernican would have found this situation immensely disconcerting. To a friend who

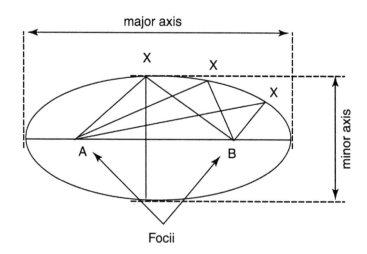

FIGURE 8.8   Definition of an ellipse; XA + XB is a constant.

protested against the replacement of the sacred circle by the *imperfect* ellipse, Kepler dismissed the circle as a[24] "voluptuous whore," enticing astronomers away from the "honest maiden of nature." Even Copernicus had been led astray by the siren of ages to chase circles around the sun. It was a pitiful miscarriage of beauty!

Defying the inherited ideals of classicism reflected so strongly in his own model of Platonic solids, Kepler was ready to break down the old order. He was among the community of innovators swept up in the turbulent transition from the Renaissance to the Baroque. A parallel metamorphosis transpired in Galileo's thinking about motion on earth (Chapter 4). The same innovative spirit emerged among artists of the Baroque. Guilo Romano tears down the orderly world of the Renaissance in *The Fall of the Giants* (Figure 8.9). As Jupiter destroys the palaces of the rebellious giants with thunderbolts, the antique universe collapses around them. With theatrical flair, characteristic of the Baroque, ceilings and walls come tumbling down to signal the collapse of classical rules of order, stability, and static symmetry. A new structure would arise.

Poised to explore planetary system dynamics that emerged from the luminous power of an eccentric sun, Kepler was among the Baroque innovators who broke out into unconstrained exploration of skewed forms that provided a new sense of vitality to describe planetary motion. Renaissance and Baroque styles stand in dramatic contrast through a comparison of *Last Supper* (Figure 8.10) by Castagno (1445–50) with *Last Supper* (Figure 8.11)[27] by Tintoretto (1592–4). With architectural clarity, Castagno paints Christ at the tranquil centre of the drama of communion. But Tintoretto abandons left–right symmetry, straight edges, and the calm of the classical to create visual tension with diagonal lines. Instead of placing the table parallel to the picture plane, the rebel skews it to thrust the dynamic of his mystical message onto the mind of the observer. In the Renaissance versions of the Last Supper, apostles sit symmetrically on either side of Christ, except for Judas. Many artists of Castagno's period adopted this classical style and arrangement. Domenico del Ghirlandaio (1480) framed his symmetric work in circular geometry (Figure 8.12) instead of Castagno's rectangular panels. But such elements are nowhere to be seen in Tintoretto (Figure 8.11). Instead, there is a new dynamism in Tintoretto's subjects. Light energy blazes from the head of Christ and the Holy

FIGURE 8.9   The Fall of the Giants, 1534. Romano, Giulio (1499–1546).[25]

FIGURE 8.10   *Last Supper*, ca. 1445–50, Castagno.[26]

FIGURE 8.11   *Last Supper*, 1518–94, S. Giorgio Maggiore, Venice, Italy, Tintoretto, Jacopo.

Ghost, portrayed as a ceiling lamp, and illuminates the apostles' haloes. Clouds of smoke transform into angels, rendering earthly to divine, natural to supernatural. A radiant Christ stands off-centre.

DREAMS COME TRUE

Apollonius' conic section replaced the circle. What an ironic coincidence! The Greek from Alexandria, who invented the ugly epicycle to account for retrograde motion, had unwittingly provided astronomers the future key to model planetary orbits. The ellipse was ready-made since antiquity to cure the circular obsession. An important feature of the ellipse is the amount by which it differs from a circle, called the eccentricity (e). See Figure 8.13 for a definition. The eccentricities

FIGURE 8.12   *The Last Supper*, 1480, Domenico del Ghirlandaio.[28]

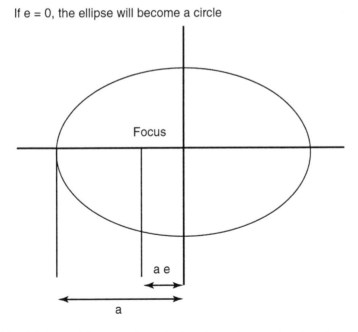

FIGURE 8.13   The definition of the eccentricity (e) of an ellipse. The semi-major axis is a. The product (ae) is the distance of the focus from the centre.

TABLE 8.1

Eccentricities of Planetary Orbits

Planet	Eccentricity (e)
Mercury	0.206
Venus	0.007
Earth	0.017
Mars	0.093
Jupiter	0.048
Saturn	0.056

Except for Mercury, the small eccentricities make it easy to mistake the elliptical orbits as circular

of all the planetary orbits are small (Table 8.1) because their paths are nearly circular, but the small differences are crucial. Mercury has the greatest eccentricity (20 per cent). Mars has the next largest eccentricity (9 per cent), which explains why it gave headaches to astronomers bent on fitting its orbit with circle combinations. Moon's orbit around earth has an eccentricity of about 5 per cent, leading to the curious phenomenon of the annular solar eclipse. When the moon is 5 per cent farther from earth and a total solar eclipse occurs, the moon does not block out the entire disc of the sun, leading to an annular shape eclipse, a dramatic visual manifestation of the moon's noncircular orbit (Figure 8.14).

Even after the radical ellipse breakthrough, Kepler was far from fully satisfied. He had just broken new ground. It was not enough to know the exact shape of the planetary paths. Was there a connection between the sizes of planetary orbits? What were the tunes of the cosmic music box? Why was the speed of a planet *nonuniform* in its elliptical path around the sun? Planets moved faster when closer to the sun. Was there any law that governed the changing planetary speeds? All were immensely fertile questions.

FIGURE 8.14   An annular total solar eclipse. When it is a few per cent farther away from earth in its eccentric orbit, the moon cannot block out the whole sun, leaving an outer ring.[30]

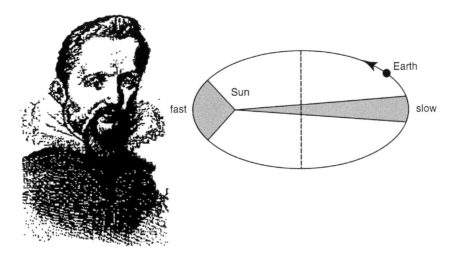

FIGURE 8.15   Kepler[31] and his equal area law. The sun is at the focus of the ellipse. When a planet is closer to the sun, it moves faster than when it is distant from the sun. As a result, the line connecting the planet to the sun traces out the same area in the same time interval.

To evaluate planetary speeds, Kepler imagined a spoke connecting the sun to the planet, like the hand of a clock. It was clear that the spoke did *not* revolve at constant speed, as did the clock hand. Was there any element of constancy in the spoke's movement? Kepler wrestled long and hard with numerous possibilities. After many years of trial and error, near-success and dismal failure, he finally discovered an unchanging physical quantity, intimately related to the speed of a planet. As the spoke connecting the sun and the planet moves, it sweeps out equal *areas* in equal times (Figure 8.15). Only a mind with Kepler's geometrical proclivity would recognize the subtle connection. Kepler's Second Law of Planetary Motion states:

Planets sweep out equal areas in equal times over their orbits.

The area-rule delighted Kepler. A simple mathematical law governed the varying speed of a planet as it orbited the sun. Kepler's original thinking about planetary motion in terms of non-circular figures and connecting spokes illustrates logician and mathematician Augustus de Morgan's perceptive observation:[29] "The moving power of mathematical invention is not reasoning but imagination."

Yet the original quest for harmony among planetary movements remained unanswered. There was still no obvious connection between orbit sizes for different planets. What would take the place of those perfect solids? And there was no quantitative relation between the orbital periods of the planets as they made their circuits around the zodiac. A separate construction for each planet was still necessary, a glaring flaw in both Ptolemaic and Copernican schemes. Kepler continued scrutiny of the data, searching for some overall connection, but his labours were fruitless for a long time. Still, he was inspired to persist by the belief that laws of nature are governed by simple, mathematical principles that were nearly within his grasp.

After many unsuccessful tries over a ten-year period, he finally hit upon a third law (1619):[32]

At last, at last, the true relationship … overcame by storm the shadows of my mind, with such fullness of agreement … that I at first believed that I was dreaming.

Kepler was ecstatic to discover a Third Law of Planetary Motion, now called the Harmonic Law. "I contemplate its beauty with incredible and ravishing delight":[33]

The square of the orbital period T is proportional to the cube of the semi-major axis R of the orbital ellipse. R^3/T^2 is a constant.

TABLE 8.2

Application of Kepler's Third Law to Six Planets

Planet	Relative Distance from Sun (R)	Period in Years (T)	Square of Period (T²)	Cube of Distance (R³)
Mercury	0.387	0.241	0.058	0.058
Venus	0.723	0.615	0.378	0.378
Earth	1	1	1	1
Mars	1.524	1.881	3.538	3.54
Jupiter	5.203	11.862	140.707	140.851
Saturn	9.539	29.458	867.774	867.977

The agreement of numbers conformed to Kepler's high standard of precision

An orbital period is the time a planet takes to complete one revolution about the sun. Earth takes one year, Saturn takes 29.5 years. The semi-major axis (R) of the ellipse is approximately the average radius of the orbit, for small eccentricity. The value of the constant in Kepler's Third Law is the *same* for every planet! Finally, a single relationship unified orbit construction for all planets. Even more thrilling, the relationship involved *both* orbit sizes and planetary speeds. Planetary synthesis was at hand.

The first two columns of Table 8.2 list the distances of each planet from the sun and the orbital period of each planet. The final two columns give the computed values of R^3 and T^2. Kepler's Third Law states that the entries in the last two columns are identical.

Kepler had at last fulfilled Pythagoras' prophecy. There was a mathematical pattern governing planetary orbit sizes. Indeed, he surpassed the Greek's original ambition. Kepler's Third Law unified orbital sizes and speeds. For the first time in history, there was a relationship to weld two planetary physical properties together into a single law. It was a watershed event in the quest for unity.

A sublime mathematical relationship finally crowned Kepler's labours. But the result was far from the simple numerical harmonies in whole number ratios that Pythagoras hoped for; one might even call the new relation contorted. It entangles the cube of a spatial quantity with the square of a temporal one. Yet R^3/T^2 was mathematically precise. It was celestial music indeed, not to the ears, but to the mind. In the rousing words of Alfred Noyes[34]

> And so in music men might find the road
> To truth, at many a point where sages grope.
> … music is the golden clue
> … Planets move
> In subtle accord like notes of one great song
> Audible only to the Artificer,
> The Eternal Artist.

Through mathematical relations, only Kepler was finally able to listen with a clear mind to heaven's sweet harmony. Despite his overwhelming successes in developing a mathematical model for the planetary system, he never relinquished his search for the literal celestial harmonies. Forever etched upon Kepler's immortal soul,[35] "The heavenly motions are nothing but a continuous song for several voices." Near the end of his life, he even tried to work out the exact musical notes sounded by each planet by relating the maximum and minimum orbital speeds to concordant intervals of the musical scale. For Kepler, earth rang out "Mi–Fa–Mi" as if to sing about the perpetual conditions of Misery and Famine dominating a planet torn apart by ravages of war.

A NEW BOND FOR THE HEAVENS

Although delighted to replace epicycles with ellipses, and supersede Ptolemy's separate contorted constructions by unifying equal area and harmonic laws, Kepler had an uneasy feeling. Something important remained missing from the new scheme. Gone forever was the scaffolding of crystalline spheres that held the heavenly bodies together. He needed a new physical model to replace them. It was not enough to describe accurately the geometry of motion. He longed to understand the physical cause of planetary motion and penetrate the rationale underlying their arrangement. What moves the planets? What moves the earth? What governs the order?

His next insight came like a bolt of lightning. A planet sweeps out equal areas in equal times because a planet's speed is inversely proportional to its distance from the sun (Figure 8.16). Kepler did not realize this was only qualitatively correct, but it was still very instructive. When a planet *is close to the sun* and the connecting spoke is short, the planet moves fast as the spoke sweeps out a certain area per unit time. When the planet is *far away from the sun* it moves slowly, but the spoke continues to cover the same area per time because it is longer.

Why does the speed of the planet increase when it comes nearer to the sun? Kepler imagined there to be a power, or soul, which emanates from the sun and drives planets in their orbits. Decreasing with distance from the sun, like radiated light from a point source, the sun's influence controls the planets. Kepler was quick to advance from the animate to a more physical concept. The sun must hold the planets together by some "mechanical bond":[36]

> For once I firmly believed that the motive force of a planet was a soul … Yet as I reflected that this cause of motion diminishes in proportion to distance, just as the light of the sun diminishes in proportion to distance from the sun, I came to the conclusion that this force must be something substantial.

With a conceptual leap from "moving soul" to "something substantial" Kepler put forth another revolutionary principle, one of his best ideas. *The sun moves the planets.* Planets orbit the sun not because a divine being guides them, not because angels push them, not because their ethereal nature ordains them, but because the sun exerts a *physical force* on them. Mechanical force replaced fictitious wheels upon wheels. There was more. If the sun rules the motions of its companion planets, the earth must influence the motion of its attendant moon, because the moon orbits the earth. The core idea was pregnant with enormous potential.

What was the nature of this mechanical force? At first, he thought the propelling force to emanate from the sun like spokes of a windmill. Hearing of his contemporary, English scientist William Gilbert's experiments with magnetism, and his spectacular demonstration that the entire earth is a giant magnet (Chapter 3), Kepler speculated that perhaps magnetism is the force which binds

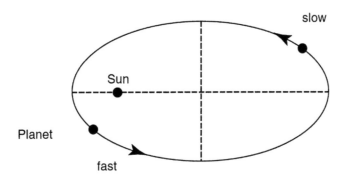

FIGURE 8.16 Kepler observed that when a planet is close to the sun it has a higher speed than when it is farther from the sun. This correlation led him to the idea that there is a mechanical force which binds planets to the sun, foretelling gravity.

planets to the sun. Not only earth but all planets as well as the sun are magnets. Perhaps earth's axis always points in the same direction because it is a kind of magnetic compass needle. Abandoning physical spokes in favour of a more intangible connection, Kepler foreshadowed the type of explanation that Newton later offered with gravitational force. Moving in this new direction, Kepler was the first to initiate thinking that the universe runs like a clockwork machine:[37]

> My aim is to show that the heavenly machine is not a kind of divine, live being, but a kind of clockwork ... insofar as nearly all the manifold motions are caused by a most simple, magnetic, and material force, just as all motions of the clock are caused by a simple weight.

Without a solar force, planets would come to a stop like the hands of a mechanical clock that someone forgot to wind. Like the first glimmer of light from a new dawn, Kepler envisioned the birth of a mechanical universe. Considering that Kepler began with a search for occult symbols underlying the fabric of the universe, it is astonishing that the road he followed eventually brought him to tangible, physical laws.

A NOVA SHINES OVER WAR-TORN EUROPE

Nature blessed avid astronomers of the time with another stunning celestial display, just three years after Brahe's passing. In 1604, during a conjunction of three planets, Jupiter, Saturn, and Mars, there appeared another nova in the midst of the planetary concourse. By itself, the conjunction of even two planets would be considered a momentous event. For astrologers, a multiple conjunction was especially dangerous, as the influences of the planets cast a combined spell. Previous multiple conjunctions were believed to have marked major historical episodes, such as the appearance of Adam, Noah, Christ, and the rise of Charlemagne, the first Holy Roman Emperor.

Conjunctions are rare, but novas are rarer still. While astronomers and astrologers all over Europe intently watched the region of the dreaded meeting, another surprising celestial spectacle greeted them. Word about the new cosmic arrival spread among the intelligentsia like wild fire. Naturally, the extraordinary appearance of a bright new star right in the middle of a conjunction gave rise to a flood of astrological predictions. Kepler diligently continued observations on the new star until it disappeared two years later. Every night it held its ground among the constellations, and so could not be a comet. At its brightest it only rivalled Jupiter, and paled compared to Brahe's nova. Like Brahe's nova, the star showed no parallax, making it abundantly clear that the new star resided not only beyond the moon but far from the limits of the solar system. Kepler put out a book about it. Today we call the apparition *Kepler's nova*. The next nova of comparable magnitude was not to be seen on earth without a telescope until 1987, a topic for the end of this chapter.

Asked to make predictions about the portents of the new star, Kepler hopefully suggested the emergence of a new republic under which the world would finally find peace. Perhaps Islam would collapse, ending forever the threat to Europe. Patrons valued Kepler's abilities in astrology far more than his achievements in science. Over the course of his productive life, Kepler wrote 800 horoscopes. During religious wars, opposing factions repeatedly vied for favourable astrological predictions from Kepler, who rose to Imperial Mathematician soon after Brahe's demise.

THE BOOK IS WRITTEN

Making astrological predictions never thrilled Kepler. His destiny was to reform astronomy, not to be a prophet. He was anxious to finish an extensive compilation of new astronomical tables. These he named *The Rudolphine Tables* (published in 1627) after the German Emperor, Rudolph, who was himself an amateur astronomer. It was a painstaking effort involving deep personal sacrifices.

Unlike Brahe, Kepler had no staff of assistant calculators. On his own, he performed all the arduous numerical computations of planetary orbits from the volumes of Brahe's detailed nightly observations. Fortunately, the recent invention of logarithms helped relieve the drudgery. Only a skilled mathematician such as Kepler would be in touch with the latest developments. Brahe's faith in the skilled mathematician had been well-placed.

For the first time in history, the error in computing the orbit of wayward Mars dropped from a bothersome 5 degrees to less than 10 arc-minutes, an improvement in precision by a factor of 30. Together with two decades of computations, Kepler included Brahe's new catalogue of stars, and a section on computations using logarithms. Decorating the work with lavish illustrations, he wrote:[38]

> The die is cast, the book is written to be read either now or by posterity, I care not which – it may well wait a century for a reader, as God waited six thousand years for an observer.

Publication involved another prolonged struggle. All through his persistent efforts, Kepler had to fight off members of Brahe's family who wished to take possession of the valuable notebooks. At publication time, Brahe's heirs disputed Kepler's right to use their patriarch's observations to contradict the great Dane's contorted cosmology. But nothing could stop Kepler now, not even when the printing press in Linz burned to the ground during a battle of the Counter-Reformation. When the Austrian emperor's treasury ran dry due to religious wars, Kepler made a deal to finance publication with his own money, provided that the new Emperor, Ferdinand II, would at least come up with payments owed. Even though Kepler finally received only a third of the arranged sum, he completed his part of the bargain. Astronomers used *The Rudolphine Tables* as the new standard for more than a century after Kepler.

Worn out by years of exertion, Kepler died in 1630 at age 60 while journeying to the Emperor's court to beg for arrears in salary promised him by an unstable ruler nearing bankruptcy. Historian Arthur Koestler compares his talents and fate to those of Mozart. The fertility of his ideas, his dogged persistence, the energy and speed of execution, and the slow decline of his personal situation thoroughly justify the comparison. Kepler's epitaph justly reads:[39]

> I measured the skies, now the shadows I measure.
> Skybound was the mind, earthbound the body rests.

Smallpox carried by soldiers fighting in the Thirty Years War killed his favourite son at the age of six. Eleven days later, he lost his wife. Kepler's aunt was burnt alive as a witch. Persecutors accused his mother of witchcraft and tried her several times. They had already burned six witches in her town over one single winter. To condemn her, a priest swore that Katherine Kepler collected herbs to concoct potions inside the skulls of dead parishioners which were made into drinking cups for her astronomer-son. Fortunately, because of his reputation as Imperial Mathematician, Kepler was able to intervene and save his mother. Through all the turmoil that surrounded his country and his family, Kepler continued the heroic quest to fathom nature's mysteries:[40]

> Let us despise the barbaric neighings which echo through these noble lands and awaken our understanding and longing for the harmonies.

UPDATES

THE BIRTH, LIFE, AND DEATH OF STARS

What was that new apparition in the firmament that Brahe named "nova" in 1572? Kepler and Galileo witnessed another spectacular one in 1604. These were exploding stars, now called "supernovae," a name coined during our age of superlatives, supermarket, superman, and others. Stars are not eternal as the ancients believed. European, Chinese, and Islamic cultures spotted other supernovae in the years 185, 386, 393, 1006, 1054, and 1181. In our galaxy a supernova goes off more than once every century. We are long due for another spectacular one that is bound to astonish us.

How did we learn that Tycho's and Kepler's novae were exploding stars? In much the same way as sky-watchers diligently observed and catalogued the sun, moon, planets, and stars over 3,000 years, astronomers over the last century scrutinized stars and nebulae, and measured their luminosity (as per Hipparchus), colour, and spectra to estimate their surface temperature. With the remarkable correlation found between luminosity and colour, stargazers ordered the vast multitude of stars into sensible families, such as main-sequence stars (our sun is one these), red giants, red dwarfs, and white dwarfs. From these connections they discovered new relationships that spoke about the death, birth, and entire life cycles of stars.

Advances in classifying and understanding stars was another far-reaching benefit that emerged from understanding the atom and the nucleus (Chapters 2 and 12). Down in the microworld, chemists ordered the multitude of elements into the Periodic Table (Chapter 2), to find number patterns ($2n^2$) in the periodic order that hinted at a deeper atomic structure. Spectroscopists discovered unique patterns of discrete colour lines in the spectra emerging from atoms excited in a gas discharge. Dark absorption lines at the same discrete wavelengths appear in the rainbow light spectrum when there is a cloud of gas between the source of light and spectrometer. When light strikes the gas atoms, not all the light makes it through due to specific absorption by these elements. Dark lines appear corresponding to the excitation energies of the absorbing atoms. The spectrum of the sun shows many such dark lines discovered by Joseph Fraunhofer (1787–1876). The dark lines reveal the presence of elements in the surface of the sun, such as hydrogen, iron, calcium, and magnesium. The strength of these lines for hydrogen and other elements vary significantly from star to star, which helps to better classify stars.

Advancing nuclear physics also played a prominent role in understanding the life and death of stars. Peering into the heart of the atom, Rutherford and Chadwick (Chapter 2) unveiled protons and neutrons in the nucleus. Nuclear physicists who followed them pursued the secret of nuclear energy through the nuclear fission of uranium atoms, and nuclear fusion of hydrogen into helium. In fusion they discovered the key to understand how a star resists the mighty crunch of gravity due to its gigantic mass, to maintain its size and equilibrium. The mass of the sun as an example star is more than 300,000 times the mass of earth. Through the comprehensive synthesis of the fields of atomic physics, nuclear physics, and astronomy emerged the astrophysics of star-birth, star-evolution, and star-death.

STELLAR CLASSIFICATIONS

The first stage in the long process of understanding stars was their systematic classification by their colour, spectrum, and absorption lines. Once again, classification and ordering preceded deeper understanding. Annie Jump Cannon (1863–1941) and Henrietta Swan Leavitt (1868–1921) worked diligently as "human computers" along with 20 other women, dutifully performing tedious examinations on tons of photographs from the copious data streaming out from the new Harvard Observatory station at Peru in the Southern Hemisphere. Chair of the Harvard astronomy department, Henry Pickering needed skilled and attentive assistants to examine thousands of photographs. The "computers" worked seven hours a day, six days a week at a meagre 25 cents an hour, less than what secretaries made. Not being housewives as expected, the women in "Pickering's Harem" were roundly criticized for being out of place at elite Harvard. Later the demeaning appellation was sanitized to "Pickering's Women" or the "Harvard Computers."

We will discuss in Chapter 9 Leavitt's vital breakthroughs in pinning down distances to remote stars in neighbour galaxies. Cannon's goal was to map every star in the sky to a photographic (Hipparchus) magnitude of 9, around 16 times fainter than the human eye can see. She also analysed the starlight spectroscopically. Cannon became a super-skilled examiner to manually classify more than 350,000 stars over her career, as compared to Hipparchus' and Tycho's meagre 1,000. At her peak she was able to record and categorize 200 stars an hour, three stars a minute just by staring at their spectral patterns and judging the intensity and frequency of absorption lines through her hand-held magnifying glass. And Cannon went one crucial step further than cataloguing. She set up a lasting classification system.

Although stars appear to be mostly white at first glance, they span a wide range of colours: blue, blue-white, white, white-yellow, yellow, orange. Cannon organized stars by their brightness (luminosity) as well as their colours. She based her classification also on the strength of different absorption lines in the spectra of stars. Accordingly, she put stars into classes O, B, A, F, G, K, M and created a helpful mnemonic, "Oh Be a Fine Girl, Kiss Me" to help remember her characteristic classification. Unfortunately, Pickering was given sole credit for the constructive classification Cannon invented.

Cannon harboured a rebellious streak, joining the budding women's suffragist movement, and became a member of the National Women's Party. Even though she was not allowed to take courses at Harvard, she became one of the first women to receive an honorary doctorate, but from a European university.

As understanding advanced about atomic structure and light-spectra, astronomers discovered that the Harvard classification system actually depends on the star's surface temperature, which in turn depends on the mass and age of the star. Stars radiate light somewhat like glowing coals. Just as a glowing red-hot coal is cooler than a white-hot coal, so a red star has lower surface temperature than a white star. The colour of light is a guide to a star's surface temperature. The continuous emission spectrum of a star shows colours over a broad range of wavelengths, but the wavelength where the emission is maximally concentrated (the peak) is related to the star's temperature. A hot object emits light at short wavelengths (blue end) and a colder body at long wavelengths (red end). The peak of the sun's spectrum is in the blue.

Our sun's surface temperature is about 6,000 Kelvin. Although it looks yellow from Earth, the light of the sun would actually look white from space. It looks somewhat yellow because our atmosphere's nitrogen molecules scatter some of the shorter (i.e., blue) wavelengths out of the beams of sunlight that reach us, leaving more of the long wavelength light behind. This also explains why the sky is blue: the blue sky is short-wavelength sunlight scattered by the gas (Chapter 12).

The coolest stars in the universe are the red dwarf stars with surface temperatures of about 2,000 K. These are stars with just a fraction of the mass of our sun (about 10 per cent). We discussed these (Chapter 7) in connection with exo-planets. The light released from their surface looks mostly red to our eyes. The hottest stars have temperatures over 40,000 K.

As mentioned, Cannon based her widely used stellar classifications on star colours as well as on distinctive features of the dark absorption lines. In 1925, Cecilia Payne (1900–79) in her PhD dissertation found a key link between colour, ionization characteristics of gases, and temperature. Payne was a child prodigy excelling in languages, literature, and mathematics, but she fell in love with astronomy. Although she earned her degree at Harvard, she was awarded it from Radcliffe, since Harvard did not allow women. Only as late as 1956 did Harvard finally accept women students and scientists. Harvard made Payne a full professor and later chair of the astronomy department, a fitting successor to the sexist Pickering who hired women computers.

Payne discovered that the light spectrum from a star shows different patterns and intensities of the dark lines because at certain surface temperatures different atomic transitions in the gas atoms (particularly abundant hydrogen) are more common, and therefore make some of the dark lines stronger. Cool stars (K and M type from Cannon's system) show weak dark lines. Light from the core of the cool star does not have enough energy for hydrogen atoms to absorb, so the dark lines in the hydrogen spectrum are dim. In medium-temperature stars (F and G type) the hydrogen absorption lines are strong and dark. But O-stars, the hottest of all stars, once again show weak absorption lines. The surface temperatures are now so great that most of the atoms at the star's surface are already highly ionized and abundantly excited to absorb any more energy to give strong dark lines.

The colour–temperature classification later revealed the evolutionary stage of a star: young, aging, burned out, or about to become supernova.

STAR BIRTH

How does a star form? All stars are distant suns, unimaginably hot, raging furnaces, only serenely beautiful from a distance. They are born from cold and sparse clouds of interstellar gas and dust that appear to naked-eye astronomers as faint white patches of light, visible only on a crisp, clear night. Ptolemy recorded a few (Chapter 6) interstellar clouds, while Messier (Chapter 9) compiled a sizeable catalogue. William and John Herschel (Chapter 9) identified 5,000, classifying some as hot gas clouds, others as globular star clusters, and still others as candidate galaxies.

Through a powerful telescope, nebulae emerge as wispy lagoons of gas of fascinating shapes and colours, glowing with the light of stars coming to new life inside majestic stellar nurseries. Figure 8.17 shows the Orion Nebula. It appears to the naked eye as a fuzzy patch in Orion's sword. But through the best telescope, the view is breath-taking. William Herschel prophetically called it[41] "an unformed fiery mist, the chaotic material of future suns." His contemporaries on the Continent, German philosopher Immanuel Kant (1724–1804) and French mathematician Simon de Laplace (1749–1827), formed the Kant–Laplace nebular hypothesis for the formation of the solar system. Since all planets revolve around the sun in the one direction, and also rotate around their axes in that same direction, they suggested that the solar system evolved from a giant nebula originally swirling around in that same direction. As the gas grew concentrated with gravitational attraction, the rotation accelerated, leaving behind outer rims of gas to eventually condense into planets. The core of the spinning nebula eventually ignited to form into the hottest part, the sun.

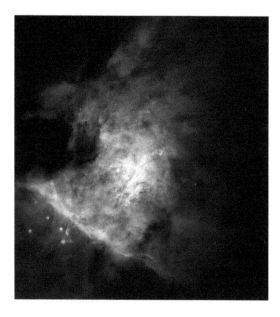

FIGURE 8.17   Orion Nebula. Credit: HST, ST ScI, NASA, WFPC2.[42]

Many a night from yonder ivied casement, ere I went to rest,
Did I look on great Orion, sloping slowly to the west …
Here about the beach I wander'd, nourishing a youth sublime
With the fairy tales of science, and the long results of Time.

[*Locksley Hall*, Tennyson][43]

Modern theories of star formation still carry a strong resemblance to the Kant-Laplace model, but with many important new features. The typical diameter of an interstellar cocoon is 100 LY, far larger than a star, and quite small compared to a galaxy, typically 100,000 LY across. (Reminder: a LY is the distance light can travel in one year, equal roughly to 10 trillion km or about 6 trillion miles.) With a mass of a million suns, interstellar clouds are mainly (70–75 per cent) hydrogen gas and 25 per cent helium, as is most of the universe, which retains the memory of its primary composition from the Big Bang. Heavier elements make up just a few per cent. Due to omnipresent gravitational tugs from outside disturbances, such as spiral galaxies swirling in the neighbourhood, smaller (less than 1 LY) and denser regions inside interstellar gas cloud start to slowly twist and turn. Sometimes, the rotation of a segment can cause it to fragment into two clouds to eventually form into a binary star system. In other cases, the cloud may fragment and evolve into a cluster of stars. A few times every century, there is a violent star-birth or a cataclysmic star-death in the galactic neighbourhood. Shockwaves rip through the cloud, ploughing up piles of matter where high densities trigger gravitational contraction and eventually more star-births. Multiple dense cores can form giving rise to star clusters, some containing hundreds of stars.

Proto-Star, Planets, and Star-Birth

As a turbulent region of the interstellar cloud starts to contract from gravitational attraction, it rotates faster, like a swift ice-skater who draws in her arms to go into a dizzying spin. A rotating, contracting cloud assumes a spherical shape, but not for long. It tends to flatten into an equatorial disc with a bulging sphere at the centre. A similar effect occurred early in its history with the rotating, spherical earth, albeit on a much smaller scale. Earth is slightly flat at the poles and bulges at the equator.

Why does a spinning gas cloud contract into a disk under gravity? On the surface of a spinning sphere, the angular speed (in degrees per second) is everywhere the same, but the linear speed (in metres per second) is highest at the equator, because the equatorial circle has the largest radius of any latitude circle. By virtue of high velocity tangential to the disc, material in the equatorial plane can better resist the inward suction of gravity. But slower moving mass at higher latitudes and especially the poles cannot counteract the inward pull of gravitation and so heads for the centre (Figure 8.18). Over millions of years, the spherical region contracts into a central core surrounded by a rotating equatorial disc. A proto-star pulls itself together, leaving behind planetary material circling the core in the equatorial plane. Solid grains in the rotating equatorial disc slowly collide and stick together, forming giant rocks that eventually accrete into planets. Hydrogen and helium escape from the low-mass bodies, leaving behind heavier elements. Thus planets begin to emerge.

When planets congeal from flying rocks, they remain in the plane of the disc surrounding the central proto-star. Looking from inside the planetary disc, ancient civilizations found all the sun's companions moving within a belt-like region they named the zodiac. As a vestige of the primordial rotation of the interstellar cloud segment, all planets revolve in the same direction around the sun, counter-clockwise as seen from the north. Most planets also rotate counter-clockwise around their axes, with a few exceptions, such as Venus, Uranus, and Pluto, which rotate clockwise. Rotation axes are roughly perpendicular to the plane of the solar system.

As matter from the stellar nebula cocoon continues to plunge inward, the central core increases in density. Gas compression raises the temperature, just as air vigorously pumped into a bicycle tyre heats up. The shrinking cloud exceeds 10 million degrees at the centre, with densities a hundred times higher than water. Under such extreme temperature and pressure conditions, nuclear reactions ignite (see section below "Star Life – Fusion of Stars and Elements"). A star is born. It lights up the stellar nursery like a thundercloud illuminated from inside. Upon ignition, the star violently expels interstellar dust in its surrounding cocoon, becoming visible to the universe at large as a "nova", later renamed "supernova." Ensuing shockwaves trigger star formation elsewhere in the nebular neighbourhood, like a wild fire through a forest. Nuclear ignition clears the low mass debris, but leaves giant rocks of high mass intact to form planets. Once the solar nebula clears of debris, the

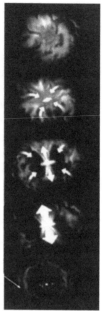

FIGURE 8.18   Collapsing nebular region becomes a cocoon for star-birth. The right panel illustrates the stages of star formation from the interstellar cloud segment.[44]

main process of planet formation comes to a halt. But some accretion continues for a long time among the circling rocks. For example, meteors striking earth's atmosphere as shooting stars show that earth is still at the tail end of its process of planetary formation. These falling bits of rock and metal still add about 40,000 tons per year to earth's mass.

Our sun began to form about 5 billion years ago. The sun is a relative newcomer to the galaxy which formed about 10 billion years ago. Earth coalesced half a billion years after the sun's formation. Soon after condensing from the debris, continents began to take shape amidst vast blue oceans. A half-billion years later, the first speck of life began to swim around and make copies of itself to evolve into humanity through countless evolutionary forms over another 3 billion years. In *The Princess*, poet Tennyson (1809–92) shrewdly hints at prevailing, Kant–Laplacian ideas about the origin of stars and planetary systems from interstellar clouds of gas and dust:[45]

> This world was once a fluid haze of light,
> Till toward the centre set the starry tides,
> And eddied into suns, that wheeling cast
> The planets.

A HAIL OF COMETS

When our sun formed from its massive fragment of interstellar gas and dust, it left behind in its remote outskirts, about 1 LY away from the core, kilometre-size chunks of rocky ice and dust that rove like exiled bandits in a spherical shell, barely hanging on to their parent star by the feeble force of the distant gravity of the sun. Called the Oort cloud, after the astronomer who conceived the idea, it is the source of erratic comets. Thus comet material is the closest approximation to our primary interstellar ingredients. Another source of comets is the Kuiper belt, miniature ice-worlds "parked" in distant outer regions of the accretion disc out of which planets formed. Occasionally, the gravity of a passing star dislodges a comet from the Oort cloud or from the Kuiper belt. Star Gliese 70, for example, will pass within 1 LY of our solar system in about 1 million years. It may well shake up a rain of comets. Dislodged icy bodies then follow trajectories that take them toward the inner sanctum of the solar system. Comets approaching us from the Kuiper belt consistently circle the sun counter-clockwise, staying in the equatorial plane. Like planets, they move among the zodiac constellations, and in highly eccentric elliptical orbits that faithfully obey Kepler's laws with periods less than 200 years. Comets originating from the remote, spherical Oort cloud, however, have paths that are not confined to the zodiac plane. They approach from random directions, some circling the sun counter-clockwise, others clockwise. Most have highly eccentric orbits with periods far greater than 200 years. Hence the confusion of Aristotle who thought comets must be random atmospheric effects, such as lightning, because some comets follow wayward paths often outside the zodiac. Every year, amateur astronomers will detect half a dozen new comets while keeping close track of recurring comets.

Most new comets launched from the nether regions of the planetary system never get any closer to the sun than Jupiter. Being a large planet, Jupiter ejects the vagrants back to the Oort cloud, or even expels unwelcome bandits completely out of the solar system. Thus the king of planetary gods protects the inner planets from deadly bombardment, not by hurling thunderbolts as in Greek mythology, but by tossing out comets. At times, Jupiter even takes a direct hit. When comet Shoemaker-Levy crashed into the giant planet in 1994, the catastrophe unleashed energy equivalent to 1 billion megatons of TNT dynamite, a hundred times the entire nuclear arsenal of our planet's contentious nations.

Some comets do escape Jupiter's protective shield and head right for inner planets. Arriving between Mars and Jupiter, solar heat evaporates ice into steamy jets. The pressure of sunlight and solar wind eject the streaming gas into spectacular tails pointing away from the sun. In ancient times, whenever one of the long-tailed celestial bodies flashed the sky, numerous frightening predictions of death and destruction followed. Indeed, comets can bring disaster to planets, but not in the form of bad weather, pestilence, or death to kings. Some comets collide with planets and moons, leaving the pock-marked, cratered appearance of these worlds. Sixty-five million years ago, a rather large comet rammed right into earth, creating the 150-kilometre-wide Chicxulub crater on the ocean floor off the coast of Mexico. The collision threw up such a mighty cloud of dust over the entire surface of the earth that it blocked out the sun for years. The interminable winter that followed wiped out dinosaurs and 70 per cent of other species of the Cretaceous period. For almost 200 million years the giant reptiles had dominated life on earth, only to vanish in an instant of geological time. Were it not for this cataclysm, mammals and their human evolutionary descendants would never have survived to inherit the earth.

How do the proponents of such a theory know that a cometary collision was responsible for dinosaur extinction? Iridium from the cometary dust cloud shows up as a 30 per cent excess in the 65-million-year-old geological layer of earth's soil, marking the end of the Cretaceous era. Will civilization be ready for the hail of comets that descends toward earth when Star Gliese shakes up the Oort cloud 1 million years from now?

STAR LIFE – FUSION OF STARS AND ELEMENTS

What keeps the sun and stars stable against further gravitational collapse? Deep inside the sun, where temperatures exceed 10 million degrees and densities 20 times that of earth, hydrogen atoms whiz by each other. When four fast-moving hydrogen nuclei (protons) fuse into a helium nucleus (Figure 8.19), about 1 per cent of their total mass converts to pure energy according to Einstein's mass–energy equivalence, $E = mc^2$ (Chapter 4). It's a tiny amount of energy, but the sun produces 10^{38} such fusion reactions per second. The energy released is spectacular.

The fusion of a helium nucleus is a multi-stage process with interesting by-products that reveal the inner workings of the sun. Two protons first come together to form a deuteron (one proton and two neutrons), the nucleus of heavy hydrogen. The reaction releases an energetic positron (an

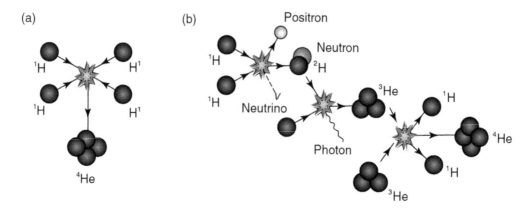

FIGURE 8.19   (a) Fusion of four hydrogen nuclei into helium to power the furnace at the core of a star. (b) Intermediate stages of the fusion process showing release of positron, neutrino, and photons.

anti-electron) and a neutrino, the most elusive of nature's fundamental particles (Chapter 2). A third proton then bombards the deuterium product to form a nucleus of lightweight-helium (otherwise known as He-3) releasing an energetic gamma ray that rattles around the surrounding gas to further heat it up. The final stage is the most powerful reaction, akin to a hydrogen bomb explosion. Two fast-moving He-3 nuclei collide to fuse into He-4 (regular atomic weight helium), liberating the colossal energy of fusion, and returning two protons to feed the raging nuclear fires. During the intermediate reactions, positrons annihilate with electrons to heat up the sun, but neutrinos fly out essentially unscathed because neutrinos react only weakly with matter. An average neutrino can travel through a block of lead 1 LY thick without interacting. Trillions of neutrinos from the sun's furnace pass through our heads every few seconds. Even at night (when the sun is on the other side of the earth) neutrinos fly through the earth and then our bodies unabated. They don't harm us because they don't interact at all with our bodies. Despite the colossal flux, the probability for a person to absorb even one neutrino over a lifetime is only about 25 per cent.

Nuclear fusion is the same process that propels a thermonuclear weapon, the hydrogen bomb. When just one gram of matter (such as uranium or plutonium) transforms completely, the energy released is equivalent to an explosion of 20 kilotons of TNT dynamite (10^{14} joules), enough to obliterate an entire city. The bomb that destroyed Hiroshima was 15 kilotons of TNT. But the sun does not blow up like a bomb because each fusion reaction causes local gas expansion, decreasing density and reducing the probability of another proton–proton fusion reaction in the immediate vicinity. Instead of a chain reaction as in a hydrogen bomb, solar fusion is a self-regulating process. Our sun transforms 6 million tons of hydrogen mass into pure energy (6×10^{26} joules, or about one trillion Hiroshima bombs!) *every second*, providing just enough outward pressure to resist the mighty inward pull of gravity on its enormous mass, nearly 300,000 times earth's mass.

Energy roars out from the fiery nuclear furnace at the core of a star. Just one part in 5 billion of the light and heat generated by the sun reaches our earth, but that is plenty to keep our planet warm and humanity alive. Having formed about 5 billion years ago, our sun will keep burning hydrogen into helium for another 5 billion years, a comforting prospect.

But stars come in all sizes, as astronomers discovered after classifying their luminosities, colours, and absorption spectra. Star masses range from a hundred times the sun's mass to only one-tenth of the solar mass. A deadly fate awaits stars which are a few times more massive than the sun. Due to excessive gravitational squeeze, a massive star develops a smaller core that burns hotter and brighter.

Astronomers' models reveal information about the star's life, life-time, and eventual fate. Smaller stars are cooler and redder. A red dwarf star with a tenth of the sun's mass will live for a trillion years. Bluer stars are intrinsically brighter because they are more massive than white or red stars. The more massive stars burn much faster and hotter than less massive stars. The bluish type-O stars, for example, are 30–50 times more massive than yellow-white stars, like our sun. But O type stars burn a million times brighter, so they have far shorter lifetimes, as they rapidly use up their hydrogen fuel. As a result, O and B type stars only last a few million years before they die in spectacular supernova explosions, while cooler and less massive K and M type stars burn steadily for billions of years. Higher mass stars are brighter, hotter, and blue. A blue star on the main sequence is superhot but relatively young, and not burnt out already.

The massive outward pressure generated by the energy released balances the contraction pressure from the enormous force of gravity. All hydrogen-burning stars belong to the "main sequence" (Figure 8.20) of the Hertzsprung–Russel (H–R) correlation diagram. A graphic way of thinking about star organization, this famous diagram is one of the most important in astronomy. The brightness (luminosity) is presented on the y-axis, and temperature on the x-axis (cooler from right to left). The main sequence is the population of stars shown on the diagonal from top left to bottom right. Star brightness along the main sequence increases with temperature, hot stars to the left (top) and cool stars to the right. Stars actually show a spread over the region of the diagonal line. As a star goes through its life, it moves its position in the H-R diagram.

The most massive (O type) and most luminous stars have solar mass of about 60 (times the sun's mass) and are very unstable, living for only millions of years. Life is highly unlikely to develop around any planet orbiting such a massive star because the star's life is too short for evolution. Main-sequence K and M type stars are very common. The least massive and dimmest stars we can observe have less than 0.2 solar mass. These are red dwarfs which use their fuel very slowly and

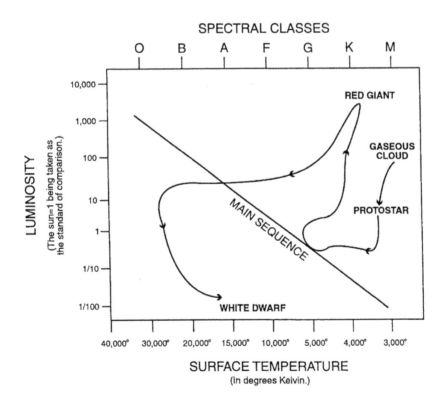

FIGURE 8.20 Hertzsprung–Russel diagram showing star luminosity (relative to sun) versus star surface temperature. Most hydrogen-burning stars fall along the main sequence diagonal. When a star exhausts its primary fuel it walks off the main sequence. The meandering line shows the probable evolutionary path of our sun over a very long timescale.

survive for hundreds of billion to trillion years. Because low-mass stars have very long lives there are many more such stars around than higher mass stars. "Stars" with smaller than 0.2 solar masses not having hydrogen to fuse and ignite are also called "brown dwarfs."

OUR SUN'S PAST AND FUTURE

About 90 per cent of the stars in the universe, including the sun, are main-sequence stars, where they spend most of their long stable lives of many billions of years. These stars range from about two-tenths of the mass of the sun to up to 200 times. Figure 8.20 shows the probable history and fate of our sun, from its birth from a gas cloud to a protostar, to its present position on the main sequence. At present the sun has about 70 per cent of its primary fuel remaining. As the core approaches the limit of fuel available, the sun will increase in luminosity to raise the temperature on earth, so that oceans will evaporate and much of the atmosphere will disappear into space. When our sun exhausts its primary fuel, it will depart from the main sequence as shown and very rapidly turn into an enormous "red giant" that will swallow up the nearby planets Mercury and Venus, and make earth so hot that it will become un-inhabitable.

As a red giant, the sun will be in its last stages of life for about 1 billion years after expelling its outer layers. The surface will become cooler. The helium core will contract and heat up enough to start fusing. In its final years it will burn helium to carbon and oxygen, but it will not be hot enough to form heavier elements, as more massive stars can. The sun will finally collapse into a "white dwarf." Such stars are very common in our galaxy and in the universe, as these are the remains of stars with masses close to the sun's mass. A white dwarf does not produce any nuclear energy and resists the inward pressure from gravity by a quantum mechanical (Chapter 13) effect called "electron degeneracy pressure." As it radiates its residual energy into space it becomes cooler until eventually (a billion years later) it likely will become very cool and dark to become a "black dwarf." Basic calculations show that a white dwarf with mass greater than 1.4 solar masses will eventually collapse into a "neutron star" (see below) with the colossal density of an atomic nucleus. As a neutron star it does not collapse any further. It is supported by the "neutron degeneracy pressure." But if a star's mass is greater than 2–3 solar masses the degeneracy pressure will not be able to resist gravity and it will further collapse into a black hole. Gravity is so strong that nothing, even light, cannot escape (Chapter 11).

Over 97 per cent of stars in our galaxy will become white dwarfs. White dwarfs are stars that have ceased nuclear fusion, but still emit light from stored thermal energy. If a white dwarf is part of a binary system (two-star) – as are many stars – it may suck in mass from its companion thereby increasing its temperature. Eventually it will get massive enough and hot enough to trigger carbon and oxygen fusion, with explosions on the surface. If the white dwarf swallows enough matter from its neighbour to become greater than 1.4 solar masses, it can no longer resist the deadly contracting force of gravity from its own mass. It will explode in a thermal runaway as a Type Ia supernova with its core collapsing into a neutron star. The peak luminosity of a Type Ia supernova exceeds all other types of supernovae. Due to their well-understood and predictable light decay patterns Type Ia supernovae are very useful as calibrated light sources with a defined intrinsic luminosity as reliable distance lampposts (Chapter 9).

DEATH-MARCH, SUPERNOVA TYPE II

After squandering its core hydrogen fuel, a large-mass star (8–10 solar masses) shrinks at the core and heats up. Abundant with heavier helium, the core collapses under its own weight. The outer envelope expands to a giant size and cools down from white hot to red, consuming any planets orbiting the outskirts of the star. We can spot with the naked eye a reddish tinge in red giant stars such as Betelgeuse. At the shrunken core, temperature and density jump higher. Nuclear fire re-ignites with fury, triggering nucleo-synthesis of elements of the Periodic Table. Helium fuses into carbon, intensifying the nuclear furnace while postponing the inevitable end. Heavier still, the core contracts further, and the furnace stokes up its intensity to create more elements. In a pattern of concentric burning, carbon fuses to neon, neon to oxygen, and up the Periodic Table to silicon and finally to iron.

Chemical elements familiar to us have not always been around since creation. Stars are their factories. Generations of stars cooked up the atoms in our bodies, from hydrogen in water to calcium in our bones, and iron in our blood. We are indeed all star stuff! And the atoms in our bodies came from the debris of several thousand stars that had to die to spread the assortment of chemical elements over the region of the interstellar cloud that gave birth to our sun. Many twisted shape interstellar clouds were probably formed during supernova explosions. Poet Albert Ahearn captures the ecstasy.[46]

> I am a scion of the Milky Way
> Wholly unique to the highest degree
> My soul is as old as light-years away
> My provenance stems from cosmic debris
> I need not religion to guide my life
> My quintessence antedates mankind's creeds
> The brief time walking beneath starlit nights
> Imbue my soul more than mankind's prayer beads
> Every thought, all that I am is akin
> To these heavenly designers birthplace
> The very essence, my soul within
> Began eons in interstellar space
> Knowing who I am and where I came from
> Is my greatest joy than what I've become.

Elements heavier than iron, such as iodine in our thyroid glands, do not form inside stars but in a more violent process that is the supernova. At each stage of a collapsing star, burning accelerates. Hydrogen and helium burn for millions of years, carbon burns for 100,000 years, and oxygen for 10,000 years. But silicon disappears in just a few tens of hours to cook up iron! As nucleo-genesis works its way up the mass ladder, the star develops an onion-like, layered structure, with light element ashes in outer shells, and heavy iron clogging up the shrinking core.

STAR-DEATH AND COLLAPSE

Consuming fuel with wild abandon, giant massive stars have relatively short lives. Stars with 100 times the sun's mass use up their fuel in several million years, incredibly quickly compared to the sun's 10 billion years. English physician and poet Erasmus Darwin (1731–1802), grandfather of the author of the theory of evolution, wondered prophetically about the ultimate fate that could befall such lilies of the heavenly fields:[47]

> Roll on, ye stars! exult in youthful prime,
> Mark with bright curves the printless steps of time …
> Flowers of the sky! ye too must yield,
> Frail as your silken sisters of the field.

An unconventional freethinker, Darwin's poetic imagination showed startling perspicacity.

When a massive star exhausts its endowment of fuel, it explodes into a supernova Type II. Suddenly it increases in brightness by a factor of 1 billion within one day, outshining all visible stars combined. The peak brightness is less than that of the Type I supernova, but the brightness can last almost twice as long (six months). The light which floods the cosmos can be seen across several galaxies. The most powerful supernovae show up over 100 MLY distance. Why does the star explode?

After all the fuel at the centre turns to iron, there is an energy crisis. Protons and neutrons in an iron nucleus are so strongly bound together that when iron burns there is a deadly consequence. Rather than producing energy, nuclear fusion of iron absorbs energy. Thermonuclear fusion in the star comes to an abrupt end. With diminishing outward pressure, the shrinking nuclear furnace can no longer resist the mighty inward pull of gravity.

In one brief second, the entire core of the star suffers a catastrophic implosion, followed by a phenomenal explosion into a vast interstellar cloud of gas and dust, laden with elements manufactured inside the star over millions and billions of years. It is one of nature's most cataclysmic events, short of the primordial Big Bang. As the eruption propagates to the star's outer layers over the period of 10–20 hours, the supernova outshines an entire galaxy of 10 billion stars. Its light output exceeds the total energy that our sun will give off over 10 billion years.

In the final crunch, 1 billion tons of iron squeeze down to into one giant dense nucleus within one second. All atoms lose their identity. Electrons combine with protons to form a rapidly spinning neutron star, a single giant nucleus only a few kilometres in radius. What stops the collapse? One factor is the neutron degeneracy pressure arising from the Pauli Exclusion Principle (Chapter 13): neutrons cannot all fall into a single energy level. Another factor is rotation. Imagine the mass of 1 million earths squeezed into the island of Manhattan. Now imagine Manhattan spinning at a dizzying speed of 100 turns per second. Skyscrapers would fly apart like mud from a spinning potter's wheel. But the gravitational squeeze on the compact yet colossal mass keeps the exotic star-nucleus together. Since each neutron is a tiny nuclear magnet, the monolithic nucleus becomes a giant magnet, with 1 trillion times the strength of the earth-magnet. From its magnetic poles, the neutron star emits an intense narrow beam of radio waves into space. As the star spins, it sweeps the beams across the sky like a lighthouse beacon, but with a period as regular as an atomic clock.

If the radio beams intersect earth, astronomers detect regular pulses and identify the neutron star as a *pulsar* (pulsating radio sources), monument to the power of gravity. When a young British graduate student, Jocelyn Bell, found the first pulsar in July 1967, the extreme regularity of its pulses opened the titillating possibility that the signal came from an extra-terrestrial civilization. It nearly earned the acronym LGM for "Little Green Men." Shortly thereafter, sensitive radio telescopes detected many more pulsars from different parts of the sky, clearly discounting the exciting prospect.

Spectacular as the light energy output of a supernova may seem, it is only 1 per cent of the total energy lost. A giant storm of neutrinos speeding through outer star shells at the velocity of light carries away most of the supernova energy during the fatal instant. Calculations from star collapse models reveal an unimaginable 10 billion, trillion, trillion, trillion, trillion (10^{58}) neutrinos escaping with the star's energy bank. The sun emits about 2 per cent of its energy in neutrinos and about 98 per cent as electromagnetic waves (photons). A supernova, in contrast, releases 99 per cent of its energy as neutrinos, and only 1 per cent as photons.

As the neutrino exodus saps life out of the core, the star pressure drops precipitously. Meanwhile the outer envelope of the star, which split up after exhausting its hydrogen fuel, is still too distant to experience any disastrous effects of the rug that has just been pulled out from under. Matter surrounding the central neutron core begins to collapse at speeds approaching 10 per cent of light velocity, and heats up from friction against the out-rushing neutrino swarm. Hitting the hard-core neutron star, the contracting shell bounces off and detonates outward, generating a powerful shock wave that finally blasts through remote outer shells. Like an overheated pressure cooker that blows up, the star expands enormously, becoming a brilliant sphere with rapidly growing luminosity. When light reaches earth, thousands of millions of years later, humans spot it as a newcomer to the firmament, a supernova.

Rippling through outer layers, the intense pressure wave triggers new nuclear reactions generating copious quantities of neutrons. Cosmic alchemy through energetic neutrons forges unstable elements all the way up the Periodic Table to uranium. When these fission they form elements heavier than iron and generate nuclear energy to prolong the outburst. Within a second, the supernova stamps out tons of gold and silver and other heavy elements in the blowout. Rich with neutrons from copious nuclear reactions, the supersonic boom continues to synthesize uranium, radium, and other super-heavy elements for humans to isolate eons later from ores and mines. Understanding the details of nuclear interactions unravels the secrets of the stars.

Stellar ejecta blast out into space, scattering gas and massive quantities of dust all over the heavens, eventually to become ingredients for new stars and planetary systems. Shock waves spreading out from the exploding star stimulate new stars that rise from the ashes of dead stars in neighbouring clouds of interstellar gas and dust. Explosive death spasms of millions of supernova forged the heavy elements necessary for the complexity of life on earth. All went off in our galactic neighbourhood, over the billions of years since our Milky Way first formed. If stars did not die we would not be here to wonder about our origins. We are indeed one with the universe, as the Buddhist philosopher Confucius dreamily contemplated.

As the cloud of debris from the furious eruption dissipates, the supernova continues to shine brilliantly for more than a year due to the radioactive decay of unstable nuclei assembled inside the belly of the beast. Even after radioactivity dies out, the supernova remnant keeps glowing from exotic radiation emanating from high-speed electrons spiralling around the intense magnetic field of the spinning neutron star. X-ray and radio telescopes can image supernova remnants of Tycho's and Kepler's supernovae (Figure 8.21), and the remnants of the more recent supernova, SN 1987A. Expanding factories of interstellar dust may at one time seed the birth of new stars. Thus matter recycles and regenerates itself. High-energy particles blown out by supernovae eventually arrive on earth as energetic cosmic rays to randomly mutate the genetic material of life. Some changes are destructive, others produce life forms better suited for success in the struggle for survival of the fittest. Over millions of years, mutations evolve to generate the infinite variety of species that inhabit the earth. Michelangelo's reverie takes on fresh significance:[48]

> Every beauty which is seen here below by persons of perception resembles more than anything else that celestial source from which we are all come.

Supernova Pair Re-Born

In 1934, two major personalities in the supernova story, Walter Baade (1893–1960) and Fritz Zwicky (1898–1974), made the first breakthroughs in understanding novae. Much like Tycho Brahe and Johannes Kepler, they formed a fateful collaboration. Educated in Switzerland, Zwicky joined the physics department at California Institute of Technology near Los Angeles, where he established a reputation as a pompous, egotistical character, calling his colleagues pedestrian astronomers. Lecturing to graduate students about extragalactic nova sightings of unprecedented intrinsic luminosity, he coined the name "supernova," putting him in league with Brahe who dubbed the first stellar nova. Just as Brahe diligently combed the skies to map the stars and chart the detailed motion of heavenly bodies in the vain hope of justifying his hybrid planetary model, Zwicky pursued supernovae with the single-minded purpose of vindicating his ideas. Like Brahe, Zwicky scoured the universe with the latest instruments to detect the first hundred extragalactic supernovae.

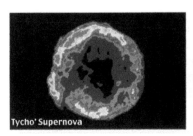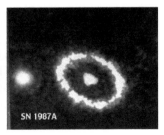

FIGURE 8.21   Supernova remnants scale.[49]

Walter Baade, like Kepler, was a gentle-mannered soul caught between the adversaries of world war. Just as Kepler fled Germany under religious persecution, Baade had to abandon Germany as it crumbled under forced reparations for World War I. Baade set his sights for Mt Wilson Observatory outside Los Angeles. But when war broke out again, the United States government declared him an enemy alien from Nazi Germany and quarantined him at Mt Wilson. It was just as well. Over skies darkened by wartime blackouts in Los Angeles, Baade took some of the clearest photographs of galaxies.

Both Zwicky and Baade were passionate observers. Here the analogy with the binary astronomers of the Renaissance fades. Kepler, the theorist, had to wait until Brahe passed on to work out the meaning of Brahe's precious data. But Baade and Zwicky formed a fruitful living collaboration to make daring predictions about supernovae.

As the last gasp of dying stars, supernovae connect intimately to the birth of other stars, inextricably binding with the birth of earth and the life of its inhabitants. Countless stars had to live and die to give humans the variety of necessary substances and life. Understanding the sagas of star-birth, star-death, and star-life-cycles is a prime example of unification between two vastly distinct arenas: the physics of the microcosmos and the macrocosmos. Nuclear physics and particle physics play crucial roles in understanding what goes on inside stars from beginning to end.

THE SMOKING GUN

With scarce data from just 20 supernovae sightings, Zwicky and Baade boldly predicted in 1934 the supernova model (described above) including the squeezing of matter into a super-dense neutron star. It was natural that colleagues were sceptical of such wild conjectures. And it was barely a year since the neutron was first discovered (Chapter 2). Zwicky's answer was to launch an intergalactic search for supernovae, and study them to death. It was vintage Brahe. When colleagues claimed Zwicky was mad to look for rare supernovae, he blasted their thinking as Babylonian, much as Brahe branded his astronomer colleagues as "blind watchers of the skies." Using the best instruments available, he doubled the number of supernovae sighted within five years. Excitement grew, and others joined the search. By the time Zwicky died in 1974, he logged more than 100 out of a full complement of over 10,000 supernovae observed. Every time Zwicky announced a new find, his faithful collaborator Baade would make careful measurements of the light curve (how the brightness increased and faded away over months) from the best telescope at Mt Wilson. Similar to the 1572 nova that Brahe meticulously observed, the luminosity would intensify for the first few weeks, then decay over a year. With evidence mounting from supernovae outside our Milky Way, astrophysicists advanced the models described earlier to fill out our present understanding. What was missing was the smoking gun, a supernova in action.

At precisely 7.35 a.m. Greenwich Mean Time on February 23, 1987, a swimming-pool-sized water tank holding 3,000 tons of ultra-pure water, deep inside the Kamiokande lead mine in Japan, picked up 12 rare flashes of blue light. Flooding the detector was a swarm of 1 billion, billion neutrinos arriving from a star in the pangs of death 160,000 years ago in our neighbouring galaxy, the Large Magellanic Cloud. Halfway round the world, inside a large salt mine under Lake Erie, seven additional neutrinos interacted with 7,000 tons of water. But no one was watching the momentous electronic signals at the time in either pool. No one could appreciate the significance of the cosmic event and alert the world.

About 20 hours later, Canadian astronomer Ian Shelton was developing photographs of the Large Magellanic Cloud from his post atop Las Campanas mountain in Chile. A puzzling dot of light appeared on his plate along with a crisp image of the familiar neighbour galaxy. After studying the unbelievable spot for a few minutes, he pulled out a plate from the previous night for comparison. The white patch on the new photograph was 600 times brighter than anything in the vicinity of the earlier image. Perhaps it was a flaw in the plate. But Shelton was closely connected to the cosmos. If there was such a bright star out there, he should be able to pinpoint it with his own eyes. Rushing out of his dark room into the clear night he caught a faint star glowing right in the heart of the hazy Large Magellanic Cloud. It was an erupting supernova. It was too good to be true. Of course Shelton had absolutely no idea of the neutrino bonanza that appeared around the world a day earlier.

As news spread around the southern hemisphere through the astronomers' telegraph network, the supernova grew brighter, over days, to the delight of eager southern observers (Figure 8.22). Aware of predictions based on astrophysical supernova models, neutrino experts at Kamiokande and Lake Erie began to pore over their data to finally identify the select few neutrinos that left behind tell-tale flashes of light. It took two weeks of checking to gain enough confidence to announce the meagre harvest of 19 total neutrinos out of the colossal initial flux of the 1 billion, billion that struck the detectors; but that was just the right amount expected from the elusive character of the cagey particle. Neutrinos hinted at the formation of a neutron star at the core, vindicating Baade and Zwicky.

After brilliant detective work through countless archival photographs, astronomers identified the parent star catalogued by Nicholas Sanduleak of Case Western Reserve University in 1969 (see Figure 8.22) – the first supernova with a known original star. It was 15 times heavier than the sun, not a behemoth, but sufficiently massive to get into fatal trouble. By supernova standards it was a feeble explosion. Yet it was the nearest to us in recent history. Astronomers have not yet detected the pulsating radio beams from the neutron star presumed to be at the centre. We need to wait for the intense debris to thin out and expose the heart. Even then the neutron star may not reveal itself if the highly directional pulses from it do not intersect earth.

SUPERNOVAE PAST, PRESENT, AND FUTURE

Providing strong confirmation for many predicted processes, SN 1987A was the first supernova visible to the naked eye after the sensational blasts that sparked Brahe's interest in 1572 and Kepler's studies in 1604. These supernovae of the High Renaissance took place within our Milky Way galaxy and were far more breath-taking than 1987A, involving even larger mass stars. By now, however, astronomers with powerful telescopes have recorded over 10,000 supernovae in other galaxies.

FIGURE 8.22   Supernova 1987a.[50] (Left) before eruption. (Right) after eruption. In *Night Thoughts* Goethe wrote: I pity you, unhappy stars, who are so beautiful and shine so splendidly.[51]

With almost a trillion galaxies all over the universe, there's always a supernova exploding somewhere. The Large Synoptic Survey Telescope (LSST) now under construction in Chile will document millions over its operating lifetime starting around 2020.

X-ray and radio telescopes have located wispy interstellar cloud remnants 10 LY in diameter from explosions observed over human history by European, Chinese, and Islamic cultures. All remnants appear at recorded locations. And all reside in constellations lying near the luminous band of the Milky Way, as to be expected from the high population of stars inside our galactic disc. Some, like the wispy Crab nebula in Taurus, remnant of the 1054 supernova, clearly reveal a blinking pulsar at the centre, and gas jets flying out at 1,400 km per sec.

Judging from the frequency of supernovae spotted on earth, we conclude that such explosions are rare astronomical occurrences. In our Milky Way galaxy, a visible supernova goes off rarely, once every few centuries. No wonder people came to believe that the heavens are unchanging. (In larger galaxies, the supernova rate is one per 30 years per galaxy. In the Milky Way the rate could be as high as a few per century. Even though a rate of 1 per 100 years may seem small, our galaxy formed 10 billion years ago. It has therefore experienced 50 million supernovae spread out over its 75,000 LY physical span. Within the 100 LY neighbourhood of the interstellar cloud from which we were born, several thousand supernovae must have gone off to brew the rich elemental mixture composing our solar system. Most likely, a supernova blast about 5 billion years ago also triggered the birth of our sun and its neighbours within our star cluster.

With the passage of four centuries since Kepler's supernova, we are long overdue for another inside our Milky Way. But it may be hidden to the naked eye by interstellar dust that obscures our view of most of the galaxy. Galactic dust dims a large proportion of the stars in our galaxy. Only about 10% of the Milky Way stars are optically discernible (Chapter 9). But we are ready with infrared and X-ray telescopes as well as sensitive neutrino detectors to see through the fog. Today we have seven neutrino detectors around the world. One likely candidate about to blow in less than 10,000 years is Betelgeuse, the brightest star in Orion. Having exhausted its allotment of hydrogen, it is a now reddish super-giant, 1,000 times larger in diameter than the sun and 20 times as massive. Even though it is 550 LY away, when it blows it should be visible for many months in daytime. Another candidate is Eta Carinae (Figure 8.23) in the Southern Hemisphere, one of the most massive stars known (100 solar masses) and 3,700 LY away. In 1843 it erupted, to become the second

FIGURE 8.23   Hubble Space Telescope images of a star that could become a supernova. Eta Carinae is about 100 times more massive than the sun. Credit: NASA, ESA, N. Smith (University of Arizona) and J. Morse (BoldlyGo Institute).[52]

brightest star in the sky, but now it is hardly visible to the naked eye, masked by its own effluence. Far advanced in its life, its detonation could appear at any time.

Fortunately for humanity, no star within 50–100 LY of earth appears ready to explode. That is a blessing. If a supernova were to occur within 50 LY from us, the human race would have to hide below ground for a few decades to avoid the deathly nuclear radiation from cosmic rain. Star IK Pegasi is about 150 LY away from us and nearing the end of its life. But in about a million years when it is ready to blow, it will have moved much farther away, so it will be even less dangerous.

The age-old fascination with our connection to the heavens has given us new meanings through the birth, life, and death of stars. Human existence and destiny are indeed bound up with celestial drama. But not as told by ancient myths and astrology. Our entire make-up was minted in stars and scattered stellar explosions towards our place of origin. Through supernovae, we may justly retain the wonder about the heavens we have inherited from the past. With our early warning system of neutrino detectors all over the earth, we may dream of spotting a Milky Way supernova in our lifetime. It should be a breath-taking spectacle.

9 A New Heaven: A Wide-Open Universe

In theology the weight of Authority, but in philosophy the weight of Reason alone is valid. Therefore a saint was Lanctantius, who denied the earth's rotundity; a saint was Augustine, who admitted the rotundity, but denied that the antipodes exist. Sacred is the Holy Office of our day, which admits the smallness of the earth but denies its motion: but to me more sacred than all these is Truth, when I, with all respect for the doctors of the Church, demonstrate from philosophy that the earth is round, circumhabited by antipodes, of a most insignificant smallness, and a swift wanderer among the stars.

Johannes Kepler[1]

In the wake of Giordano Bruno's death, controversy over the heliocentric system continued to rage over Europe. Around the same time that Bruno burned in Italy, the English bard contrasted the prevalent doubts about the cosmos with the constancy of Hamlet's love for Ophelia:[2]

> Doubt thou the stars are fire
> Doubt that the sun doth move
> Doubt truth to be a liar
> But never doubt I love

While Kepler in Austria penetrated laws of celestial motion through the power of mathematics, Galileo unveiled a new heaven in the skies over Italy through the penetrating power of the telescope. It is an irony of fate that decisive discoveries for the heliocentric theory came from Catholic Italy through the pioneering telescope. Interpreting stunning new finds within the framework of the Copernican system, Galileo overhauled our view of the cosmos. Telescopic discoveries confirmed Copernicus' revolutionary theory. But Galileo went far beyond the validation of Copernicus' reorientation. Overturning the cosmological system, he made a direct assault on beliefs cherished since the amalgamation of Aristotelian cosmology with Christian theology. The dual challenge meant serious trouble for Galileo and for science in general. With the triumph of observation and reason, came the tragedy of blind faith.

A SPYGLASS FOR MERCHANTS AND MARINES

The germ of Galileo's telescopic discoveries was an accidental find in an optical shop in Holland. Making glass for spectacle lenses had grown into a fine art. Glaziers melted sand mixtures inside moulds and stirred their brew with firebrick paddles into a perfectly uniform composition. Cooking up a batch of high-quality glass took weeks. If the glass was non-uniform, with any bits of crystals embedded, or any lingering bubbles, lenses would form wavy images. As demand for spectacles increased, the quality of lenses improved. A busy centre of activity in glass and metal polishing, Holland was the ideal breeding ground for the telescope.

An apprentice of a Dutch spectacle-maker, Hans Lippershey, was examining curious distorted images obtained from various combinations of lenses, when he made an astonishing discovery. Through a particular pair of convex lenses he could see faraway objects as if they were close by. A remote church steeple looked enormous. But it stood upside down. It was not just an annoying distortion; the steeple appeared so much closer that he could see details barely visible to the naked eye. A jubilant Lippershey showed the toy to his friends. He quickly recognized the battlefield potential of the discovery. What a great device to determine the identity of far-away ships! He tried to sell the idea to Dutch sea-captains who were engaged in a fierce fight for independence from powerful Spain. They promised him a prize of 900 florins if he could make it into a usable binocular device.

Word of the marvellous toy spread through Europe. A faint echo found its way to Padua where Galileo was teaching Euclid's mathematics and Ptolemaic astronomy, lecturing to crowded audiences on the amazing supernova of 1604. What difficulties the new star presented to Aristotle's ideas about the incorruptibility of the heavens! Within one year, the curious instrument arrived in Venice. The Doge could purchase it at a steep price.

While immersed in research on dynamics, Galileo earned income on the side as an accomplished instrument-maker, providing sailors with accurate compasses for navigation, and sundials for timekeeping. A military compass fabricated by Galileo was the best measuring and calculating device that money could buy. He engraved the arms of compasses, some with scales for multiplication and others to calculate squares, cubes, and square roots. Mathematical precision was Galileo's forte. For the richest patrons, he used gold and silver. In the same period, Galileo devised a hydrostatic balance to determine accurately the density of gold alloys, the thermoscope to measure temperature, and the pendulum to measure heart rate. Through such enterprising ventures, he grew familiar with problems in the shipbuilding yards and artillery workshops of Venice.

When Galileo heard about Lippershey's toy, he quickly started to experiment with lenses he could purchase from the local optician. Such a delightful instrument was a must for his workshop. Using an old organ pipe as a tube, he placed a convex spectacle lens at one end, and a concave lens at the other. A large convex lens converged rays of light from distant objects into a small-scale image that he could project onto a piece of white paper to form a sharp replica at the focal plane. Instead of forming the real image, Galileo examined it through a concave lens to find a magnified image. The organ pipe could easily hold both lenses and, with his nimble fingers, he could systematically change the distance between them.

Galileo's first simple telescope had a three-fold magnification and showed distant objects without inversion. Soon he improved it to about nine-fold magnification. Delighted by its power, he took friends to the highest building in Padua and showed them the sails of ships in the Venetian harbour, normally invisible to the naked eye at 20 miles distance. Surprisingly, they could even identify the insignia on fluttering flags. Merchant friends were delighted. What a great help it would be for their lookouts who kept anxious watch from towers all over the city, waiting for ships to come in, laden with exotic spices from Constantinople. Sea trade was the lifeblood of the prosperous city-on-water. The tube was truly magic. From St Mark's Square in Venice, Galileo's friends could see the facade of a tall church in far-away Padua. Across the water on the island of Murano, they could see people entering and dismounting the gondolas at the ferry. Murano glass-works made the high-quality glass Galileo used to fabricate telescope lenses.

Instead of selling the device, as Lippershey had tried, Galileo made a gift of it to the Senate of Venice. Sea-captains of the flourishing maritime power were thrilled. They could use the spyglass to better aim their artillery. Without surrounding walls to defend it against invasions from the sea, Venice would find the spyglass critical for defence. Spotting enemy ships two hours earlier, the Doge could send a fleet to repel any aggressor out in the ocean, far from homes and families. The grateful Duke granted Galileo a lifelong professorship, and doubled his salary.

THE TOY THAT PENETRATED THE HEAVENS

Was it possible to increase the penetrating power? Galileo continued to improve the telescope till he reached a magnification of 30. It was an impressive device, nearly 4 feet long with a

2 1/4-inch diameter objective lens. In a flash of inspiration, he turned his telescope to the night skies, first to the moon and the sun, then to the planets and the stars, and finally out to the Milky Way. Through the power of the penetrating tube, celestial bodies revealed a host of breath-taking new features. Into his eager eyes and receptive mind, the telescope opened the depths of space:[3]

> Since they are infinitely stupendous, I am infinitely thankful to God who has deigned to make me the first observer of things so admirable and hidden to all past ages.

By interpreting what he saw through the magical glass, Galileo overthrew astronomical beliefs staunchly held since antiquity. Through new discoveries about the sun and moon, he bridged the great divide between heaven and earth. Close scrutiny of planets revealed the consequences of heliocentric order to his astute mind.

The moon is not a smooth and perfect sphere as Pythagoras claimed. It has a rough surface, just like our earth. There must be mountains on the moon. How did Galileo infer this? Turning his lens-tube to the crescent moon, he realized that the boundary which divides light from dark is not sharp at all (Figure 9.1). It forms a wavy line. Wherever the shadow of a tall mountain falls, the dark region extends beyond the crescent defining line. Judging from the length of mountain shadows, Galileo could estimate the highest lunar mountains to be 4 miles tall (Figure 9.2). At a time when the height of mountains on earth were not yet measured, Galileo could determine the height of lunar mountains! For a heavenly body, it was remarkable how the moon's surface looked more irregular than the earth's surface. In some parts he recognized the magnified features as wide plains. One reminded him of the plains of Bohemia. How could heavenly and earthly matter bear so many close similarities? All these were troubling conclusions.

Galileo was not just the careful viewer. He was also the brilliant interpreter, with astonishing insights. The ability to observe something new does not necessarily lead to an understanding of its significance. Insight is essential. When the moon was a very thin crescent, he could see the large, dark portion of the globe bathed with a faint soft light. Many had noticed the faint ashes of light

FIGURE 9.1 Galileo's sketch of the craters on the moon. Note the wavy boundary dividing light and shadow over the quarter-cycle moon.[4]

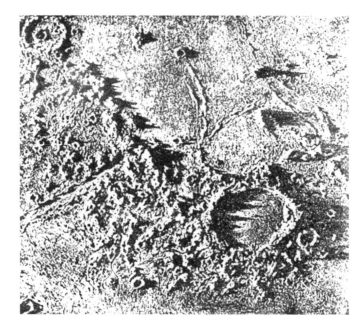

FIGURE 9.2   A modern, high-magnification photograph of the moon showing shadows cast by crater walls and mountains.[5]

surrounding the dark globe. An age-old Scottish ballad captured the pale glowing orb in a romantic vision:[6]

> I saw the new moon late yesterday evening
> With the old moon in her arms.

   Where does this dim light on the dark moon come from? It is not geometrically possible for sunlight to hit the opposite side of the moon. Galileo was the first to understand the reason (Figure 9.3). Our sunlit earth illuminates the moon. What we see is the bounce of sunlight reflected from earth

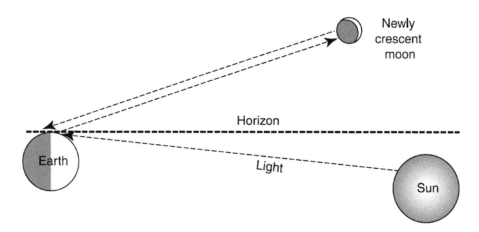

FIGURE 9.3   During the first crescent moon, the dark portion of the moon's face has a faint glow. This is the reflection of a reflection. Light from the sun illuminates the earth, which then illuminates the dark part of the moon's face.

to the moon, now called "earthshine." But there is a deeper significance to the effect. If the earth reflects sunlight, it must also appear as a shining orb (from outer space)! The earth is not a dark body. It shines like other planets.

Not long after Galileo's discovery of the cratered moon, the pock-marked image he described found its way onto the canvas of contemporary painters. The most famous rendition comes from Ludovico Cigoli, who shows the Virgin Mary (Figure 9.4) as queen of a crescent moon, realistically depicted with craters and fissures which the artist had the privileged opportunity to view through Galileo's telescope. Cigoli's moon bears a stark contrast to the more traditional representations of a heavenly smooth surface. Cigoli was Galileo's close friend and learned the techniques of mathematical perspective from the master. At the same time, Galileo learned the interplay of light and shadow from his artist friends, before he could apply these lessons to interpret the crescent shadow on the moon.

Although it was not Galileo who made the connection, lunar mountains can occasionally give the total solar eclipse the spectacular appearance of a diamond ring. Just as totality begins or ends, a small part of the sun's bright sphere can peek out through a lunar valley. Together with the ring-shaped solar corona edge, the moon looks like a ring studded with a sparkling diamond (Figure 9.5).

THE SPOTTED SUN

Turning the magical instrument to the sun, Galileo made another remarkable discovery, disturbing once more the prevailing views of the cosmos. The sun has spots (Figure 9.6)! How could there be

FIGURE 9.4   Assumption of the Virgin of Immaculate Conception (with Cratered Moon), Cigoli, Sta Maria Maggiore, Rome.[7]

FIGURE 9.5   The diamond ring eclipse occurs when sunlight escapes through a lunar valley.[8]

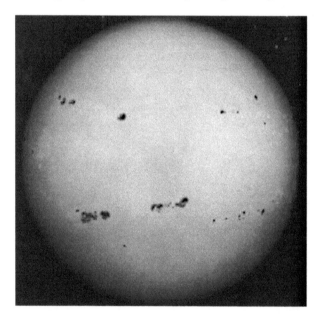

FIGURE 9.6   A modern picture of sunspots.[9]

blemishes on our eternal fountain of light and warmth? Blemishes on the moon Aristotelians could accept; the moon is just at the celestial–earthly divide. Much to Galileo's chagrin, other astronomers by this time turned telescopes to the sky in hopes of making their own discoveries. Among those who saw the sun-spots, some interpreted these to arise from shadows of planets crossing the sun's disc, a kind of partial eclipse. They were not prepared to accept that the sun could be imperfect. Divine and unalterable truth demanded perfection of the heavens. But Galileo was a much more careful and systematically regular observer with a nimble mind, free from the iron teeth of dogma. After tracking the spots for a few days, he was astonished to find that they move across the face of the sun, then disappear for some time, *to reappear on the other side* and repeat their circuit around the sun. Two conclusions emerged, both demanding courage and piercing insight:

(a) The spots are an integral part of the sun.

(b) The sun rotates around its axis with a slow period of 25 days.

If the sun rotates, earth's rotation was nothing outrageous, thought Galileo secretly. Repeated observations of the brilliant sun nearly blinded Galileo for a week. (It is always dangerous to look directly at the brilliant sun.) Indeed, his eyes were permanently affected; in old age he went completely blind. But the blinding glare did not deter his intense curiosity. Inspired to devise a new method, he darkened his room with shutters, and placed the telescope objective lens at a small opening to project the sun's image directly onto a sheet of white paper. As days passed, he carefully tracked the spots with his pencil.

THE SHAPES OF CYNTHIA IN THE MOTHER OF LOVE

Everywhere he pointed the probing telescope, the moon, sun, planets, stars, a surprise waited for interpretation. Turning to planets and stars, the telescope revealed an important difference. While fixed stars appeared as tiny blazes of light, planets appeared as moon-like discs, with well-defined boundaries. As Copernicus had so boldly conjectured, planets are not self-luminous like the stars. They only reflect the glorious sunlight. Therefore planets must have day and night, just like earth. One by one, ancient distinctions between earth and the planets were evaporating.

Examining Venus systematically every night for several months, Galileo discovered that it is not just a simple round disc. It changes its shape and apparent size, showing phases like the moon (Figure 9.7) from crescent to full. Galileo had fulfilled Copernicus' prophecy. With a telescope he could discern the phases as the heliocentric theory predicted. Throughout the history of astronomy, Venus had guided thinkers to new insights about the cosmos. Pythagoras concluded that the evening and morning star are one and the same – Aphrodite, renamed Venus by the Romans. Copernicus (Chapter 7) drew inspiration from Venus' peculiar attachment to the sun, always appearing in conjunction, to conclude that Venus must revolve around the sun, and so must other planets. Keeping systematic track of the planet of love in his magical telescope, Galileo discovered new features about Venus. When Venus sets with the sun at dusk as an evening star, it shows a crescent on the west side. When Venus rises with the sun as a morning star, the crescent is on the east side, strongly reminiscent of lunar phases.

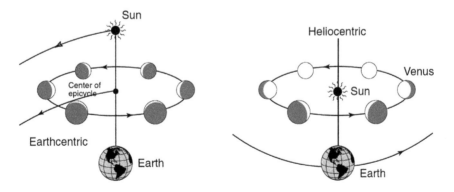

FIGURE 9.7   Galileo found that Venus shows all phases of the moon. (Right) in the heliocentric system, both earth and Venus revolve around the sun. When Venus is close to earth, it appears crescent and large. When it is far from earth, it appears full and smaller in size. Only through the telescope could Galileo observe Venus phases and correlated changes in size. In the heliocentric system, Venus is at times farther from earth than the sun. Therefore it can appear half to completely full. (Left) in a simplified Ptolemaic system, Venus has two motions. Moving along the deferent, Venus parallels the sun's motion in a large circle around earth. Hence Venus and the sun appear to move together at sunset (evening star) or sunrise (morning star). Venus also has an epicyclic motion used to account for its beguiling movements relative to the sun. Since the sun never comes between Venus and earth, Venus can never appear in full phase.

Being familiar with Copernicus' seminal work, Galileo realized that only the heliocentric system could explain the complete cycle of Venus' phases. In the Ptolemaic theory (Figure 9.7), Venus and the sun both orbit the earth. Furthermore, Venus never gets very far from the sun and always occupies the region between sun and earth. In Ptolemy's scheme, Venus should appear either as crescent or dark, *but never full.*

Venus had another surprise in store. In full phase, distant Venus appeared smaller than nearby Venus in crescent phase. The diameter of crescent Venus was six times larger than full Venus (Figure 9.8). The appearance of phases in Venus was a posthumous triumph for Copernicus and a bitter pill for the Ptolemaic diehards. With poetic flair, Galileo revealed his discoveries to the world:[10] "The mother of love emulates the shapes of Cynthia (the moon)."

Galileo romanticized Venus in the same spirit as writers and artists through the ages. Through her phases, Venus unveiled a new beauty to Galileo, an inspiring light to the heliocentric system. To the artists of the Renaissance, Venus symbolized a revitalization of the intellectual and cultural drama ringing with Greek ideas, but with newly added Christian overtones. In *Birth of Venus* (Figure 9.9),[11] Botticelli gives fresh meaning to the classical legend. Venus rises from the ocean as a gift from Heaven. Her youthful visage portrays beauty with dignity and caring, eloquently expressed by humanist Marsilio Ficino:[12]

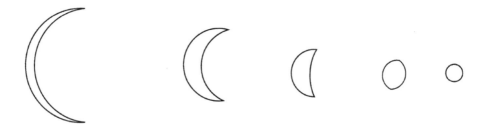

FIGURE 9.8   From left to right: when close to the earth, Venus is in crescent phase, and appears larger than when it is far from earth, and in full phase. Correlated changes in shape and size support the heliocentric theory.

FIGURE 9.9   *Nascita de Venere (Birth of Venus)*, 1486, Sandro Boticelli.

Venus is Humanitas ... Her soul and mind are Love and Charity, her eyes Dignity and Magnanimity ... The whole, then, is Temperance and Honesty, Charm and Splendour ... How beautiful to behold!

Zephyr, the god of wind, effortlessly blows Venus ashore, stirring up a storm of roses with his heavenly breath and weaving gorgeous tresses in her hair. But Botticelli turns the classical god of wind into a Christian angel by giving him wings. The spell of Dante's *Divine Comedy* continued to tie up Greek cosmology with Christian theology.

In the Christian context, the watery birth of Venus was an allegory for baptism, the rebirth of the soul. In the humanist context, her emergence from the sea was the rebirth of classical ideals. In the scientific context, Venus reborn through Galileo's telescope revealed the truth about the Copernican system of the universe.

Galileo did not yet dare declare that Copernicus was right, for fear of ridicule and accusations of blasphemy. It was not long ago that Giordano Bruno burned at the stake in Rome for proclaiming Copernicus' heresy and other sacrilegious statements. Like Bruno with his lofty conjectures, Galileo was also smitten by the incredible consequences of the Copernican universe.

FIND SOME NEW STARS TO BEAR MY NAME!

Ever since Babylonians took a special interest in planets, poets and artists through the ages imagined the temperate star Jupiter to be king of the heavens. Now the king was about to reveal his ring of followers. When Galileo turned his telescope to Jupiter, he observed something totally new and startling (Figure 9.10(a)). It was his most sensational discovery.

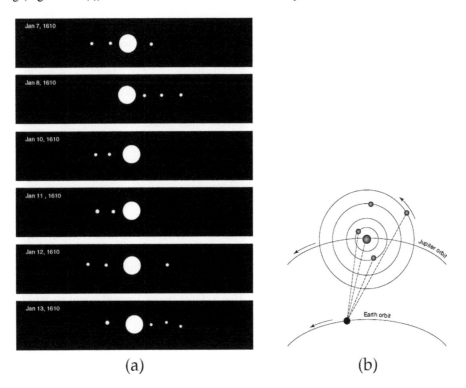

(a)   (b)

FIGURE 9.10   (a) A reconstruction from Galileo's notes on his first encounters with the moons of Jupiter which he observed faithfully every day, except January 9, when the sky was covered with clouds. Note how rapidly the moons of Jupiter change their position. Only on the sixth night was he able to see all four satellites. On previous nights one or more moons would always be eclipsed by Jupiter. (b) Jupiter's four major satellites comprise a miniature Copernican system. In this position only three moons are visible from the earth; the big planet eclipses the fourth moon.[13]

On January 7, 1610, the first night of the incredible find, he noticed three tiny companions to Jupiter's bright disc. Strangely enough, on the next night, the three new "stars" had all moved to one side of Jupiter. On January 10, there were only two stars visible, but, once again, they had changed sides. On January 11, there were again only two escorts, but surprisingly, one was bigger than the other. On January 12, there were three, but the big one had disappeared. On January 13, all four attendants were out for the first time since the new find (Figure 9.10(a)).

Slowly, Galileo began to understand the riddle of the stars that dance around the king of planets. Jupiter must have four moons of its own (Figure 9.10(b)). But these heavenly bodies are completely invisible to the naked eye. The "new stars" are not part of the firmament of fixed stars, because they change their position every few days. Like new-born offspring clinging to their mother's skirt, they dance around Jupiter, sometimes near the mother planet, sometimes farther, sometimes hidden by the planet.

Usually he could see two or three, and occasionally all four moons. By regularly following the satellites and keeping a careful timetable of their locations, he could predict where the moons of Jupiter would be on any given night. The shortest Jovian "month" is a brief 42 hours; the longest is 17 earth-days.

Galileo was quick to recognize the import of his astonishing find. New heavenly bodies were circling one of the planets, not the earth. When one of Jupiter's moons went behind Jupiter it was invisible to earth – a simple eclipse. But if everything in the heavens was supposed to revolve around the earth, why did Jupiter have its own entourage? It was a mortal blow to traditional cosmology. How could earth be the centre of the universe if at the same time there were other worlds with satellites whirling about them? The earth no longer held the singular honour of orbiting bodies. There were moons around another planet. The earth was not the centre of the universe. Jupiter and its four moons formed a miniature Copernican system, just like the sun and its attendant planets. One of the old objections against the heliocentric system raised by earth-centric diehards was that if the earth were to move around the sun, it would leave the moon behind. Now here was Jupiter dragging its four moons along as it made its regular trajectory through the zodiac.

Galileo had broken old habits of thought since the rediscovery of classical knowledge. Bubbling with fresh ideas, his mind was receptive to new phenomena. He was a product of a time when many were upsetting traditional currents by introducing new patterns of thought. In *Hamlet*, Shakespeare stresses the importance of being open to new discoveries and ideas:[14]

> Horatio: O day and night, but this is wondrous strange!
> Hamlet: And therefore as a stranger give it welcome.
> There are more things in Heaven and earth, Horatio
> Than are dreamt of in your philosophy.

Galileo named the moons of Jupiter "Medicean stars," in honour of his financial sponsor, Cosimo de Medici, a generous supporter of arts and sciences. Cosimo was delighted to be immortalized in the heavens. When he visited the Duke, Galileo personally showed him the new heavenly bodies. Soon after, Galileo received a letter from France, asking that he discover, and as soon as possible, some new heavenly body to which the name of Henry IV (king of France) could be attached. Of course Galileo tried to oblige, searching the planets, but the moons of Jupiter remained a unique find. If only Henry IV had been wise enough to fund a more powerful telescope, Galileo may have found moons around Mars and Saturn as well as more moons around Jupiter to both satisfy the king's vanity and nudge the progress of astronomy.

There is another fascinating story behind the naming of Jupiter's moons, now numbering 79 or more. For each, astronomers chose a name corresponding to one of the many lovers of Zeus: Io, Europa, Callisto, and so on. According to Greek legendary myths, Jupiter assumed many disguises to hide his numerous liaisons from his jealous wife Juno. In the sensuous painting *Jupiter and Io* (Figure 9.11),[15] the king of gods appears as a cloud embracing the willing nymph Io, planting an ethereal kiss upon her sensual lips.

A SPURT OF MILK FROM THE BREAST OF A GODDESS

Turning to the stars, Galileo unveiled a majestic new universe. In any region of the night sky, he could count ten times as many stars through his telescope than with his naked eye: instead of seven stars in the constellation Pleiades (Figure 9.12) he found an abundant 36. Instead of the famous nine stars in the belt of Orion he found an overwhelming 80 more. Despite the magnification of his telescope, stars still appeared as twinkling pin-points of light, not much different than with the naked eye, just hundreds more wherever he looked. Stars twinkle because microturbulences in the earth's atmosphere easily distort the narrow light beam arriving from a single point, causing momentary dips in light intensity. A nearby planet shines steadily, however, because light rays arriving from various points of its disc add up coherently to give a steady beam. Even at 30 times magnification, a telescope is still too weak to expand the distant stars. Planets are much closer. No wonder it was still impossible to measure the parallax of stars from different positions of the earth during its orbit around the sun.

Turning to the eternally flowing celestial river, the Milky Way (Figure 9.13), Galileo was astonished to discover that the heavenly fog is actually a vast conglomeration of countless stars:[16]

> Upon whatsoever part of it the telescope is directed, a vast crowd of stars is immediately presented to view … By the aid of the spyglass, any one may behold this in a manner which so distinctly appeals to the senses that all the disputes that have tormented philosophers through so many ages are exploded at once by the irrefragable evidence of our eyes … for the Galaxy is nothing else but a mass of innumerable stars planted together in clusters.[17]

FIGURE 9.11 *Jupiter and Io*, 1511, Antonio Allegri da Correggio.

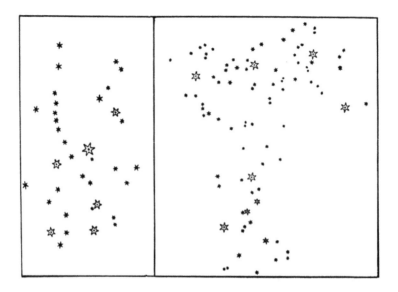

FIGURE 9.12   Through the telescope, Galileo observed many more stars than he could see with the naked eye (shown larger). (Left) in the constellation Pleiades, (right) in Orion.[18]

FIGURE 9.13   Galileo, and the Milky Way as it appears from earth through a powerful telescope.[19]

Humans have always stared at the night sky to be awed by the Milky Way. Democritus the Greek, who proposed atoms, also proposed that the milky white band might consist of distant stars. Who better to pay homage to the stellar swarm composing the Milky Way than Galileo's contemporary, Shakespeare? In his most famous love tragedy, Juliet declares:[20]

Come, night; come, Romeo; come, thou day in night;
For thou wilt lie upon the wings of night
Whiter than new snow on a raven's back.

> Come, gentle night, come, loving, black-brow'd night,
> Give me my Romeo; and, when he shall die,
> Take him and cut him out in little stars,
> And he will make the face of heaven so fine
> That all the world will be in love with night
> And pay no worship to the garish sun.

Poets were quick to incorporate Galileo's fascinating revelations into romantic visions. Integrating ancient river myths with the telescopic find, Longfellow paints the Milky Way as a[21]

> torrent of light and river of the air,
> Along whose bed the glimmering stars are seen
> Like gold and silver in some ravine.

A MESSAGE FROM THE STARS

It was time to open the eyes of the world and reveal the wonders he saw through the magical device. Galileo put out a new pamphlet, the *Starry Messenger*, to make regular reports and impressions of telescopic discoveries. It was the 46-year-old Galileo's first book. He wrote in rich Italian rather than in erudite Latin, and with a popular style, illustrating arguments with analogies. When it came to attacking old notions, Galileo did not hesitate to use sarcastic wit. Easily readable, the *Starry Messenger* was a crucial contribution to the progress of science through the process of democratization. The fascinating reports in the popular pamphlet inspired others to buy telescopes and turn towards the heavens. All over Europe educated people waited eagerly for each new edition of the magazine. What would Galileo's telescope reveal next? News from the heavens sent shock waves through the learned world. Discoveries became the talk of the town. Intoxicated by the revelations, English poet John Donne wrote:[22]

> Man has weav'd out a net, and this net thrown
> Upon the Heavens, and now they are his own.

Equally enthralled by the opening of the heavens, John Milton paid this delightful homage to the discoverer:[23]

> Before [his] eyes in sudden view appear
> The secrets of the hoary Deep – a dark
> Illimitable ocean, without bound,
> Without dimension.

In *Paradise Lost*, Milton described the Milky Way as:[24]

> A broad and ample road whose dust is gold
> And pavement stars.

In 1611, Rome invited Galileo to demonstrate his famous telescope to the eminent personages of the pontifical court. Pope Paul V showed the inventor great honour by allowing him to remain standing in his presence; all others knelt before the Holy Father. Galileo persuaded each to look through the telescope, showing them how it was possible to read holy inscriptions on a parapet 3 miles away. Such a noble invention had to have a special name. The inventor himself preferred "occhiale," for spyglass. A Greek poet and theologian present at the meeting offered the name "telescope."

The cardinals enjoyed what they saw. In hearty commendation, they proclaimed Galileo a true son of the Church. But they were totally averse to accepting Galileo's seditious Copernican interpretations. They could hardly abandon long-held views built on Aristotle's cosmology and its historically secure place in Christian theology. As the calculators of the new Gregorian calendar had so clearly declared, earth rotation and revolution were only a mathematical device to ease the burden of calculations through simpler Copernican models. Nevertheless, the cordial reception from Rome greatly encouraged Galileo.

Hostile reactions to Galileo's astronomical discoveries, and especially to his heliocentric interpretations, began to crop up among established university professors, bound by Aristotelian authority and scholastic tradition. Who would dare use a meagre toy of glass and metal to penetrate heaven's majesty? Some were incredulous, calling the results outright trickery. Others refused to peer through the tube, afraid it was witchcraft. Why should they trust such an artificial contraption interposed between God's beautiful nature and His generous bounty, the human eye? Soon, objections turned into passionate controversy. How dare Galileo spoil the celestial face of the moon, making it as ignoble a body as earth! What a travesty it was to blacken the sun with ugly blemishes! How could other worlds have satellites whirling about them? Why should there be more than seven heavenly bodies? Why would God change his Divine plan? There are seven musical notes in the scale, seven branches for the candlestick of the Jewish Temple, seven virtues, and seven deadly sins. Some laughed at the idea of adding four new planets to the set of heavenly bodies totally complete since time immemorial. Blinded by passion and prejudice, others dismissed the new discoveries as illogical. If Jupiter's satellites were invisible to the eye, they could have no influence on earth. So they were useless to humans. Therefore, they could not exist. If God created the universe for humanity why would He put so many things in the sky that are invisible to us? Staunch clergymen denounced the new discoveries as obnoxious heresy. Clinging fervently to the past, Aristotelians accused Galileo of brazen fraud.

THE EXTRAORDINARY STUPIDITY OF THE MOB

At home in Padua, Galileo's observations and interpretations received their share of scepticism and ridicule. Many of Galileo's eminent philosopher contemporaries refused to waste time looking through the glassy contraption, complaining that it gave headaches. One reported that he:[25]

> Tested this instrument of Galileo's in a thousand ways, both on things here below on earth and on those above in the heavens. Below it works wonderfully; in the sky it deceives one, as some fixed stars are seen double.

Could this have been a serendipitous "discovery" of binary stars? When an obstinately critical philosopher named Libri died, Galileo wrote to a friend that Libri would perhaps see the moons of Jupiter on his way to heaven. Libri was among those who adamantly refused to look through the telescope. Adept at rebuttal with his biting wit, Galileo wrote to Kepler in Germany:[26]

> I wish, my dear Kepler, that we could have a good laugh together at the extraordinary stupidity of the mob ... In spite of my often repeated efforts and invitations, they have refused ... to look at the planets or the moon or my glass *(telescope)!* ... What peals of laughter you would give forth if you heard with what arguments the foremost philosopher of the university opposed me ... labouring with his logic-chopping argumentations as though they were magical incantations wherewith to banish and spirit away the new planets *(moons of Jupiter)* out of the sky!

Although the two stalwarts Kepler and Galileo never met, they exchanged correspondence. Galileo knew that Kepler was a Copernican at heart. When he received a copy of *Mystery of the Universe*, Galileo was too polite to criticize Kepler's mystical connections between perfect solids and the heliocentric model, not to mention Kepler's enigmatic logic. Instead of reacting, he diplomatically thanked Kepler, saying he had not yet had an opportunity to examine his gift but was glad to hear it was based on the Copernican theory. He himself had accepted Copernicus on the basis of his own

telescopic observations. But since the Church still opposed Copernicus while learned men ridiculed him, Galileo did not dare to become fully defiant, as yet.

When Kepler first heard about Galileo's discovery of new moons dancing around Jupiter, he reacted, "I felt moved in my deepest being."[27] Delighted that Galileo supported the Copernican view, he could not wait to get his hands on the clever gadget to gaze upon the new heavenly bodies. Myopic since childhood, Kepler's vision was hardly suitable for astronomy. Besides, he could not find the glass nor the lenses of the same quality in Prague to construct his own telescope. Galileo was fortunate that the island of Murano near Venice had the best glass-making expertise in the world. And it was not trivial to reproduce the quality of workmanship at Galileo's lens workshop, where they chose the best lenses after rejecting dozens. These were the problems many experienced when they tried, without success, to look for the moons of Jupiter through their own inferior versions.

Despite Kepler's repeated requests for one of Galileo's exquisite telescopes, the master would not oblige. Although his workshops had by now made many telescopes, these were mostly for dukes and rich patrons, so as to gain monetary favours. Galileo had to pay fat dowries for his sister's marriage and to support his wayward musician brother. Having given away all his models to royalty, he had none to spare. To make a brand-new model of 30 magnification for Kepler would require much time and labour. But Kepler finally got lucky. A royal patron whom Galileo had favoured earlier with one of his high-quality instruments happened to spend a summer holiday in Prague. There he loaned Kepler a sharp telescope. At long last, Kepler could feast his eyes on the moons of Jupiter to enthusiastically exclaim:[28]

> O telescope, instrument of much knowledge … How the subtle mind of Galileo … uses this telescope of ours like a sort of ladder, scales the farthest and loftiest walls of the visible world, surveys all things with his own eyes, and, from the position he has gained, darts the glances of his most acute intellect upon these petty abodes of ours – the planetary spheres I mean – and compares with keenest reasoning, the distant with the near, the lofty with the deep.

How quickly Kepler recognized that it was not just the high magnification instrument but the sharp mind of Galileo that had the real penetrating power. And how quickly Kepler foresaw the crucial role of future telescopes in establishing a distance ladder to climb the heavens.

It was important for Kepler to confirm Galileo's discovery. As Tycho's successor, the Imperial Mathematician had earned the reputation of leading authority on astronomical matters. His enthusiastic endorsement helped crush some of the scepticism. Kepler coined the word *satellites* for Jupiter's moons, a Latin word meaning "those who stay close to someone rich in the hope of picking up favours." With fresh insight, Kepler went on to apply his laws of planetary motion to the moons of Jupiter. Indeed the periods and sizes of the new satellite orbits followed the harmonic law. But the value of the constant in Kepler's equation, $R^3/T^2 =$ constant, was totally different from that of the solar planetary system. Newton's synthesis later made clear that the different constants arose from the mass of the central body which, of course, is the sun for planets, and Jupiter for the Jovian moons (Chapter 11).

THE SATURN ANOMALY

In a new issue of the *Starry Messenger*, Galileo reported another unusual find, this time about Saturn. To ensure priority, he shrouded his discovery in the mystery of an anagram:[29]

SMAISMRMILMPOETALEUMIBUNENUGTTAURIAS

Kepler, who by now kept a sharp look-out for Galileo's enlightening pronouncements, worked on the riddle for a few days to come up with the following solution:

SALVE UMBISTINEUM GEMINATUM MARTIA PROLES
Hail, burning twin, offspring of Mars

Could there also be a satellite of Mars? Perhaps all the planets had satellites of their own. Kepler would have no trouble accepting it. But what Galileo actually meant was:[29]

> *Altimissum planetam tergeminum observavi*
> I have observed the highest planet (Saturn) in triplet form.

There appeared to be two protuberances on either side of the farthest planet, like handles on a tea-cup. Could there be satellites dancing around Saturn? The lobes presented another challenge. In a few months they disappeared. Galileo wondered: did the planet devour its kin like Saturn who swallowed up its offspring? A generation later, Christian Huygens discovered, through a more powerful telescope, that Galileo had stumbled upon the rings of Saturn. When Saturn's orientation tilts, the rings become harder to spot. Hence the temporary disappearing lobes.

The game of anagrams between Galileo and Kepler actually started even earlier with another delightful banter. Galileo first disguised his discovery about the phases of Venus:[29]

> *Haec immatura a me jam frustra legunturoy.*
> These immature things I am searching for now in vain.

In the mystery of anagrams, he hoped to protect the priority of his finds against impostors, now that many could partake of the joys of the telescope. Two years later, when Galileo was ready to go public, he provided the solution:

> *Cynthiae figuras aem ulatur mater amorum.*[29]
> The mother of love (Venus) emulates the shapes of Cynthia (the moon).

Upon receiving the first challenge, Kepler also tried to unravel this conundrum to come up with a possible solution, but again the wrong one:[29]

> *Macula rufa in Jove est gyratur mathem.*
> There is a red spot in Jupiter which rotates mathematically.

If the sun has spots and rotates, perhaps Jupiter is the same! The irony in the contest of the anagrams is that even though Kepler's guesses were way off the mark, later telescopes did in fact reveal satellites of Mars as well as a giant red spot that rotates with Jupiter! Kepler had a great gift for stumbling upon truths.

THE WAY THE HEAVENS GO

As the clashes of Reformation and Counter-Reformation continued to rage over all Europe, the optimism which characterized the Italian Renaissance began to evaporate. In the academic world, Aristotle's physics and Ptolemaic cosmology remained deeply and inflexibly entrenched, keeping the world and humanity at dead centre. Mathematicians and astronomers were not to meddle in matters of faith and discuss questions about creation. They were supposed to merely observe the stars, and use that information to compile almanacs, accurate calendars, and horoscopes for navigators and astrologers. Truth about the heavens was to be found in the scriptures or through comparison of ancient texts. He who best quoted chapter and verse was victorious in debate.

Attracted by a new offer of generous financial support from Cosimo de Medici, Galileo moved from Padua to Florence. The relocation proved to be an unfavourable turn for Galileo's personal fate. A stronghold of Aristotelian thought, professors at the University of Florence were lost in the medieval fog. The Church exerted a powerful influence. Galileo's friend, Sagredo, warned him that since Jesuits controlled Florence, he would hardly find the academic freedom he enjoyed at Padua. Ruled by a thriving merchant class, the free republic of Venice was an ideal place for free thinkers like Galileo. Copernicus had once studied at the University of Padua. Here William Harvey discovered the circulation of blood. At the *Theatro Anatomico*, daring medical faculty of the upstart university devised ways to circumvent the Church's ban against dissection. Meanwhile, in the stifling atmosphere of Aristotelian Florence, medical lecturers were hopelessly buried in the past, teaching from Galen's medical encyclopaedia, an amalgamation of procedures and superstitions that the Greek doctor had compiled 1,500 years prior.

Despite prescient admonishments, Galileo decided to move to Florence; he needed the money. At his new home in the glittering court at Tuscany, he acquired the rank and high salary of chief mathematician and philosopher. Although the grant was magnanimous, it was subject to the whim of the Duke. Galileo continued to publish the *Starry Messenger*. Heartened by his friendship with the Pope, and encouraged by the cordial reception from Rome, he grew bold enough to now openly declare his support for the Copernican theory. His observations on the phases of Venus and the moons of Jupiter testified that Copernicus was right, and Ptolemy was wrong. It was time to stop hiding behind the old facade. Claiming the earth's motion as real, not just a mathematical device, Galileo turned into a fierce advocate of the Copernican view.

Writing eloquently about discoveries and interpretations, Galileo's works reached a wide audience, making it difficult for the Church to ignore the bold proclamations. Worried over the immense popularity and excitement generated by telescopic discoveries, the cardinals of Rome admonished Galileo about the way in which he was phrasing his interpretations. He should be satisfied with the glory he acquired thanks to God's new revelations. But to use these discoveries to defend radical thoughts was a travesty against the gift of God.

Instead of obedience, an unruly Galileo embarked upon a lonely crusade, taking on the challenge of enlightening the Church hierarchy. He even wanted to convert the Pope to the Copernican view. Echoing a Roman cardinal of antiquity, Galileo proclaimed:[30]

> The Bible should not be treated as a text book of physics … the Bible shows the way to go to heaven, not the way the heavens go.

It was vintage Thomas Aquinas, the early medieval thinker who separated matters of faith from revelations of reason. Galileo was reminding the Church to respect the divorce between faith and reason. The light of reason can be an independent source of knowledge as much as the word of the Bible can guide the soul to salvation. Scriptures were not meant to teach science. One should not quote them as evidence. Rather than accept the literal meaning, a thoughtful reader must try to discern the true meaning behind the words. He who interpreted the scriptures literally was living on a stationary planet.

Interpretation of scriptures raised a constellation of thorny issues. The Catholic Church considered itself to be the sole authority with complete control over religious thought. If the Bible was wrong about one item, could it be wrong about others? Private interpretation could blossom into a full-fledged practice, threatening the entire credibility of the scriptures and thereby the power of the church.

A MESSAGE FROM THE CHURCH

Controversy over the moving earth seethed at home and abroad. By his revelations and deductions, Galileo made many enemies. Aristotelian professors, who felt their authority threatened by this upstart, united against Galileo to stop the "false apostle" from spreading radical notions. With the cooperation of preachers, they accused Galileo of blasphemy, and denounced him to the dreaded Inquisition. The Pope's council formally censured the heliocentric propositions as contradictory to the Holy Scriptures. Theologians joined together in banning the assertion that the earth can move, issuing a salutary edict as the final judgement on the subject:[31] "The sun moves and the earth is stationary." The Inquisition passed a decree prohibiting Copernicus' book. One bishop demanded that Copernicus should be arrested and jailed, oblivious to the fact that the man had been dead for 70 years. Pope Paul V called upon Galileo to give up his revolutionary views and stop teaching them, or the Inquisition would be forced to throw him into jail. On February 26, 1616 the papal commission ordered Galileo to abandon the view of the earth's motion and refrain from defending it further. Scientific freedom in Italy was truly dead. Galileo submitted reluctantly; he was a religious man. But in a confidential letter, he wrote:[32] "I believe there is no greater hatred in the world than the hatred of … knowledge." Some historians claim that his later trial hinged on the alleged violation of a written decree delivered to him personally. Others maintain that a false record of such a decree was later inserted into his files to strengthen the case against him during trial.

For seven long years Galileo held his tongue. In 1623 the Church installed a new Pope. Cardinal Barberini became Urban VIII. The new Pope was an intellectual cardinal who wished to transform Rome into a glorious city. With an enlightened Pope at the helm, the 60-year-old Galileo took new hope. He was working in secret on a new book, *On the Systems of the World*. Here was his chance to convert the Church. Cardinal Zollern, chief of the German mission, told the Holy Father how much harm had been done to the reputation of the Catholic Church in Lutheran countries by its outright prohibition of Copernican theory. Presenting lavish gifts to the respected scientist, Urban granted Galileo six audiences.

To the intellectual Pope, the idea of a moving earth went against common sense and experience. If the earth flies at astronomical speed, we would all be thrown about from place to place. How could earth be moving if its inhabitants did not feel its motion? It was lunacy to move the earth. Galileo answered by analogy with a moving ship. A stone released from the mast of a ship will always hit directly below the mast, never behind. Both mast and stone move forward together with the ship. It is impossible to detect uniform motion from inside the moving ship alone (Chapter 4). So it is with the earth.

In one private meeting, Galileo offered the Pope his opinion that tides do in fact provide a direct and visible proof of the earth's motion, both rotation and revolution. Shake a bowl of water and notice how the water sloshes back and forth. In the same way, seas rise and fall as earth moves. On the night side, water gains speed as it moves in the same direction as the earth's orbit around the sun. On the day side, water loses speed as it moves in the opposite direction to the orbital. These changes in velocity cause water to slosh about as rising and falling tides. To which the stubborn Pope countered: maybe God commands the seas to move back and forth *without* moving the earth at the bottom. God can do anything He desires for He is all-powerful. Did not Christ perform miracles?

Galileo was wrong about tides. Both the orbital motion and the rotational motion of the earth are relative to the sun, not relative to an observer on earth who moves with the earth. It was a rather grievous error, considering his own explanation of why we do not sense the motion of earth from earth. There is a second important factor. His theory of tides could not explain why there are two tides per 24-hour period for any place on earth. Nor could he explain why the time of high tide changes regularly, by about one hour, from day to day. Galileo dismissed these deficiencies as minor effects, just as the role of friction is subsidiary to the fundamental laws of motion. Here we must critique Galileo for committing a classical error. He ignored evidence that did not fit into his overall system. Would that he had taken the same stance as Kepler did for the eight-minute discrepancy in the orbit of Mars. Beautiful laws are often killed by "ugly" facts. Small deviations should not be ignored for they could be the tip of the iceberg, lurking to reveal profound truths.

THE DANGEROUS DIALOGUES

With a friendly Pope in the Vatican, Galileo felt safe to finish the definitive book in which he hoped to clearly lay down all arguments supporting the Copernican system. To allow publication,

Urban demanded that such a book refrain from theological considerations. As before, he directed Galileo to represent the Copernican system only as mathematical hypothesis. What a long shadow Ossiander had cast (Chapter 7)! For the Pope, an acceptable purpose of a new work, presented by an eminent Catholic, Italian scientist, would be to show that Rome did not act out of ignorance when it rejected the Copernican theory.

Given the finality of the salutary edict, Galileo knew he could not get away by openly supporting the heliocentric system. As in a debater's dispute, he decided to craft a *Dialogue on the Two Greatest World Systems*. Three debaters, Salviati, Sagredo, and Simplicio, "dispassionately" discuss the two world systems. Salviati is Galileo's *alter ego*, competently defending the heretical point of view. Sagredo plays the impartial listener, always asking the right questions. Representing the orthodox views, Simplicio is an imaginary character, supposedly modelled after a well-known 6th-century Roman scholar and commentator on Aristotle and Ptolemy. Galileo portrays the anti-Copernican Simplicio as a stupid ignoramus. With a mind clouded by antique cobwebs, he is unable to understand the simplest logical arguments. In a spectacular political error, Galileo put into the mouth of the simpleton some of the Pope's personal arguments in favour of a stationary earth.

The dialogues take place over a span of four days. On the first, Galileo tackles head-on the issue of whether heaven and earth can really be two separate realms. Having discovered mountains on the moon, spots on the sun, and moons around Jupiter, he argues how there can be no clear distinction between terrestrial and celestial regions. Heavenly bodies are made of the same stuff as earthly objects. Future discoveries about the composition of heavenly bodies clearly validated his claims. If we can understand laws that operate on earth, we can understand the heavens. In a daring extrapolation, he concludes that human reason can partake of the divine mind! He goes further. Heaven and earth must be subject to the same laws, accurately presaging the discoveries of Newton (Chapter 11). Such positions were bound to rankle the Church. Why should God be bound by laws deduced by humans? How then to account for the miracles of Christ and the saints, or that angels in heaven can defy gravity on earth, but mortals cannot?

On the second day, the debaters discuss the usual objections to a rotating earth and deal with counter-arguments already discussed. On the third day comes a presentation of the new order of the heliocentric system. As earth orbits around the sun with other planets, it is indistinguishable from other heavenly bodies. Yes, earth motion defies all common sense, but to recognize the truth the force of reason must prevail:[33]

> For the arguments against the whirling of the earth are very plausible ... the experiences which overtly contradict the annual movement are indeed so great in their apparent force that, I repeat, there is no limit to my astonishment when I reflect that Aristarchus and Copernicus were able to make reason so conquer sense, that in defiance to the latter, the former became mistress of their belief.

With the power of reason our mind can comprehend a richer reality that flickers beyond what our perceptions provide only a faint inkling of from our immediate environment. Finally, on the fourth day, Galileo brings in the tides as proof. It is here that he commits a scientific error, but also his lamentable political error. He puts the Pope's arguments in the mouth of Simplicio.

On reading the finished dialogues, the Pope did not come out immediately against the book, a surprising turn of events. Perhaps the crafty dialogue duped him on the first reading. Still, fearful of the broad cosmological implications, the Pope asked Galileo to consider changing the title of the finished product. When Galileo came back with the alternative, *Dialogues on the Ebb and Flow of the Seas*, the Pope became nervous. The alternative now emphasized too strongly Galileo's idea that tides provide incontrovertible proof for earth's motion. And the work still left the reader with the feeling that there was a choice in the matter. There had to be a final statement supporting the position sanctified by the Church. Anxious to get approval for publication, Galileo promised to oblige with a new introduction and a conclusion.

These he wrote over several years. When he finished the new sections, they contradicted the book in content and tone. His final version still appeared ambiguous in purpose. He recalled the salutary edict which[34]

> in order to obviate the dangerous tendencies of our present age, imposed a reasonable silence upon the ... opinion that the earth moves.

Going further in dangerous commentary, he expressed his opinion that the salutary decree "had its origin not in judicious inquiry, but in passion." Only after making these disclaimers, which were really protests, did he yield, and again only reluctantly, to the statement that the earth is motionless[35]

> for reasons that are supplied by piety, religion, the knowledge of Divine Omnipotence, and a conscious-ness of the limitations of the human mind.

There is a double meaning. To assuage his critics, the human mind may be too limited to understand the mind of God. Alternatively, in a defiant spirit, our mind is too limited to accept the truth that defies common sense.

Galileo submitted the manuscript, written in popular Italian, to the censorship of the Florentine Inquisition. Pope Urban's secretary thought that it was a brilliant work and quietly gave approval to the master of the palace for printing. But he ordered the Bishop of Florence to keep a watchful eye on revisions to ensure that the author consistently presented the earth's motion as hypothesis.

Popularity and fame quickly followed publication. All over Europe, enthusiastic readers hailed the work as a scientific and literary masterpiece. The lively, dramatic dialogue turned astronomy from a sterile, erudite, and aristocratic science into a living, stirring, popular science. In the same spirit as the *Starry Messenger*, Galileo propelled the democratization of science with eloquent vernacular.

Controversy boiled again when the Church re-examined the work. Arguments presented in favour of the earth's motion appeared all too cogent. Now the Pope's apparatus decided that the book should be banned. Copies already printed were to be handed over to the Inquisition. How did the work ever slip through the ecclesiastical censor? Powerful prelates sentenced the secretary to exile and suspended the master of the palace. Naively, Galileo prevailed upon his protector, the Grand Duke of Tuscany, to send a strongly worded protest against the ban. But it was to no avail. Jesuits of Rome spread the word that Galileo's teachings were worse than the heresies of Calvin and Luther. A crafty introduction made the book thoroughly Copernican in spirit. While Galileo bluffed the censors, printers escaped the watchful eyes of the Pope's apparatus. Galileo had even dared to hold the Pope in ridicule through the character of Simplicio. Outraged, Urban resolved to persecute Galileo. Already in political trouble from the failure of military adventures, the Pope could not afford to lose further credibility by allowing freethinkers to spread revolutionary ideas. The Church of Rome wished to stop the world from sliding farther and farther down the slippery slope of heresy.

THE NOBLEST EYES

Summoning the 70-year-old Galileo to Rome, despite his conditions of arthritis and hernia, the Inquisition held him in custody at night and put him through the ordeal of hearings by day. They insisted how clear it was that the accused still believed in the motion of the earth. Threatening him with instruments of torture, they insisted on a retraction. His grim treatment was a warning to sci-entists all over Christian Europe that spiritual obedience demanded intellectual obedience. Galileo hoped to the end that his old friend, Pope Urban, would intervene on his behalf. On the last day of the trial, he kneeled in the presence of seven cardinals and, with Bible in hand, he recanted. There is an indestructible legend that deserves to be told. As he rose from his knees after the recantation, he muttered under his breath, "And yet it moves." Unanimously quashed by historians of science, the legend is truer to Galileo's character than to history. From pulpits throughout all Italy, priests arrogantly read out the renunciation extracted from the old man.

The tragic affair was not simply a rational debate about the merits of the heliocentric system versus the traditional geocentric system. It was about the infallibility of the Church, a spectacular example of how institutions of authority abhor change and resist new ideas. Galileo's untiring onslaught posed an extreme risk to the foundation of Christianity at its home base. Already under severe threat from Protestants, the Church could not afford to tolerate new challenges. But ultimately Galileo's position prevailed. His assault proved crucial to topple the authority of Aristotle, and liberate human thought from the shackles of tradition.

Sentencing Galileo to imprisonment in the dungeons of the Holy Office for an indeterminate period, the Inquisition soon reduced the respected scientist's sentence to exile at Arcetri, one mile from Florence. Here Galileo lived out the last years of his life near his beloved daughters. Constant surveillance prevented him from ever leaving his villa. At Arcetri, Galileo recovered from the ordeal, and began to work again. His discoveries could not be buried, nor his ideas repressed. French philosopher-mathematician René Descartes, and English poet John Milton visited the ageing master. It was their unique privilege to see the wonders of the night sky through Galileo's magical telescope, guided by the discoverer's hand at the telescope and mind on the heavens. One month after condemnation, he surreptitiously arranged for publication of *The Dialogues* in Strasbourg, Germany.

With the controversial dialogues finally in print, Galileo returned to work on dynamics, his first passion. Fortunately, the strenuous tribulations had little effect on his scientific vigour. In a lively new work, *Discourses on Two New Sciences*, Galileo collected investigations made over 30 years in the science of dynamics, the laws of free fall, inclined planes, projectile motion, and the principle of the pendulum. When completed, the visiting French ambassador smuggled *The Discourses* out of Italy and published them in Holland. Italy's misfortune was that its most famous scientists could only publish in Protestant countries.

Near the end of his days in exile at Arcetri, Galileo slowly lost his eyesight, previously damaged by observations of sunspots. But his mind never failed him. There remained one last astronomical discovery: the librations of the moon. Observing the moon regularly, Galileo could occasionally see a bit more than 50 per cent of the moon's face, a few per cent of the dark side towards both east and west. Here is the simple reason. Normally the moon presents the same face to earth, because friction from tidal forces due to earth's gravity has slowed down the moon's rotation. The moon still does rotate very slowly. Because the moon also revolves around earth, the two motions compensate just enough to hide most of the dark side, almost all the time. But, as Kepler discovered, the moon does not move in a circular orbit with constant speed. Rather its orbit is elliptical and its speed non-uniform. The compensation breaks down when the revolution and rotation speeds do not exactly match to reveal a bit of the dark half on one side or the other. Unwittingly, Galileo's discovery of librations provided another item of evidence for Kepler's elliptical orbits. When he made the discovery, the nearly blind Galileo wrote to a friend:[36]

> This heaven, this earth that I have extended a thousand times beyond the limits of all past epochs by wonderful observations, have now shrunk to the narrow confines of my own body. Thus God likes it; so I too must like it!

Another friend, one-time-pupil and now faithful disciple, Father Benedetto Castelli mourned the loss of the great man's eyes:[36]

> The noblest eyes that nature had ever created grew dark … it may be said of them that they had seen more than all the eyes before them and that they opened the eyes of all those who came after him.

Galileo died in 1642. Not yet satisfied after bringing the great man to his knees by the power of the Inquisition and the dead weight of Aristotelian authority, the Pope sent his emissary to prevent the proud Florentines from burying Galileo beside their other favourite son, Michelangelo. No monument was to be erected to the heretic condemned by the Church.

Reactionaries could not suppress Galileo's glory for long. Less than a century later, Italians moved his remains and solemnly interred them beside Michelangelo's tomb. The epitaph proudly bears the famous legend: "And yet it moves." For a period after Galileo, scientists avoided the wrath of the Church by treating the new knowledge as hypothesis. Two centuries were to pass before the Church finally removed the works of Copernicus, Kepler, and Galileo from its index of prohibited books.

Remarkably, the historic controversy persists in modern conscience. The Papal position remained unchanged till 1979, when Pope John Paul II formally pardoned Galileo. At a session of the Vatican Academy of Sciences in November 1979, he conceded that Galileo[37] "suffered greatly – we cannot conceal this now – from an oppression on the part of the Church." It was a grudging admission. Still, the modern Vatican tried to rationalize the action of the Church by stating that Galileo's repentance was "divine illumination in the mind of the scientist." In the mind of the Pope, the tragedy oddly emphasized the harmony of faith and knowledge, religion and science.

Within 50 years of Galileo's work, an amazing change took place in the intellectual climate of science. The experimental method became accepted as the best way to advance science. Astronomy with the telescope flourished. Men of science became less isolated as they formed scientific societies. One by one they discarded geocentric views. Galileo had paved the way for modern science. He is truly its father.

THE SECRET RINGS

Not long after Galileo's passing, Christian Huygens (1629–95) in Paris transformed the rudimentary telescope into a powerful instrument for precise astronomical observations. Encouraged by his father and helped by his brother, the task of making ever more powerful telescopes became Huygens' lifelong passion. Like Galileo, he was a man of many talents. He hit upon a new and better method for grinding lenses. With keen mathematical ability, he advanced ray-tracing methods to build telescopes of extraordinary optical quality, and with magnifications between 100× and 150×. His largest refractor was 23 feet long. Huygens' powerful telescopes showed Saturn's anomalous "ears" pondered by Galileo to be a complete ring surrounding the remote planet.

France was to be the centre of the new intellectually liberated universe. With lavish funds from the Sun King, Louis XIV (1638–1715), a famous Italian architect was directed to design a new observatory in Paris. The king's enlightened minister, Jean Colbert, personally ordered the best lenses from Italy, authorizing French astronomer Jean Picard to attract similar talent from all over Europe and commission a new Uraniborg (Chapter 8) to continue Tycho Brahe's adventurous missions. To attract qualified astronomers from Italy and Denmark, Picard offered Jean Cassini princely living quarters, and put up Ole Roemer in his own apartment. Working in Italy with increasingly powerful telescopes, Cassini had already reported impressive new finds. After recognizing spots on Mars and Jupiter, he tracked them to deduce that both planets rotate. Mars makes a complete turn in about 24 hours and Jupiter in ten hours. Recall how Galileo used moving sunspots to determine the sun's rotation. Evidence for earth's rotation continued to grow stronger. If the sun rotates, and the planets rotate, why should rotation be so unusual for planet earth?

Stifled in Catholic Italy, Cassini jumped at the opportunity to move to the new intellectual centre, the Academy of Paris. When he arrived and saw the plans for the grand Paris observatory building, already half constructed, he immediately requested an audience with the king. A famous Italian architect had drawn up plans that would make an excellent palace, Cassini asserted, but the building would be totally useless for astronomy. It would never have the stability to make precise telescopic observations. Furious at the criticism cast upon his ornate design, the proud Italian architect refused to make any alterations. Undeterred, Cassini decided to set up his powerful instruments, but on the grounds outside the useless observatory building. From there he found that the rings of Saturn are actually two rings, separated by a gap, now called the "Cassini division." He also discovered four moons around

Saturn. Modern telescopes reveal more than 60 moons. Keeping track of surface features, now visible on Saturn, Picard, Cassini, and their colleagues found that all the planets rotate about their axes. What was particularly interesting was that they all rotate counter-clockwise (except Venus), which is also the direction of planetary orbits around the sun. What a neat organization the Creator had in mind! Later, this observation inspired French scientist and mathematician Simon Laplace (1749–1827) to propose the nebular hypothesis that our solar system evolved from one rotating cloud of gas. It was one of the first clues to the birth of stars and the origin of the solar system (Chapter 8).

THE IMMENSITY OF THE SOLAR SYSTEM

From spotting new planetary features, scientists at the Paris Academy expanded their ambition. They planned to measure the size of the solar system. Toward this goal, the French academy sponsored several scientific expeditions. The most notable foray was to Cayenne in French Guyana (near the equator) where they set up to measure the parallax of Mars against the fixed stars by simultaneous observation of Mars from two places on the earth's surface.

In the heliocentric model, Copernicus gave the relative size of the solar system (Table 7.1). For example, the farthermost planet, Saturn, is 9.1 times more distant from the sun than earth. What was still missing was a direct and precise measure of any one of the *absolute distances.* If astronomers could measure the distance between the sun and just one neighbouring planet, they could unravel Copernicus' pioneering table to give the distances between the sun and *all the planets.*

The telescope opened the possibility of accurately measuring small parallax angles. Cassini chose to make the measurement for Mars even though Venus is closer to earth. The reason is that when Mars is nearest to us, Mars is visible at night (in opposition) and therefore easy to locate in relation to stars. When Venus is nearest to us, it appears directly between the sun and the earth. With its phase like a new moon, *Venus is nearly invisible.*

The best time for measurement was when Mars came into perfect alignment with the earth and sun (Figure 9.14). Cassini chose Paris as one location. For the other vantage point (Figure 9.15), he sent French astronomer Jean Richer to Cayenne on the northern shore of South America. Thus they established the first absolute scale of our planetary system. Mars is nearly 80 million kilometres from earth. What a shock it was to discover that the distance to a nearby planet was nearly the same as Ptolemy's early estimate of the total size of the entire cosmos! Copernicus' table then revealed the earth–sun distance of 140 million kilometres. Saturn, the outermost planet known at the time, is 1.5 billion kilometres from the sun.

The Parisian cosmos completely dwarfed the universe of antiquity. For some it was more of the same disturbing news, ever since Copernicus displaced the earth from its privileged position and marvelled at the immensity of the cosmos from the absence of stellar parallax. Our home was becoming lost as an insignificant speck of dust in an infinite universe. Aghast at the upsetting news coming from all directions, poet John Donne (1572–1631) portrayed the troubled mood:[38]

> And the new Philosophy calls all in doubt,
> The Element of fire is quite put out;
> The sun is lost, and the earth, and no mans wit
> Can well direct him where to look for it.
> And freely men confess that this world's spent,
> When in the planets and the firmament
> They seek so many new; they see that this
> Is crumbled out again to his atomies.
> 'Tis all in pieces, all coherence gone.

Artists responded to the immensity of the cosmos and humanity's demotion to the insignificant and infinitesimal. In *Landscape with Church and Village* (Figure 9.16),[39] Jacob von Ruisdael paints

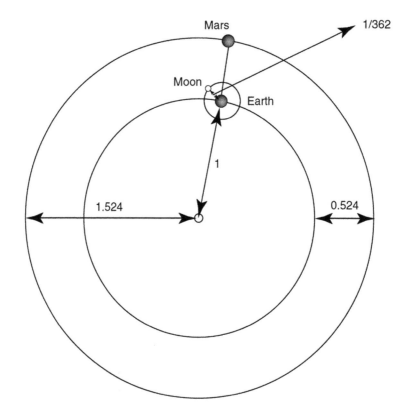

FIGURE 9.14 Paris astronomers measured the parallax of Mars when it came closest to earth. According to Copernicus' Table 7.1, the reference earth–sun distance is unity, and the Mars–sun distance is 1.524. Therefore the Earth–Mars distance is approximately 0.524. Determination of the absolute distance between earth and Mars yielded all the absolute distances between the sun and its planets in km.

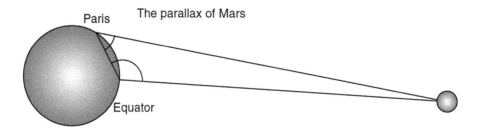

FIGURE 9.15 Location of two points on earth to determine the parallax of Mars, and from there the scale of the solar system.

the natural world as a cosmographic landscape engulfing humankind. Bright light and great floating clouds explode into an expanding sky. Is our world still a special part of God's plan, or are we floating aimlessly in a boundless cosmic sea?

THE MOONS OF JUPITER REVEAL THE SPEED OF LIGHT

Galileo did not realize it, but his sensational discovery of Jupiter's satellites gave the Parisian scientists a successful method to measure the speed of light. Constructing detailed tables for the revolution of Jupiter's moons, he proposed their frequent eclipses as a universal time-keeping

FIGURE 9.16 *Landscape with Church and Village,* 1670, Jacob von Ruisdael.

mechanism, since eclipse times of Jovian moons are the same for observers anywhere on earth who can see Jupiter's moons through their telescopes. Soon after Galileo's passing, Ole Roemer used the moons of Jupiter to make the first successful light-speed measurement in 1676. Roemer came to Paris at the urging of Jean Picard, organizer of the grand Paris Observatory, where Cassini, Huygens, and Richer gathered to make new observations with sharp telescopes. By the time Roemer arrived, Cassini had carefully worked out the exact moments at which Jupiter eclipsed each satellite.

Carefully comparing eclipse cycle times through the year, Roemer found slight shifts in their expected schedules. Precise measurements have always led to impactful new discoveries. At some times through the year, Io's eclipses came a little too early, and at other times a little too late. Not one to dismiss minute discrepancies, Roemer realized that the time between eclipses was shorter when earth was closer to Jupiter in its orbit around the sun (Figure 9.17). The difference over six months grew to roughly 16 minutes, clearly measurable and demanding explanation. In a flash of inspiration, he linked the discrepancy to the additional time light must take to cross earth orbit. Knowing the diameter, Roemer estimated the velocity of light as 2×10^8 m/sec. It was the right order of magnitude, but lower than the modern value of 3×10^8 m/sec. Such inaccuracies are typical at the raw edge of the frontier of human knowledge. Recall how Aristarchus greatly underestimated the size of the sun relative to the size of the earth (Chapter 6).

Once and for all, Roemer banished speculation that light travels at infinite speed. Now the speed of light has become the primary yardstick for the size of the solar system, galaxy, and universe by using the light year (LY), the distance light travels during one year (10 trillion km). Continuing the tradition of Eratosthenes, Hipparchus, Brahe, and Kepler, Roemer's scrupulous attention to detail led to the unthinkable. To convince his contemporaries of his incredible feat, Roemer came up with a stunt. He predicted that, in exactly two months, the eclipse of Io would be precisely 600 seconds late. Taking the challenge, Cassini and others were stunned to find Roemer's prediction accurate to one second. That level of extraordinary precision was a brilliant stroke of luck! And it turned sceptical Paris astronomers into true believers.

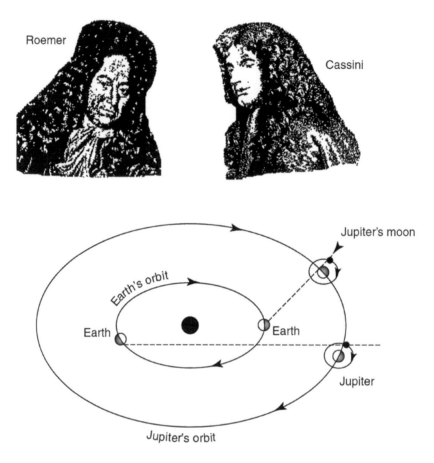

FIGURE 9.17   Roemer makes the first determination of the velocity of light from precise observations made by Cassini on the eclipses of the moons of Jupiter.

UPDATES

DISTANT NEBULOSITIES

Stars were not the only objects of astronomers' affection. Ptolemy dutifully made note of seven cloud-like blemishes in the heavens, later dubbed *nebulae*. With the help of a telescope, French astronomer and avid comet-hunter Charles Messier, in 1784, expanded Ptolemy's preliminary list to 101 diffuse patches with a hazy clouded appearance. But Messier was not really interested in nebulae. He kept track of the nuisance clouds to avoid confusion with comets he really wished to discover. Astronomers still refer to many of Messier's blotches by their M number.

Given the similarity of their misty appearance with the Milky Way, German philosopher Immanuel Kant (1724–1804) suggested that nebulae might be "island universes" densely populated with stars, much like our own Milky Way. Indeed, 20th-century stargazers later confirmed Kant's conjectures by identifying many nebulae as galaxies. Plunging deep into space with self-made telescopes to "look into the mind of God," William Herschel (discoverer of planet Uranus, Chapter 5) and his son John expanded the collection of blurry stellar patches to 5,000. They could resolve many nebulae into disc-like shapes they suspected could indeed be island-universes. Some they properly picked out to be dense, globular star clusters with millions of stars, and other nebulae to be interstellar clouds of gas and dust, later understood to be cocoons for new-born stars (Chapter 8). Overshadowing the Herschel collections, Johann Dreyer published a massive catalogue in 1908 of 15,000 nebulae.

But none of the map makers had any idea of the size of these extra galactic clouds, nor of their immense distances from earth. We will continue with the stories of the nebulae as well as distance determinations after we explore further advances in telescopes as well as their grand discoveries in the heavens. As telescopes grew more powerful, looking deeper and deeper, the universe grew increasingly immense in grandeur and size.

POWERFUL REFLECTIONS

After Galileo's telescope revealed the Milky Way to be a dense congeries of stars, German philosopher Immanuel Kant offered a striking geometric explanation for why the Milky Way stars form a ribbon of pale light around the sky. Stars in our galaxy are not scattered randomly about the heavens without design. They populate a flat disc. Our sun is one star residing inside the disc. From inside a pancake of stars, our view of the Milky Way appears as a faint belt. When we look along any direction roughly parallel to the plane of the pancake, our eyes catch a high density of stars within the disc (Figure 9.18). If we look in other directions, outside the plane of the disc, our eyes encounter relatively fewer stars because the disc is thin. Clearly we are well inside the disc because the Milky Way extends as an uninterrupted circle across the entire sky, and is visible in both hemispheres.

A similar effect takes place on a smaller scale, to explain why planets appear to move in the belt-like region of the zodiac. All planets are confined to the region of the planar disc surrounding the sun. Earth dwellers look at the planetary disc from inside and see the planets, sun, and moon travel along a belt-like path. Both the Milky Way and the solar system evolved from rotating gas clouds into a disc-like shape through the combined effects of rotation and gravitation (Chapter 8).

Though aesthetically pleasing, Kant's galactic disc idea was purely speculative. Inspired by geometry, the simple model raised fruitful questions. Is the Milky Way the entire universe? Are we and the sun at the centre of this vast planar universe of stars? Judging from their milky appearance,

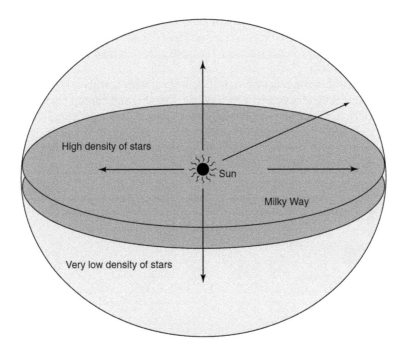

FIGURE 9.18 Kant envisioned a model for why the Milky Way galaxy appears as a belt across the sky instead as a disk. Looking in any direction parallel to the plane of a disc-shaped galaxy, we see a high density of stars in the shape of belt. But looking outside the plane, the density of stars decreases substantially. The sky looks mostly empty.

Kant wondered in 1755 whether hazy nebulae could also be distant galaxies, dubbing them "island universes." Could such a single and simple explanation account for all the nebulae that Charles Messier had carefully documented?

The refracting telescopes of Galileo, Kepler, Huygens, and Cassini were unveiling countless items of evidence for the Copernican system. But there was a nagging problem. Colours soiled refracted images, a nuisance called *chromatic aberration*. To circumvent the problem, young Isaac Newton (Chapter 11) devised a radically new type of telescope (Figure 9.19). Instead of a large convex lens, he used a 1-inch diameter parabolic mirror to collect rays of light from a distant object to focus in front of the mirror. A convex eyepiece could then magnify the image. His reflecting telescope was much shorter than refractors with the same light-gathering and resolving power. Newton's first reflector was only 6 inches long, but it had magnification of 40×, equal to a refractor 6 feet long. It could clearly display the moons of Jupiter. In a breakthrough for astronomy, Newton had conceived the principle of the most powerful telescopes to peer into space beyond the solar system. Mirrors would continue to grow in size to increase their light-gathering capability and their grasp of a rapidly widening cosmos.

When news of Newton's miniature telescope leaked out, his colleagues at Cambridge urged him to send the marvellous toy to the Royal Society. But Newton had lost the little instrument, absorbed as he was in mathematical discoveries. He quickly made another. The miniature telescope caused a sensation, evoking national pride among members of the Royal Society. Various dignitaries, including the King himself, wished to examine the English device. To recognize his achievement, the Royal Society immediately awarded Newton a membership, a high honour for such a young scientist. The enthusiastic reception caught Newton by surprise. He considered the puny gadget of little value compared to his on-going mathematical breakthroughs, topics for Chapter 11.

When Newton started reflectors, the technology of high-quality mirrors lagged considerably behind lens grinding and polishing. Musician William Herschel and his sister Caroline soon bridged the gulf (Chapter 5). They began to grind and polish mirrors to make their own telescopes. The brother and sister team became fanatical grinders. Turning their home from a music school by day into an optics shop by night, the Herschels ground and polished mirrors for up to 16 hours at a stretch. Caroline Herschel wrote in her diaries: "I see every room turned into a workshop … a huge turning machine in a bedroom."[40]

Herschel's largest home-made reflector was 20 feet long, with a mirror 18 inches in diameter. Through an earlier model, he discovered a new planet, Uranus, and earned resounding accolades from King and country (Chapter 5). Over 20 years, more than 400 mirrors came out of their workshop. Some they used and others they sold to amateurs. The proceeds financed even larger telescopes. For an astronomer of the time to acquire an Herschelian telescope was equivalent to a violinist owning a Stradivarius.

 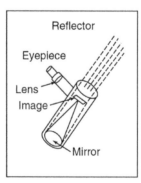 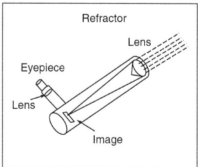

FIGURE 9.19   Newton looks through his hand-made reflecting telescope. Ray diagrams for Newton's reflecting telescope compared with a refracting telescope.

To increase their light-gathering capability beyond their best instrument, the Herschels launched construction of a mammoth telescope with a 36-inch diameter mirror, even though it was too large for their workshop. But no metal foundry would agree to undertake the giant task, nor could the Herschels afford the steep price. Going back to the basement of their house, they constructed a cheap mould out of a wagon-load of horse dung. The whole family took turns at pounding the mould into shape and shaving the surface smooth. On the fateful day to cast the giant mirror, they anxiously poured out the hot liquid metal. With a resounding crack, the mould shattered under the intense heat and liquid metal gushed onto the floor. As the party fled for their lives to the garden, a river of lava pursued them. Fortunately, it solidified rapidly before anyone was hurt.

It was clearly the end of the road for large-scale homemade telescopes. But King George III of England agreed to fund the famous astronomer's next model. Ever ambitious, Herschel planned for a 48-inch mirror. Industry offered to try. When the mammoth device was finally complete, an eager King attended the dedication ceremony, bringing along the Archbishop of Canterbury. Proud of the new instrument, the King bragged: "Come my Lord Bishop, I will show you the way to Heaven."[41]

At a time when his contemporaries were still studying planets and measuring the size of the solar system with refractors, the Herschels dived deep into space with powerful reflectors. Eager to uncover God's plan for the universe, they set out to map the heavens with ferocious intensity by making systematic star counts in every field of view of his most powerful instrument. What was the pattern to the stellar distribution? To map the shape of the galaxy, Herschel decided to survey stars both inside and outside the Milky Way. Over 20 years, Herschel counted 90,000 stars in 2,400 sample areas. From their numbers and locations, he concluded that the Milky Way is indeed a flat disc with the shape of "grindstone," about five times as broad as it was thick. An appropriate analogy for contemporary readers would a hamburger bun. With the same number of stars visible in any direction from inside the disc, Herschel concluded that the sun must be located near the centre. But, without any method to estimate stellar distances, Herschel could not determine the disc's diameter.

Through his intense search, Herschel discovered more in heaven than anyone could dream of. Collecting light from hundreds of distant nebulae, he sketched a variety of shapes (Figure 9.20). Inside some white fleeces, he could identify stars. These turned out to be dense swarms, later called globular clusters. By resolving a third of all nebulae into spherical clusters, Herschel left behind a major clue to the structure of our galaxy.

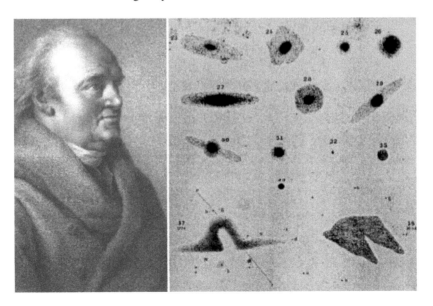

FIGURE 9.20   Herschel and his sketches of nebulae shapes. Some turned out to be interstellar gas clouds, others dense globular star clusters, and still others were galaxies like our own.[42]

Could other nebulae be star congregations, yet to be resolved? A most interesting feature was that all nebulae avoided the plane of the Milky Way. Could they be outside the galaxy? Could they be the island universes Kant dreamed about? Following in his father's steps, Herschel's son John confirmed the disc shape of our galaxy with a star count of 70,000 from the Southern Hemisphere. And his nebulae catalogue expanded to 1,700. Stars had enticed the entire Herschel family.

SCALING THE COSMOS

The parallax method (Chapter 6) was the first rung on the ladder to scale the cosmos. Measuring distances to external galaxies is even more challenging, because parallax angles are much smaller than 1 milli-arcsecond (which limits the measurable distances to about 5,000 LY). Another method to determine stellar distances is from star luminosity. A generation after Galileo, Christian Huygens (and later, Newton) assumed that if Sirius and the sun have the same intrinsic luminosity, and the star's apparent luminosity is so much smaller than the sun's it means that Sirius is much much farther from earth than the sun. Huygens looked at the sun through small holes in a piece of cardboard and decreased the apertures until he could match the apparent brilliance of Sirius. Using Hipparchus' brightness scale, if Sirius has brightness magnitude −1 and the sun −28, Sirius is dimmer by 27 orders, which is equivalent to 2.5 multiplied by itself 27 times, equal to 63 billion. Spherical geometry gives the simple conclusion (Figure 9.21) that the fraction of light arriving at earth from a distant star decreases as the square of the distance to that star. Hence Sirius is $\sqrt{63\,\text{billion}} = 250,000$ times farther than the sun. In modern terms, if the sun is eight light minutes away, Huygens' distance to Sirius would be 4 LY. Sirius is actually 8.7 LY away. Indeed Huygens and Newton made excellent first estimates, but they went wrong with the simple assumption that the intrinsic brightness of Sirius is the same as the sun's. To figure distances with the brightness method through the inverse square law, it is crucial to know the *intrinsic luminosity* of stars, for example, the number of suns that would replace Sirius' brightness. Was there some other property of stars that would reveal their *intrinsic* brightness?

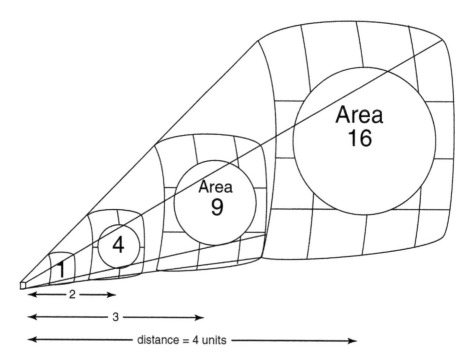

FIGURE 9.21   Light flux through a unit area at radius r decreases as r^2 because the surface area of a sphere increases as r^2. Doubling the distance decreases illumination by a factor of four.

Cepheid variables provided the new tool. In 1912, while examining stars in the small Magellanic Cloud, Henrietta Leavitt made a remarkable discovery about intrinsic luminosity of some particular type of stars. It was to have lavish implications for our ability to measure galactic distances. Leavitt's main job as research assistant was to be a human computer along with 20 other women (including Annie Jump Cannon from Chapter 8) dutifully performing tedious examinations on tons of photographs, and calculations on copious data streaming out from the Harvard Observatory station at Peru in the Southern Hemisphere. Women at the time were not allowed to earn Harvard degrees, only attend classes. Leavitt's assignment was to look for changes in brightness among the same stars on many photographs taken over a period of time. By now, it was well-known that stars are not constant and unchanging, contrary to staunch beliefs of ancient astronomers. A striking example in the constellation Cepheus was a star varying *periodically* in brightness. In 1784, English astronomer John Goodricke had spotted such a variable star after spending long cold nights at the telescope. He eventually died of pneumonia from the brutal exposure. But he left the world a magnanimous legacy, a cosmic beacon, later called a Cepheid variable.

Volumes of photos from the large Magellanic Cloud showed Leavitt a number of Cepheid-variable stars, with extremely regular behaviour. They grew dimmer and brighter with a variety of periods ranging from a few days to two months. After thousands of hours of scrutiny, her inquisitive bent of mind found an important pattern. The period of pulsating Cepheids was precisely proportional to their *observed* luminosity: the longer the period, the brighter the star. It was a sudden new light. What Leavitt discovered later evolved into a new celestial yardstick with the luminosity method. All the Cepheid stars in the Magellanic Cloud were approximately the same distance from earth, being in a single foreign galaxy, neighbour to the Milky Way. Therefore the *period of a Cepheid* served as a direct indicator of not only its observed luminosity but also its *intrinsic* luminosity. The shortest-period Cepheids were 100 times more luminous than the sun, while the longest ones had 40,000 times the intrinsic luminosity of the sun. Later models of star evolution showed that stars pulsate in brightness when they leave the main sequence band of the HR diagram (Chapter 8), and become unstable for a period as they pass through a region defined as the vertical instability strip on the diagram.

Establishing actual distances in LY to the remote Cepheids from their brightness periods, however, was not easy; there was a temporary insurmountable difficulty. All detectable Cepheids were too far away to measure parallax with earth-based telescopes. Facing up to the formidable challenge to find the distance to any Cepheid, Eljnar Hertzsprung (1873–1967) and Harlow Shapley (1885–1972) (more about him in the next section) invented indirect methods to ascertain distances to Cepheid-variable stars inside the Milky Way. One example is "proper motion" of stars relative to the sun. Star motion can be resolved into two orthogonal components: radial motion which is moving towards or away from the sun, and proper motion across the sky (perpendicular to radial motion). Just as the apparent luminosity decreases with distance, the farther a star the less its apparent "proper motion." Proper motions need time to accumulate, so comparisons must be made using a star's position over many years, or even over centuries.

We will not delve into other techniques to calibrate Cepheids, except to applaud their role as provisional lampposts along the indispensable, expanding cosmic ruler. Although remote Cepheids became accepted distance indicators, their reliability was always open to improvement. Deriving intrinsic luminosity from the period, astronomers could deduce Cepheid distances from observed brightness through the inverse square law. Every doubling of the distance decreased brilliance by a factor of four. Cepheids became the prime method to establish distances to nearby galaxies as well as the size of galaxies, including our own Milky Way.

THE GREAT DEBATE ON THE UNIVERSE

Telescope mirrors continued to grow in size. Every generation of astronomers managed to see a little bit farther than their predecessors. At his private estate in Ireland, amateur astronomer William

FIGURE 9.22   Lord Rosse's spiral nebula, which later turned out to be a galaxy outside the Milky Way.[43]

Parsons (Lord Rosse) assembled a Leviathan in 1845 with a 72-inch mirror. Precariously perched atop a platform high above the mirror, he spent many cold and tedious nights looking through the suspended eyepiece. Parsons made some remarkable sketches of nebulae shapes. Many were spiral pinwheels of light (Figure 9.22). Parsons' intriguing sketches inspired Vincent Van Gogh with heavenly paintings such as "Starry Night." In some spirals that Herschel was unable to resolve, Parsons found stars. Kant's island universes were clearly in the picture as other galaxies. But no one knew the distance to the spirals. Although parallax triangulation methods, luminosity techniques, and later Cepheid variables were yielding stellar distances, the extent of the Milky Way was yet uncertain, making it hard to assess whether spirals and other nebulae were inside or outside our Milky Way. Could the Milky Way be our entire universe?

Continuing precise star counts with the cooperation of observatories around the world equipped with the best telescopes, Dutch astronomer Jacobus Kapteyn finally put a distance scale to the Milky Way disc in 1922. Despite petty objections of nations caught up in the vengeance and aftermath following World War I, Kapteyn struggled to promote international cooperation to amass and analyse data on the brightness, numbers, and motion of stars. His three-dimensional map showed our galaxy disc to be about 30,000 LY in diameter and 6,000 LY thick. His map revealed the sun to be nearly in the centre of the galactic plane, at the unique location of highest density. Displaced from the centre of the solar system by Copernicus, humanity was now back in its special place in the cosmos.

More penetrating intellects looking through more powerful telescopes would challenge Kapteyn's extent of the Milky Way as well as the sun's position in it. A decade after assembling the largest *refracting* telescope with a 40-inch lens at Yerkes in Wisconsin, telescope enthusiast George Hale (1868–1938) went on to build a grander reflecting telescope atop the San Gabriel mountains in California. With donkeys hauling a 60-inch high-quality mirror up narrow, winding mountain roads, Hale assembled the powerful telescope on top of Mt Wilson outside Los Angeles (near Pasadena). At the University of Missouri, a young undergraduate student by the name of Harlow Shapley (1885–1972) was trying to decide his major field of study:[44]

> I opened the catalogue of course … The very first course was a-r-c-h-a-e-o-l-o-g-y, and I couldn't pronounce it! … I turned over a page and saw a-s-t-r-o-n-o-m-y: I could pronounce that – and here I am.

From the telescope on Mt Wilson, Shapley once again dethroned the sun from the centre, this time from the centre of the galaxy. He started by studying star clusters, classifying them systematically with the hope of finding some order among the hundreds he found. There were two distinct cluster types (Figure 9.23). Open clusters were groups of hundreds to thousands of stars, spread out uniformly all through the Milky Way disc. A starkly different class of clusters, the closed globular clusters, were compact swarms of 10,000 to 1 million stars, tightly clumped together in spherical clouds. Amazingly, if there is a rotating planet with an atmosphere orbiting a star in a globular cluster, night time would be almost as bright as daytime due to millions of nearby stars lighting up the night sky.

Shapley found some astonishing features of globular clusters that contrasted with the open clusters. Most globular clusters fell outside the plane of the Milky Way, whereas the open clusters were all inside the plane of the Milky Way. Were globular clusters outside our galaxy? There was another surprising property. Most globular clusters lay in or near constellation Sagittarius, rather than being randomly distributed over the entire sky. But the *open* clusters were symmetrically distributed through the entire sky. Why were there such drastic differences between open and compact clusters?

Shapley's challenge was to measure distances to globular clusters even if they were too far away to have any measurable parallax. Cepheid lamp-posts provided the needed distance markers. Shapley was the first to successfully calibrate Leavitt's famous period–luminosity relationship which brought new findings quick and fast. From the edges of compact clusters, Cepheids revealed to Shapley the distances to 86-star conglomerates. From distance measurements, Shapley confirmed that globular clusters were widely distributed in latitude and *symmetrically* above and below the galactic equator. But as he already noticed, their longitude distribution was strangely skewed. A majority of globular clusters displayed a concentrated presence on one side of the Milky Way, near Sagittarius. Once again, asymmetry rang bells to open new doors.

Shapley drew several daring conclusions. Even though they fall outside the galactic plane, all globular clusters are nevertheless physically associated with our Milky Way and bound to it via gravity. If they do not belong to our galaxy, a random and symmetric distribution over the entire sky would appear. The centre of the globular cluster halo must therefore be the true centre of our galaxy. Furthermore, the diameter of that halo of clusters surrounding the Milky Way must be equal to the width of our galaxy, 300,000 light years in size. The next conclusion was even more radical and upsetting: the reason for the asymmetry in the distribution of compact clusters. Most globular clusters appear predominantly on one side of the Milky Way because *our sun is not at the centre of our galaxy!* Constructing a three-dimensional cluster map from his distance estimates, Shapley located the sun to be 60,000 LY distant from the galactic centre (Figure 9.24). And finally, Shapley dared to make an audacious pronouncement: at 300,000 LY, the Milky Way galaxy is so large that it must be the entire universe. There are no other galaxies outside our own. Shapley was proud to have established the true size of the entire universe.

Opposition was vigorously strong, on all three counts. Shapley's size estimate for the Milky Way was too large. And, by displacing our sun from the centre to near the edge, he rendered our sun, our

FIGURE 9.23   (Left) open star cluster (right) closed globular star cluster.[45]

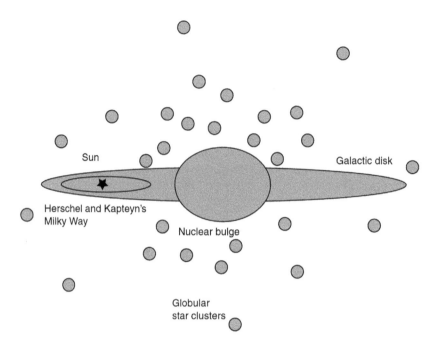

FIGURE 9.24   Viewing the Milky Way through a thick fog of gas and dust, Herschel and Kapteyn could only see a limited distance, so they thought the sun to be at the centre.

earth, and humanity peripheral – for the second time in history. Even though 300,000 LY seemed like an enormous distance, surely the universe could well be much larger.

Factions assembled to challenge each other in a dramatic debate staged in 1920, ostensibly to vie over the extent of the universe. Shapley led the charge to argue forcefully that the universe comprised just one, single mega-galaxy. Faced with the gigantic 300,000 LY extent, he was sure that our Milky Way encompassed all known structures of the universe, including spiral nebulae. How could there be hundreds or thousands of galaxies out there? Most important, he displaced the sun from the centre of the Milky Way. The sun was just another nondescript star at the edge of the galaxy. Shapley reflected over his galactic revolution:[46]

> Yes we have been victimized by the chance position of the sun near the centre of a subordinate system, and misled by the consequent phenomena, to think we are God's own appointed, right in the thick of things. In much the same way ancient man was misled, by the rotation of the earth, and by the consequent apparent daily motion of all heavenly bodies around the earth, to believe that even this little planet was the centre of the universe, and that his earthly gods created and judged the whole.

Rival Heber Curtis, from Lick Observatory in Santa Cruz, California, held the familiar galactic ground. Curtis argued that galaxies could not be as large as Shapley's estimate of 300,000 LY. Kapteyn's international collaboration had made reliable star counts to determine a diameter of 30,000 LY, with the sun at the dense centre. How could Kapteyn be off by a huge factor of ten? Curtis was also an enthusiast for the island universe hypothesis, promoting many galaxies lying outside our own. Evidence was mounting that spiral nebulae must be swirls of stars, perhaps the distant galaxies of Kant's speculations. Bright novae were showing up at spiral edges to prove a stellar composition for the nebulae. Andromeda must be a galaxy, not a gas cloud. Distant galaxies must be scattered all over the universe. The Milky Way must be just one galaxy in an enormous universe filled with galaxies.

Curtis offered evidence that spirals had to be well outside the Milky Way. Since novae in spirals were ten times less brilliant than novae in the Milky Way, the spirals were a hundred times farther away than the largest extent of the Milky Way, and therefore well outside our galaxy. If the Milky

Way was 300,000 LY as Shapley proposed, Andromeda and other spiral universes would have to be an incredible 30 million LY away. Brightness considerations led to the same staggering estimate. If Andromeda, like the Milky Way, was also 300,000 LY wide, it would have to be millions of light years away to account for its observed luminosity. (Later this was found to be exactly so.) Such stupendous distances in the universe were completely inconceivable for the time. Curtis' imagination was open to accept many galaxies outside our Milky Way, but he could not accept a cosmos of such enormity. So Shapley must be wrong about the 300,000 LY estimate.

Both Curtis and Shapley were right, each in some ways. Both Curtis and Shapley were wrong in others. As the famous adage goes, the vying contestants had just grabbed different parts of the proverbial elephant and were describing what they could see. At the great debate, there was a great disconnect. Shapley spoke mostly on the size of the galaxy while Curtis focused his presentation on spirals. There is always fruitful confusion and controversy at the raw edge of the frontiers of science. Both were guilty of a bit of Ptolemaic thinking on the unfamiliar galactic scale. But the open debate was healthy, exposing different pathways of thought.

RESOLUTIONS

Hale soon moved on to assemble a 100-inch reflector at Mt Wilson Observatory. Now builders could haul delicate parts up the mountain by trucks along the same treacherous roads travelled earlier by donkeys. Several trucks toppled over the steep edge of cliffs. When finally completed, Mt Wilson housed the best instruments of astronomy, until Hale's next gallant jump to a 200-inch reflecting telescope on Mt Palomar outside San Diego.

With eyes strengthened by the 100-inch, Edwin Hubble (1889–1953) made a joyous discovery (in 1923) of several Cepheid-variable lamp-posts after repeatedly photographing the Andromeda nebula. Now he could triumphantly extend Leavitt's pioneering period–luminosity relationship to penetrate the heavens out to an astonishing 1 million LY! That was clearly too far for Andromeda to be inside the Milky Way. It was the first piece of solid evidence that Andromeda is a galaxy outside our own. Shapley was dead wrong about the limited size of our universe.

From the giant telescope, Edwin Hubble settled the great debate of 1920. Track athlete, basketball star and coach, Rhodes scholar, high-school teacher, infantry captain in World War I, and lawyer, the brash young astronomer fixed his gaze upon the outer regions of Andromeda (M31) and its neighbour M33. He found "dense swarms of images which in no way differ from those of ordinary stars."[47] Andromeda was not a nebula of fiery hot gas, but a separate galaxy of millions upon millions of stars. Among repeated photographs of faint stars in Andromeda, a dozen showed cyclic brightness variations. These were Cepheid variables, trusted distance indicators discovered by Leavitt and calibrated by Shapley. Hubble determined Andromeda was 1 million LY from earth. That was too far away to be part of our galaxy, even if the Milky Way was as large as Shapley's grand estimate of 300,000 LY. You could fit three Milky Way galaxies in the empty space to Andromeda. More powerful telescopes were to later reveal that Andromeda is a large spiral galaxy (Figure 9.25) in a local group of galaxies of which the Milky Way is a member. Andromeda's distance is 2.5 MLY.

When Shapley first heard about Hubble's find, he described it as a "most entertaining piece of literature."[48] But others encouraged young Hubble to send his discovery to be read at the meeting of the American Association for the Advancement of Science. On New Year's Day 1925, a scant audience of 100 astronomers heard the astounding result from Hubble's paper, read out in the stunning absence of the discoverer. From that day forward, the scientific world was sure that the Milky Way was not unique. If poet Alfred Noyes (1880–1958) had been in the audience he would have resonated with Hubble's discovery. In *The Torch Bearers*, a poetic trilogy about pioneers of science, the poet chants:[49]

I see beyond this island universe,
Beyond our sun, and all those other suns
That throng the Milky Way, far, far beyond,
A thousand little wisps, faint nebulae,

FIGURE 9.25   Andromeda galaxy and Hubble.[50]

> … Faint as the mist by one dewdrop breathed
> At dawn, and yet a universe like our own;
> Each wisp a universe, a vast galaxy
> Wide as our night of stars.

As one of two best papers read, Hubble received a $500 prize. But he was not present to receive the award. Back home from Mt Wilson he was preparing to reveal one of the universe's most profound discoveries: the expansion of the universe, his greatest find (Chapter 12).

BLOCKED VIEW

What was the true extent of the Milky Way, Kapteyn's 30,000 LY or Shapley's 300,000 LY? Nature is seldom what it seems to be at first or even second glance. From Lick Observatory in Santa Cruz, Swiss-born Robert Trumpler found the core reason for the wide discrepancy. A galactic fog masks a large part of our galaxy. Our galaxy's extent is indeed much larger than 30,000 LY. Looking directly into the galaxy, Herschel and Kapteyn just could not see any farther than 30,000 LY. Trumpler ran into the fog when he was trying a new approach to determine the distance to open star clusters lying inside the plane of the Milky Way. According to the standard method of determining distance by comparing intrinsic and apparent luminosity, there were many distant open clusters but those also showed unusually large diameters. Trumpler expected far away clusters to also appear smaller in diameter. The fainter the cluster the larger the discrepancy between expected and observed cluster-size. The dimming fog made clusters seem farther than they really were. But the haze had little effect on the apparent diameter of the cluster. Trumpler boldly concluded (in 1930) that Kapteyn's 30,000 LY estimate was wrong because light absorption by intergalactic gas and dust clouds hid most of the galaxy's large extent to optical telescopes.

Trumpler's interstellar dust conceals most of the Milky Way. Our galaxy's extent is not 30,000 LY. Shapley's 300,000 LY was also a gross overestimate. But Shapley was quite right about our sun's peripheral location in the galaxy. Because Kapteyn was not looking very deep into the galaxy, he was also misled by his finding that all stars he counted were distributed evenly about the sun. The sun is not at the centre of the galaxy. Unlike Kapteyn's myopic view of the galactic plane, Shapley based his estimate on observed globular clusters well outside the galactic plane, where the fog is thin. Shapley could obtain a more complete view of the galactic halo, and therefore the larger estimate of the Milky Way extent (Figure 9.24).

Shapley had grossly overestimated the size of the galaxy because of an error in Cepheid calibration. There are two types of Cepheids, as Walter Baade (1893–1960) discovered in 1954. Cepheid variables in globular clusters are less massive than those Leavitt observed in the Large Magellanic

Cloud. Therefore they are less intrinsically bright, and globular clusters are less distant. The diameter of the Milky Way dropped from 300,000 LY to 100,000 LY after correction. The modern value is 75,000 LY. The sun's distance to galactic centre also decreased from Shapley's 60,000 LY to 30,000 LY.

In 1927 Dutch astronomer Jan Oort (1900–92) and Swedish theoretician Bertil Lindblad (1895–1965) made a new discovery that our galaxy rotates slowly. Charts cataloguing the motion of stars, over a century of diligent observations, showed a rather intriguing pattern. High-velocity stars (those moving at speeds greater than 62 km/sec) displayed an interesting asymmetry in their direction of movement. Sixty per cent were approaching the sun, while others were moving away from the sun. Oort and Lindblad unlocked the riddle. The entire Milky Way galaxy rotates, but not like a disc, where every point moves at the same angular speed and constant period. Rather, the galaxy spins like the solar system, where the nearer a planet is to the sun, the faster it moves and the shorter its period due to gravity. So too, stars rotate with different periods. Those closer to the galactic centre than the sun move faster than the sun. Just as Copernicus classified a planet closer to the sun than earth as an inner planet, Oort and Lindblad realized that stars closer to the galactic centre than the sun are "inner stars" that appear to move in one direction, getting ahead of the sun. The outer stars fall back relative to the sun and therefore appear to move in the opposite direction. This also confirmed Shapley's finding that our sun cannot be at the centre of the Milky Way since 60 per cent of stars lie closer to the centre of the galaxy than the sun. In its grand circuit around the centre of the Milky Way, the sun makes one complete revolution in 200 million years. The galaxy had finally come together. Figure 9.26 shows how the Milky Way may appear from an extra-galactic perspective.

GALAXIES MULTIPLY

From the existing nebular catalogues of Messier and others, Hubble was confident there were hundreds of other possible galaxies out there, but all were too far away and showing no Cepheids. Needing a new way to measure remote galactic distances beyond a few million LY, he turned to "supergiant stars," many times intrinsically brighter than Cepheids. Betelgeuse and Rigel in Orion

FIGURE 9.26   Galaxy M83 appears similar to our own Milky Way galaxy.[51]

are nearby examples, at less than 1,000 LY. Hubble made the bold extrapolation that all supergiant stars must have the same intrinsic luminosity due to the fundamental nature of such stars. Hubble had a reasonable basis. With distances calibrated by Cepheids, his guess was valid for several nearby galaxies making up what we now call the local group of galaxies sprawled out over a region of a few million LY. The supergiant luminosity technique turned out to be good to 80 MLY. Above this distance, available telescopes failed to resolve galaxies into stars. Hubble needed yet another approach in the absence of any visible supergiants.

With imagination fired by stupendous distances, Hubble continued to cast his vision ever farther ahead:[52] "The history of astronomy is a history of receding horizons." Emboldened by his first audacious guess about supergiant stars, Hubble postulated that all galaxies as a whole should have the same intrinsic brightness. A large cluster of galaxies about 50 MLY away in the direction of the Virgo constellation, now known as the Virgo cluster, supported his simplifying presumption to within a factor of ten. Taking the median luminosity of the cluster as the standard galactic luminosity would make his error only a factor of three. Now he could push the galactic frontier out to 500 MLY, covering 100 million galaxies with average distances between galaxies of about 1 MLY. After 1954, distances determined by Hubble need to be multiplied by ten because of new revelations about Cepheids and corresponding improved distance calibrations.

To venture out beyond our supercluster of galaxies is to truly face the cosmic void. Galaxy superclusters envelop vast empty regions like soap films over giant air bubbles. The Great Wall is a sheet containing thousands of galaxy clusters spread over several hundred MLY. Many superclusters are linked in filamentary networks like a cosmic sponge, with strings of galaxies 100 MLY long separated by equally long voids.

HUBBLE IN SPACE

In April 1990 the launch of the Hubble Space Telescope (HST) by NASA became the most significant advance in astronomy since Galileo turned his 36" long telescope to the night sky. Named after the pioneer, HST has one of the most powerful eyes humanity has ever set upon the universe with a sensitivity to detect celestial objects to a faintness magnitude of 31.

Over 25 years of operation, including five shuttle missions to service and upgrade, the school-bus sized telescope revolutionized our understanding of the universe and provided us with millions of awe-inspiring images of the cosmos of stunning beauty and mind-bending scale. Thanks to the 2.5-metre diameter mirror and absence of atmospheric haze 350 miles above the earth's surface, Hubble can see astronomic objects as small as 0.05 arc-seconds, which is equivalent to spotting a flea (2.5 mm) from a 10,000-metre distance!

In December 1995, Hubble spent ten days peering into a tiny patch of the sky in the constellation Ursa Major. The result was the Hubble Deep Field, revealing nearly 3,000 galaxies. Continuing to view further and further back in time, the Ultra Deep Field View (Figure 9.27), taken over 200 hours, is a look back at the 13.8-billion-year-old universe when it was a youthful 1 billion years old. It shows more than 10,000 galaxies.

The excitement of Hubble's cosmic discoveries has become part of our modern culture. Expanding our horizons to unprecedented distances, Hubble showed us hundreds of billions of galaxies. The centre of our home galaxy contains a black hole with the mass of about 4 million suns. Nearly every galaxy with a central stellar bulge contains a supermassive black hole. Hubble played a key role in determining the age of the universe – about 13.8 billion years. The telescope has radically enlarged its namesake's discovery of the expanding universe (Chapter 13). The expansion of the universe started accelerating about 6 billion years ago propelled by what we think is a repulsive, anti-gravity type force that has been named "dark energy," for dearth of better understanding about its true nature.

Along with the Kepler satellite and other studies, Hubble shows that exo-planets (Chapter 7) are very common. The Milky Way galaxy is teeming with many billions of which billions are in the "habitable zone" where water is likely to form. Hubble is beginning to reveal the atmospheric

FIGURE 9.27   (a) The Hubble Ultra Deep Field image is the combined total of over 2,000 separate images, and the total exposure time is 23 days! It shows over 10,000 galaxies. Every spot of light is a galaxy of billions of stars. Some of the galaxies date back to just 350 million years after the Big Bang. Some galaxies are spirals, some ellipticals, some baby galaxies in formative stages, some galactic collisions. On the Hipparchus brightness scale the faintest objects in this picture are at 31st magnitude, 10 billion times dimmer than the faintest star that can be seen with the naked eye. (b) An expanded view of a small section (marked with box on left). Credit: NASA, ESA, and S. Beckwith (STScI) and the HUDF Team.

composition of a few extrasolar planets, and has even imaged some planets in orbit around their host stars.

Poets and artists through the ages have drawn inspiration from the majestic displays in the heavens to give intoxicating and romantic notions about our place in the cosmos. Time and again astronomical discoveries have demoted humanity to the peripheral and insignificant. Yet, like Walt Whitman in *Song of Myself*, we must continue to wonder how we hold such a special place and role with our calm rational ability to understand the cosmos:[53]

> And there is no object so soft but it makes for the wheel'd universe
> And I say to every man or woman, Let your soul stand cool
> and composed before a million universes.

Out in the farthest regions of the visible universe, at staggering distances greater than 10 billion light years, lie some of the most mysterious objects in the sky, named *quasars*, for "quasi-stellar radio sources." A small percentage of them emit radio waves. With super luminosities, 10 to 1,000 times more intense than galaxies, they appear to the optical telescope like stars because they are much smaller than galaxies, only about 1 AU in size. But they are not stars. Hence the term quasi-stellar. Normal galaxies at such large distances would be invisible. Recent thinking suggests these extraordinarily brilliant celestials were the black-hole nuclei of precocious galaxies born relatively soon after the creation of the universe. We can detect the light that set out from the Big Bang, a topic for the final chapter.

In *Night Thoughts on Life, Death, and Immortality*, English poet Edward Young contemplates the immensity of the cosmos:[54]

> How distant some of the nocturnal suns!
> So distant, says the Sage, 'twere not absurd
> To doubt, if beams set out at Nature's birth,
> Are yet arrived at this so foreign world.

HST originally estimated about 100 billion galaxies out in the universe, ten times more than the human population. In 2016, the estimate was completely revised. Modelling many images collected by HST over 20 years led to an estimate of over two trillion galaxies! A hundred billion to one trillion stars populate each galaxy, leading to an incredible aggregate of 10 billion, trillion (1 followed by 23 zeros, or 10^{23}) stars in the universe. The number is staggering, even comparable to the number of atoms in a handful of salt. Can earth truly bask in the privilege of being the only habitable planet (Chapter 7)?

It will be a sad day for astronomy when Hubble ceases to function, as expected around 2020. Its scientific legacy will grow for another few years. The space shuttle programme used to repair Hubble was also terminated in 2011. Hubble is expected to stay in orbit till 2028 after which it could come tumbling down to earth, hopefully in a controlled orbit that splashes in the ocean.

Hubble's successor will be the Webb Infra-Red Telescope. With a 6.5-m diameter mirror it is about three times the size of Hubble. With infra-red visibility Webb can look inside opaque dust clouds of interstellar gas to study the birth of stars and planets. It will also be able to study the centre of galaxies where theories and data suggest supermassive (100 million to billion suns) black holes reside. Webb is anticipated to launch in 2021.

Perched atop Apache Point Observatory in Cloudcroft, New Mexico, today the Sloan Digital Sky Survey (SDSS) photographs galaxies in the northern skies with digital camera technology, now better than photographic plates. The SDSS project achieved first light in 1998. The goal of hundreds of scientists, engineers, and software experts from eight astronomy groups is to compile a digital encyclopaedia giving positions and brightness of more than 100 million celestial objects, including at least 50 million galaxies. Continuing the Hipparchus legacy, SDSS will characterize the universe in a detailed and quantitative manner. Looking for large-scale patterns in galactic formations, the results will help decide between different theories emerging on cosmic questions: how did the universe begin? How did it evolve? How did stars and galaxies form?

The world's largest earth-bound telescope is the twin Keck on Hawaii's Mauna Kea, 14,000 ft above sea level as compared to world's highest mountain, Mt Everest at 29,000 feet. Each of the main mirrors is composed of 36 hexagonal sections working together as one giant mirror of 10 m (nearly 400 inches). Compare this to Hale's 200-inch mirror at Mt Palomar. Keck's giant mirror is adjusted to the perfect shape by adaptive optics which can accurately tune the position of the segments as well as adjust for varying atmospheric conditions.

Several very large earth-based telescopes are now planned or under construction: Giant Magellan Telescope (GMT), European Extremely Large Telescope (EELT), and the Thirty Meter Telescope (TMT). These extremely large telescopes will probe the origins of the early universe, black holes, galaxy centres, search for habitable planets, and seek signatures of alien life as well as the most exciting and unexpected discoveries.

GMT's primary mirror will have seven segments. Six off-axis 8.4-metre (320-inch) segments surround a central on-axis segment, to function together as a single optical mirror of 24.5 metres (960 inches) across. Compare these numbers to the 200-inch Mt Palomar mirror and the 400-inch Keck. It should be able to see ten times farther and more clearly than the Hubble Space Telescope due to its adaptive optics. The telescope will be mounted at Campanas Peak in the Atacama Desert in Chile. First light is estimated for 2023.

The EELT is being built on top of Cerro Armazones (10,000 feet) in the Atacama Desert of northern Chile. The peak has the clearest and darkest skies in the world. Its primary mirror 1,550-inch across so that it will be able to collect roughly 13 times more light than any operating telescope. Already under construction, EELT will see first light in 2024.

The TMT will also sit atop Hawaii's Mauna Kea volcano. The primary mirror will have an effective diameter of about 30 metres (1,200 inches) from its 492 segments. With construction approval in 2018, they hope to start in 2019 to have first light by 2030.

As bigger and better telescopes reveal new regions of the universe with more galaxies, we will continue to push back the frontiers of the observable universe, continuing Hubble's enthralling quest.

Part 3

Synthesis

10 Rise of the Mechanical Universe: Unifying Space and Time

They claim that the universe is in large what a watch is in little, and that everything happens by regular movements, depending on the arrangement of the parts. Admit the truth. Had you not once a more sublime idea of the universe, and did you not give it more honour than deserved?

Bernard de Fontenelle[1]

Even though their structure was hopelessly intricate, the crystalline spheres of the Greek universe had provided a visual, physical framework for holding up the heavens. But the wayward comets which Tycho placed in the heavens shattered the crystalline construction. Kepler's ellipses spelled the final demise. Planets and earth were tracing mysterious ellipses without need for any guidance. What could replace the spheres as the new structure of the heavens?

After finding precise laws describing planetary motion, Kepler searched for a physical meaning underlying the mathematical account of heavenly motion. He proposed a "mechanical force" that originates from the sun and binds planets to the hub (Chapter 8). Galileo set down a uniform set of principles for a new dynamics (Chapter 9). Rene Descartes followed the pathways opened by Kepler and Galileo. *All natural phenomena* must be subject to mechanical explanations and cast in mathematical language. His ideas matured into a *mechanical universe* that runs like clockwork, a new scientific paradigm that filled the void left by the demolition of Greek crystalline spheres.

Progress in understanding terrestrial motion combined with the discovery of a new earth. Voyages to circumnavigate the globe intertwined with advances in reckoning the passage of time through clockworks. A synthesis emerged between circumnavigation and time-keeping, cyclic motion and circular motion. New insights into circular motion on earth opened up the understanding of "circular motion" in the heavens.

AROUND THE WORLD

With Columbus' breakthrough discoveries of new continents, the question naturally arose: Was it possible to go all the way around the world? Portuguese explorer Ferdinand Magellan dared to launch an expedition to prove the outrageous possibility. The Portuguese navy had just dismissed him for trading with the Moroccans. Angry with the shoddy treatment, he went to the king of Spain with the bold idea of circumnavigating the globe. Once again, Portugal forfeited to Spain the opportunity for glory and treasure. Taking five ships, Magellan set sail in August 1519. Crossing the Atlantic and sailing down the coast of America, he found an opening near the tip of Chile. Columbus had narrowly missed the same discovery on one of his later voyages, as he roamed up and down the coasts of the New World in a stubborn, but futile, search for the mouth of the Indian river Ganges. After a stormy passage through the strait (now called the "Straits of Magellan"), they burst into tranquil waters. Grateful for the calm after the fierce storm they had just endured, they called the new ocean "Pacific." It was the sea forecast by Amerigo Vespucci. As they ventured through the vast oceanic expanse, their ships fell apart one by one. Pioneer Magellan met a tragic death in a fight with the natives of the Philippine Islands. Pushing the frontier always places the most rigorous demands on existing capabilities, at the same time providing strong incentives for breakthroughs. In the Southern Hemisphere, Magellan's astute log-keeper documented two distinct, hazy patches of stellar clouds, now called the Small and Large Magellanic Clouds, our nearest neighbouring galaxies, pivotal to the later progress of astronomy (Chapter 9).

Despite massive losses, Magellan's last ship continued westward across the Indian Ocean and around the southern tip of Africa, back toward Spain. Leaking, but laden with spices, it returned home in 1522. The historic trip took three long years. Magellan's expedition had finally found a westward route to the East. More important, he had for the first time circumnavigated the vast globe. The last doubts about the spherical earth finally disappeared. Eratosthenes' estimate of the enormous size of the earth proved accurate after all.

MARKING TIME IN EXILE

In a climate of feverish ocean exploration, scientists and inventors launched a quest for an accurate method to determine longitude at sea. Longitude measurement demanded an accurate chronometer. Why was accurate time-keeping essential to navigation? Since earth takes 1 hour to turn an angle of 15 degrees, the local time between two places is different by roughly 1 hour for every 15 degrees of longitude separation. *Therefore longitude is both a measure of place and time.* Explorers knew how to measure the local time at sea by the position of the sun during the day, or by the orientation of the Dipper at night. Every day at sea, the navigator set the *local noon* when the sun reached the highest point in the sky. The difficult problem was keeping track of time *back home* – at the port of departure. The key to accurate longitude determination was a portable timepiece that could keep accurate track of time at port of origin. By comparing the reading of a portable timepiece with the local time (i.e., the time at sea) a sailor could tell the number of longitude degrees between places travelled. Thus cartography was linked inextricably with chronometry.

Accurate time measurement turned into one of the most significant challenges for mariners sailing out of European shores into a world of uncertain size and with expanding frontiers. It propelled advances in the science and technology of clocks and watches. In parallel, scientists of the time started to picture a universe ticking along like a perfect mechanical clock governed by gears of mechanical laws.

European mechanized clocks had evolved directly from the clepsydra, the old Egyptian water thief. Water falling from a reservoir moved a wheel which slowly turned a bar that would ring a bell to sound time. But there were no hands to point out hours because most people were unable to read numbers. European water works technology inspired mechanical clocks. Water power ran mills to crush grain and pulverize ores. As mechanized water clocks grew elaborate, they rang huge metal bells decorated by cocks crowing on the hour, half-hour, and quarter-hour.

But in the cold climate of Europe, water-powered clocks would freeze silent in winter. The solution was to replace falling water with the *motive power* of a falling weight. It was a pivotal event for the birth of mechanics. Weight-driven mechanical clocks stimulated an understanding of crucial concepts, such as motive power stored in the lifted weight (modern potential energy), work done to lift the weight, and motive power released by the falling weight (modern kinetic energy).

Rapid free fall of a large weight presented a serious problem: how to break up its descent into small steps to simulate droplets of water that measured small bits of time-flow in a clepsydra? An innovative device, called an *escapement*, was the solution. It interrupted free fall into small, regular intervals. A toothed wheel would alternately stop and permit the fall. Additional cog-wheels connected the falling weight and rotating axle to bars that rang the bells and activated display mechanisms. The measurement of time became a favourite activity for thinkers and inventors.

Galileo's discovery of the pendulum principle opened a new era in time-keeping. He was surprised to find that the duration of swings remained independent of the swing's excursion (Chapter 4). Nor did duration depend on the bob's weight, but only on string length. A year before he died in "exile" at Arcetri, he returned to the challenge of constructing an accurate clock based on the pendulum's constant period.

Here is his masterful technique (Figure 10.1). A falling weight spins a ratcheted wheel by unravelling its suspending rope, wound around the axle. The wheel also has pegs protruding from the side. The essential feature is that the pendulum swing controls two curved "fingers" to stop and to rotate the wheel in regular intervals, so that the weight descends in small steps for each pendulum swing. The upper finger starts the rotation of the wheel and the lower finger stops the wheel. When a notch in the ratchet rests against the pawl (the stop) the wheel does not rotate, and the weight stops falling. When the pendulum arrives it lifts the upper finger which lifts the stop so that the wheel rotates and the weight falls some distance until a moving peg strikes the lower finger on the pendulum, and pushes the pendulum away. As the pendulum swings out, the upper finger releases the stop, bringing the wheel to rest so that the next ratchet is blocked by the pawl. After completing one oscillation, the pendulum returns to the wheel, once again lifting the upper finger and pawl to continue the cycle.

One pendulum length defined a precise time interval to regulate all mechanical clocks. After his death, Galileo's students continued his mission for an accurate pendulum-based clock. Still missing, however, was an understanding of the physics underlying the principle of the pendulum. There was a clear connection between period and length. But what was the mathematical relationship? One of Galileo's successors figured it out (discussed later).

Before the invention of the controlling pendulum, mechanical clocks were notoriously inaccurate. A good clock would lose as much as 15 minutes per day. Galileo's pendulum-controlled clock changed all that. Clocks built by different artisans in different places could all finally keep the same time. Humanity finally held in its grasp a controllable natural cycle to regulate time, as well as to accurately measure small intervals of time.

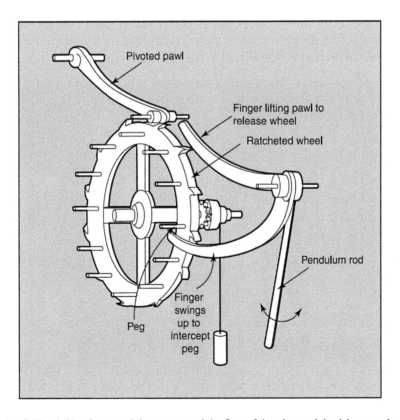

FIGURE 10.1   Galileo's idea for a pendulum to control the flow of time in a weight-driven mechanical clock.[2]

Better clocks propelled mechanical technology to produce better quality screws, gears, and ratcheted wheels machined on superior lathes and mills. Emerging technology stimulated new physics concepts of motive power, impulse, and force (Figure 10.2).

THE DIVINE CLOCKWORK

With creative imagination, Rene Descartes formally filled the void left by the demolition of ethereal spheres with the *Mechanical Universe* paradigm. Descartes followed Galileo as the scientist of the Baroque era. His best-known philosophical credo is "I think, therefore I am," often quoted in Latin as "cogito, ergo sum." Our senses are often in error, giving contradictory information on the same subject. They cannot be trusted. The existence of the external world and explanations for natural phenomena can only be built on the rock of "cogito," pure thought. Only the principles derived by the mind from absolutely logical arguments can be accepted.

Descates' Mechanical Universe was like a great cathedral clock, easy for all to picture, for such a clock was the centre-piece of every town square all over Europe. As descending weights powered the rotating hour hands, turning on their axles, mechanical toy men rang bells, and cocks crowed to tell the hour. One complex mechanism generated a multiplicity of operations to generate the infinite phenomena of an infinitely complex machine we call nature. Once set running, the divine clockwork runs forever without needing winding or repairs, as God is the perfect clock maker. Descartes was after all a religious man, brought up by Jesuits. Descartes seized Galileo's concept of inertial motion: a body set in motion continues to move forever in a straight line. It fit his mechanical picture, ensuring the eternal operation of the clockwork universe governed by physical laws.

The new paradigm captured the imagination of European scientists providing a qualitative, physical picture for the new planetary system. But poets expressed a deep sense of loss with the dawning of the mechanical era. Had the divine clock-maker abandoned the universe to run on its own? Where was his guiding hand? In the quote at the beginning of this chapter (page 349), playwright Bernard de Fontenelle expressed his disappointment at such a crude vision. How disappointing that natural philosophy had turned utterly mechanical at its core.

FIGURE 10.2   (a) Strasbourg clock. (b) Mechanisms running clockworks.[3]

In discussing the physics of motion, Descartes focused his attention on the agencies that must be responsible for changes from natural inertial motion. He submitted that bodies can only influence other bodies when they are in contact, the easiest interactions to visualize. All matter is a collection of individual particles which collide with one another producing the variety of natural phenomena. All change in motion arises from the impact of bodies upon one another. Forces arise from the impact of one particle of matter on another. A complete understanding of the laws of impact would therefore reveal the hidden workings of the clockwork universe. Though single particles may alter their motion on impact, the total "quantity of motion" during collisions must remain constant forever. Descartes had an early glimpse of the laws of conservation of energy and momentum, but he conflated the two concepts which were as yet not categorically defined. Huygens and Hooke who followed him continued his effort to understand the laws governing collisions.

PUTTING SPACE IN ORDER

Descartes is more familiar to us as the mathematician who invented the Cartesian coordinate system which bears the Latinized version of his name, Cartesius. There is a popular story about how the essential insight into the now ubiquitous rectangular system came to him. Educated in a French Jesuit school, Descartes' health was so poor that the priests allowed him to lie in bed till late every morning. But he was not lazing; his mind was on geometry and how to describe nature in mathematical language. Watching a tree branch outside his window move with the breeze across the window panes, he imagined he could specify the location of the branch-tip by two numbers corresponding to how many panes separated the tip from the corner window in the horizontal and vertical directions. So also, specifying a point on a plane needs two numbers, which give the distances of the point from a vertical and a horizontal reference line. It was analogous to using latitude and longitude on the globe. Descartes' reference lines are called the x-axis and y-axis, and their intersection point is called the origin, the starting point of order (Figure 10.3). Numbers, a and b, which specify point location, are called coordinates (a, b). Descartes' essential insight was that numbers can describe spatial location and introduce a powerful, quantitative element into geometry. Since the world is three-dimensional, the new mathematics requires three coordinates to frame our perceptions. An algebraic expression provides the distance between points (Figure 10.3).

Creating a new order for mathematical space, Descartes proceeded to fuse algebra with geometry, turning lines and curves into numbers and algebraic equations. Descartes was not alone in his inspiration to bring rational, mathematical order into space. The Baroque world of Descartes was governed by structure and order. Underlying the apparent chaos of the world Baroque artists and scientists sought rational order.

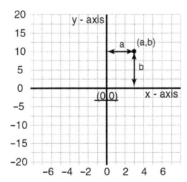 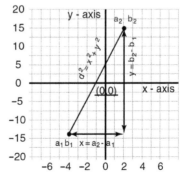

FIGURE 10.3 (Left) the Cartesian coordinate system. (Right) Cartesian distance between points. Suppose there are two points with coordinates (a_1, b_1) and (a_2, b_2). If x is the difference between their x coordinates $(x = a_2 - a_1)$ and y the difference between the y coordinates, then Pythagoras' theorem gives the distance (d) between the two as $d = x^2 + y^2$. In three-dimensions, the distance formula becomes $d = x^2 + y^2 + z^2$.

Geometrical framework ruled the design of architects commissioned to glorify the reigning French monarch, Louis XIV. Instead of towering cathedrals to praise the Lord, the king commissioned spectacular castles surrounded by extravagant gardens. Instead of the mystic zeal of religion, the power of organized intellect imposed order. In a bird's-eye view, we can see the precise symmetric layout of the sumptuous palace, gardens, and adjacent buildings of the estate at Marly (Figure 10.4). On the outskirts of Paris stood the ornate palace of Versailles reflecting the Sun King's grandeur, enhancing French prestige throughout Europe and the world. Over 20 years it took the talent of hundreds of architects and artists and the sweat of 30,000 labourers to erect the lavish palace with a capacity to house 10,000.

Baroque style found a new medium for expression. Artists and architects created spectacular visual effects to assert the wealth and dominion of the absolute monarch. Under a grand arched ceiling at Versailles Palace, architects lined the Galerie des Glaces with hundreds of mirrors set into walls opposite the windows, adding incredible depth to the narrow, enclosed corridor. Landscape artists transformed an entire forest into a park. Replete with fountains, artificial lakes, spacious lawns, and flower gardens lined with sculptures, a Cartesian framework ruled the grounds (Figures 10.4 and 10.5). Tamed and ordered, nature provided a vast palace of pleasure for king and court. Buildings, fountains, and garden paths spread outward from the centre, in precise circular symmetry. Landscape artists perfected the natural shape of each tree by trimming it to a precise cone. In the Orangerie (Figure 10.5), they laid out groves in a rigid geometrical grid. Versailles and Paris became the new centre of life – a new Athens in a Golden Age. Structure and reason controlled environment. It was the essence of a Cartesian world view.

FIGURE 10.4   Chateau at Marly.[4]

FIGURE 10.5   The Orangerie at Versailles.[5]

The sovereigns of Europe embraced Louis XIV as their model, adopting his despotism and emulating his achievements. Sumptuous estates shot up everywhere: the Würzburg Palace in Germany, and the Schönbrunn and Belvedere Palaces in Vienna, Austria, to name but a few. Everywhere Baroque artists expressed their sense of flamboyant drama in grandiose palaces capturing French splendour. In Russia and India, czars and maharajahs built their own mini-Versailles. Baroque music filled salons. As Descartes infused order into space, Johann Sebastian Bach put order into music. Canons were an integral part of Bach's prolific and structured creations. Meaning "rule" or "law," the canon is the ultimate in the skilful use of musical rules imposing a rigid structure through elaborate compositions.

BREAKTHROUGHS IN CIRCULAR MOTION

Descartes' contribution to understanding circular motion was a key ground-breaking geometric insight leading to further understanding of the operation of the mechanical universe. A moving body traces a curve only if something diverts it from a straight path. Descartes embraced Galileo's principle of inertia: a body moving with uniform velocity will not slow down, unless there is an external agency acting on it, e.g., friction or air resistance. To arrive at the underlying *ideal* motion, Galileo identified and eliminated friction and air resistance. An object at rest is inert – too "lazy" – to move by itself. A body moving with constant velocity is too "lazy" to change speed all by itself. It is also too lazy to change direction by itself. Descartes extended the concept of inertia from speed to direction. Just as any change in speed must have a cause, any change in direction must also have a cause. Galileo's principle of inertia demands that bodies which move in circles, or curves, do so through the influence of an external agency. There is no self-sustaining circular motion, not on the earth and not in the heavens. Descartes realized that there must be a *force* acting on an object moving in a circle (Figure 10.6). When legendary David whirled a stone at the end of his sling to slay the mighty Goliath, he pulled inward toward the centre of motion. A body in circular motion must experience an outside force directed to the centre of the circle, later called the "centripetal force."

Taking this small, but crucial, step toward understanding circular motion, Descartes jumped over major philosophical hurdles. Whether it takes place in the heavens or on earth, circular motion requires "outside intervention." Under the influence of the external agency, *circular motion is accelerated motion*. Acceleration takes place when either the *speed* of a body changes, or when *direction* of motion changes. But Descartes did not derive the quantitative expression for the acceleration sustained during uniform circular motion. That was a feat for Christian Huygens (1629–95) and later, Isaac Newton.

Christian Huygens was at the epicentre of commercial, cultural, and scientific activity rippling through the West. New continents were opening up through exploration and circumnavigation of the globe. New worlds were becoming visible through the increasing power of the telescope. With both the eastern route (around Africa) and western route to riches open from Portugal and Spain, Venice was now in twilight. European economic activity drifted from Italy to Spain, and to the northern countries of France, Holland, and England. Emerging political powers depended as much on economic prosperity as on military strength. They drew riches from overseas empires. Shipbuilding and exploration became principal activities of the new culture, dominating the landscape and images of artists.

From an early age, Huygens showed a mechanical bent with talent for drawing and mathematics. His mind was ripe for the dawning mechanical age. But his tutors frowned on his mechanical propensity. Such practical work did not befit a young man from a wealthy and distinguished family with a noble status. Fortunately, he ignored them. Mechanical explanations were essential to fundamental understanding of natural phenomena. A frequent visitor to the Huygens' home, Descartes was impressed by the early genius shown by the young boy's precocious efforts in geometry.

Like Descartes, Huygens realized that when a body moves in a circle at a constant speed, it does not have a constant velocity, because it continuously changes direction. Cut the string (Figure 10.7) and the stone sails forever in a straight line, tangential to the circle. Circular motion is also accelerated motion. Unlike free-fall motion where acceleration and velocity point in the same direction (down), in circular motion, velocity is tangential to acceleration, which always points toward the

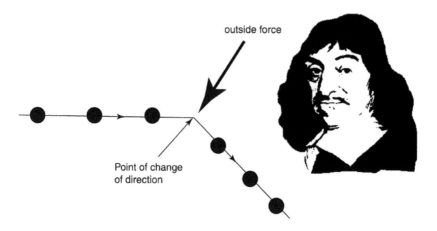

FIGURE 10.6   Descartes recognized that a change in the direction of motion requires an external cause[6] just as change of speed requires an external force.

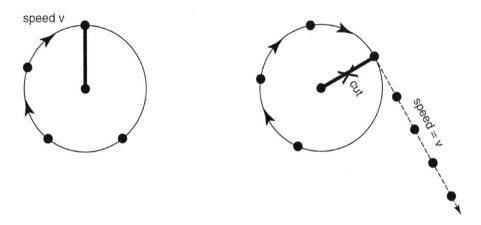

FIGURE 10.7 A body is moving in a circle at the end of a string. When the string is cut, the body flies off in a straight line and at constant velocity tangent to the circle. *v* is called the tangential velocity.

centre of the circle. In free fall, the attraction of gravity provides downward acceleration. When a weight hangs down from a string, it remains stationary because an upward pull along the string, called *tension*, counteracts the downward attraction of earth. If the same weight rotates in a horizontal plane, the string pulls in a radial direction providing *centripetal acceleration* necessary for circular motion. Increasing the vertically hanging weight to a critical value will eventually break the string when the weight exceeds the string strength. Increasing the speed of circular rotation to a critical value will also break the string. Therefore centripetal acceleration depends on tangential speed (as v^2). A larger radius of curvature means more gradual direction change; hence centripetal acceleration decreases with radius (R), for the same tangential velocity. Huygens' *tour de force* was to derive a precise mathematical formula for the acceleration (a) of an object undergoing motion in a circle of radius (R) at a constant tangential speed (v)

$$a = \frac{v^2}{R}.$$

Centripetal acceleration increases as the square of the tangential velocity and decreases as the radius (Figure 10.8). Geometrical analysis based on similar triangles and vectors played an important role in deriving the direction and the magnitude of centripetal acceleration.

THE PERIOD OF THE PENDULUM

Under the guidance of his wise minister Jean Colbert, the Sun King realized that economic prosperity depends upon technology, which in turn depends on scientific progress. Louis XIV gave a charter and financial backing to a new Academy of Sciences. Intellectuals gathered as they did at Plato's Academy of Athens. Huygens became one of the principal founders. Scientists at the Paris Academy enjoyed extraordinary freedom under the most absolute of regimes. Art and learning rose to new heights, as did achievements in technology.

Familiar with navigation issues from his home in seafaring Holland, Huygens strove to provide a method to determine longitude through an accurate chronometer. Thanks to his efforts, clockmakers everywhere used the pendulum as a time controller. Within 30 years after Galileo's death, the average error of the best timepiece dropped from 15 minutes per day to only 15 seconds per day. It was no coincidence that the proper description of physical quantities with vector properties with magnitude and direction grew mature at a time when it was crucial to have a good understanding of directional navigation in the hostile waters of open oceans.

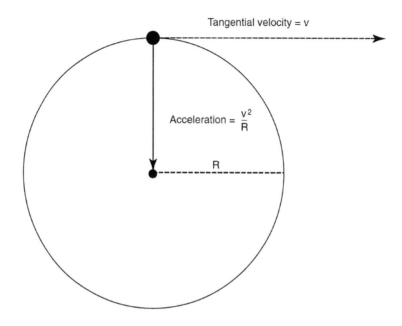

FIGURE 10.8   Acceleration for a body moving in a circle at a constant speed.

Cyclic pendulum motion carries a special name with origins in music – *simple harmonic motion*. A violin string constrained at two ends vibrates in simple patterns, with one or several humps. Any point on the string cycles to and fro, following simple harmonic motion. The violin sound box vibrates in sympathy. A violinist's finger position changes the vibrating segment length to make the string vibrate in other patterns (modes), called harmonics. Different notes are harmonious when the ratios of string length form simple, whole-number ratios, as Pythagoras determined long ago (Chapter 1).

The simple harmonic motion of vibrating strings and oscillating pendula is intimately connected with the dynamics of circular motion. Suppose a ball moves in a circle at a uniform speed in a vertical plane. Imagine a ceiling lamp above and a white table below. As the ball traces a circle, its shadow on the table executes a to-and-fro cyclic motion, identical to the oscillation of a pendulum (Figure 10.9). Simple harmonic motion is just the projection of circular motion onto what became known as the Cartesian axis.

Through the connection between simple harmonic motion and circular motion, Huygens derived a famous expression for the period of a pendulum's simple harmonic oscillations in terms of the length L and free-fall acceleration, g. The square of the pendulum period increases with length: $T^2 = 4\pi^2 L/g$.

PROBLEMS WITH THE PENDULUM

Strictly speaking, the motion of a pendulum is simple harmonic only for small excursions. As long as the pendulum makes only small-amplitude excursions from its equilibrium position, its period depends only on length. Any clock-maker in the world could use a 1-metre long pendulum to generate a 2-second period and calibrate his clock against another clock from any location. At sea, however, the weighty pendulum-driven clocks proved too inaccurate. The pitching sea disrupted the swings, while the varying temperature changed the pendulum length. The small-excursion restriction limited the pendulum's usefulness. Huygens searched for a way to make the period constant over the large swings likely on a ship at sea. He found an ideal geometric path that a pendulum bob could follow so that its swing remained isochronous over large amplitudes. Galileo's cycloid

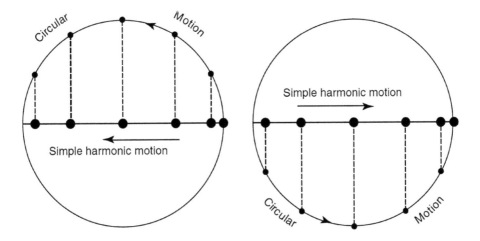

FIGURE 10.9   (Left) simple harmonic motion is a projection of circular motion, as in this shadow arrangement. (Right) when a particle moves in a circle, its Cartesian x-coordinate executes simple harmonic motion. The same is true of the y-coordinate.

(Chapter 4) proved to be the perfect path. Huygens devised a technique to force a pendulum to follow the path of a cycloid. The string had to swing between two specially curved metal plates (or cheeks). As the thread curved at the ends of its swing, the bob followed a cycloid (Figure 10.10). Huygens' cycloid pendulum clock could tell time to an accuracy of better than a minute over a day. A later model pendulum with an alternative device to restrict the size of the circular arc swings was equally accurate, and remains the preferred method today, as for the grandfather clock.

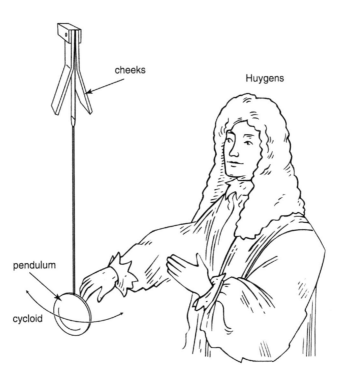

FIGURE 10.10   Huygens devised an arrangement to force the pendulum bob to follow a cycloid arc and obtain a constant period for large swings.[7]

Despite these advances in timekeeping, a more fundamental difficulty stood in the way of the pendulum clock as the basis for an accurate timepiece for voyages around the globe: the dependence of the period on the value of *g*. An unexpected connection with a major benefit cropped up during an astronomy venture. In their expedition to Cayenne to determine planetary distances and consequently the scale of the solar system in km (Chapter 9), scientists from Paris discovered a strange behaviour: a pendulum clock calibrated to run accurately in Paris would lose a few minutes per day in Cayenne. (A method for accurately calibrating clocks is discussed later.) With the grand objective of determining the absolute size of the solar system, they aimed to make two simultaneous measurements of the position of Mars – one from Cayenne (latitude near 0 degrees) and another from Paris (latitude 48 degrees N). Both measurements had to take place *at the exact same time.* Simultaneity demanded that the pendulum clock which they carried had to keep accurate time relative to the clock in Paris. Now they had a problem. In Cayenne, the pendulum clock was off by a few minutes per day, a large error. They made a correction without quite understanding the reason. To make the Cayenne pendulum have a period of one second, they had to shorten its length compared to the Paris pendulum with a 1-second period. When they returned to Paris they now found that the Cayenne-adjusted pendulum was off. They had to increase the length back to its original Paris value in order to return to a 1-second period. Clock-makers faced a serious problem. An exact-seconds pendulum needed different lengths for different locations over the globe.

At home in Paris, Huygens explained the instructive puzzle. It arises from the small but crucial influence of earth's rotation on the value of *g*, as measured by an increase in the free-fall velocity of a stone, or by the period of a "falling" pendulum. Earth's attraction provides a downward acceleration for falling objects. A major part of that downward acceleration is *g* which increases the downward velocity. But a small part also provides the centripetal acceleration (pointing downward) needed for circular motion (i.e., earth's rotation). An object at the pole has zero circular motion about the centre of the earth, and zero centripetal acceleration. The value of *g* is highest at the pole, e.g., *g* = 9.8 m/sec². At the equator, where the radius of the latitude circle is largest, the tangential velocity and centripetal acceleration due to circular motion are largest, and so is the corresponding effect on *g*, which drops to *g* = 9.7662 m/sec². Since each latitude circle on earth has a different radius (Figure 10.11), the tangential velocity of the earth's rotation depends on latitude. So does the centripetal acceleration associated with earth's rotation, and the value of *g*.

According to the famous formula for the period of the pendulum ($T^2 = 4\pi^2L/g$), if the value of *g* becomes smaller, the length of the seconds pendulum also has to be smaller to yield the same period. Therefore, scientists on the Cayenne expedition had to slightly shorten their pendulum. If *g* = 9.8 m/sec² at the pole, a 1-metre-long pendulum yields a period of 2.0071 sec for one complete oscillation. At the equator *g* reduces to 9.7662 m/sec². Re-applying the pendulum formula for a 1-meter-long pendulum yields a period of 2.01056 sec at the equator. The error in cycle time is only 0.17 per cent, which seems like a small difference on first glance. But over a day (86,400 sec) a pendulum will lose 150 seconds, or 2.5 minutes! Huygens had made great efforts to improve pendulum precision to better than 1 minute per day. Therefore his clock was sufficiently accurate to detect the effect of earth's rotation from its effect on the pendulum's period. Little did Galileo know that his pendulum invention would eventually provide a physical proof for earth's rotation. Pushing the rigours of time-keeping revealed the motion of the earth. Once again, improved precision opened a new door.

A pendulum-regulated clock with an uncorrected 2.5-minute *daily* error was totally unsatisfactory for keeping track of longitude on a long ocean voyage. At the end of a month- long journey across the Atlantic Ocean, the corresponding time would be off by 75 minutes and the longitude assessment by 19 degrees! Around the equator, the distance error is more than 2,000 kilometres, a grievous miscalculation at sea. A ship could arrive at an unfriendly port, or run aground on unexpected shallow rocks.

Another important effect accentuates variations of *g* over the surface of the earth. It arises from the slightly non-spherical shape of the earth, a topic for Chapter 11.

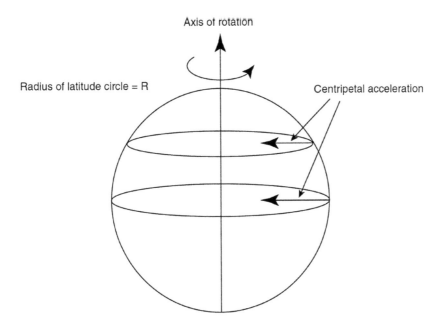

FIGURE 10.11   The change in centripetal acceleration with latitude. At higher latitudes, the radius of the latitude circle is smaller. Both the tangential velocity and the centripetal acceleration are smaller.

THE LONGITUDE PRIZE

The accurate determination of longitude became vital to empire building, trade, and economic prosperity. But the pendulum clock could never provide sufficient accuracy. Many a sea voyage continued to miss its destination, lost in the treacherous open sea. Countless ended up as shipwrecks. A catastrophe in 1707 triggered a call for concerted action. Four ships of a British fleet miscalculated their position and came to grief on the rocks off Cornwall, England, losing 2,000 sailors. It was humiliating for the greatest naval power on earth to lose men and ships so close to their home port. Desperate for a solution, the British government offered a prize of £20,000 (equivalent to several million dollars in today's currency) for a method to keep track of longitude with an error of less than half a degree. It demanded a 30-mile accuracy at the end of a 2000-mile voyage to the West Indies. A ship's chronometer would have to preserve port time to better than 2 minutes over a month-long journey. Who could devise a clock to keep time better than 4 seconds per day? Parliament appointed a Board of Longitude as judges for the fierce competition.

Anxious to ensure naval supremacy, Philip II of Spain and Louis XIV of France promised equivalent incentives. All across Europe, scientists, inventors, and amateurs dreamed of ways to solve the longitude problem. Some ideas bordered on the absurd, such as one scheme to dive down into the deep ocean, locate sunken ships, and place beacons signalling their locations.

Galileo discovery of the moons of Jupiter originated an ingenious astronomy-based idea. After discovering the moons of Jupiter (Chapter 9), he realized that their eclipse timings were as regular as a clock's. Even over the course of a single night, he could discern a significant change in the Jovian moons' positions. When Jupiter was above the horizon, anyone on earth could spot the satellites through a good telescope. Eclipses occurred predictably and often, about 1,000 times annually.

Constructing detailed tables, Galileo hoped to use frequent eclipses as a universal time-keeping mechanism for longitude determination. At first there was great optimism about the method. The Dutch awarded Galileo a gold chain for the idea. Anyone with detailed tables for Jovian satellite eclipses could read Paris time, as if from a universal clock in the night sky. But to put the method into practice on a swaying ship was another matter. Mariners had to keep steady a sensitive telescope capable of resolving Jupiter's moons. Even the slightest movement would cause Jupiter to jump out of view. Anticipating the problem, the imaginative Galileo suggested that a navigator should sit in a special chair, set in gimbals normally used to keep a large compass steady on the rolling sea. His helmet would carry a sensitive telescope. Being a competent craftsman, Galileo tried to implement his method, but failed.

On land, however, Galileo's method reigned supreme. Cartographers of world geography, such as Cassini in Paris, used it for measuring longitude to better than a quarter degree. With the availability of extensive tables, a surprising new benefit emerged. Eclipses of Jupiter's moons became the basis for Ole Roemer to measure the speed of light (Chapter 9).

A GRAVITY-FREE CLOCK

Despite steady improvements in understanding the pendulum period, and despite technological advances in time-keeping, the pendulum-regulated clock eventually failed as an accurate chronometer at sea. For global positioning, any time-regulating mechanism had to be free of gravity. Abandoning the pendulum, Huygens began to consider the next promising alternative – simple harmonic oscillations of a spring. Spring-driven clocks were already in use as compact, pocket-size timepieces. Instead of falling water, or descending weights, a wound-up coiled spring drove the inner mechanics.

Huygens' new idea was to use a spiral spring and flywheel oscillator as time controller. One end of a coiled spiral attaches to the spindle of a balance wheel (or flywheel), and the other end to an exterior fixed point, such as the clock-frame (Figure 10.12). A push on the rim sets the wheel in motion. As the spring coils up a little tighter it resists motion to bring the wheel to a stop. When the spring unwinds, it spins the wheel past the equilibrium point, uncoiling to resist motion again. Sustained by the twist-force (*torsion*), to-and-fro oscillation is simple harmonic motion with a constant period. Instead of pendulum length and *g*, the rotation period of a torsional oscillator depends on spring stiffness and the *rotational inertia* of the balance wheel. The latter depends on the mass distribution of the rotating wheel. If there is more mass at larger diameter, the inertia to rotation is greater. *But gravity plays no role.* A complete clock needs an escapement and gears to count and record the number of oscillations.

The spiral-spring clock finally broke the basic link between time measurement and earth's gravity. In the same stroke, it eliminated dependence on earth's rotation. Eventually it led to an accurate chronometer and to the first wristwatch, and remained the basis of accurate timepieces until the advent of quartz-crystal-regulated watches. In the quartz wristwatch, the natural vibrations of a tiny quartz crystal serve as the time controller. The frequency of light emission from caesium atoms serves to regulate modern atomic clocks.

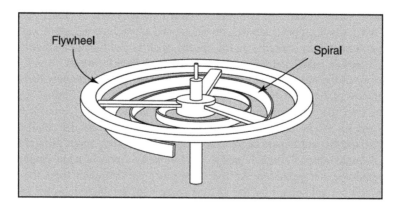

FIGURE 10.12   A spiral spring and balance wheel oscillator to replace the pendulum.[8]

PERFECT MOVEMENTS

At first, even spiral-spring clocks were unable to keep time with the accuracy necessary for mariners. The most serious problem was temperature change. The period depends on flywheel and spring geometry; in other words, how the mass distributes over the spring and wheel. Metals expand when heated. When the flywheel expands, the effective diameter of the rim and the rotational inertia increase, so that the oscillation period decreases. In addition, the length of the spring increases with temperature, increasing the rotational inertia of the spiral. Even well-made watches can lose 5–6 seconds per day for every degree rise in temperature. That would hardly suffice for the accuracy of 4 seconds per day desired, given daily swings in temperature of the order of 10 degrees at sea.

By dogged persistence, master mechanic John Harrison finally solved the problems of accurate chronometry in 1764. But it took another 75 years after Huygens first proposed the spiral-spring oscillator as a clock regulator. To eliminate thermal effects, Harrison made springs and flywheels with bimetallic strips of copper and iron (Figure 10.13). Because copper expands more than iron, the strip curves inward upon heating and outward on cooling. Inward curling reduces inertia, compensating for expansion. When the temperature falls, outward curvature increases inertia.

One of the greatest challenges in developing the chronometer was calibration. How accurate is the 1-second time interval? Harrison used an important astronomical effect. There is a small but precise time difference between the duration of a *sidereal day* (star-day) and a solar-day due to the combined effect of the revolution *and* rotation of earth. A sidereal day is the time interval between two successive appearances of a star in the exact same location of the sky. A solar-day is the time between two successive appearances of the sun at the exact same place in the sky, as viewed from one place, e.g., London. Since the earth revolves around the sun, by the time one day passes, the earth moves a small fraction (1/365.25) of its path around the sun (Figure 10.14). Because of the shift in orbital position, it takes an additional few minutes for earth's rotation to bring London into exactly the same orientation with respect to the sun as on the previous day. Harrison was well aware of the additional 3 minutes and 56 seconds precise time interval between solar and sidereal days. He used it to check his chronometers. If he kept track of the time with reference to the sun's position, the time at which a particular star rose above the horizon would change from dusk to dusk by 3 minutes and 56 seconds. Making careful observations on when certain stars disappeared behind a neighbour's chimney each night, he checked the timing of his intricate clockworks. Both rotation and revolution of earth provided the ultimate calibration of the clock.

Harrison dedicated his entire professional life to building model chronometers (Figure 10.15) of increasing accuracy to solve the longitude problem. His best models required no lubrication nor the frequent cleaning needed for most clocks to keep working smoothly during a long sea voyage. The mechanical masterpieces were impervious to rust from salty sea air. Each clock he devised was better than the one before. The last clock took him 19 years to perfect and was off by only 5 seconds after a month-long (2.6 million seconds) voyage to Jamaica. The equivalent error was only 1 mile, far smaller than the 30-mile error permitted for winning the longitude prize. Finally, time could be transported across vast oceans without noticeable loss.

But the British Parliament delayed paying the coveted prize. They only compensated Harrison for the cost of developing new models because a bitter competition arose between Harrison and proponents of the astronomical method based on carefully recording and using the moon's position

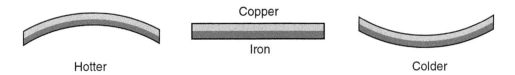

Copper

Iron

Hotter   Colder

FIGURE 10.13   The curvature of a bimetallic strip compensates the change in inertia of a spiral spring oscillator.

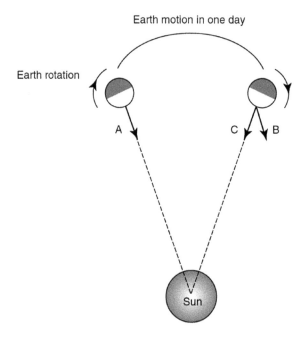

FIGURE 10.14   The motion of the earth in one solar-day takes slightly longer than the motion of the earth in one sidereal day. After one day of revolution, arrow A, which pointed towards the sun, becomes arrow B. For one complete solar-day to pass, arrow B must continue to rotate a bit farther, to the position of arrow C. Only then does earth regain the same orientation with respect to the sun.

FIGURE 10.15   Harrison's first chronometer.[9]

relative to stars. How could a provincial mechanic, the son of a carpenter, be successful where astronomers of the Royal Society failed? Admirals and astronomers on the Board of Longitude preferred to endorse the lunar-distance method pursued by the scientific elite. Judges kept changing rules to favour astronomers. Finally, in 1765, King George III persuaded Parliament to pay up. Still Harrison got only part of the promised sum.

For 150 years, the simple harmonic motion of regulated spiral springs marked accurate time for mariners navigating in a shrinking globe. With the triumph of electromagnetic waves (Chapter 3), radio communications achieved comparable precision. Today, the accuracy of satellite-based-navigation and atomic clocks has far surpassed that of radio. The accuracy of caesium-based atomic clocks is 3 parts in 10^{14}, equivalent to a drift of 1 second in 1.4 million years. This record has now been shattered by a factor 100 with ytterbium-based atomic clocks with a fantastic accuracy of 3 parts in 10^{16}.

UPDATES

UNIFYING SPACE AND TIME

Harrison's chronometer kept time to an accuracy of nearly one part in a million, allowing positioning to better than two kilometres. Leaping forward to the modern era, space-based satellite communication makes it possible to tell one's position anywhere on the globe to a precision of 5 metres, an improvement factor of 10,000 over the requirements of the longitude prize! For a few hundred dollars' investment in a smartphone, city drivers can navigate without paper maps.

Synchronous global positioning satellites (GPS) revolve around the earth in 24 hours, always staying above the same reference point on earth. A GPS determines distances to the desired location by measuring the time for electromagnetic waves to arrive from the satellite. Since light takes 33 billionths of a second to travel 10 metres, earth-clocks and satellite-clocks must be accurate to the 30 billionths of a second level for 10-metre position accuracy. Hovering at an altitude of more than five earth radii above the earth's surface, a GPS satellite moves fast enough that it becomes necessary to take into effect the relativistic time dilation effect (Chapter 4) that makes the satellite clock run slower than the earth clock. Over 24 hours, the satellite clock loses 6,000 billionths of a second, an enormous correction to keep up 10-metre position accuracy.

Starting from a purely aesthetic point of view, Einstein's special relativity (Chapter 4) demanded a drastic revision of the commonplace, intuitive notion that space and time are the same for all observers. There is no absolute space and no absolute time. Astonishing consequences follow. In order for the velocity of light to always come out constant, space and time change for moving observers. Clocks on moving vessels give different readings than stationary clocks. Moving metre sticks appear shorter to a stationary observer. Relativistic transformations provide the proper combinations of space and time.

Einstein devastated other traditional concepts to rebuild our understanding of space and time. Different observers in relative motion form different conclusions about the relative timing and separation of events they observe. Two observers moving relative to each other can disagree about the time order of events that take place at two different locations. For example, one observer in the middle of a soccer field may claim that two events take place simultaneously at entrance gates A and B across the field. But another observer moving towards gate A will claim that event A took place before event B. And a third observer, moving toward gate B, asserts that the event at B took place before A (Chapter 4).

Pronouncing the arrival of a new era in physics, mathematician, and former tutor of Einstein, Hermann Minkowski (1864–1909) declared:[10]

> The views of space and time which I wish to lay before you … are radical. Henceforth space by itself, and time by itself, are doomed to fade away into mere shadows, and only a kind of union of the two will preserve an independent reality.

Fusing the formerly separate realms, Minkowski coined the word *spacetime* to describe the physical world in motion. As one of Einstein's former mathematics professors, Minkowski thought of Einstein as a lazy student. When he later saw Einstein's special relativity paper he exclaimed in delightful surprise:[11] "Imagine that! I would never have expected such a smart thing from that fellow." Now Minkowski reformulated Einstein's theory, unifying space and time. At first Einstein resisted the hybrid treatment. He could hardly recognize his own ideas in the new formulation. But he soon embraced it wholeheartedly and made it the basis for a geometric treatment of gravity, a story for Chapter 11.

Relativistic transformations of distance and time between moving systems are analogous to rotations, but in spacetime. First consider rotations in space. When we look at a shoebox from a certain angle we see limited aspects of its apparent width, height, and depth. For example, looking exactly head on, we see no depth, just width and height (Figure 10.16 (centre panel)). If we come around to the side of the box (left panel), we see only depth and height, no width. But if we look at the box from an angle (right panel) we perceive both; at some angles we can see all three: height, width, and depth. As we move around the box we realize that height, width. and depth are just different properties of one object. Rotation mixes the spatial coordinates x, y, and z in a three-dimensional grid.

Relativistic transformations are analogous. At low velocities, we see only one face of a moving shoebox when it arrives directly in front of us (Figure 10.17). But when a shoebox moving at high speed to the right arrives at precisely the same location, we can *simultaneously* see part of the left side adjoined to the front of the shoebox! Even though we look head on, we see parts that we could not see on the static box. How? Because light rays from the left side which departed some time ago can also reach us. Light rays from the right side of the box do not reach us because the box gets in the way. Also, the front face appears contracted in the direction of motion due to relativistic effects. If the box moves faster we can see even more of the left side because light from earlier times on the shoebox "clock" can reach us. Just as spatial perception of object shape depends on orientation, spacetime "images" of events also depend on speed!

At the low speeds of present technology, our brains have become accustomed to treating space and time as independent entities. But space and time are different aspects of one unified spacetime view, just as depth, width, and height are linked aspects of one three-dimensional box. Movement at high velocities exposes the interconnectedness of spacetime. When objects move at different velocities, we can see spacetime reality from different points of view.

There is another subtle point. When we see the left side of a fast-moving shoebox, we are not looking at points on the left side at the same time as the points on the front side. Different parts are at different distances from us; therefore we see them as they were at different times. If a crack appeared on the left side at the instant the box arrived in front of us, we would not know about it; we only see the left side of the box as it appeared in the past. More generally, it is impossible to tell what is happening *now* at any point in space except at our own location. We can only tell

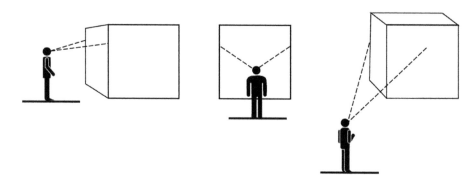

FIGURE 10.16   A shoebox as it appears from different angles.

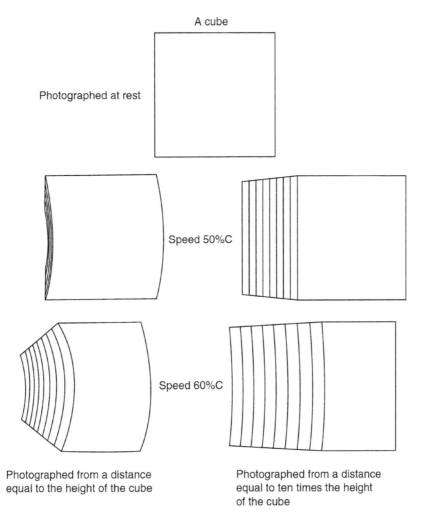

FIGURE 10.17   A shoebox as it would appear to observers at various speeds and distances from the box.[12]

what happened at other locations in the past. When Tycho saw the 1572 supernova, it had already exploded thousands of years prior. Indeed, some of the stars we see may not exist now! If an alien on an extragalactic star, 70 million light years from earth, were to possess technology powerful enough to observe earth in all its glory, he would see dinosaurs! And a moving observer would see what is happening at a different time in the past, because his clock time is different. In short, when we look at stars we look backward in time. But we cannot travel into the past.

In *Sun and Moon*[13] Escher creates an artistic metaphor for the fusion of what we normally consider as distinct. There is no sharp division between sun and moon, left and right, inward and outward. Birds moving left entwine with those moving right. Those moving radially inward mingle with outgoing ones. As Escher describes his imaginative works:[14]

> I cannot keep ... from having fun with our unassailable certainties ... It is satisfying to note that quite a few people ... are not afraid to revitalize their thought about rock-hard realities.

Throughout the quest for longitude, the accurate location of position on earth was linked inextricably with accurate time measurement. Chronometry in turn was bound intimately to motion, simple harmonic and circular, as well as the motion of earth through the heavens. The measurement of classical motion was bound inextricably with the measurement of classical space and time. But Einstein's special relativity dismissed the notions of absolute space and absolute time. Perceptions of space and time change for moving observers. Relativity demands a fusion of separate realms into the linked entity of spacetime in order to describe the physical world. Spacetime as a fused entity will play a major role in the reformulation of gravity through Einstein's General Theory of Relativity (Chapter 11).

11 Universal Gravitation: The First Synthesis

I do not know what I may appear to the world, but to myself I seem to have been only like a boy playing on the seashore, and diverting myself now and then in finding a smoother pebble or prettier shell than ordinary, whilst the great ocean of truth lay all undiscovered before me.

Isaac Newton[1]

Although Galileo set down the principles of dynamics, and Kepler found mathematical laws describing planetary motion, neither saw any connection between dynamics and astronomy. Newton brought Kepler and Galileo together, marrying the new dynamics to the new astronomy. With the law of universal gravitation he revealed the bonds holding the universe together. In that same bold stroke, he synthesized mathematical laws ruling the heavens with physical laws operating on earth. Two elegant laws governed the entire cosmos, from delicate movements of the smallest portable watch, to rising and falling tides, to earth's conical precession, to all the planets in the heavens. In the same grand sweep, universal gravitation made wayward comets predictable. Apostle of reason, Newton figured God's design in two brilliant lines of mathematics, elevating Him from clock-maker to master-engineer and mathematician.

GOOD FOR NOTHING BUT COLLEGE

The same year Galileo died (1642), a premature baby, christened Isaac, was born in England on Christmas Day to a Lincolnshire farmer by the name of Newton. The coincidence of the birth of Newton and the death of Galileo is a wonderful symbol of the continuity of thought in physics linking two intellectual giants.

Isaac Newton grew up in turbulent times. England was on the brink of bloody civil war. Exercising the divine right of kings, Charles I ruled the country at his whim and pleasure. Raising money to wage wars of territorial expansion, he imposed taxes unheard of since feudal times. But Charles' hegemony did not go unchallenged. Agitating for religious and political reforms, the Puritans were defiant. Extreme Protestants, they believed that the English Reformation had much further to go to prune obsolete Catholic doctrines. They wished to "purify" the English Church by removing all Catholic ceremonies. Rather than submit to Charles' pressure to conform to the Anglican Church, they migrated to America in huge numbers. With the express intention of replenishing his treasury through despotic measures, Charles called Parliament into session. But the Puritans in the Parliament had a different agenda. Listing royal abuses, they declared that ministers should answer to Parliament and bishops should be elected rather than appointed. Furious over such preposterous demands, the monarch personally entered the House of Parliament to arrest rebellious leaders. But they escaped before royal forces arrived.

Civil war broke out. Oliver Cromwell, a pious Puritan and brilliant general, organized an army of God-fearing men who advanced upon their enemy singing psalms. Puritans overthrew Charles I in 1649, and beheaded him for being a tyrant, traitor, and murderer. For the first and last time in English history, the country was without a monarch. Cromwell established complete control over the government and people's lives. His iron rule forbade games, dance, and theatre.

Enveloped in turmoil, mid-17th century England of Newton's childhood was full of contradictory trends. The middle class grew wealthy through commerce, but dire poverty engulfed the masses. Superstition was rampant, but the spirit of scientific inquiry thrived as never before. Conservative moral values dictated by Puritans co-existed with political revolutionaries who dared to execute a monarch.

During his early years at school, Isaac was a sickly child, backward in studies, lost in day-dreams, until a fateful event changed his destiny. One day, enraged by a bully who constantly picked on him, Isaac challenged the menace and overpowered him with new-found strength. The bully also happened to be a better student than Newton. Having won on the physical battlefield, Isaac's spirit rose. He resolved to complete his victory on the intellectual front. With hard work, he surpassed his rival to become the best student.

Young Newton had a special bent for mechanical toys and curious inventions, constructing water-wheels, kites, and furniture for a playmate's doll's house. One summer night he alarmed the citizens of Lincolnshire by launching a miniature hot-air balloon illuminated with candles attached to a wooden frame beneath a floating paper canopy. He was trying to fool the country folk into thinking it was a comet. Particularly fond of making sundials, he placed one in each room where the sun would shine. Although he mastered the art of telling time by the sun's position, he repeatedly forgot to show up for meals. All his money he spent on tools, which cluttered his room. Mice on treadmills churned wooden windmills. One stormy night, when Cromwell died, Newton stayed up all night to measure the force of the wind by pacing out the lengths of his leaps with and against the wind. Powers of concentration defied sleep. "It was my first experiment in physics"[2] he later recalled with fond nostalgia. With new-found power and understanding of wind conditions, he defeated his competitors in long-jump at school.

With his schooling complete, Newton's mother hired a tutor to teach him farming. Instead of watching sheep, Newton built model water-wheels in a brook, while sheep foraged a neighbour's cornfield, forcing his mother to pay damages. On market days, he bribed his tutor to leave him behind to build gadgets in peace. Once, as Isaac was walking his horse home, he fell into deep thought, oblivious to the world. Reaching home, he found himself holding an empty saddle. Fortunately for Newton, and fortunately for science, a family friend finally convinced the mother to abandon farming ambitions for her son, and send Isaac to grammar school to prepare for college education.

THE KEY TO THE MECHANICAL UNIVERSE

Newton entered Trinity College at Cambridge to study mathematics. At the time, Cambridge admitted about 30 students each year. As a poor student, Newton served as subsizar, earning his keep by performing menial tasks for fellows, who were the wealthy students able to pay for the privileges of education, room, and board. But they spent more time carousing than studying. Among his duties, Newton worked as a valet, rousing fellows for morning chapel, cleaning boots, dressing their hair, emptying chamber pots, waiting on tables, and eating leftovers. Kepler, the tavern pot-boy, had shared a similar past.

Despite the Copernican revolution, despite Galileo's starry messages and eloquent dialogues on world systems, Cambridge remained a stronghold of antiquated Aristotelian ideas with a geocentric view of the universe. Political and religious connections earned professorial appointments. From the wisdom of Greek philosophers, Newton learned canons of rigorous thought, and the need for a system to organize the overwhelming diversity of nature into a unifying, coherent pattern. On his own, he discovered Copernicus, Kepler, Galileo, and Descartes. Breaking free from entrenched philosophy, he wrote in his notebook, echoing the wisdom of Kepler[3] "Plato is my friend, Aristotle is my friend, but my best friend is truth."

Newton embraced Descartes' approach to seek mechanical descriptions of nature, fortified by mathematical formulations. Mastering all available mathematics within a short time, he moved on to explore new territory such as a powerful method of finding the slopes of curves. He was sowing the seeds of a new branch of mathematics, later called differential calculus. On publishing the work, he became well-known in a small circle of Europe's leading mathematicians.

For Newton, the foremost question for Descartes' mechanical universe was: what makes the divine clockwork tick? On one occasion, Newton stayed up all night intently watching the course of a comet, wondering how it would ever be possible to understand celestial motion. In the first phase of a comprehensive grasp of dynamics, Newton extended Galileo's ideas. Embracing the principle of inertia, he clearly formulated what physicist's proudly call *Newton's First Law of Motion*:

> A body remains in its state of rest or uniform motion in a straight line, unless it is compelled to change that state by an outside force acting on the body.

Without friction, without interaction with the earth, or any other body, an object will continue to move in a straight line with constant velocity in the same direction for ever. Defining when a force must be present, the first law foreshadows the concept of *force*, which Newton defined more precisely in the second law. Galileo, Descartes, and Huygens all referred to the concept of force, but with vague and different meanings. In his quintessential second law, Newton defined force precisely and quantitatively. Force is the agency necessary to realize "a change in the state of uniform motion." In modern language, force is the agency which changes velocity and therefore causes acceleration. Quantitatively, when acting on a body, a force (F) increases acceleration (a) in direct proportion to the force:

$$F \propto a \text{ (force is proportional to acceleration).}$$

There is another crucial aspect underlying the second law. When the same force acts on different objects, the acceleration is different due to a unique property of the body which Newton called *mass*. The greater the mass, the greater the resistance to change in velocity, i.e., the more the resistance to acceleration (Figure 11.1). With the concept of mass, Newton gave all matter a single nature. Newton sharpened the meaning of mass from the older and more familiar concept of weight. Put together, the two concepts of force and mass spell *Newton's Second Law of Motion*:

$$F = ma.$$

The fundamental law does not depend on the nature of the force, whether it has a mechanical origin in muscles, weights, and springs, or whether it arises from the attractions of gravity, magnetism, or electricity. Motion depends only on the magnitude and direction of force. Newton's key unlocks all motion: free fall, inertial, circular, or simple harmonic. One single law unifies all forms of motion of all bodies under all types of forces, a grand and elegant unification. Together with simplicity, the economy of its expression became the hallmark of its resounding success. Asked how he penetrated the workings of nature to reveal such a single, comprehensive, and powerful law, Newton answered,[4] "By thinking of them without ceasing," attributing the illumination to his powers of concentration.

Newton's force law is the most useful physical law ever devised. It can analyse the behaviour of levers used to build pyramids, pulleys to lift megaliths, objects sliding down mountain slopes, swinging pendula, oscillating springs, and flywheels that run mechanical clocks. Its awesome power is the basis of our engineering sciences – mechanical, civil, and hydraulic. Engineers use it to analyse the stability of bridges and tall buildings, the flow of fluids, and the drift of continents.

Newton took advantage of the law in two ways. In the first method, when knowing the forces, he determined the motion. A simple but instructive example arises when the force is zero. Acceleration is zero; therefore change in velocity is zero. Velocity remains constant. Included here is the direction and magnitude of velocity. Galileo's law of eternal horizontal motion is just a special case of Newton's force law. Rest is an even more special case, when the constant velocity is zero. Without force, there is no distinction between constant velocity motion and zero velocity. Zero is just one among an infinite number of possibilities.

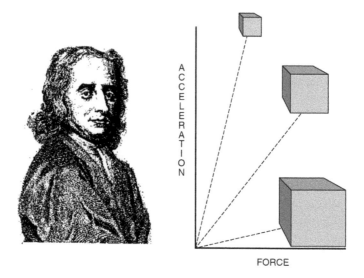

FIGURE 11.1   Young Newton and the Second Law of Motion.

In the second method, when he did not know the operating forces, Newton observed motion, and deduced acting forces through the second law. Consider centripetal acceleration of a body in circular motion. Since acceleration points toward the centre, there must be a force toward the centre. For a stone whirling at the end of a rope, string tension supplies the centripetal force. If the speed of circular motion increases, so does tension in the attached string, until eventually the string breaks when centripetal force exceeds string strength.

After penetrating the nature of centripetal force required for circular motion, Newton faced the crucial question. What supplies the centripetal force for the circular motion of the moon around the earth? The answer would turn out to be a major clue to the nature of gravitation.

MIRACLE OF FALLING APPLES

In 1665 a great plague hit London and Cambridge. With no effective treatment, people died in frightening numbers. Nearly 100,000 people succumbed to the scourge. The university closed down. Newton, then 23, returned home to Lincolnshire for 18 months. For Newton and science, it was to be a miraculous phase. During this period he conceived ideas that were to shape the development of science for the next two centuries. Among these were the theory that white light is a composite of the colours of the rainbow, the mathematics of differential and integral calculus, and the overarching concept of gravitation. Any single one of these accomplishments would be a lifetime achievement for most scientists. But Newton devoted all his powers of concentration to discovery rather than publication. The world heard nothing of his fabulous ideas. He only published them much later at age 44 in his monumental work *Philosophiae Naturalis Principia Mathematica*, commonly referred to as the *Principia*.

In the garden at his home in Lincolnshire, Newton was reflecting about the nature of the force which makes apples fall to the ground. The earth attracts the apple with the force of gravity, familiar since the time of Aristotle. Any object on the surface of earth experiences a force propelling it toward the centre of earth. What is the magnitude of that force? In free-fall motion, acceleration is g, first measured by Galileo. According to Newton's second law ($F = ma$), the force acting on an apple of mass (m) is $F = mg$. As before, mass is a property of the apple that expresses its resistance to change in motion, its inertia. All matter has inertia. Inertial mass determines how hard it is to change natural motion.

How does the earth manage to keep acceleration at the same value g for all apples, independent of their inertial mass? Galileo's epiphany in Pisa exposed Aristotle's monumental fallacy about free fall, but Galileo did not pause too long to ponder why acceleration is the same for all bodies. The universality of free-fall motion thoroughly engulfed his thinking. Satisfied with determining how, he did not jump to exploring why. But Newton had an explanation. If a horse pulls a cart with the maximum force it can muster, the cart speeds up with a maximum acceleration. If an ambitious farmer continues loading up the cart with hay to double the cart's mass, the acceleration drops to half the original value, since the horse is already pulling with maximum force. But how does the earth manage to keep up its acceleration at the same value g for an apple with twice the mass, or four times the mass? Concentrating on falling apples, Newton realized an astonishing property of earth's gravity. *Earth's attractive force on the apple must increase with the mass of the apple.* It was as strange as if a horse knew the farmer wished for it to carry more hay, and obeyed by increasing its effort accordingly! Newton had grasped a strange but important aspect about the nature of gravity. Gravitational force is proportional to the mass it acts on.

Extending thought beyond falling apples, Newton wondered whether the force of gravity was limited to a certain distance from earth's surface. Did it extend beyond the clouds? Perhaps it extended to the moon. If so, why did the moon not fall down to earth? Newton had a marvellous insight. As the moon moved horizontally, it fell just enough toward earth to follow the curvature of earth (Figure 11.2). The moon was forever falling, but its horizontal motion made it follow a curved path. It took an intellect as penetrating as Newton's to realize that the moon does indeed fall to earth like an apple!

A gravitational pull of the earth on the moon was an intriguing idea. But Newton felt it important to characterize the force quantitatively. What was the magnitude of earth's attractive force on the moon? Was it the product of the moon's mass and g? Was the moon's acceleration towards earth also g? Innovation was Newton's forte. He used the observed acceleration of the moon to *deduce* earth's force. Independently of Huygens, Newton had shown that a body moving with uniform speed v in

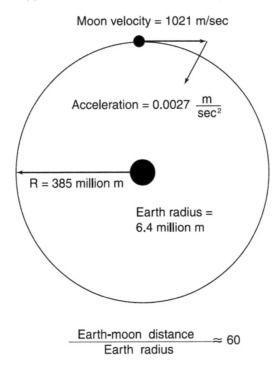

Moon velocity = 1021 m/sec

Acceleration = 0.0027 $\frac{m}{sec^2}$

R = 385 million m

Earth radius = 6.4 million m

$$\frac{\text{Earth-moon distance}}{\text{Earth radius}} \approx 60$$

FIGURE 11.2   With the available numbers, Newton determined that the moon moves in a circle (approximately) at a speed of 1,020 m/sec, and "falls" to the earth with centripetal acceleration of 0.0027 m/sec^2.

a circle of radius R had centripetal acceleration v^2/R, pointing toward the centre of motion. He used this formula to calculate the moon's acceleration, assuming a circular orbit for simplicity.

Knowing the earth–moon distance (R), as well as the 28-day lunar cycle time (T), Newton calculated the moon's speed $v = 2\pi R/T$, and from there the centripetal acceleration: $v^2/R = 0.0027$ m/sec². The result was a surprising disappointment! The tiny acceleration of the moon implied that the earth's attractive force on the moon was far smaller than earth's force on the apple which caused the familiar free-fall acceleration of the apple to be $g = 9.8$ m/sec². If earth's gravity was responsible for the moon's orbit, why was the moon's acceleration not the same as the apple's?

A lesser thinker would have quickly abandoned the idea that the moon fell toward earth like an apple. But Newton had an inspiration: Earth's force of gravity must weaken with distance. Which would make the distant moon's acceleration much smaller. Did earth's attractive force (and the resulting moon's acceleration) decrease proportionately with distance? The numbers would tell. Since the ratio of the apple's acceleration to the moon's was

$$\frac{9.81}{0.00271} = 3,609 \approx 3,600$$

the moon's acceleration was approximately 3,600 times smaller than the apple's free-fall acceleration. But the moon was not 3,600 times farther from earth-centre than the apple; it was only 60 times farther. Remarkably, there is a simple relationship between 60 and 3,600: $60^2 = 3,600$. It was Newton's supreme clue that the earth's force of gravity weakened as the *square* of the distance from earth, or "pretty nearly" to use Newton's words.[5]

It was another marvellous insight into the nature of gravitational force. Earth's force of gravity decreased inversely with distance from earth. Newton was the first person in the history of human civilization to realize why the moon orbits earth. Earth holds the moon in its clutches with the force of gravity. But, instead of falling to earth like an apple, the moon goes around earth in circular motion, like a stone in a sling. Most importantly, the numbers showed that the force of gravity decreased with the square of distance. It was the first synthesis of terrestrial and celestial motion.

Romantic poet, and fiery champion of oppressed people, Lord Byron poignantly captured the fundamental significance of Newton's first revelation:[6]

> When Newton saw an apple fall, he found
> In that slight startle from his contemplation
> A mode of proving that the earth turn'd round
> In a most natural whirl, called 'gravitation';
> And this is the sole mortal who could grapple,
> Since Adam, with a fall, or with an Apple.

But the historic discovery was still in an embryonic stage. Was a "gravity force" also responsible for binding planets to the sun? The idea flashed out of the shadows. Newton imagined a planet falling forever toward the sun (Figure 11.3), just as the moon "fell" towards earth. The horizontal velocity of a planet made it move in a curved path around the sun. But did the force between a planet and the sun also decrease with the square of sun–planetary distance? Once again, Newton determined force through motion via his second law. Knowing a planet-to-sun distance (R_{planet}), and orbital period (T) from Copernicus and Kepler, Newton calculated orbital velocities and centripetal accelerations of Mercury, Venus, earth, Mars, Jupiter, and Saturn, in terms of each R_{planet} and T. (For instance, see Figure 11.3, earth's centripetal acceleration toward the sun is 0.006 m/sec².) From the planetary accelerations, he could derive the sun's force for each planet, still in terms of both R_{planet} and T. But, to check the inverse square law, Newton needed to obtain the dependence of force purely on distance; he had to eliminate the planetary periods T. Through his extensive readings

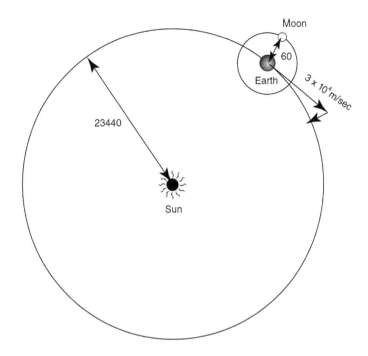

FIGURE 11.3 The earth–moon distance is 60 times bigger than the earth's radius, and the earth–sun distance is 23,440 times larger than the earth's radius. In its orbit around the sun, the earth moves horizontally at a speed of 30,000 m/sec, falling toward the sun with centripetal acceleration of just 0.006 m/sec², far smaller than $g = 9.8$ m/sec².

at Cambridge he was familiar with Kepler's laws, even though the mystical mathematician had presented them as disjointed bits of information, tucked away among hundreds of pages about the musical harmony of planets and geometric perfection of five solids. Using Kepler's harmonic law, $R_{\text{planet}}^3 / T^2 = \text{constant}$, Newton eliminated T, to derive two wonderful conclusions:

1. The force of the sun on each planet is proportional to the mass of the planet.
2. The force of the sun decreases as the square of the sun–planet distance.

From Kepler's harmonic law, Newton renewed faith in the inverse square law of gravitational attraction between the sun and its planets.

When a friend queried Newton many years later on the source of his inspiration for the inverse square law, Newton humbly offered:[7]

> It was occasioned by the fall of an apple, as I sat in a contemplative mood … and thus by degrees I began to apply this property of gravitation to the motion of the earth and the heavenly bodies.

On another occasion, Newton explained that his train of thought was to first recognize the inverse square nature of the sun–planet "bond" in Kepler's law. But he did not yet ascribe this bond to gravity. Fired by the idea that earth's gravity was the force holding the moon in orbit, he was delighted to find that this force also obeyed the inverse square law. At this point he leapt to the conclusion that the "sun–planet bond" must also be a form of gravitational attraction, just like the earth–moon bond. There was another crucial affirmation that arose when Newton derived the sun–planet force from Kepler's third law. The sun's attractive force on its attendant planets was proportional to planet mass, a property of gravity which Newton also saw in earth's attractive force on apples.

In Torch Bearers,[8] poet Alfred Noyes captures the momentous events in the Lincolnshire garden.

> Did Newton, dreaming in his orchard there
> Beside the dreaming Witham, see the moon
> Burn like a huge gold apple in the boughs
> And wonder why should moons not fall like fruit?
> Or did he see as those old tales declare…
> A ripe fruit fall from some immortal tree
> Of knowledge, while he wondered at what height
> Would this earth-magnet lose its darkling power?
> Would not the fruit fall earthward, though it grew
> High o'er the hills as yonder brightening cloud?
> Would not the selfsame power that plucked the fruit
> Draw the white moon, then, sailing in the blue?
> Then, in one flash, as light and song are born,
> And the soul wakes, he saw it--this dark earth
> Holding the moon that else would fly through space
> To her sure orbit, as a stone is held
> In a whirled sling; and, by the selfsame power,
> Her sister planets guiding all their moons;
> While, exquisitely balanced and controlled
> In one vast system, moons and planets wheeled
> round one sovran majesty, the sun.

Newton unified our understanding of terrestrial and celestial motion. Gravity makes apples fall, keeps our moon in orbit around the earth, and planets whirling about the sun. Projectile and planet move under the same laws. One might imagine that with such a powerful revelation, Newton rejoiced as ecstatically as Kepler did when he discovered the laws of planetary motion. But no. Newton was not yet totally confident about the inverse square law. There was a possible fly in the ointment. Unaware of the most accurate value for the earth's radius, he first used the "sailor's rule of thumb" that every degree in longitude spans 60 miles at the equator to calculate the earth–moon distance to be 69 times the radius of earth, not 60 times. As a result he was not completely certain of the inverse square fall off. Sixteen years later, when Newton heard of Paris scientist Jean Picard's result that 1 degree of longitude corresponds to 70 miles, he was completely reassured of the inverse square law and returned to meditate on the powerful consequences of gravitation.

ULYSSES DRAWS ACHILLES INTO BATTLE

When the university re-opened, Newton's reputation in mathematics earned him the position of Fellow. Two years later, Isaac Barrow, leading professor of mathematics, recommended that Newton succeed him as Lucasian Professor. Barrow was ready to move on to a post in theology, regarded at the time as considerably more important than mathematics.

Meeting frequently at the Royal Society in London, Newton's contemporaries, astronomer Edmond Halley, and curator of the Royal Society Robert Hooke, grappled with the question of how to prove mathematically Kepler's ground-breaking law of planetary motion – planetary orbits are ellipses, not combinations of circles. Halley had a bet with Hooke over who could solve the problem first. Hooke boasted he had found a proof for elliptical orbits, but claimed to keep it secret, so that others who tried and failed would duly realize how difficult the problem really was, and how clever Hooke was to solve it.

In an earlier correspondence with curator Hooke, Newton suggested a simple experiment to prove earth's rotation. An object dropped from a high tower does not fall exactly to the base but slightly to the east of the tower base, because its tangential velocity at the top of the tower exceeds the velocity at the foot of the tower. Newton drew a diagram showing the path. Unfortunately, going further than necessary to prove his point, he extended the line to the centre of earth in the shape of a spiral.

Newton's ideas stimulated Hooke. But Hooke was a quarrelsome character, pleased to find errors. He took the young upstart to task. The trajectory of a body falling through the earth would not end at the centre of the earth, as Newton hastily guessed. Rather, according to Hooke's intuition, the path would be an ellipse, with the body returning to its original height, provided the gravity of earth decreased with the square of the distance. Hooke was on the heels of a major discovery. With a passion for experiments, he later tried to make a measurement of the inverse square law by weighing objects at the top of mountains, and in the tall steeple at Westminster Abbey, as well as at the bottom of a deep mine shaft. But he was unable to detect any clear effect at such short distances.

Hooke's reply struck a raw nerve in Newton. It seemed that the curator was hot on the trail of the inverse square law, which Newton had deduced many years earlier in the garden. Newton did not bother to reply, claiming instead to tire of unavoidable philosophical controversy generated by publication. Later, he admitted that Hooke's reply did stir him to a challenging demonstration: when a body revolves in an elliptical orbit around a centre of attraction placed at one focus, the force of attraction must vary inversely as the square of the distance from the focus. Still the derivation remained private. But Hooke's seed had fallen upon fertile soil.

Hooke's colleague Halley kept pushing for answers. Engrossed in the problem, he mastered Arabic to devour a translation of Apollonius' conic sections. Perhaps there was a hint lurking in the Alexandrian's far-reaching enterprise. Men of learning gathered in saloons and cafes to debate the latest scientific discoveries and ideas. Although Hooke continued to boast a solution, Halley could never get a satisfying answer. Finally, he decided to go to Newton (in 1684). It was an historic meeting.

Even though Newton desired isolation, the world had grown familiar with his innovative reflecting telescope (Chapter 9) and his mathematical prowess. Without mentioning Hooke, Halley asked Newton if he knew what would be the path of a planet for an inverse square law of attraction to the sun. To Halley's astonishment, Newton replied confidently that the path had to be an ellipse, with an attracting centre at the focus. A startled Halley listened to the claim that Newton had worked out a proof (and much more) several years prior. But Newton could not immediately find the right pieces of paper on which he had scribbled his mathematical reveries.

It may seem shocking that Newton had casually mislaid the answer to the scientific question of the millennium! Reflecting upon the absent-minded Newton's temperament, it is not too surprising that while others searched fruitlessly for the force that bound the heavens, Newton had lost it. His notes were probably buried among vast sheaves of papers on divinity, theology, and alchemy, subjects which also engulfed his diverse mind. Urging Newton to supply the proof, Halley left, in excited anticipation. Three months later, Newton came through. Deriving elliptical planetary orbits from first principles, he supplied two separate proofs of Kepler's first law. On describing the historic encounter, Halley liked to say he was the wily Ulysses who lured the sulking Achilles out of hiding and pressed him to prove his mettle.

Halley received much more than expected: a nine-page treatise entitled *On the Motion of Bodies in an Orbit*. Not only did Newton demonstrate how an elliptical orbit entailed an inverse square law; he also proved the converse and much more. An inverse square law led mathematically to any of the conic sections; the ellipse was only a special case. Halley's excitement later surfaced in a tribute to Newton's *Principia*:[9]

> Lend your sweet voice to warble Newton's praise,
> Who searcht out truth thro' all her mystic maze,
> Newton, by every fav'ring muse inspir'd,
> With all Apollo's radiations fir'd;

Newton, that reach'd th' insuperable line,
The nice barrier 'twixt human and divine.

Newton's proof of elliptical orbits requires mathematics of differential calculus, which he also invented. It is a powerful method for breaking up curves into a series of small straight lines, making possible the analysis of curved motion during which attractive force, acceleration, velocity, and position all change continuously. Newton also proved Kepler's second law (the area law, Chapter 8) as a direct consequence of *central force*. Regardless of the varying force intensity, a planet sweeps out equal areas in equal time intervals of travel, as long as the sun's force acts along a line joining it to the planets.

THE POINT OF SPHERICAL SYMMETRY

These were giant strides. Starting from general principles of dynamics, i.e., the first and second laws, Newton's primary treatise on motion was succinct and powerful. Recognizing immediately the momentous significance, Halley visited Newton again, pressing him to publish. But Newton held back, bothered by an asymmetry between heavenly and earthly motion. It was reasonable to treat both the earth and moon as single points, since the distance between the two was very large compared to the radius of earth. The earth–sun and earth–planet distances were also large enough to approximately treat planets and sun as point-masses. But he could not honestly treat the earth and apple as separate points, remote from each other. He had to prove that when earth exercised its force on an apple, it behaved as though all earth's mass was concentrated at a single point at earth's centre. Again, mathematics of the time fell short of the task. Inventing an entirely new branch of mathematics, integral calculus, Newton combined the attractive force of all the particles of earth, to show to that a sphere of uniform density exerts its force of attraction as though all its mass were concentrated in one single point at its centre. It was another crucial consequence of the inverse square law. No other force law could produce such a fundamentally elegant result. The spherical symmetry of the cosmos found new expression. Through the inverse square laws, symmetry reduced the action of earth's gravity to a single point.

To reach this stage, Newton became completely immersed in mathematical labours, forgetting to eat and sleep. When he walked through the gardens outside Cambridge, he would stop suddenly to draw diagrams in the gravel. Colleagues would reverently walk around the sacred figures; Newton could return at any moment to stare at them again. Once he ran back from the gardens to his desk and started writing furiously while still standing, not wanting to disrupt his train of thought even for an instant to pull up a chair. At another time he set out to church in his dressing gown, until a neighbour stopped him in time. Wondering why he could not penetrate a problem, he once realized he had forgotten to sleep for days; so he finally went to bed.

FROM APPLE TO MOON TO SUN TO UNIVERSE

Newton did not trace his exact path of thinking to the renowned universal law of gravity, but we may well imagine how the lamps of symmetry and unity guided him, much as they illuminated the way for intellects through the history of scientific thought.

There was a bothersome asymmetry in the force of gravity. Why should gravitational attraction be unidirectional? For example, the earth attracts an apple, or the sun attracts the earth (Figure 11.4). Does an apple also attract the earth? Does a planet also attract the sun? If so, why does the sun remain stationary?

From observations on billiard-ball collisions, it is clear that an isolated single body can neither experience nor exert force by itself. Forces emerge only as a result of interactions between two or more entities. A fundamental symmetry governs such interactions. Newton phrased it as follows:[10] "If you press a stone with your finger, the finger is also pressed by the stone." In the *Principia* he

FIGURE 11.4   There is an asymmetry in the idea that the earth attracts the apple, or the earth attracts the moon, or the sun attracts the earth.

formalized the law as the final pillar for the science of mechanics. *Newton's Third Law of Motion* reads:[11]

> To every action there is an equal and opposite reaction: or the mutual actions of two bodies upon each other are always equal, and directed to contrary parts.

In modern terms:

> Whenever there is an interaction between two bodies (A and B) the force exerted by body A on body B, F_{AB} is equal and opposite to the force exerted by body B on body A, F_{BA}. Therefore $F_{AB} = -F_{BA}$.

Each force acts on a *different* body, so the forces do not cancel each other out. For example, gravitational force is mutual: earth attracts the moon with the same force as moon attracts earth (Figure 11.5). But the resulting motions of the moon and the earth are very different, because their masses are very different; hence the observed asymmetry in their motions.

Musing over the nature of gravity, Newton had recognized a very peculiar property embodying a gross asymmetry. Gravity increases with the mass of the body it acts on. For example, earth's attractive force on an apple increases with the mass of the apple. Similarly the sun's gravity force on a planet is proportional to *planet mass*. Reciprocity leads to another inescapable consequence. The

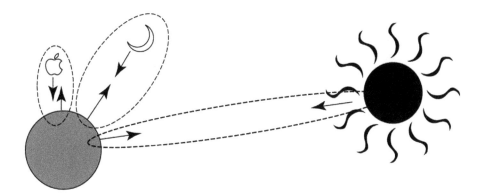

FIGURE 11.5   Gravitational interaction between two bodies must be symmetric. The earth attracts the apple with the same force as the apple attracts earth, but in the opposite direction. Similarly for the earth–moon and sun–earth systems.

reciprocal gravitational force which a planet exerts on the sun is also proportional to the *sun's mass*. Now here is the asymmetry. The planet mass and sun mass are vastly different, considering that the sun has nearly 1 million times the volume of earth. How can the earth's force (proportional to the sun's mass) be equal to the sun's force (proportional to the earth's mass)? To restore symmetry, the magnitude of each force must be proportional to the *product of both masses*. By using a product, the expression for the mutual force becomes symmetric with respect to both masses.

And the final giant leap: if the earth attracts the apple, does the apple attract the earth? Does the moon also attract the apple? Newton's deep insight was that gravity must be universal. With one final vault, he reached the summit. Every object in the universe attracts every other through the innate force of gravity. Aesthetics of universality demanded the sublime principle of gravitation:

> All matter moves as if every particle attracts every other particle with a force proportional to the product of their masses, and inversely proportional to the distance between them. That force is universal gravitation.

Newton's law of gravitation has been called "the greatest generalization achieved by the human mind."[12] With the law he brought a wealth of physical phenomena into a single theoretical framework. Gravity applies to all objects, everywhere and anywhere: between apples and apples, between the moons of Jupiter and Jupiter, planets and planets, stars and stars. It was an arduous climb, the heroic struggle of a solitary mind to master the universe. Through pure thought, Newton penetrated a universal law, in mathematical form:

If a distance R_{1-2} separates any two masses, $mass_1$ and $mass_2$, the force of gravity between them is proportional to both masses, and to the inverse square of the distance between them (Figure 11.6). To replace proportionality by equality demands a constant, now called the gravitational constant G. Assembling the inspired conclusions, Newton's law of universal gravitation reads:

$$\text{Universal Law of Gravitation}: F_{grav} = G\,\frac{mass_1\,mass_2}{R_{1-2}^2}$$

G should not be confused with $g = 9.8\ \text{m/sec}^2$, acceleration due to earth's gravity.

Newton's universal law joined dynamics and astronomy with a mathematical bond. Cannonballs, earth, moon, planets, and stars: all move under the same laws. Newton's epiphany culminated 2,000 years of observations, questions and answers, many false trails and blind alleys. Poet Francis Thompson (1858–1907) captured the universal and all-encompassing nature of gravity:[13]

> All things by immortal power
> Near or far
> Hiddenly
> To each other linked are
> That thou canst not stir a flower
> Without the troubling of a star.

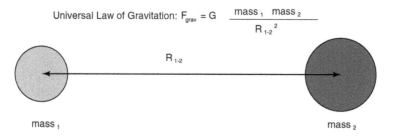

FIGURE 11.6   Newton's Universal Law of Gravitation applies to any pair of masses.

Gravity explained mysteries that baffled the best thinkers of all time. All heavenly bodies and earth are spherical in shape because gravity pulls their mass together. Poet Samuel Rogers (1763–1855) eloquently expressed the pervasive manifestation of gravitation:[14]

> That very law which moulds a tear
> and bids it trickle from its source
> That law preserves the earth a sphere
> and guides the planets in their course.

Newton answered the gnawing philosophical objection raised by sceptics when Copernicus introduced the heliocentric system. If earth is not the true centre of the universe, why does everything always fall toward earth? If the sun is the legitimate centre, should not everything fall toward the sun? With the unifying force of gravity, planet earth, and everything on it, does indeed fall toward the sun. Because of its tangential motion, earth circles the sun. Circular motion is equivalent to falling into the centre.

Stepping back to view the visionary Newton's universal law, mutual attraction between two bodies depends on a "gravitational property" of objects, which Newton identified as mass. To be explicitly clear, the mass which appears in universal gravity is called "gravitational mass." Newton's gravitational force is a natural attraction, like the force between magnets or electrical charges. A priori, *gravitational mass* is not the same as the *inertial mass* in Newton's force law, $F = ma$. Here inertial mass is the property of bodies which resists motion. Through observations on motion followed by powerful reasoning, Newton reached an incisive conclusion. Gravitational mass is the same as inertial mass. The identity of the two masses is the fundamental reason underlying Galileo's epiphany that acceleration due to gravity is independent of mass.

But there is no fundamental reason from any general principle that *gravitational mass*, the "property" which determines the strength of attractive force, should be identical to *inertial mass*, the property of matter that resists acceleration. Just as there is no reason to expect that magnetic force depends on inertial mass. The two masses have quite distinct meanings at this stage. Inertial mass determines how hard you have to push to get a cart to move faster. Gravitational mass is how hard you have to lift to raise the cart. Later we will see that the manifest identity of the two masses is due to the intrinsic nature of gravity which Einstein ascertained.

THE MISSING PIECE

Newton did not know the value of G, but that hardly stopped him from using his law to make predictions. First, he deduced the value of G in terms of earth's mass, for which he had to make a clever but excellent guess, since he had no idea of the earth's mass either. Only the earth's radius, and therefore volume, were known, thanks to Eratosthenes (Chapter 2) and confirmations from Picard in France. Newton approached the unknown earth mass in terms of earth's density, thanks to Archimedes (Chapter 2): *Mass = Density × Volume*. But Newton did not know the density of the earth either. So he made an educated guess:[15]

> The common matter of our earth on the surface … is about twice as heavy as water … in mines (matter) is found about three or four, or even five times more heavy, it is probable that the quantity of the whole matter of the earth may be five or six times greater than if it all consisted of water.

Guessing the average density of the earth to be 5.5 times the density of water, Newton's estimate of earth mass turned out to be 6 trillion trillion kilograms (6×10^{24} kg). In the face of huge unknowns, scientists often make educated guesses to set the scale. Newton was extraordinarily lucky to come so close to a good answer for earth's mass and from there a value for G!

Take the case of the gravity force between the earth and the apple. By sequential applications of the force law ($F = ma$) and gravitation law ($F = Gm_1m_2/R^2$), the apple's mass can be eliminated to express

G in terms of earth's properties: mass, radius, and g. All he knew was earth's radius, as well as the value of g, thanks to Galileo (Chapter 4). With his clever guess for the earth's mass, Newton determined the value of G as 6.67×10^{-11}, when mass is measured in kilograms and distance is in metres.

Because gravitational force is extremely weak, G is an extremely small number, and among nature's most difficult fundamental constants to measure accurately. Only when one of the masses in the law of gravitation is gigantic, such as the mass of a planet or the mass of a star, does gravity assume a familiar strength. The attractive force of gravity between two masses separated by 1 metre is nearly 1 trillion times smaller than the weight of each mass (which is equal to the earth's attractive force). But Newton's reasoning was sufficiently powerful to deduce the existence of even a minuscule gravitational force between two apples! As Parmenides and Plato taught in antiquity, the eye of the mind can see much further than the senses.

A hundred years after Newson, Henry Cavendish (1731–1810) devised an exquisite method to measure minute gravitational forces on kilogram masses in a laboratory experiment, and directly determined the value of G. Conceptually, Cavendish's method was very simple (Figure 11.7). Attaching two masses to a rod, he suspended this dumb-bell assembly from the middle with a wire holding a mirror. A ray of light striking the mirror cast a spot on a remote scale. Slight dumb-bell rotation twisted the mirror by a small angle, but produced a large deflection of the light spot due to the long arm of the reflected ray. Bringing two large masses close to each end of the dumb-bell, he upset the delicate torsion balance. From the amplified twist, Cavendish measured the force. It was the first direct confirmation of Newton's law of gravitation, and the first direct measurement of G.

Cavendish never viewed his experiment as one to check Newton's theory of gravity. Fully accepting Newton's law, he claimed instead to measure the value of G, and so to determine earth mass. Cavendish claimed his experiment was designed to weigh the earth for the first time. Cavendish confirmed Newton's guess for earth's density, and obtained a confident number for earth's mass.

BRIDGE TO THE HEAVENS

Newton's overarching theory encompassed previously understood phenomena. Recall how Newton was inspired by Kepler's third law, $R^3/T^2 =$ Kepler's constant (Chapter 8), to establish the inverse square distance dependence in his law of gravitation. As a truly successful theory, Newton's treatment went even further. He could derive all of Kepler's laws from the more general principles of his gravitational theory. Now the law of gravity combined with Kepler's law led directly to toward unanticipated results, predicting incisive new consequences. Sitting alone in his study at Cambridge,

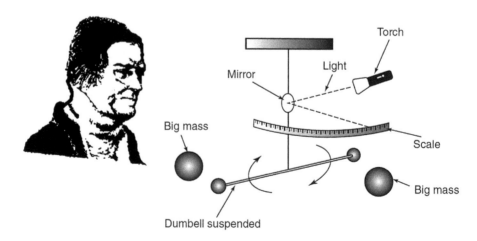

FIGURE 11.7   Cavendish measured the force of gravity in the laboratory to determine the value of G, and so weighed the earth. Cavendish determined the mass of the earth.

Newton could determine the mass of the sun! Newton revealed a new physical property of the solar system: *Kepler's constant = G Mass$_{sun}$/4π^2*. The mass of the central body in the solar system is a crucial part of Kepler's constant. Newton extended our knowledge of the solar system to determine the solar mass to be 300,000 times earth's mass.

When R is measured in astronomic units ($AU = 1$ for earth–sun) and T is measured in years (the earth takes one year to orbit the sun), Kepler's constant $R^3/T^2 = 1$. To determine the mass of the sun in kg we have to use R in km and T in seconds (MKS units). In this case $R = 1.5 \times 10^8$ km and $T = 3.15 \times 10^7$ sec. Using these constants, the sun's mass becomes 2×10^{30} kg.

Just as Kepler's third law applies to the sun and its orbiting planets, the law also applies to Jupiter and its satellites as an independent orbital system, with Jupiter as the central mass but a different Kepler's constant which depends on the mass of the central body, in this case, Jupiter. Newton determined Jupiter's mass to be about 300 times earth's mass. More generally, from the motion of the moons of any planet, his method provided the central planetary mass for that system. The motion of Saturn's moons (Chapter 9) yielded the mass of Saturn, and similarly for Mars.

The earth–moon system could be used to confirm the mass of the earth, determined earlier by density guesses. But how to determine the mass of moon, which has no satellites? It was a difficult problem but Newton devised a brilliant method; symmetry in the force of gravity made it clear that our moon indeed has a very familiar satellite, the earth. The moon attracts the earth, so that the earth does indeed revolve around the moon, but the centre of that small circle is near the centre of the earth, and too difficult to assess. But Newton realized that the attraction of the moon is responsible for the tides on earth. We will return to this topic later. Newton estimated the moon's mass from the height of tides on earth.

Uniting earth and heaven with one set of laws, Newton constructed a new mental bridge to make an earthly body into a celestial one! Propelling Galileo's projectiles with successively greater initial velocities, Newton imagined how projectiles launched with higher and higher velocity would describe ever larger parabolic trajectories that would eventually become large enough to orbit earth (Figure 11.8). A powerful enough cannon would make it possible to propel an earthly object into orbit around earth. The canon (or rocket) would need to provide a launch velocity of 18,000 mph! Newton conceived of artificial satellites 300 years before rocket technology!

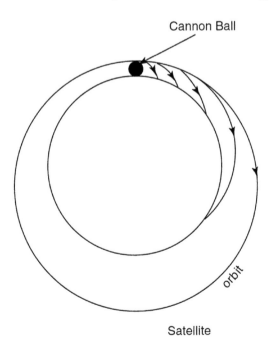

FIGURE 11.8   Newton's sketch showing how an earthly body can become a celestial object.

Today geosynchronous satellites routinely orbit high above earth at precisely the same rate as earth rotates, staying exactly above one spot all the time. Such satellites are essential for global communication, positioning, and weather monitoring. Just a few in orbit permit radio transmission from any point on the earth to any other point. Direct satellite television antennae point directly to synchronous satellites. Kepler's third law applied to such an earth-satellite system yields an enormous orbiting height of 36 million metres, about 5.5 times earth's radius.

A perfect circular orbit right above earth's surface occurs at a fantastic launch speed of 18,000 mph (8 km/sec). At this tangential velocity, earth's gravity force provides the exact centripetal acceleration for circular motion directly above earth. In comparison, earth's rotation speed is a mere 0.5 km/sec. But if earth were to miraculously speed up and spin 16 times faster, turning once every one-and-a-half hours, Newton's apples would no longer "fall" to the ground. Acceleration due to gravity would be entirely centripetal, nullifying the familiar consequence of gravity. There would be no more downward-falling apples.

Defying customary notions, René Magritte paints an imaginative scenario in *Castle in the Pyrenees* (Figure 11.9), floating effortlessly above a hypothetical earth! Even though it is unlikely that he knew how the physics of gravity and rotation would permit such an incredible levitation, he had the intuitive foresight to impose such a vision of the world. With suggestive pictures of reality, Magritte always tries to disturb our thought habits and outsmart our certitudes.

Every time we take a joyride on a loop-the-loop roller-coaster, we experience the illusion of defying the force of gravity for an exhilarating instant, when the centripetal acceleration of the cart and its excited passengers speeding over a tightly curved track becomes exactly equal to *g* (Figure 11.10). Of course, the radius of the loop is small (e.g., 10 m) and the necessary speeds are correspondingly less (e.g., 10 m/sec = 22 mph). In such an image, Newton visualized how the moon on its heavenly roller-coaster ride never falls down to earth.

FIGURE 11.9 *Castle in the Pyrenees*, René Magritte.[16]

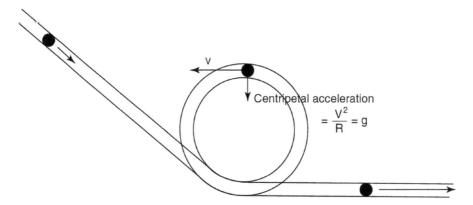

FIGURE 11.10   When centripetal acceleration of the ball at the top of a loop-the-loop track becomes equal to free-fall acceleration g, the ball traverses the entire track without falling off. A minimum release height from the top of the slide is necessary for the ball to gain enough speed to fulfil the condition.

But Newton went further. A satellite's path does not have to be a perfect circle around the earth; it can be any one of the conic sections (Figure 11.11). When its launch speed is greater than 8 km/sec, centripetal acceleration *exceeds* g, and the orbit becomes elliptical. Gravity is too weak to provide the exact centripetal force necessary to maintain precise circular motion, a simple way to under-stand how planetary orbits turn elliptical from circular. So also, planets move a bit too fast for sun's gravity to keep them in perfect circles. With even greater velocity, elongation increases, as centrip-etal force increasingly overwhelms gravity force. At a launch speed of 11.2 km/sec, an earth satel-lite can entirely escape earth gravity in a parabolic orbit. At still higher speeds, the orbit becomes hyperbolic, barely disturbed by gravity. At the other extreme, for speeds less than 8 km/sec, gravity wins over centripetal force; a satellite will fall toward earth, also in a parabola.

How far Newton had progressed from simple reflections in the garden at Lincolnshire! How radically changed was Newton's dynamic universe from Aristotle's static heavenly architecture of concentric spheres! Newton's imagination liberated artists to mingle heaven and earth.

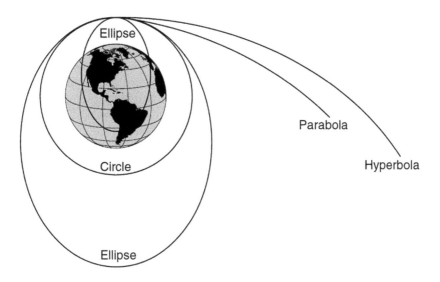

FIGURE 11.11   Under an inverse square law of gravity, it is possible to have any one of the conical sections as an orbit.

Extending gravity from the planetary system out to the firmament, Newton wondered what hindered fixed stars from falling into each other to collapse into one giant fireball. At first he thought symmetry would avert cosmic collapse. Each star would feel equal forces in all directions from other stars, balancing to a net zero force, and so a merciful equilibrium. But he quickly realized that even the slightest unevenness would trigger a local collapse, and eventually a mega-collapse. Perhaps God would intervene from time to time and spread the stars apart. Believing that the universe is static, Newton did not consider that stars do not fall into each other because they move, orbiting the centre of our galaxy, just as the moon does not fall onto the earth because it orbits the earth. The discovery of star motion was around the corner, but the discovery of the spinning Milky Way was far in the future (Chapter 9).

Is there any evidence for the prevalence of gravity beyond the planets? After Newton, John Adams, and Jean Leverrier discovered the eighth planet Neptune through shrewd application of gravity to minute wobbles of Herschel's seventh planet, Uranus (Chapter 5). Beyond the solar system, Herschel discovered binary stars, dancing around each other, held together by the force of gravity. As it turns out, more than half the stars in the sky are binary. Bessel, who first discovered annual stellar parallax, observed anomalies in the motion of Sirius to deduce the existence of a companion star, Sirius B, later detected through more powerful telescopes. From gravitational wobbles of stars, modern astronomers deduce the presence of planets circling stars (Chapter 7). Gravity holds globular star clusters together and keeps our whirling galactic spiral from flying apart. At 30,000 LY from the centre of the Milky Way, our sun orbits the sea of suns with a period of 240 million years. Newton's version of Kepler's harmonic law reveals the mass of the Milky Way inside the sun's orbit to be 100 billion solar masses. Thus galactic mass leads to an estimate of the 100 billion star population in our galaxy. Galaxy clusters and superclusters (Chapter 9), out in the far reaches of the observable universe, hang together by the universal force. New England Puritan, Cotton Mather, part mystic, part lover of science, wrote:[17] "Gravity leads us to God and brings us very near to Him."

TAMING OF THE COMET

From observations of remote antiquity to Newton's time, cometary tales provided fascinating fodder for astrologers and doomsday prophets. Newton's understanding was to gradually bring an end to cometary superstitions.

Most scientists of Newton's period held the opinion that irregularly appearing comets must be foreign to the solar system, not governed by its laws. At first, Newton was also uncertain whether laws of gravitation applied to such erratic apparitions. In 1682 a bright new comet appeared in the night sky, moving towards the sun. By the end of the month it disappeared. But in the following month another comet appeared, this one moving away from the sun. It was even more spectacular than the first one, its tail four times as broad as the moon. English astronomer John Flamsteed (1646–1719) proposed that the two comets were one and the same. First the comet was attracted by the sun, disappeared behind it, and then it was repelled by the sun. Like Kepler, he was erroneously thinking in terms of the magnetic forces of attraction and repulsion.

Applying observations from London, Avignon, Nuremberg, Boston, Maryland, and Jamaica to his purely attractive force of gravitation, Newton determined that the complete cometary orbit must be a highly eccentric ellipse with a major axis enormous compared to any planetary orbit. Comets were not erratic apparitions, but periodic and predictable, subject to laws of motion and universal gravitation.

Once again, Newton and Halley joined forces. As a young boy, Halley developed an avid interest in comets when two appeared during dire events, one associated with the Great Plague of 1664, and another with the Great Fire of London of 1665. Diving into historical records, he studied the paths of 24 comets appearing between 1337 and 1698. On a sharp lookout for a pattern in their appearances, he found remarkable similarities in the sightings of 1456, 1531, and 1607, and the most recent 1682 comet. Most gratifying was the 75–76-year recurrence period. Could all these be one and the same comet? From Kepler's law and the 75-year period, the average orbital radius came out an incredible 17.7 astronomical units, nearly twice as large as the average orbital radius for Saturn (9 AU), the farthest planet known at the time (Figure 11.12).

If cometary orbits were closed, as his dear friend Newton suggested, Halley forecasted that the 1682 comet should re-appear in 1758. If only he could live as long to see his prophecy fulfilled! It was the first time anyone ever *predicted* a comet to appear in a specific year. Halley had the temerity to forecast that the comet would return precisely at the end of 1758, around Christmas, from a particular part of the sky on a particular orbit. If Newton's laws of gravitation and motion governed unruly comets, humans could finally banish the superstition of heavenly daggers.

Unfortunately, Halley died 16 years too early to witness his scientific forecast. As the year of the expected comet approached, a fierce competition erupted among astronomers who wished to be first to sight it. On sharp lookout, eager comet hunter Messier mistook a nebula for a comet. Determined not to repeat the error, he started a detailed nebular catalogue, now the famous Messier catalogue which confers a name and number upon galaxies, globular clusters, and nebulae. These turned out to be even more fascinating members of our vast universe than the simple comets he yearned to find. Messier spotted a dozen other comets in the process.

Returning almost exactly as foretold on Christmas Day 1758, Halley's Comet celebrated Newton's triumph, presenting to the entire world the first visible proof for Newton's laws. Comet Halley confirmed the cosmic order of God's celestial clockwork, running with laws penetrated by the human mind. And yet the superstitious retained their strange beliefs. People sealed their windows and doors to prevent the comet's advertised poison gas from entering their homes. An intimidated mob in Oklahoma was ready to sacrifice a virgin to assuage the hairy star, but fortunately was stopped by police.

Even after the crowning triumph of understanding, comet fears have never fully subsided, even to this day. The 1997 Hale–Bopp Comet triggered the suicides of 39 members of a cult in Southern California. The cult fervently believed that hiding behind Hale–Bopp was an enormous extra-terrestrial spaceship ready to destroy the planet. In Night Thoughts on life, death, and immortality, 18th-century English poet Edward Young refers to new scientific ideas about comets, but the old superstitions remain alive.[18]

Hast thou ne'er seen the comet's flaming flight?
Th' illustrious stranger, passing terror sheds
On gazing Nations, from his fiery train
Of length enormous, takes his ample round

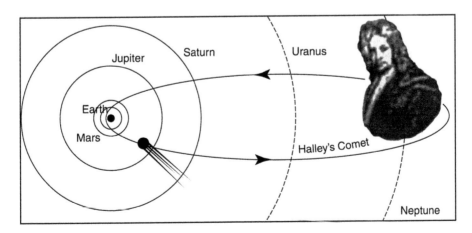

FIGURE 11.12　The path of Halley's Comet as compared to the orbit sizes of planets.[19]

FIGURE 11.13   The Bayeux Tapestry showing the appearance of Halley's Comet in 1066.[20]

> Thro' depths of ether; coasts unnumbered Worlds,
> Of more than solar glory, doubles wide
> Heaven's mighty cape; and then revisits the Earth,
> From the long travel of a thousand years.

Looking for new comets and calculating orbits became a popular activity for astronomers. The heavens captivated the world. In the United States, the predicted appearance of Halley's Comet spurred the government to grant financial support for the first time to astronomical observatories. The laws of gravity could foretell the future. Halley's Comet continues to return every 75 years, the last time in 1986; it will re-appear in the middle of 2061.

With equal agility, Newtonian mechanics could examine past cometary appearances. Halley's Comet was the one which marked the downfall of a Roman emperor in AD 218. In 1066, it hailed the victory of William the Conqueror over King Harold, commemorated as the Norman Conquest of England. A group of women artists stitched the story of the battle in 75 embroidered scenes into the Bayeux Tapestry. (What a curious coincidence they used 75 panels!) The detail (Figure 11.13) shows English King Harold, ensconced on his throne inside the castle, listening intently to a messenger who bears evil tidings. William has ordered a fleet of ships to be built to cross the English Channel. On the left, a group of men marvel at the comet [*isti mirant ur stellam*], which artists brandish as fair warning to Harold to prepare for battle or to concede. Comets were similarly believed to have foreshadowed the birth and death of Julius Caesar, and the fall of Jerusalem.

TIDES MAKE GRAVITY VISIBLE

Returning exactly as foretold at Christmas 1758, Halley's Comet was compelling evidence for Newton's theory, 30 years after the master's passing. But Newton did not wait for posthumous affirmation! He challenged his own theoretical premise to work out other powerful consequences from the inverse square law and mass dependences of the gravitational force, to be verified by direct observations. Through new physics insights, he proceeded to account for ocean tides that baffled thinkers for 2,000 years. Since tides are especially high during the full and new moon, Poseidonus of Alexandria first speculated (around 100 BC) that the moon must be responsible for tides. The lunar phase also appeared to play a role. Another connection was that high tide occurred about 50 minutes later each day, just as the time of moon-rise slipped by 50 minutes each day.

To Newton, tides provided tangible manifestation for action-at-a-distance, astronomical forces operating without need for contact. Here is his explanation for the main factors causing tides. The tidal forces arise from the imbalance of the moon's attraction across the diameter of the earth (Figure 11.14). When the rotation of the earth brings the ocean under the moon, the moon's force on the water at the surface (being closer to the moon than the earth's centre) is larger than the average force on the earth below, and the water there accelerates more (to form tides). At the same hour, on the opposite side of the earth, the average force on the earth is larger than the force on any water present at the surface, causing the earth to accelerate toward the moon more than the water. Here the water is left behind to cause a simultaneous tide. Thus the moon's gravity perturbs the shape of the earth. Any water arriving at these positions suffers bulges, giving rise to two tides per day. Since the moon changes its position as it orbits earth, a particular point on the ocean has to rotate 50 minutes longer than the previous day to return to a spot directly under the moon (or to a spot diametrically opposite) for the next high tide to occur. Hence high tides repeat every 12.8 hours instead of every 12 hours, a simple explanation for the mysterious correlation between the 50-minute time slippage of both high tide and moon-rise.

The sun plays a similar role in tide formation (Figure 11.15 (top)). Even though the sun's gravitational force on earth is 180 times greater than the moon's, the solar effect on tides is only about half that of lunar influence, even though the sun's mass is much greater than the moon's. Because the sun is much farther away, its force does not change as much across earth's diameter, the main reason for tide formation.

When the sun and moon align, both at new moon and full moon, their tidal pulls reinforce each other (Figure 11.15 (bottom)). Resulting tides are especially high, over 4 feet in the middle of the ocean. These are called "spring" tides from the German word "springen" which means to rise up. (It has nothing to do with the season of spring.) Neap tides occur when the sun and moon are about 90 degrees apart, as viewed from earth; their gravitational attractions partially cancel. Hence the intensity of tides depends on lunar phase, an intriguing correlation observed centuries ago. Neap tides are under 2 feet. Both spring and neap tides occur regularly, about once every 14 days. Due to scattering and reflections of the water off the shore, details of the shoreline can intensify the normal height of tidal waves. In some regions, tides can rise over 20 feet.

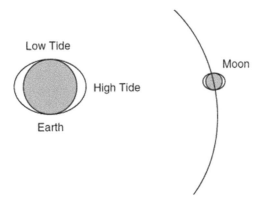

FIGURE 11.14 The moon's force on the ocean under the moon is larger than the average force on the earth, giving rise to a bulge (tide). Similarly on the opposite side, the moon's force on the earth is larger than the (farther) water, giving rise to a simultaneous second tide. Both the earth and moon assume a prolate shape. Tides occur when oceans arrive at the bulge positions with the rotation of earth.

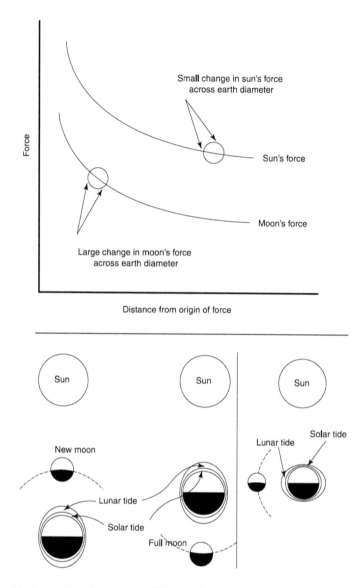

FIGURE 11.15 (Top) even though the sun is 400 times farther away than the moon, its gravitational force on earth is much larger than the moon's force, because the sun is nearly 30 million times more massive than the moon. On the other hand, because the moon is much closer to the earth, the *change* of its gravitational force across earth's diameter is much larger than the sun's. Therefore the moon has a much larger effect on producing tides. (Bottom) both sun and moon cause tides. (Left) the effect of solar and lunar tides add during a full moon and (middle) a new moon to form spring tides. (Right) the combined effect is less during the quarter-cycle moon.

From the observed height of spring and neap tides out in the ocean, Newton separated the lunar and solar contributions to obtain approximately 3 and 1.5 feet each. Having previously determined the sun's mass, he could independently assess the solar tide height from his law of gravitation to find gratifying confirmation of quantitative aspects of his far-reaching theory. Going further, Newton estimated the mass of the moon from the height of the lunar tide alone. Thus Newton provided the first estimates of the masses of the earth, moon, Jupiter, and the sun.

SHAPE OF THE SPINNING EARTH

Another puzzling aspect of earth's motion provided fertile ground for Newton to harvest quantitative evidence for his law of gravitation. Why does the tilted axis of the spinning earth precess in a cone over an immense period of 26,000 years, as Copernicus (Chapter 7) so skilfully explained? Hipparchus was the first to make the sharp discovery of equinox shifts (Chapter 6). Not only does earth spin, but it gyrates like a top (Figure 11.16). Newton provided the explanation for the top-like precession of the earth's axis of rotation. A top precesses because gravity tends to tip its bulgy half. But, since the top spins, its axis sweeps in a conical motion, rather than fall over. A perfectly spherical earth would not show such wobbly behaviour.

Could solar or lunar attraction also account for the conical precession of earth's axis (as for the tides)? Newton reasoned that if earth precesses, it cannot be a perfect sphere. Pythagoras would turn in his grave at the thought. Just as a spinning sphere of clay on a potter's wheel has a tendency to bulge at the centre and flatten at the top, the rotating earth also bulges out. Compared to earth's gravitational acceleration ($g = 9.8$ m/sec^2), centripetal acceleration due to rotation is a very small 0.0027 m/sec^2. Nevertheless, earth has gained a small equatorial bulge, and flattening at the poles since its formation, claimed Newton. Astronomer Cassini in Paris had already noticed that Jupiter, which rotates once every ten hours, has a noticeable bulge at the equator. From pole to pole, Jupiter's diameter is 6 per cent smaller than its equatorial diameter. Newton verified that the observed oblateness of Jupiter is consistent with its rotation speed, provided Jupiter's matter can yield to stretching forces. Since earth rotates, it also assumed an oblate shape earlier in its history when the substance of earth was still plastic. But, because of its lower rotation speed, once every 24 hours, earth is much less oblate than Jupiter. Newton estimated a corresponding asymmetry of about 0.4 per cent (Figure 11.16).

Newton still needed a mechanism that tugs asymmetrically at the bulge. The moon attracts the nearer side more strongly than the far side, providing an asymmetric pull that tries to straighten up the earth's axis (Figure 11.17). Its line of pull also does not act exactly through earth centre. Similarly for the sun's force of attraction. With extra lunar and solar tugs, earth's axis wobbles in a conical motion we observe as precession.

In one swift stroke, Newton solved the problem that tormented astronomers for 2,000 years. And he was quantitative. Precession rate depends on earth's spin, and mass distribution. If the earth's equatorial diameter is 55 km larger than its polar diameter, the extra gravitational tugs lead to a precessional period of 26,000 years, in agreement with observation.

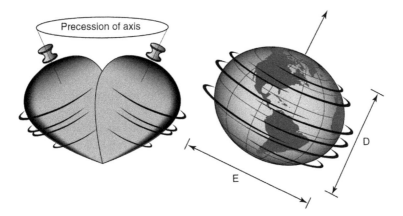

FIGURE 11.16 A spinning top precesses because of asymmetry in its mass distribution. A similar effect could occur for the tilted axis of the spinning earth, if the earth were not a perfect sphere. Newton estimated that the earth's equatorial diameter (E) is larger than its polar diameter (D) by 55 km or by 0.4 per cent.

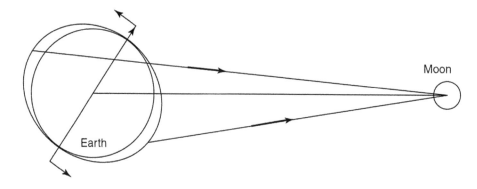

FIGURE 11.17   Since earth is oblate in shape due to rotation, the moon's attraction on the near-side of the equator is larger than for the far side, as shown by the length of two arrows. Earth's protuberance provides a lever for the sun and moon to tug at, disturbing the simple revolution and rotation, causing precession as well as other perturbations. There is also a periodic change in the obliquity of the axis due to the perturbation effect of the planets on the earth's protuberance.

Once again, nature breaks symmetry, this time due to the interaction of gravity and rotation, the same symmetry-breaking mechanism responsible for the disc shape of the solar system and the galaxies. Earth's departure from spherical symmetry also has a subtle but profound effect on weight, another opportunity for Newton to extend the reach of the law of gravitation. Because gravity force varies with distance from the planet centre, *weight is a variable quantity*. Mass, an intrinsic property of matter, remains the same. Since earth swells at the equator and shrinks at the poles, a 100-kg mass weighs 0.2 per cent more at the pole than at the equator. So also a climber's weight at sea level is slightly greater than her weight at a mountain-top. Since gravity also depends on planetary mass, weight on earth is quite different from weight on the moon, or on Mars. Weight is just one of the manifestations of gravitational mass.

A related consequence of earth's prolate shape is that g (acceleration due to gravity) varies over earth's surface. If g is 9.832 m/sec^2 at the poles, it is 9.814 m/sec^2 at the equator, a small but significant effect for accurate timekeeping (Chapter 10). Since the oscillation period depends on g, a 1-metre-long pendulum clock set accurately at the pole loses 80 seconds per day at the equator, due to the earth's prolate shape and the inverse square law. When Paris academician Richer went to South America to measure the parallax of Mars and determine the extent of the solar system, he found his time-keeping pendulum to swing a bit more slowly at the equator than in Paris. Huygens cleared up only part of the mystery: rotation of the earth slows down the pendulum by altering the value of g due to the centripetal acceleration. But Newton's gravitation law added another contribution. A pendulum loses a total of 230 seconds (nearly four minutes) per day between the poles and the equator. Earth's spin alone is responsible for two-thirds of the change in g with latitude, while the inverse square law is responsible for one-third.

Well after Newton's passing (1727), Paris scientists remained sceptical about remote gravitational forces as well as the earth's bulging shape. Followers of reputed French astronomer Jean Cassini formed an anti-Newtonian natural philosophy group. Cassini and his assistants were busy drawing a detailed earth map by careful surveying with telescopes. From some erroneous measurements, they believed the earth was shaped more like a tall watermelon than a squat pumpkin, to exaggerate the shape differences. If polar regions were flat, as Newton suggested, a degree of latitude near the poles would span a larger distance than a degree of latitude near the equator. To settle the controversy over the exact shape of the earth, French academicians sent out two expeditions, one to Peru near the equator, and another to Lapland near the North Pole. Reaching their destination in a year, the polar expedition under Pierre Maupertuis confirmed Newton's predictions for the earth's shape. One degree of latitude near the pole covered a 1 per cent longer distance along earth's surface than at the equator, vindicating Newton. A decade had passed since his death.

When news of Newton's posthumous victory reached France, the leader of the Lapland expedition received the following taunt from French philosopher François Voltaire:[21]

You have confirmed in the lands full of boredom what Newton knew without leaving home.

In biting wit, Voltaire congratulated the expedition on having flattened both the poles as well as the disciples of Cassini!

THE BIBLE OF CLASSICAL PHYSICS

In his monumental treatise *Principia Mathematica*, Newton synthesized terrestrial and celestial motion into one coherent system, an harmonious, symphonic composition embracing 2,000 years of observations and theories. In the preface he lauded the mechanistic philosophy:[22]

I wish I could derive the rest of the phenomena of Nature by the same kind of reasoning from mechanical principles.

Through gravity, the unity of heaven and earth produced explosive results. Any single one of his revelations would have been a spectacular achievement all by itself. Poet Alexander Pope captured the *Principia*'s overwhelming impact:[23]

Nature and Nature's laws lay hid in the night:
God said, Let Newton be! and all was light.

When it came time to publish the *Principia*, Robert Hooke, now a prominent figure in English scientific circles, claimed priority for the inverse square law and demanded credit. But the inverse square nature of force was just one piece of the master jigsaw puzzle. While Hooke made crude guesses, Newton formulated overarching principles, proceeding to crucial mathematical demonstrations of powerful consequences. However, as prestigious curator to the Royal Society, Hooke held significant power, which he wielded without hesitation. Halley tried hard to mediate between the two luminaries, but it was hopeless. Newton finally relented, including a passage to acknowledge how Hooke inferred certain conclusions which Newton had explained in proper measure. Even then, the Royal Society was reluctant to publish. Driven to the edge of bankruptcy after printing a lavish and unsuccessful *History of Fishes*, the Society timidly announced a lack of funds. Finally, Halley, who had inherited a small fortune from his father, decided to finance the publication himself, frequently going back and forth between printers and author. The first edition sold out quickly. After that, eager students had to copy the precious manuscript by hand. As payment, Halley accepted 75 copies of the *History of Fishes*. Determined to promote the *Principia*, he presented a copy to King James II and sent others to leading philosophers and scientists in Europe.

When the *Principia* appeared in 1687, England was embroiled in religious and civil strife. Parliamentarians executed an absolute monarch in a brutal civil war. When Cromwell died, chaos took over. Restoration of the monarchy seemed to be the only solution. Charles II returned to the throne, but with diminished powers, marking an end to the only gap (11 years) in English monarchy over a fifteenth-century history of England. Unopposed in 1685, Catholic James II succeeded Charles. Religious and political battles continued between Anglicans, Puritans, and Catholics. Because of foolhardy attempts to restore his country to Catholicism, James' reign lasted only four years. Just when the *Principia* appeared in print, the king decreed that Catholic professors be guaranteed appointment to Cambridge. To defend their right to select professors on the basis of merit alone, Cambridge sent Newton to London. Eventually Parliament overthrew James II in the Glorious

Revolution, and placed two Protestants, William and Mary, on the throne. But first they agreed to sign the Bill of Rights, limiting the power of royalty and ensuring the supremacy of Parliament. From that moment forward, the divine right of kings gave way to people's rights; monarchy slowly dissolved before rising democracy. Newton was present at the historic coronation.

In contrast to the confusion surrounding passionate issues of politics and religion, Newton's work offered welcome relief in rigorous orderliness through mathematical language and precise predictions. In the *Philosopher Giving a Lecture at the Orrery* (Figure 11.18),[24] Joseph Wright captures the intense curiosity of the lay public through the entranced expressions on the faces of children as they witness the marvels of the clockwork cosmos, following planetary paths around the magic lamp at the centre. Surrounding the eager children are thoughtful elders in rapt attention, caught up in the miracle of scientific knowledge.

THE MAN AND THE LEGEND

Like Kepler and Pythagoras before him, Newton was a mystic. Even while he wrote the most carefully reasoned sections of the *Principia*, he kept alchemical fires burning in search of some wonderful substance, like the philosopher's stone to turn mercury into gold. Lavoisier's work (Chapter 2) was decades in the future. For a man of science, it is surprising how many frustrated years he lost on alchemy, writing many books without making a single contribution to the science of chemistry. On biblical and theological topics, he wrote over a million words. Believing that God had left secret clues in the natural world, he aimed to unravel God's secrets through pure thought. But he also hunted for signs in mystical tradition, the floor plan of Jerusalem's ancient temple, and papers handed down since Babylonian times.

For a man of exalted stature and acknowledged intellectual prowess, it is amazing how many acrimonious battles Newton fought with his contemporaries over issues of priority. In later editions of the *Principia*, he vindictively deleted credit grudgingly conceded at first to Hooke for the inverse square law. Newton became embroiled in a notorious quarrel with Wilhelm Leibniz in France over who was the first to invent calculus, feeling sure that Leibniz had plagiarized from him.

FIGURE 11.18   *Philosopher Giving a Lecture at the Orrery.*

Newton waged another bitter quarrel with England's budding astronomer John Flamsteed. Newton was busy trying to solve the intractable three-body problem, the motion of the moon due to the combined gravitational pull of the sun and earth. A problem that seriously challenged his supreme intellect, the moon gave Newton frequent headaches. If successful, he would have a scheme to track longitude by predicting the moon's precise position against the stars. An astronomically based method for longitude would prove a big bonus for English merchant fleets, rapidly becoming the largest in the world. But Newton needed Flamsteed's data to check his calculations. Flamsteed was in the process of preparing better star charts to improve navigation after persuading King Charles II to build a National Observatory at Greenwich to match the fame and prestige of the Paris observatory. Methodical and meticulous, Flamsteed was too slow for Newton. From his prestigious post as President of the Royal Society, Newton kept harassing the Astronomer Royal, insisting on quick publication, ordering Flamsteed around. But Flamsteed remained adamant. Finally, working on behalf of Newton, Halley managed to acquire some of Flamsteed's data and hastily printed it. Flamsteed was furious. Observations painstakingly taken with instruments he financed belonged only to him. Newton and Halley protested that an Astronomer's Royal work was public property. Getting hold of 300 copies of the premature publication, Flamsteed burned every one.

England heaped honour after honour upon Newton, the patriarch of English science: President of the Royal Society, Warden of the Mint, Member of Parliament, first scientist to earn knighthood. The Royal Society became Sir Isaac's parliament. From his modest lodgings at Cambridge he moved into a palace in London, rode around the city in carriages, commanded servants and received a luxurious income of £500 a year. As Warden of the Mint, a position he coveted for some time, Newton took the job so seriously and was so effective that he became a terror to counterfeiters, whose hangings he personally attended. A popularly elected Member of Parliament, Newton served for several years. But he never spoke. During one session, when Newton rose, the House fell hushed, eager for a chance to hear from the great man. But all Newton desired was that someone close a window to stop a draught!

When Newton died at 82, a grand funeral preceded burial in Westminster Abbey with all the pomp and ceremony accorded kings, queens, and heroes of England. A museum preserves a lock of his hair and a piece of the legendary apple tree's trunk. Newton's epitaph reads:[25]

Let the mortals rejoice that such and so great an honour of mankind ever existed.

But the man who culminated the scientific revolution of the 17th century graciously recognized the foundations laid down by his predecessors with modest words:[26]

If I have seen further than other men, it is because I stood on the shoulders of giants.

Newton eloquently expressed his diffidence for nature's wonders in the quote at the beginning of this chapter.

THE AGE OF REASON

With accompanying advances in telescopes, astronomers detected small shifts from ideal elliptical orbits. Jupiter's orbit appeared to be shrinking ever so slightly, and Saturn's orbit was expanding. Halley noticed that lunar speed was slowly increasing over accepted values. An interesting anomaly cropped up over the short period of a comet, called Encke, orbiting the sun once every 3.3 years. With every pass near the earth, it arrived a few hours earlier. Was it defying the laws of gravity?

Perturbations from ideal orbits were all extremely small, but nonetheless worrisome. What would be the cumulative result over millennia? If such disturbances were typical for all heavenly bodies, would the moon eventually fall to earth, and earth into the sun? The stability of the entire solar system was in jeopardy if perturbations were any larger than calculated. When Newton discussed deviations arising from the influence of planets upon each other, he expressed concern over their continuous accumulations. Could planetary irregularities eventually throw the entire solar system into total chaos? Perhaps the solar system might need occasional correction by God. From Paris, Leibniz attacked Newton for picturing the Creator as a clumsy watchmaker. Whereupon Newton countered:[27]

> God ... not only composes ... but is himself the author and continual preserver of their original forces ... and consequently 'tis not a diminution, but the true glory of his workmanship, that nothing is done without his continual government and inspection.

Nearly a century later, French mathematician Simon Laplace tackled the knotty problem of planets cross-pulling each other. Determined to understand whether the solar system is stable or headed towards ultimate catastrophe, Laplace examined new deviations in planetary motion to show that perturbations are not cumulative but periodic. Changes of the solar system nearly repeat themselves at regular intervals (roughly every 1,000 years), and never exceed a moderate amount. There was no risk of the moon crashing into the earth. Obeying Newtonian gravitation, the solar system was perfectly stable. Newton's gravity was immortal. Precise regularity of the clockwork universe restored absolute faith in Newtonian mechanics. With determinist vision, Laplace boasted:[28]

> Given ... a mind which would comprehend all forces by which nature is animated and the respective situation of the beings who compose it and a mind sufficiently vast to submit these data to analysis, it could embrace in the same formula the movement of the greatest bodies of the universe and those of the lightest atoms; for it nothing would be uncertain and the future as the past would be present to its eyes. The human mind offers in the perfection which it has been able to give to astronomy, a modest example of such an intelligence.

Assembled in a giant treatise, Laplace's *Mecanique Celeste* became the *Almagest* of the 19th century. On receiving many volumes of the great work, Emperor Napoleon protested:[29]

> Newton often spoke of God in his book. I have looked through yours, but I did not see his name once.

Whereupon Laplace arrogantly replied that he had no need for that hypothesis! Natural law was supreme.

Mechanics emerged as the first exact science. Newton forged the sun, moon, and stars together with the hammer of powerful reason, shaping the course of the heavens through precise mathematics. Authoritative triumphs generated supreme optimism. The universe is fundamentally orderly. There is a rational author of nature with a grand design which the human mind can penetrate and cast in precise mathematical terms. Guided by the clear and resounding voice of pure reason, the tenor of questions asked by succeeding generations of scientists changed sharply. Newton's approach became the prototype method for all scientific inquiry; mechanics and mathematics were the models to emulate. Could other fields of human endeavour eventually yield to intellectual understanding, perhaps even a Newtonian solution? Pioneers attempted to measure up to the new standard of mathematical rigour introduced by the *Principia*. In *Wealth of Nations*, Adam Smith grappled with laws of economics so that policies could be defined on a rigorous basis, approaching Newtonian science. In writings such as a *Sketch for an Historical Picture of the Progress of the Human Mind*, Marquis de Condorcet argued that culture was governed by laws as exact as those of physics.

Rationalists argued that if scientific laws, not divine command, governed natural motion in the universe, there must be natural rights for humanity and natural laws that governed society, not arbitrary rules of monarchs. Freethinkers, like John Locke, declared that men have the natural right to life, liberty, and property. The Age of Reason engendered bold and exciting prospects of freedom and equality, calling into question the divine right of kings to rule people. The king was human. Since government was established not by the power of God through kings but by people, it was meant to protect their *natural* rights and to rule according to *natural* laws. If government failed to fulfil this fundamental role, people had the right to revolt, to overthrow their government, and establish a new one. Freeing themselves from the yoke of absolute monarchy, the English had successfully adopted a constitution which held the king responsible to the nation by written contract. France was poised to do the same. Everywhere rationalism propelled restless men to demand change.

Locke and Newton were personal acquaintances. Locke's ideas about natural law and natural rights inspired Thomas Jefferson when he drafted the Declaration of Independence:[30]

> When in the course of human events it becomes necessary for one people to dissolve the political bands which have connected them with one another, and to assume among the powers of the earth, the separate but equal station to which the *Laws of Nature* and of Nature's God entitle them, a decent respect to the opinions of mankind requires that they should declare the *causes which impel them* to the separation. – We hold these truths to be self-evident, that all men are created equal, that they are endowed by their Creator with certain *unalienable Rights,* that among these are Life, Liberty and the pursuit of Happiness.

Ideas of immutable natural laws, natural rights, and rational causes clearly influence Jefferson's eloquent expression and defence of principles of freedom, equality, and happiness. With new-found confidence, he invokes causes that impel men, just as forces impel planets. Around the same time, Benjamin Franklin published an essay entitled "On Liberty and Necessity; Man in the Newtonian Universe."[31] Political dynamics was mirroring Newtonian dynamics. Closer to our time, Hermann Bondi (1919–) reflected on the effusive optimism generated by Newton's triumphs:[32]

> His solution of the problem of motion in the solar system was so complete, so total, so precise, so stunning, that it was taken for generations as the model of what any decent theory should be like, not just in physics, but in all fields of human endeavour. It took a long time before one began to understand – and the understanding is not yet universal – that his genius selected an area where such perfection of solution was possible.

JUST THE FACTS

Not all scientists of Newton's time were ready to accept the sweeping theory. Just a simple mathematical recipe asserting a force binding planets to the sun seemed barren. Cassini, Huygens, and Leibniz had profound conceptual difficulty with the concept of action-at-a-distance, akin to a spiritual force. How could forces transmit without contact? Newton's gravity brought occult qualities back into physics. What was the cause of gravitation? Capturing the uneasiness accompanying the mysterious force, one of Newton's many portraits carries the disquieting caption:[33]

> See the great Newton, he who first surveyed
> The plan by which the universe was made;
> Saw Nature's simple yet stupendous laws,
> And proved the effects, though not explained the cause.

At the end of the *Principia*, Newton stuck to the facts:[34]

> Hitherto we have explained the phenomena of the heavens and of our sea by the power of gravity, but have not yet assigned the cause of the power ... I frame no hypothesis; for whatever is not deduced from the phenomena is to be called an hypothesis; and hypothesis whether metaphysical or physical, whether of occult qualities or mechanical, have no place in experimental philosophy ... And to us it is enough that gravity does really exist, and act according to laws which we have explained, and abundantly serves to account for all the motions of the celestial bodies, and of our sea.

Satisfied to present the facts and show how the theory worked, he decided to "feign no hypothesis" for the vexing problem of gravity's origins. Realizing the limits of his science, he was not ready to speculate further about matters beyond proof, a position strongly reminiscent of Galileo's at the base of the legendary tower of Pisa, when he realized that free-fall acceleration is independent of mass. Why, he did not know. He was glad to take one significant step toward ultimate understanding. Yet, despite his humble protests, Newton did indulge in much creative speculation, guided by aesthetics, to reach the law of universal gravitation. But he successfully followed these hypotheses with testable predictions.

Privately, Newton expressed doubts about action-at-a-distance in correspondence with a friend:[35]

> It is inconceivable that inanimate brute matter should, without the mediation of something else, which is not material, operate upon, and affect other matter without mutual contact.

Eventually some agency might be found to explain universal gravity.

With mutual planetary perturbations properly factored in, Newton's laws gave a thorough account of all planetary orbits. One notable exception remained. Mercury's orbit continued to show a minute anomaly defying gravitational calculations. Rather than trace the same path on each revolution, Mercury's ellipse rotates slowly in space (Figure 11.19) precessing around the sun to make a complete cycle in 226,000 years, more precisely at a rate of 574 arc-seconds per century. Planetary perturbations account for 93 per cent of the observed precession, leaving an error of 43 arc-seconds per century. Mercury always runs ahead of gravity's predictions. It was a small discrepancy, but Kepler had taught astronomers that such deviations could be just the tip of a massive iceberg lurking beneath the surface to tear apart old conceptions. Earlier, a small anomaly in Uranus' orbit led Adams and Leverrier to predict an eighth planet, Neptune. And errors in Neptune's orbits were symptoms of Pluto's presence. Confident that the law of gravity would prevail once more, Leverrier tackled the conundrum of wayward Mercury. Perhaps there was an invisible planet hiding between Mercury and the sun; Leverrier made the calculation and picked out the name Vulcan in anticipation. Even as astronomer's telescopes improved, no Vulcan was to be found. Was Mercury opening cracks in Newton's powerful theory?

UPDATES

A Happy Thought – The General Theory of Relativity

Combined with the force law, Newton's law of gravitation explains the astonishing result of Galileo's legendary experiment at Pisa: as the mass of an object increases, so does the force of gravitation.

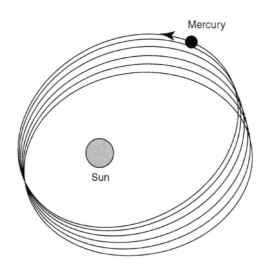

FIGURE 11.19   Slow precession of Mercury's elliptical orbit.

Hence acceleration remains constant, a strikingly simple account. Yet it holds a crucial clue to deeper questions about the nature of gravity and mass. According to Newton, gravitational mass, which appears in the law of gravity, and inertial mass, which appears in the force law, are *identical*. But the two laws describe very different physics. Inertial mass is an innate property intimately connected with motion, describing how a body moves under the influence of a general force. Gravitational mass determines the intensity of gravity force. Little did Newton know what a monumental hint he left behind, one which every scientist continued to ignore, until the arrival of Einstein.

Meditating on how Newtonian gravity might be adapted to the framework of special relativity, Einstein had an inspiration which he called[36] "the happiest thought of my life." In 1907 he recognized a deep significance to Galileo's epiphany that all bodies fall with the same acceleration in the presence of gravitation. It was not a quirk of nature. In that euphoric moment, Einstein glimpsed an epoch-making synthesis: the effects of gravity and acceleration cannot be distinguished from one another.

Einstein imagined a person inside an elevator in free fall who conducts Galileo's experiment (Figure 11.20). From the passenger's point of view, an apple he releases remains stationary. Elevator, occupant, and apple, all fall freely with exactly the same acceleration. Gravity is in effect turned off. In free-fall acceleration, there is no weight. If the elevator passenger stands on a scale, it registers zero.

An astronaut in a satellite orbiting earth is also weightless. Objects inside the satellite seem to float around freely, as space-shuttle astronauts playfully demonstrate on television. According to Newton's epiphany in the garden, both astronaut and satellite are effectively in free fall, like

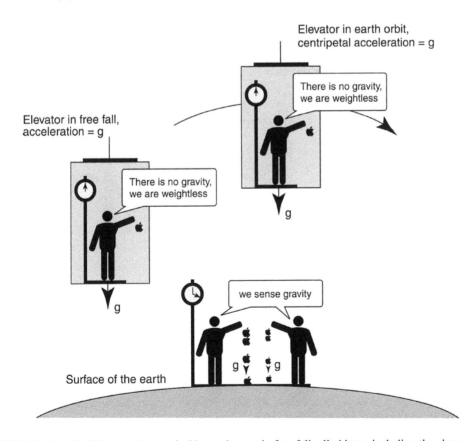

FIGURE 11.20 (Left) for an observer inside an elevator in free fall, all objects, including the observer, are weightless. (Bottom) on earth, all objects have weight and fall freely with the same acceleration (*g*), independent of mass. (Top) for an observer inside an elevator in orbit around earth, all objects are again weightless as for the elevator in free fall.

apples on earth. But they maintain a constant distance from earth because of their circular motion. Gravity can cause either vertical free fall or centripetal acceleration for circular motion. In a creative impulse, Einstein recognized there is no way to tell *from inside a free-falling elevator or from inside a satellite* whether gravity has been turned off or whether the satellite and all its occupants are accelerating toward the centre of earth.

We have seen earlier hints of an intimate link between acceleration and gravity. Huygens' glimpse of a key connection between free-fall acceleration and centripetal acceleration explains why pendulum clocks run slower at the equator than in Paris. Since the value of g becomes less due to higher centripetal acceleration at the equator than at the pole, the oscillation period of the pendulum increases. At the edge of imagination, Magritte's *Castle of Pyrennes* becomes weightless if earth speeds up its rotation by 16 times.

What if there is no earth? Einstein imagined an elevator out in free space, *accelerating up* (towards the ceiling of the vessel) at $g = 9.8$ m/sec². Inside is a volunteer carrying out Galileo's famous experiment. Before departure from earth, she is drugged into stupor, so she does not know she has left earth. When she wakes up in outer space and releases an apple, she sees it accelerate to the floor. She thinks she is still on the surface of planet earth. If she agrees never to look outside her capsule, she cannot tell that there is no planet. She thinks the apple falls because of gravity. But a stationary observer outside the elevator knows there is no earth under the elevator, and no gravity for her: the floor accelerates up to the apple. Gravity for the elevator astronaut inside is equivalent to acceleration for an observer outside. Einstein elevated this delightful conclusion to a general *equivalence principle*: constant acceleration is equivalent to uniform gravity at every point in space.

Here is the pleasant consequence of equivalence which Einstein appreciated. From the point of view of an outside observer, for every object which the elevator astronaut releases, independent of its inertial mass, acceleration will always be g; because g is the elevator's acceleration (Figure 11.21). From the point of view of the elevator occupant, who thinks she is on the surface of earth, all objects

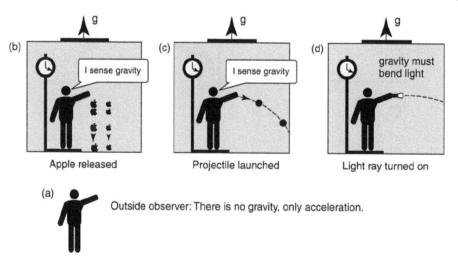

FIGURE 11.21   (a) An outside observer sees an elevator accelerate upward in free space with acceleration g. (b) All objects, irrespective of their mass, approach the floor with acceleration g, because the elevator moves upward with acceleration g. But an observer inside the elevator (who agrees not to look outside) thinks she senses gravity, as if she is on the surface of earth. All apples fall with the same acceleration g because of the nature of gravity. (c) Projectiles follow parabolic paths. According to the outside observer the parabola results from a combination of the elevator's upward acceleration and the horizontal projection velocity. According to the elevator passenger, the parabola results due to a combination of gravity fall and horizontal motion. (d) A light beam also follows a parabolic orbit. According to the outside observer, the light path is a combination of upward motion of the elevator wall and horizontal light velocity. According to the elevator passenger, the light bends because gravity must bend light, a new prediction from the equivalence of gravity and acceleration.

fall with the same acceleration g because that is a quirk of gravity which Galileo discovered and Newton explained. Einstein's inescapable conclusion was that there was no need to add a postulate that gravitational mass is identical to inertial mass. If gravitation is fundamentally equivalent to acceleration, gravitational mass must be the same as inertial mass. Einstein unified inertia and gravity. Once again, aspects of nature that we perceive as distinct are really alternate manifestations of the same reality.

Are there any new consequences of such an elegant unification? Remember that a better theory must not only embrace known behaviour but must make testable new predictions. Instead of dropping a stone, if the elevator astronaut throws it horizontally, she sees it follow Galileo's curved parabolic path due to a combination of gravity fall and inertial horizontal motion. The stone hits the opposite wall at a point lower than her hand. For a high projection velocity the landing point is only slightly lower than the launch point, because the time of horizontal flight is less. An outside observer (no gravity) interprets the parabola to result from a combination of the stone's inertial horizontal velocity and the upward acceleration of the elevator floor. But what if the astronaut flashes a pin-point sharp ray of light to the opposite wall? Again, the outside observer sees a light spot at a point very slightly below, since the entire elevator accelerates up during the time of flight of the light ray.

Here is the brand-new consequence of the equivalence principle. If gravity is equivalent to acceleration, gravity must also deflect light. The elevator astronaut, who thinks she experiences a planet's gravity, must also see light follow a bent path, like a stone. Gravity bends light! According to Einstein's special relativity, light energy is equivalent to mass through $E = mc^2$. If light has mass, then gravity attracts light, since gravity acts on all masses. But it takes a very large gravitational force to make a significant deflection. Einstein made the bold prediction that the sun's enormous gravity should bend starlight by 0.87 arc-seconds when light grazes its surface. It is a small angle, but comparable to annual stellar parallax of the nearest stars, and well within an astronomer's detection capabilities. Einstein's deduction about the bending of light was just a stepping stone to something far grander.

CRUMPLED SPACE AND WARPED TIME

A short review of special relativity is in order. Einstein took Galileo's profound idea about the invariance of physical laws to its ultimate conclusions. The laws of physics must be the same on a moving vessel as they are on a vessel at rest, as long as the ship moves with constant speed in one and the same direction. Einstein adopted the invariance of light speed as a physical law to uproot deep-seated notions about space, time, energy, and mass (Chapter 4). Space and time must change for moving observers, not the speed of light. Moving clocks run slower. There is no absolute flowing river of time; the only time we know is the time we measure by the clocks we carry. Moving rulers become shorter in the direction of motion. Perpendicular to the direction of motion, rulers do not change. Although participants describe observations differently, they agree to one set of rules. Under special relativity, concepts never before enjoined come together. Special relativity links mass and energy. Einstein's former mathematics teacher, Minkowski, knitted space and time inextricably together as aspects of a single reality, called spacetime.

In the framework of special relativity, Einstein found much that was disconcerting about Newton's law of gravity, stipulating instantaneous action-at-a-distance. If you stir a flower, in the words of poet Francis Thompson, you cannot immediately change the course of stars. Gravity forces cannot be transmitted instantly. Gravity force cannot travel faster than light speed. If the sun ceases to exist, everything on earth will still continue undisturbed for eight glorious minutes, until the devastating effects arrive at light speed. Gravity has to conform to consequences of special relativity. The masses and distances in Newtons gravity equation are both relative; they depend on the velocity of the observer.

Generalizing the idea that laws of physics must be the same in all frames of reference, Einstein wished to extend special relativity to the case where observers can be accelerating with respect to each other. Through acceleration he set out to make gravity consistent with special relativity. Acceleration is easier to understand than mysterious gravity. Einstein proceeded to examine the effect of special relativity on a rotating system as a typical accelerated frame of reference (Figure 11.22). A clock on the rim of a spinning disc moves at a higher velocity than a clock at centre, and should therefore lag behind in time according to the conclusions of special relativity. Therefore acceleration warps time. By the principle of equivalence, gravity must also warp time, a startling consequence of general relativity. A clock at the bottom of a tower, where gravity is stronger near to earth, should run slower than a clock at the top. Satellite clocks far above earth's surface should run faster than earth clocks. Stunning new conclusions.

Einstein advanced to relativity effects on space. Spin a disc of diameter D to rotate at high speed (Figure 11.22). A metre ruler on the outer rim is shorter due to relativistic contraction, but one placed anywhere along any diameter remains 1 metre long, since the ruler is perpendicular to the direction of motion. If an inhabitant on the spinning disc measures the diameter, he will report the value D, the same as for a stationary disc. But when he marks out the outer circumference, he will report a value larger than πD, since his ruler is shorter as he goes around the circumference. How can the ratio of circumference to diameter be larger than π, the sacred number characterizing the purity of a circle? *Because Euclidean geometry breaks down in a rotating system, Einstein realized.* From the equivalence principle, the same novel effect should be valid for gravitation. In the presence of gravity, geometry becomes non-Euclidean. Gravity must curve space to give non-Euclidean results. With general relativity, Einstein breathed life into Newton's rigid space and rigid time.

Einstein needed to calculate exactly how gravity curves space. He realized from the rotating disk example that gravity distorted Euclidean geometry. But what would the new geometry be? About a century earlier, German mathematician Carl Friedrich Gauss had studied one type of non-Euclidean geometry for two-dimensional curved surfaces, such as the surface of a sphere, to find remarkable properties that Einstein became aware of. According to Gauss, distance and angle measurements on the surface make it possible to determine the curvature of a two-dimensional surface without hopping out into the third dimension. Euclidean space, like a flat sheet of paper, has characteristic defining features: the circumference of a circle of radius R is always $2\pi R$, the distance between parallel lines is always constant, and the sum of angles in a triangle is always exactly 180 degrees. But on the surface of a sphere, none of these features hold true. Longitude

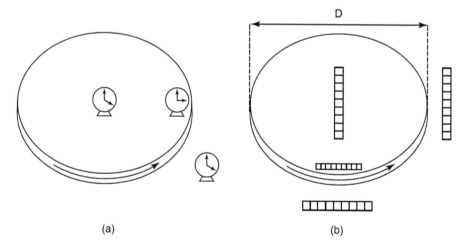

(a)   (b)

FIGURE 11.22 (a) A rotating disc is an accelerating system. Due to relativistic time dilation, a moving clock on the outer rim runs slower than a clock at the centre, or a stationary clock outside the disc. (b) Due to relativistic distance contraction, a moving metre ruler parallel to the outer rim is shorter than a metre ruler placed along the diameter or a stationary metre ruler outside the disc. Because the moving ruler is shorter, a measurement of the circumference of the moving rim will give a value larger than πD. Acceleration warps space.

lines appear parallel at the equator, but the distance between them gets shorter, and they all meet at the poles. The sum of angles in a triangle (Figure 11.23) is greater than 180 degrees. And the circumference of circles can be less than $2\pi R$. Suppose a person on the North Pole, who does not know about the curvature of earth, walks a short distance south to latitude 80 degrees, and calls that distance along the surface R. His value of R is larger than the true radius of the latitude circle as measured on a three-dimensional sphere. From flat-earth, Euclidean geometry, he expects the circumference of any circle to be $2\pi R$. But when he travels around the latitude circle and measures the circumference, he finds a shorter result. He concludes, from his distance measurements, that he must live on a curved surface.

Embracing Minkowski's hybrid spacetime as the basis of a geometric treatment of gravity, Einstein recognized that without the inclusion of matter, gravity, or acceleration, Minkowski's spacetime is flat and Euclidean. Everything moves at constant velocity in straight-line paths. Such paths also appear straight to those who live in a Euclidean world. But matter curves spacetime, just as a person who sits on a mattress deforms it locally. Cannon-balls and planets still follow inertial paths, called *geodesics*, through curved 4D spacetime country. Thinking about 3D space as flat, Euclidean folks see geodesics as curved paths, and attribute them to an invisible force, such as gravity. For visualization in lower dimensions, consider two objects moving along the surface of a sphere, following "straight" lines of longitude. They slowly come together until they meet at the pole. People who live on such a surface, but only recognize flat-earth-Euclidean geometry, are bound to come up with a theory that there is a gravity-like, attractive force pulling objects together.

Paraphrasing cosmologist Archibald Wheeler's succinct description of general relativity, mass tells spacetime how to curve, and curved spacetime tells mass how to move. There is no reason to invoke mysterious forces, like mutual gravitation. With a leap of imagination, Einstein banished action-at-a-distance. The inverse square law is simply a geometric property of spacetime. One mass moves freely along contours through the curved spacetime of another mass. Geometry reigns supreme once again, but it is non-Euclidean. The sum of three angles of a triangle drawn around the sun is less than 180 degrees!

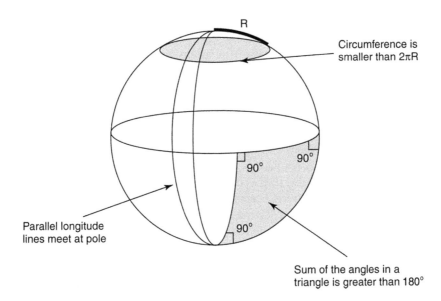

FIGURE 11.23 Non-Euclidean geometry on the surface of a sphere. Two longitude lines are parallel at the equator but meet at the poles. The sum of three angles on the triangle shown is greater than 180 degrees. The measured circumference of a circle turns out to be less than $2\pi R$, if R is the distance *along the surface* from the North Pole to the circle.

Space and time are not just a defining static framework. Both participate in dynamics. Space curves around massive earth so that a near-satellite follows an inertial path through warped space. Occupants feel no gravity because they follow inertial paths. No force is necessary to maintain orbits because satellites follow their natural states of motion through curved spacetime. But we see the satellite path as a circle. Similarly a planet follows a natural, inertial path, obeying Newton's first law of inertia, but through curved spacetime. There is no gravity binding planets to the sun. Planets travel "straight" paths in 4D spacetime but they appear to us to move in ellipses in 3D Euclidean space. On earth, a cannon-ball drops to the ground from a tower because space around massive earth is curved; the ball follows its natural path, like following the curvature of a bowl to the bottom. *Gravity is a fictitious force.*

As he started to tackle the knotty geometry of curved spacetime, Einstein pleaded with his mathematician friend[37] "Grossman, you must help me, or else I'll go crazy." As a measure of the complexity, while it takes only the radius to specify the curvature of a circle, curved 3D space takes six numbers, and 4D spacetime takes ten numbers. Being an excellent mathematician, Marcel Grossman knew that Bernhard Riemann had constructed a non-Euclidean geometry 30 years after Gauss to deal with curved higher dimensional spaces, now called Riemann spaces. Minkowski's spacetime is just one of the Riemann spaces. Curvature of 3D space is also an intrinsic property that can be detected by measurements in space without jumping out into the fourth dimension. All the mathematics Einstein needed already existed, like Apollonious' abstract conic sections were ready-made for Kepler to apply to real planetary orbits. Grossman taught Einstein the mathematics he needed. It was a veritable feast of new techniques for encapsulating Einstein's soaring ideas. With that bit of help, Einstein was on his way to show that gravity is a manifestation of curved spacetime geometry. The resulting equations of general relativity work out how much curvature a massive body produces. But it was a long hard struggle demanding eight years of intense and agonizing work. As for Newton, the gravity problem consumed Einstein's phenomenal powers of concentration. And all around him, instead of civil war surrounding Newton, the entire world was at war (World War I).

At the end of the exhilarating voyage, Einstein's mathematical law gave the same predictions as Newton's, except for very small corrections. All planetary ellipses rotate very slightly because of perturbations from other planets. But Mercury does not keep time with Newton's law of gravity. Its orbit precesses 574 arc-seconds per century, whereas planetary perturbations account for 531 arc-seconds. Einstein's general relativity triumphantly explained the extra 43 seconds of arc. On clearing up the mystery that baffled astronomers for nearly two centuries, Einstein felt such tremors of excitement and heart palpitations that he could hardly sleep for days.

The general relativity correction is largest for Mercury because it is closest to the sun and has a high velocity associated with its large orbital eccentricity. Modern studies show that ellipses of the next two planets after Mercury, Venus and earth, also advance due to the curvature of spacetime in agreement with general relativity predictions. Newton's law remains an excellent approximation, expedient for use in all practical cases, such as detecting extra-solar planets from the slight wobble of parent stars. General relativity effects become dominant only when the mass of astronomical bodies becomes enormous, such as for a neutron star.

Light was the primary actor in the special relativity drama. Now gravity was the primary actor in the general theory. With the complete theory, Einstein eliminated the contradiction between gravity's instantaneous action-at-a-distance and special relativity's light speed limit. When a person gets up from a box-spring mattress, the springs vibrate for a short time before returning to a flat profile. Rapid movements of mass in spacetime create undulations that ripple through space. Just as Maxwell's equations predict electromagnetic waves from oscillating charges, Einstein's equations lead to gravity waves that originate from vibrating masses and propagate like electromagnetic waves at light speed. The search for gravity waves was on.

NEWTON ECLIPSED

Revisiting the bending of starlight grazing the sun's surface, Einstein calculated a new deflection of 1.75 arc-seconds, twice the value from special relativity alone, due to the deformation of space from the sun's enormous mass. Such an angle could be measured, Einstein predicted, during a total solar eclipse, when the moon blocks out the blinding sunlight. It was time to put general relativity to a definitive test.

Eclipses have revealed startling aspects of nature throughout the history of science. Lunar eclipses first revealed to Aristotle solid evidence for the curvature of earth. Lunar and solar eclipse geometry helped Aristarchus to put a first scale to the earth–moon and earth–sun distances. Now a total solar eclipse was about to reveal the curvature of space. Newton's laws predicted the exact time on May 29, 1919 and the best location for the spectacular appearance of stars. Fortuitously, one of the background constellations was Hyades, a dense cluster of stars bright enough to see in eclipse twilight. It would be a formidable task, for totality was to be a tantalizingly short minute or two. British astronomer Sir Arthur Eddington organized an expedition to Principe Island off the Atlantic coast of Africa in 1919. Comparing correct star positions in the absence of the sun with apparent positions during the eclipse, Eddington's photographs measured the predicted shifts (Figure 11.24). Einstein's eclipse prediction came true. Newton's world eclipsed in the shadow of Einstein's.

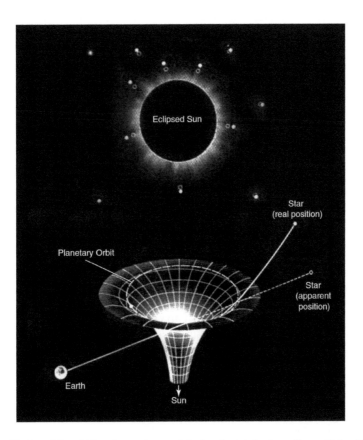

FIGURE 11.24   (Top) position shifts of stars detected during the total solar eclipse of May 1919. Stars closer to the sun show larger shifts. (Bottom) a two-dimensional model of how mass curves space. Light beams and planets follow "straight" paths through curved space, but paths appear curved to us in flat Euclidean space. Planets follow their natural curved paths in spacetime. Gravity is a manifestation of spacetime geometry.[38]

A few months later, Eddington announced the epoch-making result to the Royal Society as one of the greatest achievements in the history of human thought. Present at the meeting, Alfred North Whitehead reports:[39]

> There was dramatic quality in the very staging – the traditional ceremonial, and in the background the picture of Newton to remind us that the greatest of scientific generalizations was now, after more than two centuries, to receive its first modification … a great adventure in thought had at length come safe to shore.

The discovery hit front pages of the world's newspapers. "Light's All Askew in the Heavens," headlined the *New York Times*.[40] Upon receiving the congratulatory telegram, Einstein was surprisingly calm. Puzzled by the mild reaction, his secretary asked "What if the experiment had failed to confirm your predictions?" To which Einstein quipped without hesitation[41] "Then I would have felt sorry for the Lord, for the theory is correct."

Eddington's sensational announcement seized the scientific world by the collar. But very few scientists could comprehend the appallingly difficult mathematics. At one lecture, a dumb-struck listener asked Eddington if he was one of the only three people in the world who really understood general relativity. After a long and dramatic pause, he retorted[42] "I am wondering who the third person is." Since then, general relativity has withstood the intense glare of experiments, some testing its accuracy to a few parts in 10,000. Radio telescopes detect bending of radio waves from stars as they graze by the sun, but eclipses are no longer necessary for such experiments since the sun hardly emits any radio waves. On its way to earth from the edge of the observable universe light from dazzling quasars bends as it passes by massive galaxies. A single quasar splits into multiple images, the most spectacular forming four ghost images, known as Einstein's cross.

A light-emitting atom is a very precise clock since it emits light of a very specific frequency (Chapter 12). Intense gravity slows down a clock and lowers the frequency, shifting the wavelength of a characteristic spectral line towards the red end of the spectrum, an effect known as *gravitational redshift*. An atom on the surface of the sun would emit slightly longer wavelengths than the same atom on earth. Atomic clocks are sensitive enough (Chapter 10) to measure minute time differences due to the stronger gravity on the street as compared to the penthouse of a skyscraper. On global positioning satellites (GPS) hovering above earth's surface, atomic clocks run *faster* than earth clocks by 46 microseconds per day, a huge effect, compared to special relativity (Chapter 10) which *slows down* satellite clocks by 7 microseconds per day due to their orbital velocity. If GPS systems ignored general relativity, they would fail within one hour.

Still missing was direct detection of gravity waves (see last section). Within a few years after Maxwell's triumphant theory, Hertz produced electromagnetic waves and Marconi beamed them across the Atlantic (Chapter 3). But physicists have been looking for the elusive gravity waves for a long time. There is indirect evidence discovered 20 years after Einstein's passing. Ultra-massive neutron stars (Chapter 8) rotating around each other lose energy through gravity waves; their pulse rate slows down just the right amount predicted by general relativity.

Guided by the lamp of symmetry illuminating the long road from Pythagoras to Newton, Einstein made laws of physics the same for all inertial observers to unify space with time, and matter with energy. Inspired to generalize the aesthetic principle to accelerated systems, he weaved matter-energy into the fabric of spacetime. Galileo's inertia, free-fall acceleration, and Newton's mutual gravitation, all became fundamental properties of space and time. Grasping reality with pure thought, Einstein realized the dream of ancient philosophers. Left undisturbed, heaven and earth continue natural, inertial motion in fundamental beauty. One rule governs all.

I Told You So, Gravity Waves

1.3 billion light years away, two black holes ran into each other in a violent and destructive collision generating strong gravitational waves that weakened to miniscule distortions as they travelled the vast interstellar distance to reach earth on September 14, 2015. The amount of spacetime wobble they generated was thousands of times *smaller than the nucleus of an atom*! This is a change in length of approximately one part in 10^{21}. Such inconceivably small deformations of space are what the Laser Interferometer Gravitational Wave Observatory (LIGO) was judiciously designed to observe. As the space shivers from the massive black-hole encounter journeyed to earth, they stretched space in one direction and squeezed it in the orthogonal direction. LIGO had just finished upgrading its detectors by a factor of three over five years and come back on line just two days earlier. The distortions matched Einstein's century-old predictions of general relativity for the cosmic merger and subsequent "ringdown" of a single black hole. If he were alive today, he would have the absolute right to say: "I told you so" (Figure 11.25). A few months later in December LIGO observed a second superb event. In 2017, the discoverers received the Nobel Prize in physics. LIGO's discovery will remain one of humanity's greatest scientific achievements.

The LIGO "interferometer" shoots powerful lasers down two long (4 km) tunnels. The beams reflect back and forth 280 times before exiting the tunnels to arrive at a sensitive detector. As long as space is perfectly undisturbed the waves exactly subtract (destructive interference) to give zero light signal at the detector. Any passing gravitational waves will distort spacetime, making one tunnel slightly longer than the other, throwing off the exact coincidence between the two returning beams and producing a signal. The pattern of this signal provides information about the incident gravitational wave and its source. LIGO has two observatories separated by 3,000 kilometres to act as a check on each other, one in Hanford, Washington state, and the other in Livingston, Louisiana. Since gravity waves travel at the speed of light, the difference in arrival times at the two locations is about ten milliseconds.

FIGURE 11.25   Someone placed an appropriate banner on Einstein's famous statue at Princeton (*The Shirk Report* – Volume 357).

A third, independent instrument called Virgo, near Pisa, Italy, became operational in 2017.

By January 2019 LIGO and Virgo reported new ripples that bring the total gravity wave detection counts to 11. One of these came from the merger of two neutron stars confirmed by the orbiting Fermi telescope which saw a burst of electromagnetic radiation at the same time as the LIGO and Virgo detectors picked up a signal.

Once again, Einstein was completely vindicated!

12 Quantum Wonderland

ASCENDANCY OF DETERMINISM AND CAUSALITY

Newton replaced the supernatural order in the heavens with a mechanical one. He captured the laws of motion mathematically with moving masses and forces, expecting to penetrate the grand design of the universe on all scales and in all regions. Applying Newton's laws allow us to send people to the moon, and spaceships to the edges of the solar system.

Newton's success dismissed belief that animistic forces and occult powers are responsible for natural phenomena. The universe is not ruled by ghosts and phantoms that pop up and disappear unpredictably. Newton's world of simplicity, elegance, and intelligibility made it reassuringly beautiful. The universe is certain and predictable. Laplace's boast (Chapter 11) captured the spirit of determinism inaugurated by the successes of Newtonian mechanics. With determinism extrapolated to the extreme, everything has been predetermined from the beginning of time. The present state of the universe is the effect of its antecedent state. The present state of the universe is the determined cause of all that is to follow. 11th century Iranian poet Omar Khayyam[1] captured the powerful notion:

> And the first Morning of Creation wrote
> What the Last Dawn of Reckoning shall read.

The universe is just playing itself out. Even the result of coin flip should be predictable. If all the starting conditions are *precisely* known, and outside interferences, such as air temperature and air resistance included, the laws of physics can certainly predict the result of a coin toss. Nothing occurs at random. Everything has a cause. The cosmos is a machine moving in predictable ways.

Determinism explains any process as being determined by certain causes which make the process predictable. Causality is the bedrock of determinism. Ours is a world of cause and effect. Causality is self-evidently true like mathematical truth. All science follows the principle of determinism with cause and effect, without exception.

COLOUR, DIFFRACTION, AND INTERFERENCE

If Newtonian physics could model the movement of the solar system, could it do equally well to pattern the behaviour of atoms and molecules? As we will see in the development of Quantum Mechanics in this Chapter, Newtonian mechanics fails in the world of atoms. We now continue the journey to the heart of the matter, to the structure of atoms, that we left unfinished in Chapter 2. As we saw, Rutherford's model for the atom raised a host of troubling questions. A central nucleus carrying all the positive charge is surrounded by negative electrons buzzing around like planets. But how exactly are the electrons distributed in the relatively enormous empty space around the nucleus? And why do not the negatively charged, whirring electrons simply radiate away their energy, and spiral into the positively charged centre? The circular motion of charged electrons around the nucleus cannot work to keep the electrons in orbit like planets. It was well-known by this time from Maxwell (Chapter 3) that charged particles moving in curved paths must radiate away their energy as electromagnetic waves. Rutherford's model was heading for a disaster.

Clues from the nature of light and emission of light from gas discharges provided fascinating hints to the riddles about how the electrons distribute around the nucleus. The search for answers opened new doors to nature's wonders. Many disjointed areas of the world came together, the nature of light and colours, the nature of electrons, the nature of atoms, and the interaction of light with matter (atoms and electrons).

What is light? What is colour? What are the reasons for nature's fascinating colours? Why is the sky blue? How does the rainbow arch acquire its gorgeous colours? Why does the sky appear reddish near sunset? Is light a wave or a stream of particles? The struggle to answer these questions gave rise to the quixotic science of Quantum Physics.

We go back to Newton and his investigations into the nature of light and its colours. During his private studies at home in Lincolnshire (while Cambridge University shut down due to the terrible plague) Newton carried out experiments with prisms to make the pioneering discovery that white light is a mixture of colours. Light bends (refracts) when it travels from one medium to another, as for example from air to glass, or glass to air, as manifested by an apparently bent pencil in a glass of water. Different colours bend by different amounts. Blue light bends more than red light. When white light passes through a glass prism, the refraction by glass splits light into its natural colours. In these famous optics experiments, Newton isolated the red rays to show how the prism only bends the red beam, but without any further colour splitting. He then recombined the full coloured band back into white light by focusing the coloured beams with a lens and passing the band through a second prism. White light is not pure light; it is a mixture of the pure colours of the spectrum.

The spectacular rainbow arc forms overhead in the sky when sunlight passes through rain-drops. The drops of water acting like a prism refract the different colours and separate them (see Figure 12.1). The back surface of the droplet reflects the coloured band to make it appear that the rainbow comes from the opposite side of the sky as the sun, just as your reflected image appears opposite you in a pail of water. The angles of a rainbow are determined by the angles of refraction. Red light bends at an angle of 42 degrees from its original direction from the sun (See Figure 12.1). Violet light bends by an angle of 40 degrees. We do not see a rainbow if the sun is at an angle larger than 42 above the horizon, because in this case the rainbow forms below the horizon. A lower sun's position will cause a higher rainbow, with the maximum angle of the rainbow reaching 42 degrees. A rare second rainbow, higher than the first, will form when the bright sunlight also enters the drops from the lower half. This light reflects twice inside the drop (getting dimmer) and then exits with dispersed colours from the lower half, but with the order of the colours reversed in the second bow.

A band of many, many raindrops sends red to the eye at one angle, as another band of drops (below the first band) delivers yellow and so on. The coloured bands appear curved because the

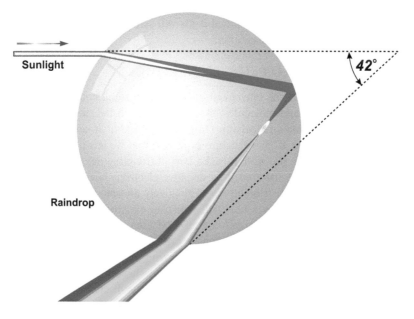

FIGURE 12.1   Sunlight enters a raindrop from the upper half, disperses into colours, reflects on the opposite surface of the drop, then exits from the lower half, further dispersing the colours. Credit: Shutterstock ID 440189362; photo by Fouad A. Saad.

raindrop bands delivering each colour are at a fixed distance from you, forming a circle with you at the centre. The circle of drops delivering green light is smaller in radius, so the green circle of the rainbow lies inside the red. Each person sees a slightly different band of coloured light from his position formed by a different arch of raindrops. The mist from a waterfall can similarly form a beautiful rainbow. You can create an artificial rainbow with a garden hose if you stand with your back to the sun and deliver a fine spray into the air.

A most fundamental aspect of light is its wave-like properties. The waves are oscillating electric and magnetic fields propagating in a vacuum, as firmly established by Maxwell (Chapter 3) through the unification of electricity and magnetism. Light of different colours is light of different wavelengths. Short wavelengths (for example, blue) bend the most during refraction. Compared to familiar ocean waves, or even sound waves, light waves have a very small wavelength, defined as the distance between two succeeding wave crests, and symbolized by the Greek letter lambda (λ). The wavelengths of light are less than one-millionth of a meter, about a hundred times smaller than the width of human hair, compared to wave-lengths of sound-waves, between few centimetres and several metres. For any type of wave, there is a simple relation between wavelength and vibration frequency (f) via the speed (v) of wave propagation, $v = \lambda f$.

Besides refraction, waves have other characteristic properties, diffraction and interference. Sound waves can bend around obstacles that are comparable or smaller in dimension to the wavelength of sound. Thus you can easily hear all the instruments in a concert band even if you are sitting behind a pillar in the auditorium, and cannot see the instruments. Light waves also diffract around obstacles, provided the dimensions are comparable to light wavelength. The corner of a sharp razor blade (Figure 12.2) has razor edge dimensions which are comparable to the short wavelength. The brightest fringe in the diffraction pattern is called the first order of diffraction.

Colours can also be dispersed by a diffraction grating, which consists of many closely spaced grooves such as on those on the surface of a compact disc (CD) or digital versatile – or video – disc (DVD). The diffraction grating separates the colours of white light to produce the spectrum from the diffraction of transmitted light (or reflected light). The direction of the diffracted beams will depend on the wavelength. The first order beam for red light travels at a greater angle than the first order beam for blue light (shorter wavelength) to form a spectrum. Multiple dimmer spectra can form from the other orders of diffraction.

When sound waves from two nearby sources meet they interfere with each other to give a beating effect with loud periods interspersed by soft (or silent) periods. Loud periods occur when the peaks of the two waves coincide, and soft periods occur when the trough of one wave cancels out the peak of the other. Similarly, two light beams emerging from two nearby (compared to wavelength) sources interfere with each other to form light and dark bands on a screen where the light beams impinge.

The wavelengths of visible light range from 400 nm (violet) to 700 nm (red). Shorter waves form ultra-violet light and longer waves give infra-red light. Ultra-short light waves are the same as Roentgen's X-rays, and even shorter waves are gamma-rays (Chapter 2) that can emerge from

enlarged view of the area outside the geometric shadow of the razor blade's edge

interference fringes

FIGURE 12.2   Light diffraction and interference. The first diffraction fringe is brightest.

radioactive decays. On the other end, waves longer than infra-red overlap with radio, TV, and microwaves.

The colours that appear in nature are primarily due to the effect of the atmosphere on light of different wavelengths. The sky appears blue because the air molecules scatter blue light the most, and in all directions. When the sun is near the horizon at sunrise or sunset, the light travels through a longer path through the atmosphere so that blues and greens are absorbed leaving behind the reddish to orange waves to colour the sky and reach your eye to give a reddish hue to the setting sun. During a total lunar eclipse (Chapter 1), the moon appears with a slightly reddish color. The earth blocks all the light from sun from reaching the moon, except some of the red light bent by the earth's atmosphere which illuminates the eclipsed moon with a faint red glow.

MUSIC OF THE ELEMENTS

Colourful electrical discharges in gases are responsible for one of nature's breath-taking wonders, the Aurora Borealis in the northern latitudes (also the Aurora Australis in the south). Usually auroras emit greenish-yellowish lights, and sometimes reddish-violet hues. When charged particles emitted by solar flares stream toward the earth's atmosphere they are trapped near the poles where the earth's magnetic field is strongest. As the particles traverse the atmosphere the current excites and lights up the nitrogen and oxygen about 100 km above the earth's surface.

When electric current (a stream of electrons) is forced with a high voltage (kilovolts) through a low-pressure tube of gas, the remaining gas lights up with different colours which depend on the gas present inside the tube. Lighted neon tubes are commonly used for bright and colourful advertisement displays. Neon lights gives out blushing red light, sodium a harsh yellow, mercury is ghostly bluish, and other gaseous elements produce a stunning variety of colours, such as hydrogen (red), helium (yellow), and carbon dioxide (white). JJ Thomson (Chapter 2) took the lightning (discharge) tube to the extreme by pumping out nearly all the gas, so that there was no light in the chamber. But surprisingly, electrical current still flowed through the tube, which led to his discovery of electrons as carriers of electricity.

Why do different excited gases emit different colours? When the light from the gas discharge is resolved into its constituent colours using a prism or diffraction grating, sharp bright lines of colour stand out on a dark background. These are called emission lines. Each element emits a unique pattern of lines (Figure 12.3) which can be used to identify the element, just like a fingerprint is used to identify a person. When white light passes through a gas and is resolved by a prism or grating, the continuous rainbow spectrum is interrupted by numerous dark lines which fall in the same position as the bright emission lines from the same gas when excited by an electrical discharge. The bar-code like appearance of the dark lines can similarly be used to identify the element. Dark lines appear

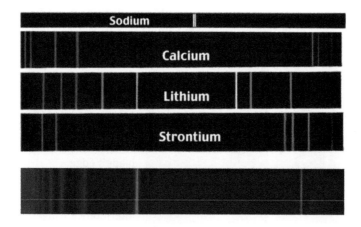

FIGURE 12.3   Discrete emission spectral lines of hydrogen, and a few other elements, strontium, lithium, calcium, and sodium. The spectrum of hydrogen at the bottom forms the Balmer series.

because the element absorbs impinging light at the same wavelengths as the hot element's emission lines. The dark lines are called absorption lines.

The spectrum of sunlight formed by a high-quality prism or diffraction grating, upon careful observation, also shows many fine dark lines (called Fruanhoffer lines) in a characteristic pattern. These dark absorption lines are caused by gases and impurities in the slightly cooler outer regions of the sun as the light from the inner hot region passes through the cooler surface layers. Thus we identify the elements present in the sun as hydrogen, calcium, sodium, magnesium, iron, and others. Spectroscopy of the dark lines proved to be a very important tool in astronomy (Chapter 8) in determining the composition, temperature, size, and age of stars.

A key discovery that led to the arrangement of electrons in the atom was the understanding of why there are so many *discrete* emission or absorption lines in the spectra of elements. Is there any order among the array of lines? Is there any number pattern? Just as Dalton found a simple number relationship in combining the weights of two elements that formed a variety of chemical compounds, Johann Balmer (1825–98) searched for a simple expression to capture the striking regularity in the spectral lines. He was extremely successful for the simplest element, hydrogen (Figure 12.3).

Balmer noticed that the most intense lines of the discrete spectrum of excited hydrogen were in the red, the next intense was blue, and the remaining lines fell between blue and violet (bottom of Figure 12.3). The shorter the wavelength (violet region), the more the lines converged, as if looking up at a tall ladder from the bottom. Rungs appeared to grow increasingly closer together at the distant end. Was there a pattern to the wavelengths, wondered Balmer?:[2]

> The wavelengths (in nanometers) of the first four H lines are obtained by multiplying the fundamental number $b = 3645.6$ in succession by the coefficients 9/5, 4/3, 25/21, and 9/8. At first sight these four coefficients do not form regular series.

Balmer had no idea where the number $b = 3645.6$ came from. But he realized that if he multiplied both numerator and denominator of the second and the fourth terms by 4, the series of puzzling fractions turned into a pleasing pattern: 9/5, 16/12, 25/21, and 36/32. With vintage Pythagorean delight, he recognized how the numerators were just 3^2, 4^2, 5^2, and 6^2, while the denominators were $3^2 - 4 = 5$, $4^2 - 4 = 12$, $5^2 - 4 = 21$, and $6^2 - 4 = 32$. In short, the wavelength series had a simple algebraic expression:

$$\lambda = bn^2 / (n^2 - 4).$$

Indeed, there was mathematical harmony in spectral lines. But it was pure numerology without any underlying physical reasoning. Balmer only knew about the visible series of spectral lines from hydrogen. In 1908, spectroscopist Louis Paschen discovered a very similar expression to fit another series of lines from excited hydrogen in the infra-red region, beyond the visible red end of the spectrum of sunlight:

$$\lambda = bn^2 / (n^2 - 9).$$

Besides the order of the Periodic Table, mathematical order in the spectral lines was pointing to an underlying structure to light-emitting atoms. But what was that structure? How did an atom emit light in mathematically precise and discrete spectral patterns from the movement of electrons within a neutral atom?

The First Quantum Revolution, Wave or Particle?

At the turn of the 20th century, Max Planck (1858–1947) and Albert Einstein (Chapter 4) introduced the *quantum theory* which revolutionized our understanding of the nature of energy, in particular the nature of light energy. Planck had climbed to the reputation of prince of German physics. He was one of the editors of the most prestigious physics journal, *Annalen*. At the time he was working on improving the efficiency of light sources, which required a better understanding of the nature of light emission from hot bodies such as wire filaments in a light bulb.

As temperature rises the wavelength of the maximum emission intensity shifts from red, as in red-hot poker, to yellow to blue. But the most troubling question for Planck was why the intensity of light emitted always tapered off at shorter wavelengths (higher frequencies). Prevailing theory predicted the opposite: that the intensity of radiation should increase without limit for shorter and shorter wavelengths (or higher and higher frequencies). In the simple picture of light emission at the time, there was a limit to how long the wavelength of light could be in order to fit inside the dimensions of a hot oven, so the intensity tapered off at very long wavelengths, as expected. But there was no physical limit to how much light energy could be present at short wavelengths. It should be possible to fit more and more short wavelengths of energy, from ultra-violet to X-rays, into a hot object. In fact, an infinite number of short wavelengths could reside in the oven. All the wavelengths from a hot body should emerge as ultra-violet and X-ray waves. If such a trend did exist, the super-hot sun would emit mostly ultra-violet rays, killing off all life on earth. It would be an ultra-violet catastrophe!

To explain the puzzling spectra of light emission from hot bodies, Planck introduced a brand-new idea. Energy is absorbed or emitted from hot objects in tiny discrete bundles called *quanta*. Each quantum carries a precise quantity of energy which is related to the frequency (f) of light waves by $E = hf$. Here h is a fundamental constant of nature called Planck's constant ($h = 6.63 \times 10^{-34}$ m^{2}kg/sec). A blue photon for example carries 4×10^{-19} joules of energy, and a red photon carries 3×10^{-19} joules. Planck's equation launched the quantum revolution to become as famous as Einstein's $E = mc^2$.

Since the wavelength and frequency of light waves bear a simple relationship via the speed of light (c), Planck's energy relationship can also be cast in terms of wavelength (λ), as $E = hc/\lambda$. The energy of absorbed quanta is inversely proportional to wavelength, and directly proportional to the frequency of the waves.

Using the creative discretization principle for energy emission, Planck could accurately account for the curious shape (intensity versus wavelength) of the emission spectra from hot bodies at different temperatures.

If light emission and absorption takes place in discrete bundles of light (quanta), and the energy of these quanta depends on frequency, then the energy in the high frequency (short wavelength) quanta becomes quite high. Hence the output of light at higher and higher frequencies drops off with higher frequencies, as observed. Planck's quantum saved the world from the ultra-violet disaster.

Planck insisted that light by itself is continuous, as Maxwell had shown (Chapter 3), but absorption and emission take place in discrete energy steps, quanta by quanta. He was not too concerned about why energy emerged in this chopped-up fashion. He was happy to come up with the much-needed mathematical trick.

Einstein took the next bold step. He generalized Planck's quantum theory. As we saw in Chapter 4, Einstein disliked special conditions such as Planck hypothesized. Why should light bundles be involved only in absorption or emission? Light quanta spoke to him about the intrinsic nature of light. Light by its very nature must be quantized. Einstein took the daring leap. Not only light but *all energy* is quantized. By nature, energy is not smooth and continuous. Energy breaks up into discrete quanta. For light, the amount of energy a quantum carries depends on the frequency of light via Planck's constant.

Einstein's inspiration for "atoms of light" came from the newly discovered "photo-electric" effect, now the basis of operation of solar devices to produce electricity, or plant leaves to produce photosynthesis. Philip Lenard (1862–1947) in Heidelberg made a curious observation about electricity from light, how electrons emerged from metals irradiated with light. Here the high frequencies were playing the dominant role, by contrast with emission from hot bodies, where high-frequency emissions die off. In the photo-electric effect, only high-frequency (ultra-violet) light can eject electrons out of a metal. The higher the frequency of impinging light, the higher the kinetic energy of the emerging electrons. But low-frequency (red or green) light is not capable of releasing any electrons at all from a metal. Turn up the intensity of red light as much as possible, but no electrons emerge. The extra energy in high-intensity red light waves had no effect. Einstein used Planck's idea that only high-frequency quanta of ultra-violet light carry enough punch ($E = hv$) to eject electrons from the metal. The higher the intensity of ultra-violet light, the more electrons emerge, but with the same kinetic energy that each quantum of ultra-violet can deliver.

Einstein's quantum theory removed a grating asymmetry in nature, between energy and matter. Why should light energy be continuous if matter is atomic and grainy? At this time in his career, he was already homing in on the equivalence between mass and energy via $E = mc^2$. The quantization of energy would be a genuinely unifying principle. The world of matter and the world of energy are no longer separate in their fundamental nature.

But there were many upsetting questions making it difficult for most scientists to accept such a crazy jump. How can light be a stream of *particles* (quanta) if Maxwell, less than 50 years prior, firmly established that light is a smooth electromagnetic *wave* which demonstrates diffraction and interference characteristic of waves? How can two nearby streams of light particles interfere with each other to give a region of no particles (dark bands)? Light waves bend (diffract) to penetrate darkness around razor sharp corners (Figure 12.2). Only waves can do that. Maxwell's light and radio waves were gospel truth. Waves spread out all over to bring music and news to everyone, *all at the same time*, even if we are sitting at different places. How can a narrow stream of particles do that? A particle must always be somewhere. You can pick it up. It is here, but it is not there. Now Einstein was killing Maxwell's waves with quantum bullets. Planck argued to make the quanta go away. But Einstein continued to maintain that the very nature of light is quantized, much to Planck's chagrin.

In 1922, a brand-new American discovery by Arthur Compton confirmed that X-rays (also a form of light) and electrons collide like billiard balls. Compton used the laws of energy and momentum conservation for particle collisions to explain the scattering of X-rays. So once again, for Compton scattering, as for the photo-electric effect, light behaves as particles, not waves. But when the American informed the world about his marvellous discovery, the information reached everyone at the same time via radio waves that spread everywhere at once, and not like a stream of quanta. The mystery of the particle–wave quandary deepened.

Bohr's Atom Explains the Periodic Table

Arriving from Denmark as a young post-doc to work with the atomic masters of the electron and the nucleus (Chapter 2), Niels Bohr (1885–1962) first studied under Thomson before gravitating to Rutherford where he felt his ideas were more appreciated. Theoretical aspects of the new atomic structure fired his imagination and tinted his rhetoric with Pythagorean optimism:[3]

> Rutherford's model of the atom puts before us a task reminiscent of the old dream of philosophers: to reduce the interpretation of the laws of nature to the consideration of pure numbers.

Bohr showed a youthful zest for innovation. How could negative electrons orbiting a positive nucleus account for the discrete light emission spectra from elements, with unique patterns for each element? In his model for the atom, Bohr introduced several creative inventions (postulates) based on the discreteness of energy in Einstein's quantum picture. Of all possible orbits, only certain ones with the right energy are allowed for an electron around a nucleus (see Figure 12.4). He dubbed these "stationary orbits" or "stationary states," because an electron does not emit energy while it resides stably in one energy state. When an excited electron jumps from a higher orbit to a lower one, it radiates energy. This energy must be an integral number of energy quanta, as given by the Planck–Einstein quantum theory. When an energized atom emits light energy, it drops discontinuously from

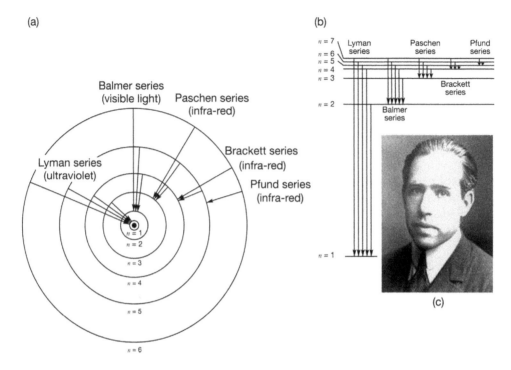

FIGURE 12.4   (a) Energy levels explain how the hydrogen atom emits (b) discrete light spectra that follow exact mathematical patterns as determined by (c) Niels Bohr.[4]

a "high energy state" to a "low energy state," releasing a quantum of light energy corresponding to the energy difference between upper and lower levels. Bohr grasped the discrete nature of the elemental spectra and transformed the discreteness to electron orbital states.

An instructive analogy for an electron moving inside a simple hydrogen atom is with a steel ball-bearing rolling inside a special hemispherical bowl with the inside surface stepped regularly like the rows of discrete steps in an amphitheatre. The ball can remain stationary on one of the steps or glide along the step; but it cannot reside in between steps. The geometry of the allowed space, as a stadium-ring, forbids the ball from residing in between two levels. As long as the ball stays on one ring at one level, its energy remains constant. If the ball falls one level due to the attraction of earth's gravity, it releases a certain quantized amount of energy. To raise the ball, the same quantum of energy must be supplied by an outside source.

In the atom, the positively charged nucleus electrically attracts the negative electron, just as gravity attracts the ball to the bottom of the bowl. Like stadium steps, *atomic levels* in which the electron can reside are discrete with well-defined energy. Electrons are not free to roam anywhere inside an atom; they must reside in well-defined levels. The level with lowest energy corresponds to the stable, unexcited atom. When heating with a flame or an electrical discharge excites a hydrogen atom, the electron absorbs a quantum of energy to jump up to a higher energy level. When an energized atom drops from a high to a low energy level, it releases a discrete amount of energy in the form of quanta of discrete light, as Planck hypothesized about emission. Excited hydrogen atoms display a characteristic series of light energies that form discrete colour lines on a photographic plate (Figure 12.3). Every element in the Periodic Table when excited also displays a unique series of colour lines which serve as a signature for that element depending on how the energy levels in that atom are distributed. The line spectrum identifies the presence of an element in familiar substances, such as in ores from the bowels of the earth, or in the sun and the stars. Indeed, Balmer's mathematical pattern governing the spacing of characteristic lines for excited hydrogen led Bohr to the model for atomic structure described above.

It electrons are not free to roam anywhere inside an atom, and must stay confined to stable orbits, which orbits are allowed? As soon as he saw Balmer's formula for the line spacing, the whole picture became clear. From Balmer's spectral series, Bohr worked back to conclude that the radii of the permitted circular orbits must increase as the squares of the integers. If r_0 is the radius of the smallest allowed orbit (now called the Bohr radius of an atom), the radius of the next orbit must be $2^2 r_0$, $3^2 r_0$, $4^2 r_0$, and so forth; in general, $n^2 r_0$. When an excited electron drops from a higher orbit (e.g., $n = 3, 4, 5 \ldots$) to a lower orbit (say $m = 2$), it emits discrete quanta of light with wavelengths that fall in the special Balmer series of spectral lines (Figure 12.3).

From the mathematical character of the electrical force (decreasing as the square of the distance from the central nuclear charge), he found electron energy to decrease as the inverse squares of the integers. For example, falling from orbits with $n = 3, 4, 5 \ldots$ to the particular orbit $m = 2$, the energy difference is proportional to $(1/2^2 - 1/3^2)$ or $(1/2^2 - 1/4^2)$, or $(1/2^2 - 1/5^2)$ and so on. The energy of emitted light is therefore proportional to $(3^2 - 2^2)/3^2$, $(4^2 - 2^2)/4^2$, $(5^2 - 2^2)/5^2$, $\ldots$, $(n^2 - 4)/n^2$. According to the quantum relation the wavelength is proportional to the reciprocal of the energy. Therefore the emitted wavelengths are proportional to $n^2/(n^2 - 4)$, exactly as Balmer observed for the visible light series. And if an electron should fall to orbit number $m = 3$, there should be another series of lines in the infra-red part of the spectrum, as Paschen noted in 1908 (Figure 12.4).

In Bohr's imagination the atom behaves like a musical instrument. The emitted colours of light are the notes. Balmer's spectrum is the musical scale. It was the first time anyone was able to link the magical atomic spectral colours with the atom's structure, an exact match between each colour, and each electron jump. Bohr waxed poetic about his atomic vision.[5]

> When it comes to atoms, language is like poetry.
> The poet too is not concerned with describing facts
> As with creating images and mental connections.

Bohr freely adopted Einstein's picture for quantized energy without at first being bothered by the puzzling dichotomy of the nature of light (quanta? Or waves?). His atomic model and accompanying postulates raised a host of questions. What causes the excited electron to suddenly jump up or down a rung of the energy ladder? Could it possibly be just by chance? Bohr's jumping electrons seemed to be choosing their own timing.

But physicists did not like to leave things to chance. There always has to be a reason for something to happen, the dominance of cause and effect established by the prevailing determinism paradigm that emerged from the spectacular successes of Newtonian mechanics. Phenomena must proceed from identifiable causes. The sudden hop between two levels was discomforting. How can the electron be in one level, then suddenly in another level, without going in between? This new quantum world of the atom was getting "spooky." Bohr returned to such troubling questions later in life to further develop the deeper meanings of the quantum theory (discussed later).

Bohr's model came from a mix of reasoning and inspired guesswork. One hallmark of a good theory is its ability to embrace the known. Bohr's success in connecting fragmented domains of atomic structure, spectral lines, and the quantum theory of light confirmed he was on the right track. And a powerful theory needs to guide science into the unknown. Although Bohr was inspired to shape his model to fit Balmer's formula, the real power of his new atomic structure was to predict a brand-new series of spectral lines. When an electron jumps from a higher orbit ($n = 2, 3, 4 \ldots$) to the first orbit ($m = 1$), the energy difference from the calculated radii predicted a new series of lines in the ultra-violet part of the spectrum. Bohr's prescience proved rapidly fruitful. Harvard spectroscopist Theodore Lyman found the ultra-violet series (for $m = 1$) in 1916. Frederick Brackett found another sequence in 1922 pertaining to $m = 4$, and August Pfund discovered spectral lines corresponding to $m = 5$ in 1924, all precisely as predicted by Bohr's theory (Figure 12.4).

As a spectacular result Bohr could calculate Balmer's constant ($b = 3645.6$) from the fundamental properties of electron mass, electron charge, velocity of light, and Planck's constant. In the same bold stroke, he determined the size of the hydrogen atom from the smallest allowed orbit, as 50-trillionths of a metre ($r_0 = 5.3 \times 10^{-11}$ m). For the first time there was a precise value for atomic size that came from first principles and not from statistical estimates that Einstein provided earlier in his thesis (Chapter 2). Einstein expressed both shock and amazement at Bohr's preposterous success in modelling nature through mathematical patterns:[6]

> That this insecure … foundation was sufficient to enable a man of Bohr's unique instinct and perceptiveness to discover the major laws of the spectral lines and of the electron shells of the atom as well as their significance for chemistry appeared to me like a miracle and appears as a miracle even today. This is the highest form of musicality in the sphere of thought.

THE SECOND QUANTUM REVOLUTION, WAVE MECHANICS

Despite Bohr's bold conjectures about the electronic structure of the atom, the trouble persisted about why electrons remain confined to Bohr orbits without energy loss, or why they do not whiz about in any other region. At this stage, French physicist Louis DeBroglie (1892–1987) dropped a bomb into the quantum world with a brilliant idea for his PhD thesis in 1924. He proposed to look at the nature of electrons in a different way. If light waves behave as particles, electrons, normally thought of as particles, ought to reciprocally behave as waves! DeBroglie proposed that the wavelength of the electron waves should be h/p, where p is the momentum of the electrons, and h is Planck's constant. Always open to radical insights, Einstein admired the idea very much. A missing symmetry was restored. Photons make light waves into particles. And electron particles become waves of matter. DeBroglie's advisor was ready to fail him for such a ridiculous thought. But Einstein enthusiastically recommended its potential. If electrons behave like waves, they should show interference effects, which were observed in electron diffraction experiments only three years later. DeBroglie went on to win the Nobel Prize in 1929 for his brave new idea. Figure 12.5 compares the modern diffraction pattern of an electron beam with the diffraction pattern of an electromagnetic wave X-ray beam, confirming from the similar ring patterns the wave-like nature of the electrons.

Confined to the nucleus by the electrical attraction, electron waves can only take on certain discrete wave patterns. An analogy from mechanical waves is fittingly illustrational. A string confined at two end points can only vibrate in discrete modes to emit the familiar discrete musical notes, as from a violin string. Or the skin of a drum confined at the circumference only jiggles in discrete wave patterns (Figure 12.6). As waves, electrons also take on discrete wave patterns – which

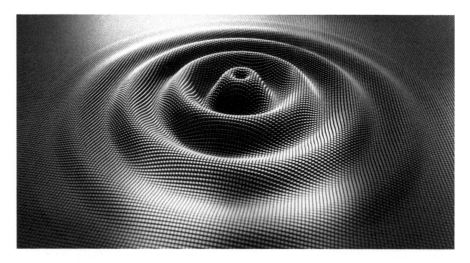

FIGURE 12.5 Comparison of diffraction patterns of electron beam with that of an X-ray beam. Since X-rays diffract as waves, electrons also behave as waves, judging from the similarity of the diffraction ring patterns.

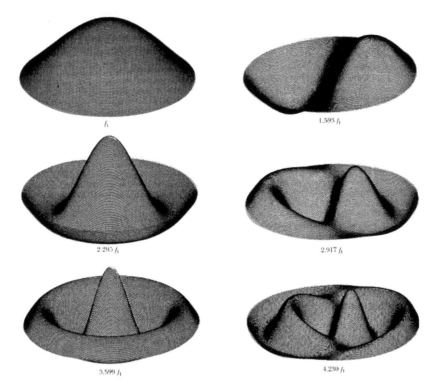

f_1

$1.593\,f_1$

$2.295\,f_1$

$2.917\,f_1$

$3.599\,f_1$

$4.230\,f_1$

FIGURE 12.6   Vibration patterns of the drum skin for different modes.

explains why there are only certain discrete levels. DeBroglie provided a crucial missing piece in the quantum puzzle of discrete Bohr orbitals.

Erwin Schroedinger seized DeBroglie's brain wave. For particles to be described by waves, a wave equation was necessary. Maxwell derived the wave equation for electromagnetic waves of light from the mathematical description of electric and magnetic fields. Electron waves must similarly spring from a wave equation of "something" waving.

The romantic physicist was skiing in the Swiss Alps with the latest of his many lovers, feeling the exhilaration of surfing waves. With matter waves on his mind, and a beautiful woman on his arm, Erwin Schroedinger came down from Mount Arosa with a mathematical wave equation for 3D electron waves around the atomic nucleus. Electrons must occupy special 3D wave shapes around the nucleus. Schroedinger was also a devotee of Indian mysticism. He searched for a universal reality, an undivided unity between matter waves and the universe. Feeling close to Brahman, the ultimate truth, Schroedinger believed the 3D wave patterns revealed how the electron's charge is distributed in space as smeared out waves.[7]

> Who sees the wave dwelling alike in all
> Perishing not as they perish
> He sees indeed
> … this is the highest way

Schroedinger developed a "wave mechanics" to describe the quantum world. The solution of his wave equation gives the wave function, famously symbolized as Ψ, to describes the wave for a particular electron system. Like a vector which has magnitude and direction, the wave function has two components, amplitude and phase. Both amplitude and phase change with time which gives Ψ the wave-like nature. Phases can provide the interference and diffraction effects of electrons and waves.

The wave equation describes how the wave function evolves with time. The wavefunction became the central ingredient of quantum physics, an instrument of prediction of how the quantum state evolves. Since the wave function can be negative in places, and even has an imaginary component of phase, Schrodinger declared the square of the amplitude of the wave function $|\Psi|^2$ to be the electron charge density, when multiplied by the electron charge.

But the core questions persisted. Light *is* a wave. Electrons are waves. Everything is waves! If everything is wave-like, how can well-known particles of physics exist? For Schroedinger, particles are just an "illusion" of many waves. Like a quantum magician, Schroedinger turned waves into particles. An infinite number of waves, superimposed upon each other, add up to a narrow packet of waves which gives us the illusion of a particle.

Ψ contains all we know, and all we *can* know. *Any* question about atomic behaviour can be answered by consulting the wavefunction. For Schroedinger, there need be no more particles, no more quanta, no more quantum jumps – only smooth transitions from one wave-form into another, beautiful notes on a violin. Like Goethe's Faust opening the Book of Nostradamus,[8] Schroedinger felt he was gazing at God given signs:

> Did some god inscribe these signs
> that quell my inner turmoil,
> fill my poor heart with joy,
> and with mysterious force unveil
> the natural powers all about me?
> Am I a god? I see so clearly now!
> In these lines' perfection I behold creative nature spread out before my soul …
> How all things interweave as one … and live each in the other.

Another magic of the quantum waves is that they can spread out everywhere. Schroedinger's wave equation applied to an electron confined to a box (potential well) showed that a small part of the electron wave can leak out of the box. So, a little bit of the wave-particle exists everywhere! The wave nature of particles allows particles to tunnel out of the box. The effect has many applications, such as a quantum-tunnelling microscope. Electrons tunnel out of the silicon surface to provide a current to image atoms (Chapter 2).

ELEMENTARY ELECTRON CLOUDS

The allowed spatial configurations for electron waves in a hydrogen atom cannot be as simple as DeBroglie's circular standing waves mimicking Bohr's planetary orbits. Since space is three-dimensional, electron wave configurations can take on a variety of 3D elegant shapes, called electron clouds, governed by the three degrees of freedom for the motion of electrons in three-dimensional space. With complete symmetry, the lowest configuration is a spherical shell of the Bohr radius around the nucleus. The next levels governed by the rotational degree of freedom are also spherical shells farther away from the nucleus with radii that correspond to the increasing Bohr orbital radii. These spherical symmetric shells are the rotationally symmetric states. But in 3D there are also angular degrees of freedom for the electron's movement, which allow multi-lobe and doughnut-like configurations for the wave configurations (Figure 12.7), called *spherical harmonics*. Proper stimulation by heat, light-bombardment from a gas discharge, or other exposure to electromagnetic fields can shift the electron of a hydrogen atom into any of the many 3D configurations, which Bohr originally called *electron states*.

The many electrons of atoms heavier than hydrogen occupy similar wave configurations (or states) around their central nuclei. But how to arrange those numerous electrons of higher atomic number elements into the various possible wave states? One might ask, for example, why all ten electrons of neon do not simply pile into the lowest energy, spherically symmetric state. Two newly recognized fundamental properties about electrons prevent this. The first is that *no two identical*

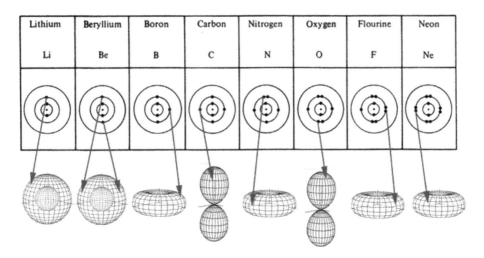

FIGURE 12.7   Electron configurations for the first two rows of the Periodic Table, which reflects the occupation plan for electrons among the states.

electrons can reside in the same configuration, as if a federal restraining order forced identical twins to live in separate countries due to their mutual dislike of each other. This is called the *exclusion principle*, recognized by Wolfgang Pauli (1900–58). We met Pauli in Chapter 2 as the scientist who proposed the existence of the neutrino to save the law of conservation of energy. Another salient intrinsic feature, also proposed by Pauli, is that electrons possess a special property, like a hidden rotation or spin, that makes them behave like tiny magnets. Magnetization, like energy, must take on discrete values according to quantum theory. An electron's intrinsic magnetization can have only one of two discrete values. Therefore, two electrons with different spins can populate each electronic state of an atom without violating the exclusion principle because they have different spins and are not identical.

In the familiar macroworld, a magnet arises from circulating electric charges or loops of electric current (Chapter 3). Therefore, it is convenient to idealize an electron-magnet as a spinning sphere of charge. Although the magnetization is real, the notion of *electron spin* should not be taken in the literal mechanical sense, since the electron is a point-like particle and not a sphere. Spin is just another attribute of an electron, technically called a quantum number. In the language of spin, electrons can take just two discrete spin values ($\frac{\bar{h}}{2}$, or $-\frac{\bar{h}}{2}$, where $\bar{h} = h / 2\pi$). For example, electrons moving in one direction can have two spin-states, clockwise or anti-clockwise, around an axis which is the line of motion. Right- and left-handed screws provide a good analogy. Most bottle caps carry righthanded screw tops where clockwise rotation (positive spin) closes the bottle. For a left-handed screw top, it takes anti-clockwise rotation (negative spin) to close the bottle.

Both the Pauli exclusion principle and electron spin originate from deeper mathematical symmetries in the quantum waves of matter, topics beyond our scope. Suffice it to say at this stage that the wave function of two identical electrons reverses sign if the positions of the two electrons are interchanged, so that by putting the two identical electrons in the same state, the wave function has to become zero – the state is not allowed and hence the exclusion principle. By contrast, the photon is ruled by an opposite type of symmetry that allows many photons to exist in the same quantum state, giving rise to fabulous possibilities such as the sharp laser beam which carries all its photons in one state, with a single wavelength. Because of the two distinct types of symmetry, the electron is categorized as a "fermion" (after Enrico Fermi). Protons, neutrons, and neutrinos are also fermions and obey fermion symmetries. Fermi (1901–54) built the first nuclear reactor at University of Chicago, and participated in the Manhattan Project to develop the first atomic bomb. The photon (ruled by the other symmetry) is a "boson" (after Satyandrenath Bose). Bose (1894–1974) worked with Einstein to classify the particles with Bose type symmetry. We will meet other bosons later in this chapter.

Without the exclusion principle and without spin, all elements would show the same chemical properties, eliminating the fantastic variety of substances we encounter in nature. In such a boring universe, humanity would not exist either. Electron spin and the exclusion principle provide key rules to populate the many electrons into the possible wave-states. The structure and patterns of the Periodic Table (Figure 2.18) have been known since Mendeleyev's discovery in 1869. Now finally the Bohr atom, Schrodinger's wave functions, the Pauli exclusion principle together with electron spin provide the rubric of organization for the masterful table, as well as explaining the distinct chemical properties of the elements arising from the electron configurations.

When it is an unexcited atom, hydrogen's lonely electron resides in the spherically symmetric ground state. Since the ground state of hydrogen can accommodate two electrons, hydrogen can capture an electron to complete its shell. Therefore, hydrogen combines readily with other elements, making it chemically very reactive. Helium, with atomic number 2, has two electrons surrounding its nucleus. Both electrons (with opposite spins) occupy the first level, in the spherically symmetric state. With two electrons occupying the lowest energy level, this level is full, and so cannot accommodate any more electrons without violating the exclusion principle. As a result, helium has no tendency to either acquire or relinquish electrons, making helium the stuck-up, noble element that it is. Since this first state can accommodate just two electrons, there are only two elements in row one of the Periodic Table.

Elements of row two, starting with lithium, assemble according to electron configurations for the second level of states of an excited hydrogen atom. There are four possible states (Figure 12.7) or four configurations for 3D electron waves – one spherical shell, two doughnut states, and one double-lobe configuration, all solutions of Schroedinger's wave equation. Each state can accommodate two electrons. Thus, the second row of the Periodic Table can accommodate 2×2^2 elements, i.e., eight elements. Lithium has three electrons. Two electrons fill up the lowest level, spherical shell state, forcing the third electron to move up into the second level to obey the Pauli exclusion principle. Now the outer electron layer of lithium which can accommodate one more electron possesses a lonely electron. Hence lithium chemically resembles hydrogen which also has one electron in the outermost layer. For both hydrogen and lithium, the lonely electron is free to leave and join partially filled outer electron clouds of other elements, making both elements very chemically active.

Moving across the second row, the number of electrons increases from three to 11, populating the doughnut- and lobe-shaped clouds in sequence, filling these up two by two, and generating a variety of chemical behaviour across the second row of the Periodic Table. At fluorine (atomic number 9), there are two electrons in the first level and seven filling the kaleidoscope configurations of level two. With an opening for one more electron in the second level, fluorine becomes a particularly reactive element. Hydrogen and lithium react strongly with fluorine making very stable compounds, such as hydrogen-fluoride (hydrofluoric acid) and lithium-fluoride.

Finally, in row 2, neon with ten electrons has two completely full levels, one of two and the other of eight electrons. Neon does not have any electrons to share with other elements, nor does it have an empty slot in its outer shell to accommodate electrons of any another element. Like helium, neon is an anti-social element.

To recap, row 1 of the table can accommodate $2 \times 1^2 = 2$ elements, and row 2 can accommodate $2 \times 2^2 = 8$ elements. Continuing this mathematical progression, succeeding rows can accommodate $2 \times 3^2 = 18$ elements, and $2 \times 4^2 = 32$ elements. A simple number pattern rules the organization of the Periodic Table. Pythagoras would be humming with delight in his grave to the music of the spherical harmonics of electron cloud configurations.

The Periodic Table introduces order among the large number of elements, taming the bewildering variety of chemical properties. When that order links to a number pattern, i.e., $2n^2$, the numerical relation points to an internal structure. Underlying that structure is the three-dimensional spatial symmetry of various electron cloud configurations of the excited hydrogen atom. Copper, gold, and silver appear drastically different in our everyday world. But in their electronic shell structure they are very much the same. Hence they exhibit similar chemistry and fall in the same column of the table. Nature reveals a number of different faces with the fascinating variety of elemental chemical properties, but they are all facets of one and the same gem! The new quantum physics provides a strong basis for a complete explanation of all the features of the Periodic Table.

Like Mendeleev who predicted ekasilicon, ekaaluminium, and ekaboron (eventually discovered as germanium, gallium, and scandium), Bohr used his electron organization to predict that the electronic and chemical properties of element 72, not yet discovered, should be similar to those of zirconium. When discovered by a Danish scientist from zirconium ores it appropriately earned the name hafnium for Copenhagen, Bohr's hometown.

DUALITY AND COMPLEMENTARITY

But the nature of Schroedinger's waves remained a mystery. What exactly is waving? For a wave on a string, the wave function describes the displacement of the string; in the case of water waves, water displacement. In the case of Maxwell's equations, the wave function describes the behaviour of oscillating electric and magnetic fields. Schroedinger believed that the electron was not a particle but a smear of charge across space in the shape of a wave, and that the square of the wave function times the electron charge gave the "charge density" of the electron wave at any particular location.

The pioneers of the quantum, Planck and Einstein, were bothered by the contradictory nature of light. How can light be both wave and particle? How can an electron be a particle and a wave? Physics was impaled on the horns a nasty dilemma. Should physics professors now teach the wave theory on Mondays, Wednesdays, and Fridays, and the particle theory on Tuesdays, Thursdays, and Saturdays? When does light behave like a particle, and when does it behave like a wave? Can an electron suddenly crystallize out of the Schroedinger mist, like a genie emerging from a bottle? It was as impossible to imagine as a sea-wave mysteriously summoning its total energy, and colliding like a cannon-ball to capsize a ship, just as an ultra-violet light wave knocks out an electron confined to a metal in the photo-electric effect?

Bohr remained undaunted. For him, contradiction and truth were allies. Light is a wave, and light is not a wave. The electron is a particle, the electron is not a particle. Waves and particles are each part of the whole truth. We must no longer speak of waves and particles separately. Waves and particles are two sides of the same coin. There is duality. In duality there is wholeness. He would no doubt find comfort in Alice's conversation[9] with the Caterpillar in "Alice in Wonderland."

Caterpillar: "Who are you?"
Alice: I hardly know, Sir, just at present—at least I know who I was when I got up this morning, but
        I think I must have changed several times since then.
Caterpillar: Explain yourself.
Alice: I can't explain myself, because I'm not myself, you see.

FIGURE 12.8   *Discovery*, by Rene Magritte. ©2019. Photothèque R. Magritte/Adagp Images, Paris/SCALA, Florence. © ADAGP, Paris and DACS, London 2019.

For the philosophical Bohr, duality rang true and deep. Particle and wave natures are not contradictory. They are complementary. But the metaphysical duality concept was hardly sufficient. Einstein asked how duality brings us closer to nature's secrets.

The artist Rene Magritte (1898–1967) paints a helpful picture (Figure 12.8) to show how apparently different objects are alternate manifestations of the same underlying entity:[10]

> I found a new potential inherent in things; the ability to gradually become something else, one object merges into another. For example, in certain areas, sky reveals wood. It seems to me something other than a combined object since there is neither a rupture nor a limit between the two materials. In this way I arrive at paintings which the eye "must think" about in a completely different way than usual, things are tangible and nevertheless, real planks of wood become imperceptibly transparent in certain places or certain parts of a naked woman become an entirely different material … it was enough to merge things into one another, things which earlier appeared separately.

GOD DOES NOT PLAY DICE

German physicist Max Born (1882–1970) in Gottingen suggested that the electron wave is not a real wave of any form, not like a sea-wave, or an electromagnetic wave of oscillating electric and magnetic fields, not Schroedinger's smeared charge wave. Instead the electron wave is an abstract "probability distribution" that predicts the point electron's position with time. There is no exact location of the electron particle in space, only the probability of finding it at different locations, which is expressed by Schroedinger's wavefunction, a *probability wave*. The electron wave leaking (tunnelling) out of the confining box comes simply from the probability of finding the electron outside the box.

The probability wave is unobservable. It only gives a superposition of possibilities. It does not mean that the system is really in all these possible states, only a probability that each one will become real. Schroedinger's equation is a projection into the future to tell us how the probabilities evolve with time. It gives the needed information about what will happen next to the system, how the probabilities will evolve with time.

The probability interpretation opened vexing new questions. How does an electron, stuck in the quantum haze, go from a host of probabilities to the actuality of the particle's location? How does a multifaceted potentiality become a single actuality? It was another way of asking Einstein's question about how a wave can turn into a particle.

Schroedinger, the triumphant inventor of the wave function was reluctant to accept Born's probabilistic interpretation. He could not believe that the underlying laws of physics are statistical, involving pure chance. Chaos cannot be at the heart of nature.

Einstein was greatly upset by the need to bring in probability interpretations into physics. He felt we must search for real laws that allow us to predict the future behaviour with certainty, not merely the probability of occurrences. Chance and probability are normally used to express our ignorance of the details of the underlying causes. A complete scientific theory must reveal order and predictability.

He fully accepted the determinism of Newton's world in his mind, in his heart, and in his very bones. His theories of relativity were fully deterministic. He could never renounce the ideals which guided science so well and so far. We need to know with certainty where the electron must be. If you raise an apple from the ground and let it go, it will surely fall. Gravity acts with certainty. But the introduction of probability into quantum physics says if an electron is excited from its ground state in an atom, there is a probability that the electron will stay in that excited state, and a probability that it will fall back to its ground state. Why is there such a great difference between the behaviour of the apple and the electron? How can determinism fail so miserably in the quantum world?

Bohr always tried to come to grips with the weirdness of the quantum world's ideas. Every great difficulty bears in itself its own solution. It forces us to change our thinking in order to find it. He latched on to the idea that probability was the way to unite the two disjoint views of wave and particle! Waves and particles co-exist through the probability wave-form which gives the exact odds that the particle will be found here, or found there.

Einstein always challenged Bohr's new quantum ideas. In fact, Einstein ended up pushing Bohr to propose new ways of understanding. In this way Einstein became an unwilling contributor to the advancement of quantum physics, despite his strong scepticism. Strangely enough, it was Einstein who earlier introduced the probability idea into the quantum domain, although with great reluctance. Here was his reasoning. You can see through glass. But you can also see a faint reflection of yourself on the other side of the glass. All in agreement with classical laws of light wave transmission and wave reflection. But, let us suppose you turn down the intensity of light, more, and more, and more, so that all you have left is a single, solitary photon. Will it reflect, or will it transmit? It is not possible for a single photon to partly reflect and partly transmit. The single photon has no parts. It must be completely transmitted or completely reflected. So how can there be both reflection and transmission? Einstein was forced to the conclusion that we need to introduce a probability that a photon will transmit, and a probability that it will reflect. Nature rolls the dice! Probabilities may have to be allowed as fundamental, physical properties in the atomic world.

Nevertheless, Einstein continued to insist (till his death) that quantum probability clouds were just philosophical smoke. The universe does not run like a casino! He strongly believed that even for the quantum world of the atom, we must search for a picture that describes the actual stuff of nature, not gambler's odds. "I am completely convinced that *He* does *not* play dice with the universe." To which Bohr responded: "But still, it cannot be for us … to tell *God* how to run the world." He probably really meant to say: "Einstein should stop telling God what to do!"

QUANTUM MECHANICS

After the end of World War I, Germany reeled on the edge of an abyss. With extreme right, and extreme left political movements, anarchy prevailed. People sought emancipation from the burnt-out conventions that dominated culture before the Great War. European cities, Berlin, Munich, and Paris, bubbled with excitement and new ideas. The Dada movement claimed that underlying reality would be found through chance and irrationality. The very name of the movement was established by stabbing a dictionary and picking the French word Dada for hobby-horse. Bohr, Schroedinger, and Einstein flirted with wild ideas of duality, probability, and spacetime warps just as did the artists, writers, musicians, and dancers of the period. James Joyce wrote *Ulysses*, inaugurating modernist literature with an unruly mix of high brow and low brow, merging time and space using radical literary forms. Schoenberg introduced atonal compositions, Picasso explored cubism, and Marlene Dietrich reinvented singing and dancing.

Werner Heisenberg (1901–76) worked closely with Max Born in Gottingen and Niels Bohr in Copenhagen. He took long hikes in the mountains with Bohr discussing new theories and experimental results on the quantum structure of matter while clearing his stuffed nasal passages from debilitating attacks of hay fever. The quantum world appeared to violate the known laws of classical physics. Heisenberg craved bold steps to create a whole new way to carry out calculations for atomic behaviour and discrete spectral lines. He strived for new quantum principles that could remove every difficulty in a single stroke. Quantum theory was a hodgepodge of hypothesis and computational recipes rather than a logical consistent theory.

After suffering a severe attack of hay fever that plagued him for days, he retreated to the island of Heligoland in the summer of 1925 where he could safely escape all trees and pollen. There were no flowers, no fields, and no hay fever on the island. The crystal-clear ocean air freed his nasal passages as well as the cobwebs in his mind. Infinity lay within his grasp as he contemplated the open sea. He felt he had nowhere to go, and all the time in the world to get there.

The source of the trouble with Bohr's atomic structure and Schroedinger's wave mechanics was their desire to construct a visualizable solution. According to Heisenberg we never can know what really goes on in the atomic realm. We must abandon all attempts to construct pictures. All that we can legitimately work with is what we observe directly, which are the atomic spectral lines and intensities of those emitted lines. Heisenberg refused to subscribe to Bohr's mysterious orbits or Schroedinger's picturesque waves as the appropriate way to think about what electrons did inside an atom.

Heisenberg only thought about how atoms transition from one state to another, like the knights on the chess board, hopping into different squares. We only know what we have at the beginning and at the end of each hop. To try to describe what happens in between was pure, unnecessary speculation. Fairy-land pictures to capture the quantum state only led to paradoxes, like an electron in many places at once! How can an electron in one state, then the other, be here, then there, without being in between? Beautiful pictures are illusions.

It was time for a new language with entirely new concepts. He devised a new quantum theory very different from descriptions based on Newtonian concepts of forces, motion, position, velocity, and trajectories. The fundamental objects are no longer particles, occupying "positions" in space and moving with "velocities." Classical concepts only brewed trouble in the quantum world. Instead he constructed the atom's states, their transitions, and the ensuing spectra by rows and columns of numbers (matrices). To calculate and predict quantum behaviour he established mathematical rules to multiply those tables in a special way. It was a new map of the quantum domain. Only mathematics prevailed. He had no need to make up arbitrary rules, like Bohr's postulates. He did not need a God's eye view into the atom's insides to construct wavefunctions. Schroedinger and Born had concocted a fuzzy mixture of here and there.

Heisenberg named his re-formulation "quantum mechanics" to describe the world of the atom, as opposed to Newtonian mechanics which exquisitely describes the familiar world. The new quantum mechanics of electron states was a far superior plan. The beauty of the mathematics said it all. It was fiendish new mathematics, later recognized by Max Born as matrix algebra, rarely used in physics at the time.

Schroedinger's *wave mechanics* and Heisenberg's *quantum mechanics* were two beautiful but completely distinct theories, like chalk and cheese, but they both worked to give exactly the same answers, as Schroedinger subsequently demonstrated.

UNCERTAINTY

When they met in Berlin, Einstein provoked Heisenberg to go further in understanding his own developments. Sticking only to observable quantities and sorcerer's mathematics does not help the essential conceptualization process. Einstein challenged Heisenberg's determination to discard classical entities of position and velocity, and to abandon the idea of visualizing the path of an electron in an atom. Einstein pointed out that in a cloud chamber, which experimentalists used routinely to track the motion of charged particles (and often identify new particles), the electron's track can in fact be observed quite directly, through the tiny droplets of moisture that form in the chamber along the track. (The path of an airplane high in the sky can be similarly picked out by its contrails, the tracks left by line-shaped condensation clouds formed by the plane's exhaust.) It was too strange then to accept that the electron takes an exact path with position and velocity in a cloud chamber, but there is *no path* for an electron inside an atom. Einstein dumbfounded Heisenberg, but only for a brief period.

Back in Copenhagen at Bohr's institute, Heisenberg contemplated electron tracks in a cloud chamber. Do we really observe the *precise* path of the electron in the cloud chamber? What we essentially observe is something more spread out than the actual path. All we see in the cloud chamber are individual, tiny droplets, which are much, much larger than the electron itself, trillions of times larger. What we can see is a series of *ill-defined* spots through which the electron passes. The electron only finds itself approximately in a given place, and it moves approximately with a particular velocity. The position and the velocities cannot both be determined with absolute precision, and so proved useless to Heisenberg as the appropriate physical attributes of the quantum world. We can look at the world with the position-eye, or we can look at the world with velocity-eye. But when we try to open both eyes, position and velocity get fuzzy together. We get dizzy. The electron can be here or there within a certain spread. And its velocity can be between this much or that much, within a certain range. The fuzziness of the two entities is related by the fundamental Planck's constant. If we determine the velocity with complete precision, there will be an infinite amount of uncertainty in the measurement of position. The electron could be anywhere. It behaves like a wave which can be everywhere, even though Heisenberg despised Schroedinger's wave description. And if we know the electron's position with precision, then we have absolutely no knowledge of its velocity.

The uncertainty principle burst forth with ferocity in Heisenberg's mind, setting all of physics ablaze. A particle cannot have a precise position and a precise velocity at the same time. The more precisely we determine the position of a particle, the more imprecise is the determination of its velocity, and vice-versa. At first, he called it the *indeterminacy principle*. Heisenberg originally conceived of his immortal principle as a fundamental limitation of measurement. But Bohr straightened him out. The uncertainty relation places a limit on what is knowable, not just what is measurable. There are fundamental limits to the *knowledge* of both together.

With Bohr's suggestion Heisenberg changed the name to *uncertainty principle*. The uncertainty in position (Δx) and uncertainty in velocity (actually momentum) (Δp) are related via Planck's constant h, as per Heisenberg's original formulation:

$$\Delta x \, \Delta p > h$$

The uncertainty principle is at the heart of many physical circumstances that cannot be explained using classical (non-quantum) physics. By classical logic, we might expect the negative electrons in an atom to collapse into the positive nucleus due to the attraction of opposite charges. Bohr had to deny this catastrophe by an arbitrary postulate. But the uncertainty principle provides the missing foundation for why the atom does not collapse. If an electron got too close to the nucleus, then its position would be known to the size of the nucleus, and, therefore, its momentum (and, by inference, its velocity) would become enormous. In that case, the electron would be moving fast enough to fly out of the atom altogether. Therefore, the electron maintains a fuzzy position – a symmetrical spherical shell with the radius of first Bohr orbit – around the nucleus.

Heisenberg's principle can also be expressed in terms of related physical quantities, energy and time. Recall from Chapter 1 how translation symmetry (in position) leads to conservation of momentum, connecting position and momentum as the two quantities in Heisenberg's uncertainty relation. Similarly, time translation symmetry is related to the principle of energy conservation, showing the physical connection between energy and time, and a corresponding uncertainty principle:

$$\Delta E \, \Delta t > h$$

One consequence of the energy–time uncertainty relation is that for very short periods of time Δt (e.g., 10^{-21} sec) a quantum system's energy can become uncertain δE ($= h/\delta t$) leading to enough energy creation out of vacuum to manifest as a virtual particle–anti-particle pair. Vacuum is not empty space but teeming with very short-lived virtual electron–antielectron pairs that momentarily pop in and out of existence. Another uncertainty coupled pair is angular momentum and angular position, similarly connected by rotational symmetry.

Einstein was greatly troubled by the new-fangled uncertainty principle. Our ability to simultaneously and accurately know both position and velocity has always allowed physics to exactly predict the future course of events, like the course of the planets, and the comings and goings of comets. Precise predictability of the paths of moving objects has been the physicist's central credo. In the classical picture of the natural world all the working parts can be defined with limitless precision. But if we cannot determine the particle's position and velocity precisely, how can we determine its exact trajectory? How can we determine its future position? In the quantum world Heisenberg killed determinism, the business of determining with certainty what has happened, what is happening, and what will happen. For Heisenberg there is no precise trajectory for an electron in the atom or in the cloud chamber. Uncertainty and probability demolished hopes of returning to a deterministic atomic world.

As Bohr perceptively concluded, uncertainty is built into the quantum world, so that an electron cannot possess a precisely defined velocity and position. The uncertainty principle rules the atomic world as an inherent principle. It turns particles into waves by preventing an electron from being located at a specific place. Instead there is only a probability that the electron might be here, and a probability that it might be there. So the particle turns into a wave of probabilities.

Finally, the uncertainty principle shows how the two pictures, wave and particle, can co-exist. Schroedinger's probability waves give the very same odds as the uncertainty principle that an electron will be found here or found there. Probability unites the two views of Heisenberg and Schroedinger, even though the two stalwarts detested each other and disagreed with their distinct approaches! Both uncertainty and probability are required to describe the world of the atom.

The uncertainty principle fit flawlessly into Bohr's emerging views of how the quantum world could reconcile the contradictory natures of wave and particle. We cannot know position and velocity simultaneously any more than we can know wave nature and particle nature simultaneously. Position and velocity are complementary. Waves and particles are complementary! Simultaneous expressions of wave-nature and particle-nature are not possible. Precise measurements of the two complementary entities, position and momentum, are also not possible – which is the heart of the uncertainty principle.

For Bohr, complementarity was the centrepiece of uncertainty. Both particle and wave natures of light and of electrons are valid manifestations of electrons. But the two different natures cannot be revealed at once. The two pictures are not contradictory. They are complementary. Bohr had found the way to make the two seemingly opposing theories (wave or particle) account for nature with equal elegance and success. Bohr's and Heisenberg's quantum interpretations of uncertainty and complementarity soon became known as the Copenhagen interpretation.

DEMISE OF DETERMINISM AND CAUSALITY

Quantum probability smashed the foundations of the deterministic world. The uncertainty principle rung the death knell for causality. We can only say "if we do this" then that will happen or not happen, with certain probabilities. When an atom emits light there must be a cause which determines the direction the light will take. Einstein found it quite intolerable that an electron should "choose of its own free will" the moment to jump off, and the precise direction too. Such answers cannot be left purely to chance. Einstein was not ready to abandon determinism or causality in any part of physics. Uncertainty and probability could have no proper place in physics for Einstein. Even God would subscribe to cause and effect.

If there is uncertainty, he encouraged quantum scientists to look up nature's sleeve for the hidden cause for that uncertainty. They were just missing something that must be found in the future. They must continue to believe in, and keep looking for, certainty. We must be able to restore causality once some new, hitherto hidden, properties were to be uncovered that would allow cause to be traced precisely to effect. Otherwise the poison of the quantum paradox could continue to hang in the air.

There was no way to tame the unruly quantum child and find a place for it in the elegant Newtonian universe. Unlike Isaac Newton's clockwork universe, where everything follows clear-cut laws on how to move, and prediction is possible if you know the starting conditions, the uncertainty principle enshrines a structural level of fuzziness into quantum theory.

Quantum uncertainty of classical concepts of position and velocity does NOT mean that quantum physics can no longer come up with precise answers and make precise predictions about quantum entities. The evolution of the Schroedinger wavefunction describing a physical system is perfectly deterministic. If nothing ever interrupts the Schroedinger evolution of a system, the wavefunctions governed by the equation develop according to deterministic laws to tell the complete physical story of the system. Thus quantum wave mechanics is a perfectly deterministic theory which can make exact predictions about atomic behaviour. Experiments have confirmed quantum theory's predictions with phenomenal accuracy – to a more than a dozen decimal places. The quantum theory offers a renewal of confidence in the same kind of causal deterministic laws as found in Newton, Maxwell, and Einstein's theories. Newton's world and the quantum realm provide us with precise answers to different questions. Exact predictions are still possible, just not for classical quantities, like the positions and velocities of electrons.

The power of the quantum theory has vastly expanded our ability to manipulate the atomic scale to bring us the high-tech worlds of computers, smart phones, lasers, internet, medical imaging, communications from the digital revolution, and much more. What is happening is still deterministic; quantum states are predictable with exactness.

OBJECTIVITY DEMOLISHED

At the Fifth Solvay International Conference on Electrons and Photons held in 1927 in Belgium, the principle inventors met to discuss the many aspects of the newly founded quantum theory. Planck, Einstein, Bohr, Schroedinger, Heisenberg, Born, Curie, and Pauli were all present, among 29 major contributors to advancing physics. The conference was dominated by disputes between the stalwarts Einstein and Bohr. Einstein maintained that a quantum theory based on uncertainty can never be complete. We only need to invoke probability when we have incomplete knowledge, as for example in card games. For Bohr, quantum probabilities are different from the probabilities of card games. It is not chance. It is based on the inherent *un-knowability* in the quantum world of the familiar (classical) concepts of position and velocity.

In the first quantum revolution, Einstein's genius and insight established the early foundations for the new quantum physics by recognizing the particle nature of light. But now, he surprisingly transformed into quantum physics' most determined critic. With relativity, Einstein boldly rejected the classical ideas of Newton's absolute space and time, but now he could not accept the revolutionary views of the atomic world. Uncertainty leads us to reject the classical concepts of wave and particle. Every evening at the conference, Einstein presented paradoxes arising from the new quantum interpretations, and by morning Bohr devised clever solutions to each one. And in the process of inventing the solutions, Bohr advanced quantum concepts even further to uncover new fundamental aspects of the quantum world. Einstein the relentless doubter instigated fundamental progress in many directions.

In the first salvo, Einstein offered the following challenge. We know well from the wave nature of light, that if a light wave moves towards a wall which has two openings, one at A and one at B, the wave will travel through both holes to form an interference pattern of light and dark bands. But what if a single photon replaces the light wave. If light is indeed a wave, will the photon travel as a spread-out wave, and go through both holes, or will it travel as a singular particle, and go through a single hole?

Bohr was only flummoxed for one night. The next morning he had a clever answer with a brand new insight. *The measurement will decide!* If you make a measurement you can determine which hole the photon went through (A or B), which means the photon will propagate like a particle through *one of the holes*. But if you leave the photon alone without making any measurement, the photon will behave like a wave, to *go through both holes*! In other words, if you send many photons and monitor which path the photons take, the pattern on the screen will be two separate bands of light (impinging photons) behind each hole. But if you do not monitor the path, you will get an interference pattern of light and dark bands as the two waves interfere with each other.

At first this sounds like a ridiculous answer. How can a single photon go through *both* holes, without splitting into two, violating its elementary nature? Bohr invoked Schroedinger's probability wave description for photon wave propagation. Left alone, the electron wave will pass through both holes, with a *probability* of going through each hole. The probabilities take on values such that if you repeat the experiment with many, many photons, they will impinge on a screen in an interference pattern, exactly the same as a passing wave composed of many, many photons. Quantum probability decides which hole the single photon goes through. It was similar to the answer that Einstein reluctantly proposed earlier for his own paradox about whether a single photon reflects or transmits when it encounters a pane of glass. Einstein himself had half-heartedly injected probability into the quantum world, and then fought valiantly against it.

Bohr's brand-new insight was that *measurement decides* if the photon (or electron) behaves like a particle or wave. If you make a measurement, the photon flies like a particle through hole A or B. The way to realize the photon as a particle from the propagating probability wave is measurement. So the experimental test determines which side of the photon's dual nature (particle or wave) actually appears.

Einstein was appalled by this new, disturbing aspect of quantum physics. Quantum theory exposed a new level of weirdness. Bohr was sacrificing objective reality. Reality must exist, independent of any observer. An electron should have the properties of a particle or a wave, whether you measure the travel path of the electron, or leave it alone. Nature must be cleanly separable from human beings who study it. The external world exists independently and apart from us. It does not depend on us. We do not need to test for the moon to know if it is really there. The moon is always the moon, even if no one is looking at the moon. It keeps on exerting its force on the oceans to form the tides whether we are at the ocean or inland.

Bohr's answer was that the moon belongs to the classical world. The electron belongs to the quantum world. In the quantum world, reality remains the multi-faceted probability wave, until perceived. Things hover in a probabilistic haze, until the measurement collapses the wave. By choosing what to observe (for example, particle), we destroy our ability to know and measure the complementary property (for example, wave). The unwatched electron is not actually anywhere, but is potentially everywhere!

Quantum theory represents an object differently depending on whether it is being observed or not. The properties (position, velocity, spin) of an object (for example, electron or photon) in a quantum state can take on a range of possible values that oscillate in a wavelike manner. The unobserved object is a probability wave of possibilities, not a particle with definite properties. The unrecorded quantum world exists only as an intermingled realm of possibilities.

When an object is observed, it always is at one particular place, with one particular velocity and one particular spin. It is no longer a smeared-out range of physical properties. With the measurement, the description abruptly shifts – from a range of possible attributes to single-valued actual attributes. This sudden measurement effect is called "the collapse of the wave," or also "the quantum jump." The collapse ushers one of the many potentialities into reality. Such peculiar aspects become less and less noticeable the larger the system. The exact boundary between quantum and classical is unclear at this stage. The vanishing of the wave of probabilities is no more strange than the vanishing of the probabilities that John, Dick, or Harry will win the lottery once a winner is chosen.

QUANTUM ENTANGLEMENT

After Solvay, Einstein grudgingly accepted that quantum theory is enormously successful in explaining and predicting experimental observations of the microworld. But at the fundamental core, he still held that the theory must be incomplete. A complete description of physical reality must be possible, eventually. In 1935, he thought (along with collaborators at Princeton, Nathan Rosen and Boris Podolsky) that he could put a stake through the heart of Bohr's and Heisenberg's quantum interpretations, a final disaster for the quantum braves. Bohr had truly laid a quantum egg. It was time for Einstein to finally subdue the quantum weirdness. The position and velocity of a particle do not have to be uncertain. They can both be determined definitely. He showed a way. A particle can indeed possess a well-defined velocity and position. His brilliant thought experiment is now called the EPR paradox. Here is a very simple form of his new and final challenge.

Suppose two electrons A and B emerge from an explosion with equal and opposite velocities. Einstein's goal was to show how to determine *both the exact velocity of B and the exact position of B*, in violation of the uncertainty principle. If at any moment we know the velocity of A, then the velocity of B will be the negative of that number. This flows from an undeniable law of Newton's physics (conservation of momentum). If an experimenter (Albert) measures the exact velocity of A at a particular time, then Albert immediately knows the exact velocity of B, *but without measuring the velocity of B* (it is the exact opposite of A's velocity.) Albert does not care that the position of A is now completely unknown. And then, since Albert has not actually performed any measurement at all on B, there is nothing to stop him from determining the exact position of B. So, then Albert knows by measurements … *both the exact position of B, and the exact velocity of B.* Electron B indeed does possess an exact position and velocity, and Einstein showed how to determine both.

On receiving the dare, Bohr was very upset. It would be the end of the quantum world (as he knew it), if Einstein was right. He had to find a way to clear this up at once. He searched a long time for the right way to speak about it, but failed many times. Einstein posed an ingenuous and subtle argument. Finally, Bohr understood how to answer.

Once again, Einstein unwittingly led Bohr right into a new and exciting part of quantum physics. Albert's pair of particles A and B formed *one single* quantum wave. When Albert measured the exact velocity of A, he immediately determined the exact velocity of B. There was no denying that. But with that one measurement of velocity, the position of A became completely uncertain. *And also the position of B became totally uncertain*, because the two particles are coupled into a single wave. So yes, Albert knows the exact velocity of B, but the positions of *both* A and B become totally uncertain.

Einstein fought back. How the devil did the position of the far away particle B become uncertain by Albert's measurement of the velocity of A? B could certainly have travelled too far away to have any physical communication with the partner A. Bohr declared that from their initial interaction A and B formed into *one* single quantum wave that spreads out over the universe as a single entity, as long as the pair is not disturbed (by interacting with the environment). Whatever Albert does to A *instantly* happens to its partner B. No physical information has to pass from one to the other. Indeed, physical information can only travel as fast as the speed of light, but this linkage between A and B is not subject to the limitation of information travel because of the one-wave nature of AB.

Einstein was once again appalled, calling this behaviour "spooky, action at a distance"! How can something here be instantaneously linked to something there? How can a quantum state abruptly change all the way at the other end of the entire universe after you observe it in one place? Something here cannot instantly be linked to something there! How can quantum linkage have no spatial limit?

A more modern situation demonstrating the EPR paradox involves the quantum coupling of two electrons (A and B) with opposite spins. If two such coupled electrons travel over a large distance, their spins remain coupled as a single wave. Their spins will always be opposite, but the spin of each electron will remain in the unknown, indeterminate wave state (up or down) until a measurement is made. Once a measurement determines that electron A has spin up, then electron B will instantly have – and show – the opposite spin (down). Or if the measurement determines A to be spin down, then B will turn out to be spin up. Again, it is puzzling how the far away electron B instantly "knows" that A's spin has been determined to be one way. Bohr's answer was that the two electrons form one coherent system whose wave spreads out over any distance. The information about one spin is always linked to the information about the other spin. A measurement of up spin collapses the wave to turn the probability from (50 per cent up and 50 per cent down) to 100 per cent up, and correspondingly turns the probability of the other spin to 100 per cent down.

THE QUANTUM CAT

Schroedinger became Einstein's ally in clinging to the deterministic universe. He was sceptical about Heisenberg's uncertainty principle, and the entire Bohr–Heisenberg interpretation of quantum physics (also called the Copenhagen interpretation). He was certainly very happy that Einstein had caught these dogmatic clowns by their coat tails with the EPR paradox. Einstein had shut down Heisenberg's uncertainty principle and Bohr's fuzzy reality in one sweep. A real world, with precise properties, does exist – without the need for any observer!

Again, Einstein stimulated Schroedinger to devise the famous "cat paradox" to ridicule Bohr's wily answer to the EPR paradox, using a *reductio ad absurdum* approach. Like Einstein, he could not accept that a measurement on one particle would instantly affect the other particle since that would require information to travel faster than the speed of light, in violation of special relativity. He coined the phrase "entangled particles" to express Bohr's description about the two particles as *entangled* into one quantum wave. Now "quantum entanglement" has become a famous demonstrated and useful property of the quantum world, as we will soon see.

Schroedinger thought he could devise a way to entangle an unfortunate cat with an atom, and bring about an absurd result. Suppose that a poor cat is trapped in a sealed chamber along with a diabolical atomic device that releases a deadly poison gas from a vial according to the "crazy" laws of quantum probability. Present in that same box along with the cat and the poison vial, imagine a small batch of Bohr's famous excited atoms that have absorbed photons. Let's say that in that first hour, there is a 50 per cent chance for one of the excited atoms to release a photon. There is also a 50 per cent chance that no atom will release a photon in that first hour. If a photon is released it activates a trigger mechanism to discharge poison gas from the vial. According to quantum laws, there is no way of telling in that first hour when an atom will decay from its excited state to release a blue photon. In that first hour, there is no way of knowing whether one of the atoms is decayed or not without making a measurement of the quantum state of the atom. The atoms remain in a mixed state of excited and decayed. Now here is the dastardly puzzle: in that first hour, there is no way of knowing whether the cat is dead or alive, without looking into the box. If the atoms are in a mixed state of excited and decayed, the cat must also be in a mixed state of dead *and* alive at the same time. And that is patently absurd! Cats have to be either dead or alive with certainty.

Using the laws of quantum superposition, Schroedinger came up with a cat that is both dead and alive in the first hour! We can have no knowledge about whether there has been an atom-decay or not. Only an act of observation can decide. And so, the cat is consigned to quantum purgatory, a superposition of states in which it is neither dead nor alive, or is it both? Is it really logical for observation to be the trigger that collapses a wave of possibilities into an actuality? Wouldn't the cat be certainly either dead or alive, even if not observed? Thus, he reasoned that the Copenhagen interpretation of quantum physics must be inherently flawed.

Bohr responded that Schroedinger's cat paradox was ridiculous. Quantum superposition does not work with large (classical) objects, such as cats coupled to a single or few atoms. The cat cannot form a quantum entangled wave with the atoms because the cat is composed of more than 10^{25} atoms. Any entanglement would be instantly destroyed by the cat's molecular activities, such as breathing in air, which will involve interaction with 10^{23} molecules. Bohr argued that macroscopic objects never achieve any fuzzy superposition. But how big does an object have to be to be classified as classical? Where is the quantum–classical changeover?

Recent experiments on entangled systems have been extended to molecules with multiple atoms (such as NH_3), large collections of 3,000 Rubidium atoms, and even coupled vibration in objects as large as sub-mm crystals of diamond visible to the naked eye, although for a miniscule amount of time (picoseconds, 10^{-12} s). The diamonds were about 15 cm apart. The vibrations involved 10^{16} atoms. Whether entanglement can extend to a system (i.e. cat) with 10^{25} atoms is in principle possible, but never achievable.

Quantum physics certainly has an interesting history with fascinating philosophical underpinnings. But are any of the esoteric concepts of probability, superposition, uncertainty, and entanglement of any use? Indeed, quantum physics plays a big role in our everyday activities, transforming our lives in profound ways. Quantum mechanics is at the foundation of transistors, lasers, smartphones, personal computers, the internet, lasers, and much much more. The next section explores the applications of quantum principles and effects to the quantum computer, which can solve a host of problems intractable for the most powerful computers today.

QUANTUM COMPUTING, QUANTUM COMMUNICATION

A fusion of quantum physics and computer science, quantum computing harnesses quantum superposition and quantum entanglement to process information. In analogy with classical computing circuits called "bits," the unit of quantum computing is a "qubit." A bit can be set to a 1 or a 0 since it is just a switch with two possible configurations, "on" or "off." A classical bit is like a coin which is *either* heads (0) or tails (1). A set of n coins can be described as a probabilistic mixture of 2n states, but the set of coins is actually in only one of these states.

A quantum bit or "qubit" can take on a quantum superposition of an infinite range of values between 1 and 0, behaving like a spinning coin for analogy. Each coin can be *simultaneously* heads or tails, with each state having a probability (of 50 per cent) when landed. If two coins are spinning at the same time they would represent a superposition of four states. Three coins would represent eight possible states. Fifty spinning coins would represent 2^{50} (10^{15}) states, which is more states than possible with the largest supercomputer today. Three hundred coins would represent more states than there are atoms in the universe. Because of quantum superposition, qubits can represent and process far more information than binary bits. For successful quantum computing it is necessary to keep the coins spinning properly in the superposition of states for a long time in order to process their information, but if the system is accidentally disturbed it will introduce errors.

The classical computer accomplishes two main tasks using transistor-based switches, storage, and processing. Computers store numbers in memory. A transistor can either be on or off. It stores 1 if it is on, and 0 if it is off. Long strings of 1s and 0s store any number, letter, or symbol using the binary code. A string of eight bits (byte), can store 255 different characters (such as A to Z, a to z, 0 to 9, as well as common symbols).

Computers process the stored numbers with operations like "add" and "subtract." They do more complex calculations by stringing together simpler operations into a series called an algorithm. Multiplying can be done as a series of additions, for example. To perform calculations, computers use transistor-based logic gate circuits which turn patterns of bits (stored in temporary memories called registers) into new patterns which correspond to the result of operations. A modern classical computer microprocessor is a single-chip which packs hundreds of millions (up to 30 billion) transistors onto a single chip of silicon. As transistors get smaller the computer power increases. But transistor miniaturization is approaching limits imposed by the laws of physics. For example, 5 nm line-width circuits will lose electrons due tunnelling discussed earlier. Since most conventional computers can do one thing at a time, the more complex problems take many steps and much more time. Problems become intractable if they need more computing power and time than any modern machine computer.

A quantum computer has the analogous key features of a conventional computer, qubits, registers, quantum gates, and algorithms. But it also processes all the states simultaneously. It can encode an exponential amount of information with qubits using superposition, and it can manipulate the superposition in parallel. This quantum parallelism allows some types of problems to be solved exponentially faster than with existing classical algorithms on a classical computer. Estimates suggest a quantum computer could be millions of times faster than a conventional computer. To carry out such complex calculations and exceed the capability of a classical computer, the qubits also need to work together, as two-qubit operators. Quantum entanglement is the key for the success of multiple-qubit operators. Entangled qubits affect each other instantly when measured, no matter how far apart they are. Without entanglement, quantum algorithms may not be able to show ultimate supremacy over classical algorithms.

Quantum computing needs new types of algorithms to solve problems faster than a classical computer. But there are many problems for which quantum computing is not cost-effective. Quantum computers and classical computers will coexist and complement each other.

Many types of qubits (about a dozen) are under development. The two advanced versions are the ion-based and the circuit-based. In the first, positively charged calcium ions are trapped in an array with electromagnetic fields, and the second uses superconducting electronic circuits. Information is stored by putting qubits into one of two well-separated energy states. Information is processed by converting qubits between states. The answer is read out by measuring the states after completing operations. Trapped ions have longer coherence time compared to a superconducting circuit but the gate operation time is faster in the superconducting circuit.

In the ion-based qubit, a string of ions floats between electrodes. Laser beams encode and read-out information to and from individual ions by causing transitions between an ion's electronic states. In a quantum logic gate, the ion's internal energy is used for the first qubit. The lower-energy state represents (0) and a higher-energy state represents (1). A second quantum bit is the atom's external motion: (0) represents less motion and (1) represents a greater amount of motion.

When two bits are entangled, the pair can serve as a quantum version of a CONTROLLED NOT (CNOT) gate. This is a gate which produces an output signal only when there is NOT a signal on its input. The CNOT gate is used to entangle (and disentangle) qubits. It operates on 2 qubits. It flips the second qubit (the target qubit) if and only if the first qubit (the control qubit) is in the (1) state. When the ion's internal energy state is put into a superposition of 0 and 1 the entanglement simultaneously puts the ion's motion into a superposition of 0 and 1. Thus the gate can process multiple possibilities simultaneously.

The superconducting circuit-based qubit simulates atomic states with two quantum energy levels. Quantum operations are performed by sending electromagnetic impulses at microwave frequencies (many GHz) to the resonator coupled to the qubit. In more detail, the superconducting qubit is a resonant circuit with a superconducting inductor and a superconducting capacitor that can resonate together in evenly spaced energy levels at microwave frequencies. (A vibrating string is a familiar resonator which resonates in many different modes to make different musical notes.) Instead of using wire-wound inductors and metal plate capacitors, the resonator is a microwave superconductor metal box (for example, niobium), with one part of the box serving as the inductor and another part as capacitor. By using superconducting materials, losses due to electrical resistance can be greatly reduced, which is necessary to preserve long coherence (survival) times of superposition.

To operate as a qubit, two adjacent resonant energy levels must be made uneven. These levels will then serve as the two states of the qubit, (0) and (1). With the uneven levels, the control signals can only make changes between the two chosen quantum states, and avoid accidentally promoting the excited quantum state to a higher level. To achieve the uneven levels, the superconducting inductor is replaced by a special (Josephson) type of superconducting junction. The Josephson junction contains two aluminium superconducting electrodes weakly coupled through a thin (sub-micron) insulator. The capacitor remains part of niobium superconducting cavity resonator. Quantum operations are performed by sending pulses at microwave frequencies to the resonator. The resonant frequency corresponds to the energy separation between the two selected energy levels. The duration of the pulse controls the angle of rotation of the qubit state. Quantum computers use superposition and entanglement. Two qubits are entangled by using a superconducting bus which stores or transfers information between independent qubits, or it entangles two qubits.

Quantum information is vulnerable to disturbance from the environment, and suffers decoherence (loss of superposition). Qubits can interact with their environment to cause their quantum behaviour to decay and superposition to disappear. Qubits need to preserve their superposed state long enough to perform the large number of calculations before everything collapses. To minimize decoherence from thermal interactions, a quantum computer based on superconducting circuits needs a helium-dilution refrigerator to cool down the quantum processor to a hair above absolute zero, about 15 milliKelvin. The longest coherence time in such computers today is of the order of milliseconds. Superconductors are being improved to increase these times to seconds.

Conventional computers based on silicon chips and integrated circuits have been doubling in power and processing speed nearly every two years for decades. But they still are unable to solve certain types of problems. To quickly find the prime factors for very large integers is a function of exponential growth with the number of digits. Other equally difficult problems are molecular modelling and mathematical optimization problems which grow exponentially demanding as the number of atoms in the systems increases. These are the types of problems to which the quantum computer is best suited.

Quantum computers will allow us to study, in remarkable detail, the interactions between atoms and molecules. For example, nitrogenase enzymes (produced by bacteria) pull nitrogen from the air to make ammonia for fertilizers to feed the world. A very small portion of the precious enzyme contains only four atoms of iron and four atoms of sulphur. Simulating the interactions between even this small portion of the enzyme is barely possible with the most powerful classical computer today. But as the number of atoms increases, the interactions grow exponentially making it too difficult for the classical method. Since a quantum computer can encode an exponential amount of information on qubits using superposition, and then manipulate the superposition in parallel, it can solve such problems exponentially faster than existing classical algorithms (in principle).

Simulating the interaction between molecules could help design new drugs and new materials. Pharmaceutical companies can analyse and compare compounds to create new drugs. The automobile industry anticipates using quantum computers to simulate the chemical composition of batteries for electric cars. Airlines can calculate fuel-efficient paths for aircraft from an exponentially growing list of possible paths. Cities can calculate optimal routes for buses and taxis to minimize congestion.

A quantum computer can efficiently factor large integers which are the product of a few prime numbers (e.g., products of two 300-digit primes) using Shor's algorithm to find the factors. Since integer factorization underpins the security of public key cryptographic systems, a quantum computer could break many of the cryptographic systems in use today. New cryptographic systems would need to be developed. In another powerful application, quantum computers could efficiently search for a specified entry in a large, unordered database using Grover's algorithm, greatly speeding up all search engines.

It will be many years before we get quantum computers that will be broadly useful. The primary reason is that the number of qubits is small (<100) and qubits struggle with high error rates when quantum states decohere on interaction with the environment. In 2018, Google demonstrated 72-qubit information processing. Rigetti Computing announced plans for a 128-qubit quantum chip. D-wave claims a 2,000-qubit computer for large search algorithms, and has announced a road map for a 5,000-qubit version by the year 2020. These companies use superconducting circuit-based qubits due to their prior semiconducting circuit expertise. IonQ and many university groups prefer ion-trap qubits due to their expertise in atomic physics and lasers.

QUANTUM COMMUNICATION

Thanks to entanglement, qubits physically separated by long distances could potentially create a channel for information transfer between sender and receiver. As the sender manipulates the photons at one end it results in an instantaneous change in the corresponding photons at the receiving end, thus creating ultra-secure communication network. Any attempt to eavesdrop or intercept the information would disentangle the particles which would alter the message and make it immediately obvious that a hacking attempt had occurred.

Another area of experimental advance is the increasing distance over which entangled systems are established. The distances are increasing with new attempts, one of the more recent ones having transmitted entangled photons over 1,200 km!

ANTI-PARTICLES

Like Dalton's atoms, Rutherford's neutron, and Pauli's neutrino, pure theory gave birth to another class of particles, *anti-particles*. From the symmetry of equations combining quantum physics with special relativity to describe electrons, Paul Dirac (1902–84) triumphantly predicted in 1928 that an anti-particle accompanies every particle of matter in the universe. It is equal in mass and equal in spin to its partner, but it carries an opposite electric charge. Its anti-nature describes a crucial property. If a particle collides with its anti-particle, they annihilate each other into pure energy. The anti-electron is called the positron. The proton has an anti-proton, the neutron has an anti-neutron, and so also for every fundamental particle. Soon positrons were discovered in cosmic rays, and later used for medical diagnosis in positron emission tomography (PET) to detect tumours.

Anti-particles remained Dirac's mathematical fantasy until nature provided evidence for the existence of anti-electrons (positrons) in cosmic rays. Then came anti-protons produced in collisions of high-energy protons with particles in a photographic emulsion. A particle accelerator provided the high-energy protons. High-energy accelerators now routinely produce small quantities of anti-protons, not more than billionths of a gram.

By theory, it is possible to have anti-atoms, where anti-protons replace protons, anti-neutrons substitute for neutrons, and positrons displace electrons. Anti-hydrogen has been produced and stored in the laboratory. Symmetry allows the existence of anti-planets, or even antigalaxies, although none have ever been detected in searches that extend over distant galaxies, and look for the energy output for collisions between large regions of matter and anti-matter. There are strong reasons to hold that most anti-particles disappeared soon after the universe came into being with the Big Bang. Due to a slight asymmetry that exists in the nature of the weak force, mostly matter particles survived to form the observed prevalence of matter over anti-matter in the universe.

PANDORA'S BOX

With the atom dismantled, Rutherford's nucleus raised a constellation of troubling questions. If the nucleus is so very small and positively charged, what holds it together? Why do protons not fly apart from the ferocious mutual repulsion within the positively charged core? Is there a force much stronger than electrical operating within the nucleus? If atoms are not basic entities, are protons, neutrons, and electrons the new building blocks of matter? Or can we shoot at protons (or neutrons) to find even smaller constituents?

At today's level of understanding, electrons are elementary, but protons are not. Indeed, a "strong force" operates within the nucleus to bind protons and neutrons together, a force more than 100 times stronger than the repulsive electric force between positively charged protons. When particle physics experimenters emulated Rutherford 30 years later to determine the structure of the proton by bombarding it with energetic projectiles of electrons and other protons, a veritable riot of new particles emerged. But none of these was a direct proton constituent, only combinations of other primary constituents. The proton could never be a conglomeration of such a large variety of particles. Some new particles were even more massive than protons! It was as if physicists opened Pandora's box to reveal a frightening number of new entities. To members of the proliferating zoo, particle physicists ascribed the names of Greek letters, such as *pi (π)*, *lambda (Λ)*, *sigma (Σ)*, and *xi (Ξ)*. Frustrated by the abundance of new particle species being discovered, Enrico Fermi once complained that he might as well be a botanist!

By 1963 particle physicists were drowning in the prolific Greek alphabet soup. A *Review of Modern Physics* article described 41 new particles. By 1970 the list exploded to several hundred. Most particles were unstable, decaying within incredibly short times into other particles. Like the wealth of chemical elements, there was a bewildering array, with various masses, charges, and spins. The particle harvest for experimenters became a nightmare for theorists. With each exciting announcement of a new particle or anti-particle came the painful realization that the fundamental nature of matter was getting further and further from the grasp of ultimate understanding. As the beauty and simplicity of atoms began to evaporate, physicists searched once again for order among the zoo of exotic particles. It was an effort to organize and classify, in much the same way as Aristotle sought to classify the elements of antiquity or Mendeleyev's struggle to organize the chemical elements into the masterful Periodic Table (Chapter 2).

Three broad classes emerged among the new particles: baryons, mesons, and leptons, named from Greek words: *baryos* for heavy, *meso* for medium, and *leptos* for light. For example, the proton and neutron are baryons while the electron and neutrino are leptons. Because they both feel the strong force, baryons and mesons are lumped together and called *hadrons*, from the Greek word for thick. Leptons, however do not interact through the strong force, only through the electromagnetic and weak forces.

A whole new language emerged to describe the new properties, behaviours, and interactions. In addition to charge, mass, and spin, physicists recognized new features and gave them names such as isospin, strangeness (hypercharge), baryon number, and lepton number. Underlying the first hints of order, particle theorists found mathematical patterns connecting the new properties. For hadrons, a fascinating integer relationship emerged between charge, isospin, strangeness, and baryon number. The simple connection pointed towards a more basic substructure, just as Dalton's Law of Multiple Proportions pointed to the atom, or the $2n^2$ period of the Periodic Table and the Balmer series pointed to atomic structure.

Once again, symmetry showed the way to an underpinning substructure. Murray Gell-Mann (1929–2019) succeeded in elucidating from the chaotic subnuclear landscape certain underlying patterns well hidden among the new properties. Mathematical symmetry piloted him to organize hadrons into a kaleidoscope pattern of *octets*. Eight low-mass baryons, including the proton, fit neatly into an octet family. Eight low-mass mesons fit into another octet group (Figure 12.9). With a magical flair, Gell-Mann dubbed the organization scheme *the Eight-fold Way*, likening them to

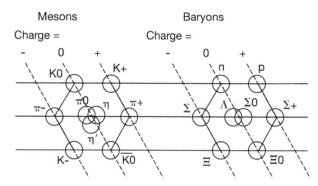

FIGURE 12.9   Gell-Mann's Eight-fold Way: subatomic particles, mesons, and baryons, arranged in octet families.[11]

the eight commandments of Buddha whose ancient scripture laid down the Eight-fold Way to guide monks through the appropriate path to Nirvana:[12]

> Now this, O monks, is noble truth that leads to the cessation of pain; this is the noble Eight-fold Way; namely, right views, right intention, right speech, right action, right living, right effort, right mindfulness, right concentration.

From the confused nightmare of exploding number and complexity, Gell-Mann's Eight-fold Way finally showed the road to the cessation of pain in particle physics. But the symmetries embedded in the arrays were barely obvious, and hopelessly intertwined as the eight heads in Escher's painting (Figure 12.10). By rotating the picture 180 degrees, one can readily recognize groups of heads not immediately obvious from one orientation. Symmetry transformation guides the eye to find concealed patterns. Order does exist, although at a deeper level than immediately recognizable. In 1965, around the time that Gell-Mann proposed the Eight-fold Way, Escher told his audience:[13]

> The laws of phenomena around us – order, regularity, cyclical repetitions and renewal – have assumed greater and greater importance for me. The awareness of their presence gives me tranquillity and support. I try to testify in my prints that we live in a beautiful, orderly world, and not in a formless chaos, as it so often seems.

THE QUIRKS OF NATURE

What was the central principle operating behind the octet patterns and integer relationships among the new properties? For Gell-Mann, the tumblers in the eight-fold lock clicked into position one by one to open a new door. Octet patterns were characteristic of a mathematical symmetry involving permutations of three labels. He could build octet groups out of three basic particles and their corresponding anti-particles. He needed a memorable name for the three new entities. A one-time student of exotic languages like Gaelic and Swahili, Gell-Mann happened to be reading James Joyce's *Finnegan's Wake*, a difficult work full of plays on words and intricate literary metaphors. The novel describes the life of Mr. Finn, who sometimes transmutes to appear as a Mr. Mark, just as a neutron transforms into a proton via beta-decay. He has three children whom he calls three "quarks":[14]

> Three quarks for Muster Mark!
> Sure he hasn't got much of a bark
> And sure any he has it's all beside the mark.

FIGURE 12.10   *Eight Heads*, MC Escher.[15]

Quark was just the whimsical name Gell-Mann sought to dub those peculiar objects composing baryons and mesons. The established number relation among the properties of hadrons demanded that members of the quark triplet have the quirky property of fractional charges of 2/3, –1/3, and –1/3. It would take inordinate courage to violate the sacred idea that charge always has to be an integer multiple of the fundamental unit of charge, as for electrons and protons. But Gell-Mann's soaring imagination was up to the task. When he called his friend Victor Weiskopf, the director of the European Nuclear Research Centre (CERN), about his dalliance with fractional charges, the shocked response was[16] "Please Murray, this is an international call!" It was strongly reminiscent of equally incredulous reactions to the concepts of atoms, electrons, and transmutation.

According to Gell-Mann's construction (Figure 12.9), all baryons are composed of three quarks, while mesons are pairs which comprise quarks with anti-quarks, like married couples. The proton consists of two up-quarks (each with charge 2/3) and one down-quark (charge –1/3). Therefore a proton's total charge is 4/3 – 1/3 = 1. The neutron consists of two down-quarks and one up-quark. Therefore the neutron's total charge is –2/3 +2/3 = 0 (Figure 12.11).

A handful of quarks could explain hundreds of baryons and mesons. Just as Mendeleyev's pioneer arrangement of chemical elements into rows and columns of the Periodic Table led to the prediction of gallium and germanium, the quark model predicted new particles, new combinations, with very specific properties. Immediately, experiments got underway to search for them. Quick discoveries sealed the credibility of the Eight-fold Way, increasing the wide acceptance of quarks.

Could quarks finally be the Holy Grail of fundamental elements? At the present level of understanding, there are six different types of quarks altogether in three generations of pairs. Their names are plucked from everyday language to describe new and exotic properties: (up, down); (strange, charm); and (bottom, top). Quark mass increases with each generation. The heaviest, top quark has the same mass as a gold atom with atomic weight 197! The lightest (up) quark is a fraction of the

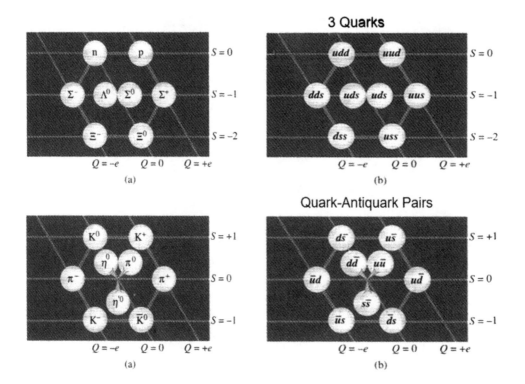

FIGURE 12.11   Baryon and meson octet compositions. Baryons are composed of three quarks/anti-quarks. Mesons are composed of quark–anti-quark pairs.

mass of a proton. Each quark has an anti-quark partner, as established by Dirac's general discovery. The leptons also span three generations in pairs: (electron, e-neutrino), (muon, muon-neutrino), and (tau, tau-neutrino). The muon is 200 times heavier than the electron, and the tau is heavier than a proton! The neutrino mass is extremely small, 500,000 times smaller than the mass of an electron, which in turn is nearly 2,000 times smaller than the mass of the proton. Why there is such an enormous range in the hierarchy of the masses of the elementary particles is an on-going quest. The quarks and leptons organized in Figure 12.12 all fall in the symmetry type of fermions.

"Familiar" matter is composed only of protons (up- and down-quarks), neutrons (up- and down-quarks), electrons, and e-neutrinos, which all lie in the first generation of fundamental particles. So why do the other two generations exist, even if only for a very short time when created in the laboratory using particle accelerator collisions? From in-depth understanding of the Big Bang creation of the universe (Chapter 13), we know that the other generations were created in the first moments after and died away in a short period. We can re-create the conditions of the Big Bang in powerful accelerators that smash electrons with anti-electrons (positrons) or protons with anti-protons in "mini-bangs," and so re-create the other generations of quarks and leptons. Accelerators and colliders have become the best way to probe deeper and deeper into the structure of matter.

To the best of our knowledge, quarks and leptons have no substructure. The modern "Periodic Table" of quarks and leptons shows a simplified arrangement of the new "elements" in the edifice of the Standard Model (Figure 12.12). The figure does not show the symmetric partner table of anti-particles discovered theoretically by marrying relativity and quantum physics, and later experimentally found in cosmic rays, and particle collisions. Perhaps in this millennium, the table will become the Standard Model for our schools in addition to the Periodic Table of chemical elements.

The Standard Model also incorporates three of the four known forces, electromagnetism, the weak force that operates in the nuclear domain to govern the radioactive decay of atomic nuclei, and

STANDARD MODEL OF ELEMENTARY PARTICLES

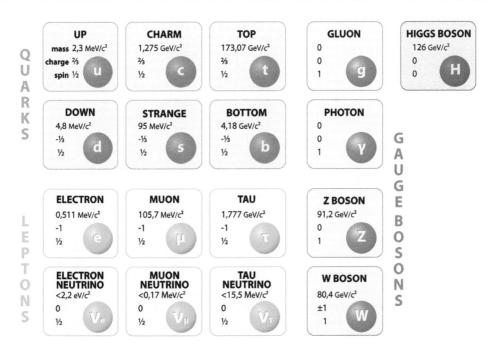

FIGURE 12.12   Modern Periodic Table of quarks, leptons, force carrier bosons, and the Higgs define the Standard Model of high-energy physics.[17]

the strong force which assembles quarks into hadrons in atomic nuclei. But it does not integrate the force of gravity, familiar since antiquity. We understand gravity separately via the general theory of relativity.

Maxwell unified the forces of electricity and magnetism (Chapter 3), while Einstein and Planck quantized the electromagnetic field to invent the photon. Advanced quantum physics of fields (quantum field theory) showed that the photon is the "carrier" of the electromagnetic force. Charged particles interact with each other by emitting and absorbing virtual photons, exchanging energy and momentum. Einstein searched in vain for a unified theory of forces, in particular gravity and electromagnetism. But he did not include quantum theory. A century later the electro-weak theory unified electromagnetism with the weak force to show that along with the photon which conveys electromagnetic forces, the conveyers of electro-weak forces are the carriers W-boson and Z-boson, particles discovered later at CERN. The carrier of the strong force is the gluon, named because of its crucial role in gluing the protons and neutrons in the nucleus together. The Standard Model (Figure 12.12) incorporates the four carriers of the forces: photon, W, Z, and gluon. Like the photon, the force-carrier particles all fall into the symmetry category of bosons.

But there was a major hiccup in the process of unification of the electro-weak force. Due to the symmetry requirement of the theory, the force-carriers, the photon, W, and Z all emerged from the theory with zero mass. This is certainly acceptable for the photon. But the W and Z which are responsible for the short range of the weak force have significant mass, nearly 100 times that of a proton. Peter Higgs (1929–) (and others) proposed an innovative solution with broad-ranging impact. Shortly after the Big Bang creation of the universe, as the universe cooled below a critical temperature, the electro-weak symmetry which zeroed all the masses was "broken," and a field, called the Higgs field, evolved spontaneously so that *any* particle interacting with it acquired a mass. Another example of symmetry-breaking is the formation of ice crystals from water molecules below a certain temperature. At high temperatures, the molecules can be aligned in any direction. But when ice forms, the symmetry of alignment breaks and the molecules make hexagonal crystals (Chapter 1).

The more a particle interacts with the Higgs field, the more its mass. The photon does not interact with the field, giving it zero mass. But the other two bosons, W and Z, of the weak force interact with the field and acquire mass. In fact, all particles in the universe must interact with this invisible Higgs field, which pervades the universe and gives particles their mass. The mass depends on the strength of the interaction of the particles with the Higgs field. This is somewhat analogous to white light interacting with the glass of the prism to separate into various colours, depending on how much the glass bends the light.

Just as the electromagnetic field has its quantum particle, the photon, the Higgs field has an associated particle – the Higgs boson. The largest accelerator operating today (LHC at CERN) discovered the Higgs boson in 2012. Long after Newton originated the concept of mass, the origin of mass, or resistance to motion, finally acquired a profound meaning connected to the origin and history of the universe. We are all immersed in the all-pervasive Higgs field, and our resistance to motion, which expresses as mass, is due to the interaction of our constituent particles with the Higgs field. A crude analogy would be fish swimming in water, encountering the resistance of moving in the fluid. If a fish discovered the H_2O molecule as the particle which constitutes water through which it must swim it would be like discovering the Higgs boson.

Is the model complete? Could six quarks, six leptons, and four force carriers along with the Higgs boson (Figure 12.12) be the fundamental entities of the entire universe? Cleary no. The Standard Model only accounts for a very small percentage of what makes up the universe, leaving gravity, dark matter, and dark energy all out of the picture. Dark matter (Chapter 8) makes up nearly 25 per cent of the mass of the universe, and dark energy (Chapter 13) makes up nearly 70 per cent of the universe's mass-energy.

GUTS AND STRINGS

As an extension of the Standard Model, theorists linked the electro-weak force with the strong force using an overarching symmetry, called supersymmetry, or SUSY. Several SUSY models have been proposed, called grand unified theories (GUTs). According to GUT, supersymmetry ruled in the very early moments after the Big Bang to unify the strong and electro-weak forces. Supersymmetry spontaneously broke down as the universe cooled to result in separated strong and electro-weak forces of vastly different strengths. The weak force is 100,000 times weaker than the strong force, and the electromagnetic force is 137 times weaker than the strong force. The separation of the strong and weak forces was similar to the precipitations of the electromagnetic and weak forces from electro-weak symmetry breaking, but at a later time after the Big Bang.

GUT theories are as yet untested and unconfirmed, because the energies necessary far exceed the capacities of the most powerful particle accelerators that can be built with known technology. One possibly testable prediction of GUT is that the proton has a finite but very long lifetime (more than 10^{34} years), so that experiments with a quarter million tons of water (10^{34} protons) are looking for rare decays of a few protons.

Another tantalizing supersymmetry prediction is that for every particle in the Standard Model (for example, quarks) and for every force-conveying particle (for example, photon and Z) there is a supersymmetric partner, called a sparticle, along with other exotic names. So electrons have selectrons, quarks have squarks, photons have photinos, gluons have gluinos, and so on, all as super-partners. The supersymmetric partners are supposed to have the opposite symmetry in the nature of their wave functions, so that since the electron is a fermion obeying the Pauli exclusion principle, the selectron is a boson, allowing a selectron laser, probably with amazing properties! And since a photon is a boson, the photino would be a fermion which obeys the exclusion principle. The early universe was filled with these heavier sparticles (if SUSY is right), but these all decayed into particles and perhaps dark matter.

None of these particles have yet been found, although there is great hope that the most powerful accelerator (LHC at CERN) may yet reveal one. The lightest, stable, supersymmetric particle is the neutralino which is a partner of the bosons, as a superposition of the photino, zino, and Higgsino. It is electrically neutral and interacts weakly with other particles of the Standard Model. Neutralinos would have the characteristics required to qualify as a dark matter candidate. Dark matter holds galaxies together and makes up 25 per cent of everything (Chapter 8). There are of course other theories about what dark matter candidates could be.

Even with supersymmetry, the Standard Model leaves out gravity. Gravity is ignored on the atomic scale, just as quantum physics is ignored on large scales. Missing is a theory of quantum gravity. A super ambitious attempt to unite gravity with quantum mechanics and bring together a unification of all the forces with gravity comes from *string theory*. The main idea is that the basic constituents of matter are not zero-dimension points but are tiny strings of energy vibrating in 11 dimensions – three spatial dimensions and time – plus another six (or more) dimensions that are invisible because they are curled up, or "compactified." Electrons and quarks merely correspond to different ways for the strings to vibrate in higher dimensions. Each vibrating string of energy is 100 billion, billion times smaller than the proton. To visualize it would be to read text from 100 light-years distance! The maths and physics of string theory are horrendously complex.

Most intriguing of all, string theory predicts (postdicts, really) the existence of a massless particle called a graviton. Gravitons are the presumed quantum messengers of gravity, the force carrier of gravity that is missing in the Standard Model. Supersymmetry has been incorporated into the quantum theory of strings, to create superstring theory, yielding impressive mathematical results to unify all four known forces in nature into a single consistent quantum field theory. String theory has the grand potential to become the Theory of Everything.

The greatest difficulty of string theory is that its extra compactified dimensions could have many different shapes and sizes, and be shuffled in billions of ways. Since the geometry of the dimensions determines how the strings vibrate and thus the particles and forces that they can manifest as, the theory allows for many combinations of physical laws and physical constants, not just the ones we are familiar with. Perhaps each mathematical possibility in string theory is realized in nature as a more expansive physical reality than our universe. Perhaps our universe is merely one of a boundless variety of multiverses, other universes that exist along with our own, but with different properties. Now of course we have left the realm of physics into wild speculations.

We continue to approach but never capture the Holy Grail. And many questions prevail. Why are there so many particles in the Standard Model? Why do quarks and leptons come in pairs? Why are there three generations with increasing masses? Is there a hidden unity between quarks and leptons, perhaps an even more fundamental particle which physicists sometimes call the "leptoquark"? It would exhibit the properties of leptons under certain circumstances and the properties of quarks under others. Could strings be yet be another layer of underlying structure?

We do not know any of the answers, but there are tantalizing hints! Physicists continue to devise and construct mammoth higher energy accelerators to probe the structure of matter, and powerful mathematical theories to further synthesize the understanding of their behaviour, as well as to unify the forces with which they interact among each other.

Perhaps the generations following us will think of our theories in the same light as we view Aristotle's primeval model of the universe as composed of shells of ether, fire, air, and water surrounding our terra firma.

13 Unity in Physics

All are but parts of one stupendous whole
Whose body Nature is, and God the soul ...
All Nature is but art, unknown to thee;
All chance, direction, which canst not see;
All discord, harmony not understood;
All partial evil, universal good:
And, spite of pride, in erring reason's spite,
One truth is clear, whatever is, is right.

Alexander Pope, *Essay on Man*[1]

Through the evolution of modern physics, the first synthesis of heaven and earth transformed into the quest for unity between the macrocosmos and the microcosmos, two vastly distinct regimes of the universe. Contemporary physics updates in previous chapters come together here. The first hint of a profound connection between the two regimes already came from the cosmic turmoil brewing during the birth, life-cycle, and death of stars. Starting from basic hydrogen, high temperatures and pressures in the nuclear furnace of stars forge the elements of the Periodic Table from helium to iron (Chapter 8). In violent star deaths, the aftermath of a supernova synthesizes heavier elements, and scatters freshly minted substances far and wide over the galaxy to provide matter for new stars to form.

Where did the starting hydrogen atoms in the universe come from? The answer involves an understanding that connects physics at the microscopic scales to properties of the universe and its contents on the largest macroscopic scales. This fascinating tale of the ultimate synthesis of space, time, matter, and forces, still under construction, will be our final story to unify the universe.

AN EXPANDING UNIVERSE

Einstein's general theory of relativity opened the door for physics to enter a cosmological phase to make predictions about the history and future of the universe. Since matter curves space and space-warps move stars and planets to change matter distribution, the universe continuously changes with time, an incredible prediction. The universe is dynamic. The cosmos expands or contracts.

Einstein was shocked to discover that his equations made such a troubling forecast. A dynamic universe was either expanding forever or contracting to nothingness. But the prevailing evidence of the behaviour of the universe (as of 1920) suggested that the universe as a whole appeared to be static, as most scientists believed. Stars moved only very slowly and randomly in all directions. The Milky Way was neither expanding nor contracting. All this confirmed Einstein's philosophical predilection that the cosmos cannot collapse under its own weight.

Stars in our Milky Way galaxy were considered the most significant members of the universe. Harvard's eminent Harlow Shapley (Chapter 9) even alleged the Milky Way was the entire universe. Whether other galaxies existed remained a matter of great debate.

But the equations of general relativity did not demonstrate a static universe. They showed that only a universe devoid of all matter could be a static universe. Even though it was[2] "gravely detrimental to the formal beauty of the theory," Einstein felt compelled to modify his formidable equations by arbitrarily adding an extra new term that would cancel out dynamic effects from the presence of matter. Calling it the *cosmological constant*, his new component kept the radius of the universe constant with time, bringing theory in line with allegedly prevailing observation. Einstein believed he stabilized the universe with a simple mathematical fix. The cosmological constant represented a hypothetical mysterious energy that repels gravitational contraction due to the presence of matter.

What a bitter pill for Einstein to swallow! After relying on pure thought to overhaul conventional concepts of space and time, energy and matter, after uprooting with pure force of mind the age-old notion that Euclidean geometry rules space, Einstein finally felt compelled to yield to the force of evidence. "All knowledge of reality starts from experience and ends in it" he had said in homage to Galileo.[3] Theory cannot be the ultimate word. Something had to be wrong with his theory, and he had to fix it, to conform to reality.

Had Einstein stuck to his aesthetic judgement, had he trusted his dependable guide through so many upheavals of physics, had he continued to believe that something so unifying as general relativity must necessarily contain a very deep truth, he would have grasped a higher reality and been the first to predict the expansion of the universe. In addition to special and general relativity, it would have been another outstanding, revolutionary discovery.

Unaware of general relativity or its aborted predictions, Edwin Hubble (Chapter 9) set out to track the motion of spiral nebulae, which he fervently believed were other galaxies like our Milky Way. At a conference in 1914 he heard about Vesto Slipher's (1875–1969) pioneering work from Lowell Observatory in Arizona. Slipher observed incredible speeds for the motion of spiral nebulae using the Doppler effect, more than 2 million miles per hour! How fast did other nebulae move? Perhaps all nebulae moved much faster than stars.

A long time prior, Christian Doppler (1803–53) discovered an important effect of sound waves. Sound emitted at one pitch by a moving observer is detected at a different pitch by a stationary observer. The pitch is sharper for an approaching train and flatter for a receding train, dropping one semitone if the car travels at 25 mph, and a whole tone at 40 mph. Fast-moving ambulance sirens show the same effect. If the source is stationary, the frequency of vibrations arriving at the ear (equivalent to pitch) is the same as the frequency which the source produces. If the sound source moves toward the ear, more waves enter each second, increasing pitch. The change in pitch can be used to determine the source's velocity. Doppler boldly predicted that since light propagates as a wave, the colour of light (which depends on the frequency of light waves – Chapter 12) should also change according to source velocity. The Doppler effect is the principle underlying radar measurements for police to check for speeding automobiles. Stellar spectra carry lines of known frequencies due to atomic transitions. If a star recedes from earth, all the spectral lines of its elements shift toward the red end, an effect called the *Doppler redshift*.

At the 1914 conference, Hubble was excited to hear Slipher announce how several spiral nebulae were travelling away from earth at astounding speeds of 900 to 1,800 km/s (several million miles per hour!), as compared to most stars which move at a relatively leisurely pace of about 200 km/s. And most nebulae were receding from the sun and from each other. The faintest showed the largest redshifts, about half a per cent in wavelength. Was there a systematic motion of matter in the universe?

Faint nebulae were Hubble's special domain (Chapter 9). From the 60-inch and 100-inch telescopes atop Mt Wilson outside Los Angeles, he settled the island universe controversy by finding stars and Cepheid variables which showed that Andromeda is our neighbour galaxy, 1 million light years from the Milky Way. Hubble was already convinced. All spiral nebulae must be distant galaxies. Inspired by Slipher, Hubble wished to accumulate spectral shifts and from there the recession speeds of spiral nebulae. But he had little expertise in the delicate techniques of stellar spectroscopy to attempt Doppler shift measurements. Fortunately, his night-time assistant, Milton Humason, a janitor-turned- astronomer, helped him out. Starting as one of the muleteers who helped George Hale lug heavy telescope parts up the treacherous slopes of Mt Wilson, Humason stayed on the cleaning staff. Soon he advanced to part-time photographer and assistant astronomer, eventually becoming expert at recording and analysing spectra. At the same time, he introduced Hubble to the pleasures of moonshine during the dry time of prohibition in the USA.

By 1929 Hubble had access to systematic shifts of spectral lines for 46 distant nebulae. Was there any pattern to the corresponding speeds of the rapidly receding nebulae? Hubble had the insight to interpret nebulae velocity data in terms of distances to nebulae. Measuring galactic distances was Hubble's forte. Unlike Andromeda, however, there were no Cepheid lampposts visible in the remote island universes to help pin down distances. But he picked up another clue. The very brightest stars in galaxies with known distances showed similar intrinsic luminosities. Now Hubble leapt to the assumption that in all galaxies, the very brightest stars must have a common intrinsic luminosity, a reasonable, but bold, assumption. He could use the inverse square law method to determine distances of remote galaxies after calibrating the intrinsic luminosity of the brightest stars in nearby galaxies using visible Cepheid markers. By 1929 he estimated distances to 24 galaxies out to 7 million LY. Plotting a graph of receding velocity versus distance, he obtained a linear relationship, now the famous Hubble law: the farther away a galaxy is, the faster it recedes! The slope of Hubble's linear relationship is called Hubble's constant, which Hubble determined to be 160 km/s per million light years. Hubble's constant turns out to play a major role in forecasting the history and destiny of the universe. The modern value is 238 km/s per million light years.

Here was the first piece of evidence that the universe is expanding. With Hubble's paper, it soon became clear that expansion and general relativity in its original form were indeed compatible. Besides the bending of starlight by the sun and the precession of Mercury's orbit (Chapter 11), yet another fundamental prediction of Einstein's powerful theory had come true, despite Einstein's unfortunate and untimely modification. In another two years Hubble and Humason extended their relationship from 7 to 100 million LY, and velocities up to 7 per cent the velocity of light.

Why are galaxies running away from us? Actually, we see galaxies receding because space itself is expanding at high speeds and carrying galaxies with it. If earth was an expanding balloon, every city on its skin would be receding from every other city. But the cities would not be sliding over the balloon surface. Only the space between them would increase. In a universe filled with galaxies, every galaxy recedes away from every other. But we must not fall into the erroneous conclusion that we are back in the special place at the centre of expansion. The view is the same from everywhere. Any observer in any galaxy will see other galaxies speeding away, and measure the same Hubble's relationship for an expanding universe without a centre. The universe is expanding. And the universe does not expand into something else, because all space expands.

The expanding universe electrified all of astronomy. But what drives the expansion? As we will see in the next section, the universe was energized to expand rapidly by the Big Bang that created the universe.

If the cosmos is expanding, we can conclude that the universe was much smaller at earlier times in the history of the expanding universe. Visible galaxies today are millions of light years apart, but earlier they were closer together, and at the beginning of the universe they must all have been so close together as to be in intimate contact with each other. So the universe of space, time, and matter began as a tiny point that suddenly exploded to create everything, everywhere, and every-when.

From Hubble's relationship between speeds and distances, it is possible to determine how long the expansion has been going on, and therefore a wonderful property, the age of the universe. Suppose a galaxy is 3.3 million light years away (in modern units one mega-parsec), and moving at 480 km/sec ($1.6 \times 10^{-3}c$), as Hubble measured. How long did it take to get there? Since speed equals distance over time ($v = d/t$), the time taken is just distance divided by speed ($t = d/v$). For Hubble's example galaxy, the time is about 2 billion years. From his expansion relation, Hubble estimated that the universe's age must be 2 billion years.

It was an embarrassingly low estimate by modern standards. Geological factors had already put the earth's age at 3.6 billion years. Rutherford, the father of the nucleus, arrived at a similar number from the cooling rate of the earth after taking into account heat generated by radioactive minerals inside the earth. (The modern estimate for the earth's age is 4.6 billion years.) How could the universe be younger than the earth and other planets? There was a serious problem with Hubble's expansion rate of 480 km/s per megaparsec. Once again we see healthy confusion at the raw edge of the frontiers of knowledge and understanding. More extensive data over time would bring the value of Hubble's constant down.

Hubble's colleague Walter Baade (Chapter 8) resolved the awkward contradiction in 1952 when he discovered, from the new 200-inch telescope on Mt Palomar, that Hubble's Cepheid-based distance calibration needed significant revision. There are two types of Cepheid lamp-posts with different period–luminosity relationships. Hubble's calibration Cepheids were larger, intrinsically brighter, and therefore much farther away than Hubble estimated. Overnight the distance scale expanded by a factor of 5 to 10, making the universe 10–20 billion years old. Scaling the distance ladder of the cosmos has always been fraught with calibration difficulties.

The modern value for Hubble's constant is 72 km/s ($2.4 \times 10^{-4}c$) speed for a galaxy 3.3 million light years away, which yields 13.8 billion years for the age of the universe, consistent with the estimated age of the oldest stars as 10 to 14 billion years. Hubble's constant also remains difficult to measure accurately because of other uncertainties in galactic distances. Determining the age of the universe from the present expansion rate also assumes that galaxies have always been receding at the same velocity since the beginning of time. For example, if galaxies receded faster in the past and are moving slower now, the Hubble time is only an upper limit to the age of the universe.

A BIG BANG SYNTHESIS

In 1930 Einstein and his (second) wife Elsa toured Mt Wilson and conferred with Hubble. When Hubble's wife, the tour guide, proudly explained to the illustrious visitors how Hubble and others used the giant telescope to determine the structure of the universe, Elsa Einstein remarked "Well, well, my husband does that on the back of an old envelope."[4] By 1931, faced with Hubble's overwhelming data, Einstein was ready to withdraw his unfortunate adulteration of his pure theory, throw out the ad hoc cosmological term, and call it his biggest blunder. Blessing the expansion of the universe, he wrote:[5]

New observations by Hubble and Humason concerning the redshift of light in distant nebulae make it appear likely that the general structure of the universe is not static.

A relatively unknown Russian mathematician, aeronautical engineer, and meteorologist, Alexander Friedman, had already challenged (in 1922) the world-renowned theorist's contrived static universe. With the proper treatment he even showed that the cosmological term did not actually succeed in stopping the dreaded expansion. At first, Einstein proudly rejected Friedman, mistakenly pointing out a mathematical error. Six months later he acknowledged his own error and magnanimously accepted Friedman's solution, but not yet the expanding universe. Friedman had come to another crucial realization. An expanding universe led to a brand-new understanding for the origin of the universe.

Quite independently, in 1927, a Belgian Catholic priest, Georges Lemaître, began developing the idea that there must have been a creation moment, a "time zero, a day without yesterday, when all space was infinitely curved and all matter and all energy were concentrated into one single quantum of energy."[6] This state he called "primordial atom." As a single atom, the "atomic weight" was equal to the mass of the entire universe. Being infinitely heavy, it must have been an infinitely unstable, giant nucleus! Lemaître called the atom's eruption into the newly born universe the "big noise." Today we call it the Big Bang. Perhaps the mysterious high-energy cosmic rays recently observed at the time were just "glimpses of the primeval fireworks." He hoped to invoke nuclear physics to explain the primary explosion of the universe. Lemaître's idea was the first attempt to link cosmology with nuclear physics, the physics of the very large and the very small, the macrocosmos with the microcosmos.

At Lemaître's time, nuclear physics was still in its infancy; the discovery of the neutron was still five years in the future. Astrophysicists such as Arthur Eddington (Lemaître's teacher) and Fred

Hoyle (who played a major role in the understanding of stellar nucleosynthesis) disliked the abrupt singularity of the "big noise" event. Hoyle was promoting the idea that the universe is in an eternal steady state. How would it ever be possible to calculate the evolution of the universe from any such dreadfully colossal cataclysm? The infinities alone would stall the mathematics. Besides, Hubble was probably wrong, since his relation yielded a ridiculous estimate for the universe's age. Hoyle deridingly labelled the idea the "Big Bang" on a BBC radio programme. The name stuck.

For the priest Lemaître, the singularity was an essential sign of God's creation of the universe. In 1927 Einstein praised Lemaître's mathematics but expressed great scepticism about the concept of an exploding universe. Why would God want to blow up his creation? Only after he met Hubble in 1930 and began to catch up with the latest astronomical evidence, did he nobly accept the foreign concepts of expansion and explosion.

In Russia, Friedman's student George Gamow picked up Lemaître's trail. Compared to the staid Belgian scholar and servant of God, Gamow was a flamboyant character who carried on like a member of the Hell's Angels gang. He roamed over Europe on a giant motorcycle when he went to study nuclear physics under Niels Bohr at Copenhagen, and drove pink Cadillacs when he eventually emigrated to the United States in 1934. Soon after Chadwick discovered the neutron in 1932 (Chapter 2), Gamow set out to determine what happened at the earliest moments of the Big Bang from the physics of the nucleus, as Lemaître hoped. Neutrons, protons, and electrons were thought to be the only fundamental constituents of matter at Gamow's time. The wild zoo of elementary particles (Chapter 12) was yet to be discovered. Extrapolating backward in time, he estimated the universe to be so small at creation that all protons and electrons had lost their identity to form a primordial soup of only neutrons. He dubbed the wild conglomerate *ylem*, from an old Greek word for the *substance* of the cosmos prior to birth of *form*. Gamow knew that neutrons are unstable and beta-decay into protons and electrons in a very short time (15 minutes), just the two components needed to pair up and form hydrogen atoms, the basic ingredient of the cosmos. Perhaps the Big Bang was a vast thermonuclear explosion which created all the hydrogen, Gamow wondered in 1948. It was at the same time as the first fearsome detonation of nuclear bombs on our planet. Perhaps the infinitely more cataclysmic Big Bang created all the elements. The theory of nucleosynthesis in the stars and supernovae for the elements of the Periodic Table (Chapter 8) was still a few years in the future.

Gamow proposed Big Bang nucleosynthesis as a doctoral research topic to Ralph Alpher at the Applied Physics Laboratory of Johns Hopkins University. With general relativity enthusiast Robert Hermann (from the same laboratory) as a fateful collaborator, they introduced a crucial new feature to Gamow's model. The neutrons had to be moving at high velocities and crashing into each other, turning the early universe into a supremely hot broth at a temperature of many billions of degrees. Radiation (photons) emitted at such high temperatures would be so energetic that it would instantly destroy any compound nuclei that tried to form. Continuous collisions would also keep photons from travelling very far, remaining tightly bound and confined to the primordial stew. Only when the universe expanded and cooled would the energy of photons drop below that necessary to form the first and lasting combinations of neutrons with protons. Today we recognize this first combination as the *deuteron*, nucleus of *heavy hydrogen*, also called *deuterium*, present in trace amounts of heavy water in our oceans.

Later calculations for Big Bang models showed that nuclei formation started a few minutes after the primary creation event when the hot cosmic soup "cooled down" to a billion degrees. Below that temperature, the deuterium nucleus (hydrogen with an extra neutron) could just survive the onslaught of "cooler" photons, and hold together. As the temperature and radiation energy continued to fall, deuterium nuclei fused together to form helium-4 nuclei which survived the assault of rapidly cooling radiation. By this stage, 25 per cent of all the mass of the universe converted to helium-4. But further nuclear synthesis ran into a hard barrier. No elements heavier than helium could form because a nucleus of atomic weight 5 is intrinsically unstable. It does not hold together long enough for another proton to stick to it, quickly breaking up into nuclei of hydrogen and helium. Yet another nuclear instability shows up at atomic weight 8, so that two helium atoms could not stick together either. As the universe continued to cool rapidly, to a temperature of 100 million degrees in a few hours after the Big Bang, element synthesis just stopped. Less than one part per billion of the cosmic mass turned into lithium, atomic weight 7. Quantitative understanding of nuclear physics and its inherent instabilities makes it impossible to reconstruct the evolution of the cosmos and the richness of the Periodic Table elements. There was a serious hiatus in element formation after helium in the Big Bang era. The familiar elements of the Periodic Table formed in other ways, later in the life of the universe.

After 25 per cent of all the mass in the universe formed helium nuclei, the rest remained as protons, the nucleus of hydrogen, with a few parts per 10,000 of deuterium and a few parts per 10 billion of lithium-7. The oldest stars in the cosmos indeed show these primordial cosmic composition fractions. According to Gamow,[7] the relative abundance of the light elements, hydrogen, helium, and lithium, is the "oldest document pertaining to the history of our universe." It is one of several items of evidence supporting the Big Bang idea, in addition to expansion discovered by Hubble. But Gamow was disappointed that the Big Bang could not synthesize all the elements of the Periodic Table, as he hoped to prove. Later, Fred Hoyle, William Fowler, and others showed that elements up to iron formed inside stars (Chapter 8). But, according to the stellar synthesis model, stars cannot make enough helium. Stellar synthesis only accounts for about 10 per cent helium, instead of the observed 25 per cent. Most of the helium in our universe was left over from creation. Every time we fill a child's balloon, we should appreciate that most of it was created in the Big Bang at the very beginning of the universe.

FIRST LIGHT

Consisting predominantly of hydrogen nuclei (protons), helium nuclei, and free electrons, the early universe was still too hot for atoms to form. Energetic photons continuously tore apart any electron–nucleus pairs. For the same reason, photons could not travel very far and remained locked tightly inside the cosmic hearth. Finally, when the universe cooled to a temperature of 3,000 degrees, nuclei and electrons could successfully link together to form atoms of hydrogen, and photons could finally escape. It was the first light.

Modern estimates give 300,000 years of cooling necessary after the Big Bang for hydrogen atoms to condense out of the fiery maelstrom of electrons, nuclei, and photons. At this stage, the universe was 1,000 times smaller than it is today. Once the photons escaped, gravity began to congeal matter into interstellar gas clouds which in turn gave birth to stars, chemical elements, galaxies, planets, life, and most recently, humans.

Gamow realized in 1948 that proper instruments should be able to detect the first light emitted from the Big Bang after the universe cooled to 3,000 degrees. Big Bang radiation should fill the entire universe since the bang took place simultaneously everywhere in the universe. It would be yet another cosmic signature of the creation event, besides Hubble's expansion and the observed abundance of light elements. Gamow yearned to see the manifestation of the first light:[8]

> In the beginning of everything we had fireworks of unimaginable beauty. We come too late to do more than visualize the splendor of Creation's birthday.

As the universe expanded and space stretched out, the wavelengths of the radiation from the 3,000-degree furnace also shifted to larger and larger values with progressive cooling. The wavelength of that initially hot light got stretched all the way from gamma rays through visible light and into the microwave portion of the spectrum. The primordial radiation redshifted by a factor of 1,000, an enormous redshift. For example, some of the most distant galaxies (Chapter 9) at 10 billion LY, moving at phenomenal speeds show a redshift of only 1, whereas quasars (Chapter 9), the farthest objects visible to us, show redshifts of between 5 and 6. Instead of seeing light characteristic of a 3,000-degree furnace, observations show wavelength shifts of a factor of 1,000, making the radiation from the hot atom-forming clouds of the Big Bang appear as if it comes from a very cold and very dark place. Before reaching earth the primary radiation travelled a distance of about 45 billion light years over 14 billion years, the expansion of the universe having enormously elongated its route.

In a back-of-the-envelope calculation, Gamow estimated that the universe today should be permeated by an ocean of fossil photons with a temperature of 50 K. (That is 50 degrees above the absolute zero of temperature where all motion ceases. Named after Lord Kelvin, the absolute scale of temperature measurement has the lowest possible temperature as zero. On this scale, room temperature is 300 K, and water freezes at 270K.) But Gamow's colleagues Alpher and Herman quickly corrected an arithmetical mistake to revise the temperature down to a super-cool 5 K. The corresponding wavelength of the radiation is 10 centimetres, like the radiation which heats food inside a microwave oven. But the grand prediction made so early by Gamow and his cohorts for the "relic" radiation from the Big Bang was completely forgotten for nearly two decades.

In the early 1960s Arno Penzias and Robert Wilson at Bell Laboratories in New Jersey were testing a microwave receiver for satellite communications. For successful communications application, it was important to identify all sources of microwave noise in the sky. They commandeered a movable horn-shaped antenna and started collecting radio emission from supernova remnants in our Milky Way. An irritating hiss in the microwave spectrum quickly upset their well-laid plans for developing advanced systems. The noise did not vary with day or night, or with the direction in which they pointed the horn. Was there something wrong with their equipment? Rebuilding the receiver had no effect on the constant buzz. When they took a look inside the horn antenna they found several pigeons roosting. Even after driving out the birds and carefully cleaning out the droppings, the background noise remained unchanged. Lesser experimenters would have abandoned the quest in the face of unstoppable noise, but Penzias and Wilson doggedly pursued the source. Perhaps the mysterious radiation was coming from the ground? So they flew the antenna around on a helicopter, but still failed to pin-point the persistent source. It was ever present and coming from all directions.

From nearby Princeton University, they heard that Robert Dicke and his colleagues had set out to build a sensitive microwave receiver to detect Gamow's faint light left over as the after-glow from the creation of the universe. The Princeton team predicted that radio telescopes should detect such radiation coming equally and symmetrically from all directions, since the first light should permeate all space.

Finally, the Bell Laboratory researchers realized their annoying hiss was really the cosmic whisper the Princeton team had set out to find. They were already "seeing" the ancient light from the birth of the cosmos 14 billion years ago. It is mind-boggling that everyone can literally see some (about 1 per cent) of the relic radiation from the fires of creation when they tune an analogue television set in between channels to experience the random static noise on the television screen. As much as we often share the thrill of the glorious sunset, or sunrise, or the Aurora in the Northern skies, we should partake of the hallowed experience of microwaves lingering in the sky from the primordial Big Bang fireball. We are all immersed in this cosmic bath of microwaves bestowed on us by the fiery birth of our universe.

The Bell scientists published a two-page paper in 1965 alongside another paper from the Princeton group which interpreted the result. More precise measurements revealed that the cosmic microwave background (CMB) whisper left over from the Big Bang corresponds to an even lower temperature of 2.725 K. It is a snapshot of the universe when light first escaped the primordial soup, and when the universe was only 380,000 years old. It was long before the first stars and galaxies started to form. The radiation is extremely uniform in every direction of the sky, as one would expect from the Big Bang taking place everywhere. Improved snapshots with more advanced probes (Figure 13.1) resolve ever more minute temperature fluctuations of less than one hundred-thousandths of a degree observed in different directions to show a micro-anisotropy in the cosmic microwave background. One can think of the CMB as the earliest baby picture of the universe. The CMB is the third

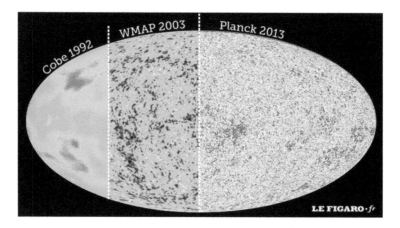

FIGURE 13.1 Progressively detailed baby pictures of the universe from advancing probes (COBE, WMAP, and Planck) of the cosmic microwave background (CMB). Regions with the greatest temperature variations subtend just one degree across the sky, or nearly twice the size of the full moon. Credit: https://archive. briankoberlein.com/2015/06/15/science-in-the-raw/index.html.

evidential validation for the Big Bang creation of the universe, in addition to the primordial element distribution, and Hubble's discovery of expansion.

The patterns of ever so slightly hotter and cooler patches in the CMB arise from tiny gravitational variations that were present in the very young universe to provide us exquisite details of the Big Bang explosion. The cosmos baby picture stunningly reveals fundamental aspects of the universe. A rapid expansion of the universe in the first moments after the Big Bang triggered shock waves that alternately compressed and rarefied regions of the primordial soup to create multiple peaks in the CMB. By studying the tones and overtones of these embryonic peaks, cosmologists are able to estimate the geometry, age, and overall composition of the universe.

It is clear that the universe is very close to spatially flat, obeying the laws of Euclidean geometry, bring us full circle to the mighty geometer of ancestry. Our cosmos consists of roughly 5 per cent familiar atomic matter, protons, neutrons, electrons, the stuff which all of us are made of, 27 per cent strange dark matter, and 68 per cent mysterious dark energy that speeds up cosmic expansion (see next section). We have naively thought all along that the universe is mostly made up of galaxies, stars, planets, black holes, comets, and asteroids. But all of that is just an astonishingly tiny part of the universe. Surprisingly, we have little or no idea of the true nature of 95 per cent of the universe's composition, which comprises dark matter and dark energy.

Dark matter was first postulated in the 1930s (Chapter 8) to explain the enormous concentration of mass in galaxy clusters to account for the unexpectedly rapid motion of the galaxies, as well as the exceptional bending of light passing by galaxy clusters to give us optical illusions of rings of galaxies. Einstein first unwittingly introduced the concept of dark energy in 1917 when he added the cosmological constant to his equations as a mysterious form of energy that prevents gravity from collapsing the universe. He later disavowed the constant, but it was resurrected in the 1990s, when observations of distant supernovae showed that the expansion of the universe is accelerating (see next section). But we still do not know much about the nature of dark energy, this major, mystifying constituent of the universe.

BEFORE THERE WAS LIGHT

Gamow and his collaborators tried to recreate the earliest moments of the universe from their understanding of microscopic physics, which at the time was nuclear physics. Their early universe was a hot broth of neutrons and protons from which they could predict the cosmic abundance of light

elements as well as the cosmic microwave background. Along with Hubble expansion, the CMB and signatures of light element abundance provide strong evidence that a Big Bang event took place at creation. The Big Bang creation of the universe is not a theoretical hypothesis.

There have since been substantial advances in the physics of the microcosmos through discoveries from high-energy particle accelerators that collide protons with anti-protons, and electrons with positrons. Protons and neutrons are not elementary, they are built from quarks interacting through the strong force (Chapter 12). Leptons, such as the electrons, muons, and neutrinos, interact via the weak force. All charged particles interact through the electromagnetic force. The strengths of these forces are vastly different. At one end of the spectrum, the strong force is 137 times more powerful than the electromagnetic force, and 100,000 times stronger than the weak force. Way out on the extreme of the spectrum is the familiar force of gravity, a 100 trillion, trillion, trillion times more feeble than the strong force.

Instead of relying on early nuclear physics of a neutron dominated ylem, strong and weak interactions allow physicists to better work their way back to moments in the Big Bang even earlier than the times that Gamow and his collaborators could evaluate. Before the ylem of neutrons, protons, and photons, the universe was even hotter. It was comprised predominantly of quarks, leptons, and a wild zoo of particles which high-energy experimenters now create and study in their detectors with particle–anti-particle collisions. Cooling to a temperature of 10 trillion degrees within a microsecond (10^{-6} seconds) after the Big Bang, quarks combined with super unbreakable bonds to form protons and neutrons in what is called the Hadron Era.

Before the Hadron Era, before a trillionth of a second (10^{-12} seconds), at temperatures above 1,000 trillion degrees, in what is called the Electro-Weak Epoch, the weak and electromagnetic forces displayed equal strength (Chapter 12). Physicists make such calculations because they understand how the electromagnetic and weak force unify, allowing them to make testable predictions in terms of new particles. These particles were subsequently discovered in the highest-energy accelerators. High collision energies in modern accelerators corresponding to trillions of volts of charged particle acceleration translate to temperature that prevailed in the early moments of the universe.

And well before the Electro-Weak era, before 10^{-35} seconds, the universe was an infernal cosmic broth at an incredible temperature of 100 trillion, trillion degrees. This is called the Grand Unification Epoch (Chapter 12), when the strengths of all interactions, strong, weak, and electromagnetic, were in a state of complete symmetry. Equal amounts of matter and anti-matter were present, potentially ready to annihilate each other completely into pure radiation with infinitesimally little matter left over to produce a hospitable universe. How did matter we are familiar with today survive? The physics of this state is not completely understood. There are several powerful, grand unification theories and predictions, but no conclusive evidence has yet emerged from the many experiments in progress. At some time in the Grand Unification Epoch, a slight asymmetry appeared, and became the genesis for the crucial imbalance between particles and anti-particles which exists in the universe today. For roughly every billion pairs, one excess particle survived the symmetric matter–anti-matter distribution, and absolutely balanced annihilations. In the subsequent evolution, most anti-matter disappeared from annihilations, and the minute asymmetry allowed matter to dominate the make-up of the universe.

We may view the wide distribution of interaction strengths of the different forces we experience today as the relics of the cooling of the Big Bang after the Grand Unification Epoch, just as we view the light element (H, He, and Li) abundances as the signatures of the early era of nuclei formation. But we do not yet fully understand how the quark–anti-quark soup evolved into the matter–radiation mix around us. We do not know for sure how the strength of the strong force differentiated from that of the other forces to give us the wide disparity in their potencies. We do not know for sure why three generations of quarks and leptons developed with their wide spectra of masses. All we are sure of is that the physics of the highest energies dominated the earliest moments of the universe. As we expand our horizons ever further, understanding high-energy physics will improve our grasp of the beginning of space, time, and matter.

What was there before the cataclysm of the Big Bang? There was no "before." Time started with the universe. The question of what went on before the universe began makes no sense because time evolved with the universe. Asking what happened before time began is similar to asking what lies south of the South Pole of the earth.

What lies outside the expanding universe? There is no outside. The Big Bang created all space, time, and matter and forces. Artist Paul Klee's insight[9] is relevant to the sum total of the universe at the time of creation. "The interior is infinite, all the way to the mystery of the inmost, the charged point, a kind of sum total of the infinite." One final time, we draw inspiration from MC Escher. In *Butterflies* (Figure 13.2),[10] he shows how nature takes form from a formless mass, breaking the perfect but structureless symmetry from the centre to evolve into a universe of lively and exquisite patterns.

ACCELERATING UNIVERSE

The universe is not only expanding, the speed of expansion is accelerating. Two independent teams of astronomers, one led by Adam Riess and the other by Saul Perlmutter, looked vigilantly at Type Ia supernova explosions (Chapter 8) in remote galaxies as their reliable distance lampposts. Type Ia supernovae explode with a prescribed absolute brightness, so that their intrinsic luminosity is well-determined wherever they take place. Nature rewarded the explorers with a spectacular find worthy of the coveted Nobel: the more distant supernovae appear dimmer than expected (show lower apparent luminosity). Expanding space had swept these supernovae unexpectedly farther away. The expansion of the universe is speeding up. Why?

The expansion is accelerating under the influence of an unfamiliar anti-gravitational force, now named dark energy, echoing Fritz Zwicky's dark matter (Chapter 8). Einstein's century-old fudge

FIGURE 13.2   *Butterflies, Circle Limit, 1959,* MC Escher.

Ten billion years before now
Brilliant soaring in space and time,
There was a ball of flame, solitary, eternal.
Our common father...
It exploded and every change began.

[*Collected Poems*, Levi][11]

factor of the cosmological constant is back! Einstein's biggest blunder was not a blunder after all. Even though he tried to put the genie back in the bottle, it has now sprung out majestically.

Dark energy is a form of "antigravity" that pushes galaxies apart by stretching space at increasing pace. Since energy and mass are related according to special relativity (Chapter 4) dark energy also has a gravitational effect. Peak analyses of CMB show that the universe is nearly flat, which means that the mass-energy density of the universe is completely balanced. If mass-energy and its associated gravity were dominant, the universe would show a curvature of spacetime as expected from general relativity. If the total amount of matter in the universe including dark matter accounts for 30 per cent of its mass, the balance of mass-energy in the form of dark energy must account for the remaining 70 per cent.

Under the influence of dark energy, the cosmos is now doubling in size every 10 billion years. Which can lead to a philosophically depressing consequence. If the mysterious dark energy remains constant, space will eventually be expanding faster than the speed of light so that everything outside our galaxy eventually will no longer be visible. The universe will become lifeless and utterly dark over the next 10 billion years, a state that has been branded the Big Freeze.

Perhaps general relativity needs to be replaced by another theory of gravity on cosmic and quantum scales. The unification of quantum physics with general relativity could bring about a deeper understanding of the trigger for the Big Bang, together with the fundamental nature of dark energy. At present, the two theories encompass completely distinct domains, one for the macrocosmos and the other for the microcosmos. With a future unification, space and time may take on completely new forms and meaning at the core of the physics-defying black holes and during the earliest moments of the Big Bang.

Such grand unknowns call for further studies on the detailed nature of dark energy, which have been in progress for the last six years by Dark Energy Survey (DES) and other studies. Results from the first two years using a sample of 207 supernovae confirm that 68 per cent of the universe's make-up is primarily dark energy, as reported by CMB. Which would mean that our universe will continue to expand with its current acceleration indefinitely to the eventual Big Freeze. But it is still too early to be certain, considering the upcoming analysis of data from 300 million galaxies measured.

Or perhaps the cosmological "constant" is dynamic in nature. If the cosmological term increases over time, the universe will end in a Big Rip. Approximately 20 to 50 billion years in the future all matter, from stars and galaxies to atoms and subatomic particles, even spacetime itself, will be torn apart by the expansion of the universe. And if the constant decreases in time, the universe will reach a maximum size, reverse course, and start contracting to eventually collapse in a Big Crunch, sometime about 100 billion years from today. It would be a reverse Big Bang and may even give rise to another Big Bang in a cyclic repetition.

How much stranger is reality than any of us could imagine!

FINAL REMARKS

Our century's revelations of unthinkable largeness and unimaginable smallness have fully captivated us. At the two extremes, the very small and the very large, between the quarks and the cosmos, the nature of the universe is inextricably entwined. The universe was around for 14 billion years before a species evolved with a deep curiosity to ask penetrating questions, and the ingenuity to come up with some answers. In *Four Quartets*, poet TS Eliot exhorts us to the eternal quest, with an uncanny glimpse of the final goal:

> We shall not cease from exploration
> And the end of all our exploring
> Will be to arrive where we started
> And know the place for the first time.

TS Eliot, *Four Quartets*[12]

The development of science has painted a magnificent picture through an unending voyage of imagination, revealing piece by piece the secreted truths about the workings of our world. Each discovery has contributed to a small piece of the giant puzzle. Each breakthrough has gathered a wider range of physical phenomena under fewer theoretical umbrellas, elucidating the underlying laws, the lenses to view and organize our world. Each breakthrough answers some questions but then gives rise to a host of new questions that could not even be imagined before the leap forward. Clearly, we have only touched the edge of the new world stretching before us, as Newton mused on the seashore.

Through our first synthesis of celestial and terrestrial motion, followed by the ultimate synthesis of matter, forces, space and time, microcosmos and macrocosmos, the unification of the universe has been a glorious adventure of the human intellect. Growing out of our innate curiosity about the world around us, science has become an integral part of our culture. Physics concepts mould our self-perception and the perception of our role in the cosmos, shaping cosmological ideas of the non-scientist as much as those of the scientist. The famous historian of science, George Sarton, wrote:[13]

> Science is not distinct from religion or art in being more or less human than they are, but simply because it is the fruit of different needs or tendencies. Religion exists because men are hungry for goodness, for justice, for mercy; the arts exist because men are hungry of beauty; the sciences exist because men are hungry for truth.

Notes

PREFACE

1. King James Bible, Genesis I, v. 6, 7.
2. Hobson, p. 91.
3. Von Carolsfeld graphic, bilwissedition Ltd. & Co. K/Alamy Stock Photo.
4. Gaposchkin, p. 255.

CHAPTER 1

1. Chesterton, p. 2.
2. *Iliad* by Homer, Book XV, line 281.
3. Bronze Zeus or Poseiden from Arternisium, copyright Gianni Dagli Orti/Shutterstock.
4. Left: Catharine Page Perkins Fund, 97.360, Museum of Fine Arts Boston/Bridgeman Images. Right: Gift of Edward Perry Warren, 92.2736/Museum of Fine Arts Boston/Bridgeman Images.
5. Hobson, p. 28.
6. https://www.thefamouspeople.com/profiles/pythagoras-504.php.
7. Anavysos Kouros, copyright Gianni Dagli Orti/Shutterstock.
8. Hoplites and Cavaliers, Attic black Figure Amphora, Photo (C) RMN-Grand Palais (musée du Louvre)/ Hervé Lewandowski.
9. Burger, p. 356.
10. Hofstadter, p. 25.
11. © 2019 The M.C. Escher Company-The Netherlands. All rights reserved. www.mcescher.com.
12. Parmenides, fragment b.
13. Whitfield, p. 28.
14. National Geographic, Vol. 196, No. 6, Dec. 1999, Supplement.
15. Falkenstein/imageBROKER/Shutterstock.
16. Burger, p. 174.
17. Raphael (1483–1520). The School of Athens, copyright Universal History Archive/UIG/Shutterstock.
18. Hobson, p. 11.
19. Trefil (1978), p. 47.
20. The Kritios Boy (Young victorious athlete), copyright PRISMA ARCHIVO/Alamy Stock Photo.
21. Riordan, p. 122.
22. Ferris (1988), p. 22.
23. Cooper, p. 119.
24. Aristotle, p. 25.
25. Boorstin, p. 94.
26. Raphael (1483–1520). Plato and Aristotle, detail from *School of Athens*, copyright Godong/UIG/ Shutterstock.
27. Ferris (1998), p. 204.
28. Hargittai, p. 4.
29. Adapted from Hilderbrandt and Tromba, p. 255.
30. KM Reinish et al., Nature, vol. 404, p. 960 (2000).
31. Ferris (1998), p. 8.
32. Steven Strogatz, New Yorker, November 19, 2015: https://www.newyorker.com/tech/annals-of-technology/einsteins-first-proof-pythagorean-theorem

CHAPTER 2

1. Burger, p. 13.
2. Shakespeare, *Troilus and Cressida*, Act I, Scene III.
3. Sagan, *Cosmos*, p. 19.

4. Meidias Painter (fl. 410 BC). Hydra with Phaon and the Daughters of Lesbos. Greek Vasepainting. 5th BC, Copyright Trustees of the British Museum.

5. Discobolus, Myron of Athens (5th c. BC), Copyright Adam Eastland Art + Architecture/Alamy Stock Photo.

6. Szabadvary, p. 38.

7. Szabadvary, p. 163.

8. Adapted from Whitfield, pp. 187, 188.

9. Farber, p. 70.

10. Tilden, p. 72.

11. Sunrise over Silicon, by L. J. Whitman et al., Alamy stock photo B71199.

12. P. Zeppenfeld, C. P. Lutz, and D. M. Eigler, Ultramicroscopy 42–44:128 (1992).

13. Helen Birch Bartlett Memorial Collection, 1926.224. © 2019. The Art Institute of Chicago/Art Resource, NY/ Scala, Florence.

14. Tyndall, p. 4.

15. Hobson, p. 29.

16. Adapted from Ronan (1982), p. 464.

17. https://iupac.org/what-we-do/periodic-table-of-elements/.

18. https://franklyandjournal.wordpress.com/2016/07/18/hydrogen-spectrum/.

19. Adapted from Segré (1980), pp. 16–17.

20. Gaposchkin, p. 341.

21. Bent, p. XXXIX.

22. Keller, p. 55.

23. Early History of X-Rays by Alexi Assmus in SLAC Beam Line (Summer 1995) p. 21, from Wilhelma, Electrical Review, April 17, 1986.

24. Adapted from Segré (1980), pp. 20–1.

25. Watson, p.181.

26. Science History Images/Alamy Stock Photo.

27. Science Photo Library.

28. Blackboard/Shutterstock, Image 784253578.

29. Mukherjee, p. 158.

30. Mukherjee, p. 160.

31. Keller, p. 99.

32. Keller, p. 99.

33. Hecht, p. 1026.

34. Riordan, p. 24.

35. Gaposchkin, p. 341.

36. Bernstein, p. 11.

37. Fritzcsh, p. 53.

38. Pagels, p. 265.

39. Guth, p. 144.

CHAPTER 3

1. Riordan, p. 8.

2. Caldwell, p. 173.

3. FerrisMW, p. 42.

4. Motz, p. 18.

5. Wilkins, p. 200.

6. Goldstein, p. 57.

7. Peter Barritt/Alamy Stock Photo.

8. Barberini Vase, Syria, middle of the 13th century, Brass inlaid with silver, H 45.9 cm; O 37 cm, AO 4090, Photo (C) Musée du Louvre, Dist. RMN-Grand Palais/Claire Tabbagh/Collections Numériques.

9. Wilkins, p. 200.

10. Averroes, p. 5.

11. Aquinas, pp. 135–6.

12. Aquinas, p. 150.

13. Bruno, p. 11.

14. Caldwell, p. 100.
15. See art historians Wilkins and Gardner.
16. Copyright World History Archive/Alamy Stock Photo.
17. The Picture Art Collection/Alamy Stock Photo.
18. Gombrich, p. 250.
19. Dürer, Albrecht (1471–1528). *Young Hare*, 1502. Copyright Foto Marburg/Art Resource, NY Graphische Sammllmg Albertina, Vienna, Austria. (Right) Dürer, Albrecht (1471–1528). *The Large Turf*, watercolour and gouache. Copyright classicpaintings/Alamy Stock Photo and Archivart/Alamy Stock Photo.
20. De Humani Corporis Fabrica Libri Septem, by Vesalius, Andreas, Bibliotheque de la Faculte de Medecine, Paris, France/Archives Charmet/Bridgeman Images.
21. (a) Westend61/Shutterstock (b) Design Pics Inc/Shutterstock.
22. Perugino, Pietro (1448–1523). *Christ giving the Keys to Saint Peter*, copyright Vatican Museums and Galleries, Vatican City/Bridgeman Images.
23. Leonardo da Vinci (1452–1519). Drawing of ideal proportions of the human figure according to Vitruvius' 1st cent AD. treatise "De Architectura." (called "Vitruvian Man"), ca. 1492, copyright Universal History Archive/Shutterstock.
24. Mortars Launching Missiles from "The Atlantic Codex," Pinacoteca, Ambrisiana, Milano. Copyright Sowa Sergiusz/Alamy Stock Photo.
25. Shakespeare, *Hamlet*, Act 2, Scene II.
26. Ronan (1982), p. 366.
27. Bruno, p. 277.
28. Lemay, p. 94.
29. Faÿ, p. 487.
30. Act II, Scene I.
31. Cohen Bernard, I, p.171.
32. Jesse, p. 294.
33. Tilden, p. 98.
34. Ronan (1982), p. 451.
35. Hecht, p. 605.
36. March, p. 73.
37. Segré (1984), pp. 159, 161.
38. Segré (1984), p. 208.
39. Everitt, p. 101.

CHAPTER 4

1. Segré (1984), p. 32.
 2. Crew, p. 60.
 3. Chronicle/Alamy Stock Photo.
 4. Peter Horree/Alamy Stock Photo.
 5. Granger/Shutterstock.
 6. Michelangelo (1475–1564) Delphic Sibyl, copyright Heritage Image Partnership Ltd/Alamy Stock Photo.
 7. Guercino (1591–1666). Aurora, ART Collection/Alamy Stock Photo.
 8. Bernini, Gian Lorenzo (1598–1680). David, copyright Alfredo Dagli Orti/Shutterstock.
 9. Kaplon, p. 85.
10. Shlain, p. 119.
11. Granger/Shutterstock.
12. Holton, p. 103.
13. Thiel, p. 150.
14. Holton, p. 181.
15. Riordan, p. 335.
16. Ferris (1988), p. 185.
17. Cassidy, p. 45.
18. Burger, p. 76.
19. Burger, p. 76.
20. Adapted from Hewitt, p. 663.

21. Pagels, p. 265.
22. Hecht, pp. 985–6.
23. Salvador Dali, *The Persistence of Memory*, copyright Martin Shields/Alamy Stock Photo, © ADAGP, Paris and DACS, London 2019.
24. Hecht, p. 990.

CHAPTER 5

1. Griffith, pp. 215–16.
2. Hawkins, p. 45.
3. Thurston, p. 46.
4. Shakespeare, *Tempest*, Act 2, Scene II.
5. Shakespeare, *Romeo and Juliet*, Act II, Scene II.
6. Credit: US Geological Survey.
7. Aveni, p. 31.
8. Shakespeare, *Julius Caesar*, Act III, Scene I.
9. Griffith, p. 16.
10. Griffith, p. 16.
11. Hobson, p. 5.
12. Thiel, p. 13.
13. Homer's *Iliad*, Book XXII.
14. Ferris (1988), p. 21.
15. Gaposchkin, p. 9.
16. Gaposchkin, p. 7.
17. Kippax, p. 47.
18. Chartrand, pp. 379–89.
19. Crooke, p. 200.
20. Gaposchkin, p.134.
21. Pindar Paen, p. 9.
22. ncamerastock/Alamy Stock Photo.
23. Kippax, p. 407.
24. Rowan-Robinson, p. 4.
25. Lodge, p. 16.
26. Cossa, Francesco del (1435–78). The Month of April: Triumph of Venus and astrological symbols. Fresco. Copyright The Picture Art Collection/Alamy Stock Photo.
27. Iliad, Books XXII and XXIII.
28. Gaposchkin, p. 203.
29. Aveni, p. 38.
30. Gaposchkin, p. 183.
31. Birth of Venus, from the Ludovisi Throne. Copyright Azoor Photo/Alamy Stock Photo.
32. The Picture Art Collection/Alamy Stock Photo.
33. (Left) Il vivace giovani Mercuro, Giovanni Bologna, Su concessione del Ministero dei Beni e le Attivita Culturali, (Right) Base and part of a pilaster with Saturn, third chapel on the right. 15th century. Copyright Nazionale del Bargello, Florence, Tuscany, Italy/Alinari Archives, Florence/Bridgeman Images and Photo © Raffaello Bencini/Bridgeman Image.
34. Thiel, p. 22.
35. Thiel, p. 23.
36. Gaposchkin, p. 222.
37. Gaposchkin, p. 226.
38. Peterson, p. 110.
39. Holton, p. 154.
40. Credit: STScI, NASA, and USGS.

CHAPTER 6

1. Koestler, p. 134.
2. James, p. 113.

3. Gaposchkin, p. 1.
4. Ferris (1988), p. 19.
5. Wilkins, p. 102.
6. Praxiteles (c. 400–330 BC). The Cnidian Venus, c. 350–330 BC Roman copy after Greek original. Copyright Jiri Hubatka/Alamy Stock Photo.
7. Crowe, pp. 64, 65.
8. Browning, Andrea Del Sarto (The Faultless Painter) lines 97–8.
9. Sarton (1959), p. 105.
10. Boorstin, p. 21.
11. Peterson, p. 21.
12. Kuhn, p. 85.
13. Adapated from Taylor (1956), pp. 24, 25.
14. Kippax, p. 218.

CHAPTER 7

1. Ficino, p. 30.
2. Thiel, p. 57.
3. Psalms 18, 92, 104, and Isaiah 40. The Holy Bible: King James Version.
4. Hecht, p. 1130.
5. Trefil, p. 55.
6. Ross, p. 100.
7. The Granger Collection/Alamy Stock Photo.
8. (Top) Pictures Now/Alamy Stock Photo, (Bottom) Christianson, p. 106.
9. Gaposchkin, p. 255.
10. Divina Comedia Swedish, Dante, Division of Rare and Manuscript Collections, Carl A. Kroch Library, Cornell University, Ithaca, NY.
11. Ferris (1988), p. 49.
12. Byzantine (476-1453). Creation of the stars. Byzantine mosaic. Copyright Granger/Shutterstock.
13. (Left) Pasachoff, p. 68. (Right) Rowan-Robinson, p. 132.
14. Wilkins, p. 295.
15. Gingerich, p. 199.
16. Kuhn, p. 178.
17. Ball, p. 33.
18. Kuhn, p. 179.
19. Kuhn, p. 189.
20. Photo (C) Musée du Louvre, Dist. RMN-Grand Palais/Angèle Dequier.
21. Copyright The Picture Art Collection/Alamy Stock Photo.
22. Copyright Scuola Grande di San Rocco, Venice, Italy/Cameraphoto Arte Venezia/Bridgeman Images.
23. ART Collection/Alamy Stock Photo.
24. Copyright Art Collection 3/Alamy Stock Photo.
25. Dobrzycki, p. 18.
26. Dobrzycki, Preface.
27. Hoyle, p. 12.
28. Psalms 18, 92, 104, and Isaiah 40. The Holy Bible: King James Version.
29. Hobson, p. 32.
30. Kesten, p. 173.
31. Hecht, p. 208.
32. Holton, p. 26.
33. Thiel, p. 87.
34. Crowe, pp. 78–9.
35. Cooper, p. 241.
36. Thiel, p. 111.
37. Kippax, p. 69.
38. Hobson, p. 307.
39. Granger/Shutterstock.

CHAPTER 8

1. Koestler, p. 59.
2. Hobson, p. 130.
3. Boorstin, p. 305.
4. Shakespeare, *Henry VI*, Part I, Act I, Scene I.
5. Shakespeare, *Julius Caesar*, Act I, Scene II.
6. Ferris (1988), p. 61.
7. Kolb, p. 19.
8. Wilkins, p. 248.
9. Shakespeare, *Henry VI*, Act III, Scene II.
10. Shakespeare, *Julius Caesar*, Act II, Scene II.
11. Ronan (1982), p. 353.
12. James, p. 146.
13. Peterson, p. 85.
14. Ferris (1988), p. 76.
15. Copyright John Dambik/Alamy Stock Photo.
16. Copyright Art Collection 2/Alamy Stock Photo.
17. Koestler, p. 79.
18. Koestler, p. 111.
19. Koestler, p. 278.
20. Ferris (1988), p. 78.
21. Cohen, p. 137.
22. Koestler, p. 135.
23. Westfall (1971), p. 11.
24. Westfall (1971), p. 12.
25. The Fall of the Giants, copyright DEA /M. CARRIERI/Getty Images.
26. The Picture Art Collection/Alamy Stock Photo.
27. Copyright Art Collection 2/Alamy Stock Photo.
28. Peter Horree/Alamy Stock Photo.
29. Burger, p. 328.
30. Association of Universities for Research in Astronomy, Inc. AURA.
31. Ronan (1982), p. 354.
32. Holton, p. 44.
33. Ferris (1988), p. 81.
34. Noyes, p. 111.
35. Wilczek, p. 12.
36. Koestler, p. 56.
37. Holton, p. 46.
38. Lodge, p. 75.
39. Hobson, p. 24.
40. Ferris (1988), p. 82.
41. Ferris (1988), p. 157.
42. C.R. O'Dell and S.K. Wong, Rice University.
43. Rowan-Robinson, p. 63.
44. (Left) Jeff Hester and Paul Scowen, Arizona State University, STScI and NASA. (Right) Rowan-Robinson, p. 67.
45. Gaposchkin, p. 250.
46. Poem by Albert Ahearn (Albee), https://www.poemhunter.com/poem/star-stuff/.
47. Gaposchkin, p. 312.
48. Gardner, p. 492.
49. (Left) National Radio Astronomy, Greenbank, (Middle) Fred Gerard, Center for Astrophysics, Cambridge, MA, (Right) NASA.
50. Anglo-Australian Telescope Board, David Malin.
51. Rowan-Robinson, p. 15.
52. Credit: STScI, NASA.

CHAPTER 9

1. Koestler, p. 159.
2. Shakespeare, *Hamlet*, Act II, Scene II.
3. Sobel (1999), p. 6.
4. Pasachoff, p. 335.
5. Lodge, p. 98.
6. Kippax, p. 337.
7. Cigoli, Ludovico (1559–1613), Assumption of the Virgin, detail, dome, Pauline Chapel, copyright Santa Maria Maggiore, Rome, Italy/Bridgeman Images. See also Morrison, p. 37.
8. National Optical Astronomy Observatory.
9. Lodge, p. 113.
10. Koestler, p. 199.
11. Su concessione del Ministero dei Beni e le Attivita Culturali.
12. Wilkins, p. 283.
13. Lodge, pp. 103–5.
14. Shakespeare, *Hamlet*, Act I, Scene V.
15. Archivart/Alamy Stock Photo.
16. Lives in Science, p. 8.
17. (Left) Rowan-Robinson, p. 111, Astronomical Society of the Pacific (Right) Chartrand, Plate 307, Rev. Ronald E Royer.
18. Galilei, pp. 60–1.
19. (Left) Rowan-Robinson, p. 111, Astronomical Society of the Pacific (Right) Chartrand, Plate 307, Rev. Ronald E Royer.
20. Shakespeare, *Romeo and Juliet*, Act III, Scene II.
21. Rowan-Robinson, p. 110.
22. Peterson, p. 45.
23. Koestler, p. 190.
24. Kippax, p. 4.
25. Boorstin, p. 315.
26. Lodge, p. 106.
27. Ferris (1988), p. 90.
28. Ferris (1988), p. 95.
29. Koestler, pp. 199–200.
30. Holton, p. 58.
31. Gindikin, p. 60.
32. Christianson, p. 297.
33. Kolb, p. 103.
34. Gindikin, p. 62.
35. Gindikin, p. 63.
36. Kesten, p. 393.
37. Gindikin, p. 43.
38. Kuhn, p. 194.
39. Ruisdael, Jacob Isaacksz.v. 1628/29–1682, Ansicht von Ootmarsum, Art Collection 2/Alamy Stock Photo.
40. Herschel, p. 53.
41. Ferris (1988), p. 156.
42. (Left) Royal Astronomical Society. (Right) British Library. B.L.c.144.1.3.
43. Berendzen, p. 15.
44. Berendzen, p. 37.
45. (Left) NGC3293. (Right) Globular Cluster 47 Tucunae, Anglo Australian Telescope Board, David Malin.
46. J. Hist. Astron., 7, 69–182, The 'Great Debate', What Really Happened by Michael A. Hoskin, Editor.
47. Berendzen, p. 105.
48. Ferris (1988), p. 172.
49. Rowan-Robinson, p. 124.
50. (Left) Bill Scheeningand Vanessa Harvey, NSF REU, NOAO/AURA/NSF. (Right) People and Discoveries A Science Odyssey, PBS.

51. M83, Anglo-Australian Telescope Board, David Malin.
52. Hubble, Science News Letter V.II, no. 306, p. 1.
53. Rowan-Robinson, p. 170.
54. Kippax, p. 12.

CHAPTER 10

1. Weaver, p. 305.
2. Adapted from Way, p. 47.
3. (Left) Shapin, p. 35. (Right) Ronan (1982), p. 289.
4. Martin, Pierre Denis (1663–1742). Child Louis XV travelling in an open carriage in view of the Abreuvoir and the Chateau de Marly, 1724, Copyright Gianni Dagli Orti/Shutterstock.
5. Cotelle, Jean (1642–1708). View of l'Orangerie and the Swiss ornamental lake at Versailles. Gouache on vellum. Copyright Gianni Dagli Orti/Shutterstock.
6. Shakespeare, *Richard II*, Act 5, Scene V.
7. Segré (1984), pp. 36–40.
8. Adapted from Way, p. 69.
9. National Maritime Museum, Greenwich, London.
10. Shlain, p. 132.
11. Hecht, p. 997.
12. Mook, p. 131.
13. Cordon Art, Holland.
14. Schattschneider p. 239.

CHAPTER 11

1. Shlain, p. 142.
2. Thiel, p. 180.
3. Ferris (1988), p. 105.
4. Peterson, p. 84.
5. Ferris (1988), p. 108.
6. Lord Byron, X Canto.
7. Hecht, p. 223.
8. Noyes, p. 193.
9. Speyer, p. 105.
10. Holton, p. 127.
11. Mook, p. 32.
12. Feynman, 1965, Character of Physical Law, p. 14.
13. Hecht, p. 227.
14. Kippax, p. 259.
15. Hecht, p. 215.
16. René Magritte, The Chateau in the Pyrenees, copyright The Israel Museum, Jerusalem, Israel/Gift of Harry Torczyner, New York/Bridgeman Images, © ADAGP, Paris and DACS, London 2019.
17. Taylor, p. 65.
18. Proctor, p. 37.
19. Ronan (1982), p. 358.
20. The Bayeux Tapestry - 11th Century, Gianni Dagli Orti/Shutterstock.
21. Cooper, p. 66.
22. Hall (1992), p. 223.
23. Cooper, p. 31.
24. Pictorial Press Ltd/Alamy Stock Photo.
25. Segré (1984), p. 45.
26. Hobson, p. 91.
27. Park, p. 221.
28. Hecht, p. 1092.
29. Bruno, p. 41.
30. Ballif, p. 151.

31. Franklin, *A Dissertation on Liberty and Necessity*.
32. Trefil, p. 265.
33. Ferris (1988), p. 120.
34. Holton, p. 163.
35. Koestler, p. 154.
36. March, p. 138.
37. Ferris (1988), p. 199.
38. Adapted from Pasachoff, p. 144.
39. Frank, p. 141.
40. Pasachoff, p. 146.
41. Hobson, p. 298.
42. Frank, p. 262.

CHAPTER 12

1. FitzGerald, p. 197.
2. Cooper, p. 305.
3. Cooper, p. 322.
4. Hecht, p. 1057.
5. Baggot, p. 47.
6. Segré (1980), p. 124.
7. Moore, p. 349.
8. Goethe, p. 14.
9. Carroll, p. 56.
10. Zeri, p.25.
11. Riordan, p. 87.
12. Riordan, pp. 90–1.
13. Schattschneider, p. 239.
14. Riordan, p. 102.
15. © 2019 The M.C. Escher Company, The Netherlands. All rights reserved. www.mcescher.com.
16. Riordan, p. 120.
17. https://www.physik.uzh.ch/en/researcharea/lhcb/outreach/StandardModel.html, Image from Shutterstock.

CHAPTER 13

1. Hecht, p. 1093.
2. Ferris M W, p. 207.
3. Riordan, p. 335.
4. Berendzen, p. 200.
5. Berendzen, p. 200.
6. Ferris (1988), p. 211.
7. Ferris (1988), p. 277.
8. Ferris (1988), p. 212.
9. Burger, p. 219.
10. © 2019 The M.C. Escher Company-The Netherlands. All rights reserved. www.mcescher.com.
11. Primo Levi, *Collected Poems*, In the Beginning.
12. Riordan, p. 355.
13. Bruno, p. 7.

Sources and Bibliography

Agassi, Joseph, *The Continuing Revolution: A History of Physics from Greeks to Einstein* (New York: McGraw-Hill, 1968).

———, *Faraday as a Natural Philosopher* (Chicago, IL: University of Chicago Press, 1971).

Aquinas, Thomas, *Selected Writings* (New York: Penguin, 1998).

Aristotle, *On the Generation of Animals* from an English translation by A L Peck (Cambridge, MA: Harvard University Press; London: Heinemann, 1942).

Armitage, Angus, *John Kepler* (London: Faber, 1966).

Arons, A, *A Guide to Introductory Physics Teaching* (New York: Wiley, 1990).

Asimov, Isaac, *Asimov's Biographical Encyclopedia of Science and Technology: The Lives and Achievements of 1195 Great Scientists from Ancient Times to Present, Chronologically Arranged* (Garden City, NY: Doubleday, 1964).

———, *Asimov's Chronology of Science and Discovery* (New York: Harper & Row, 1989).

Aveni, Anthony, *Stairways to the Stars – Skywatching in Three Great Ancient Cultures* (New York: Wiley, 1997).

Averroës, *On the Harmony of Religions and Philosophy* a translation, with introduction and notes of Ibn Rushd's *Kitāb fasl al-maqāl* by George F Hourani (London: Luzac, 1961).

Baggott, J E and Jim Baggot, *The Quantum Story: A History in 40 Moments* (Oxford University Press, 2011).

Ball, Sir Robert S, *Great Astronomers* (Plainview, NY: Books for Libraries Press, 1974).

Ballif, Jae R and William E Dibble, *Conceptual Physics* (New York: Wiley, 1969).

Bent, Harry A, *Molecules and the Chemical Bond* (Trafford Publishing, 2011).

Berendzen, Richard, Richard Hart and Daniel Seeley, *Man Discovers the Galaxies* (New York: Science History Publications, a division of Neale Watson Academic Publications, Inc., 1976).

Boorstin, Daniel J, *The Discoverers* (New York: Random House, 1983).

Brecher, K and M Feirtag, editors, *Astronomy of the Ancients* (Cambridge, MA: MIT Press, 1979).

Brodrick, James, *Galileo, The Man, His Work, His Misfortunes* (London: Chapman, 1964).

Browning, Robert, *Men and Women, and Other Poems* edited, with an introduction and notes, by J W Harper (London: Dent; Totowa, NJ: Rowman & Littlefield, 1975).

Bruno, Leonard C, *The Tradition of Science; Landmarks of Western Science in the Collections of the Library of Congress* (Washington, DC: Library of Congress, 1987).

Burger, Edward B and Michael Starbird, *The Heart of Mathematics, An Invitation to Effective Thinking* (Emeryville, CA: Key College Publishing, 2000).

Burke, James, *Connections* (Boston, MA: Little Brown, 1978).

Burke-Gaffney, S J and Michael Walter, *Kepler and the Jesuits* (Milwaukee, WI: Bruce Publishing, 1944).

Butterfield, Sir Herbert, *The Origins of Modern Science, 1300–1800* (New York: The Free Press, 1965).

Byron, George, *Don Juan*, Tenth Canto, John Murray, London, 1859.

Caldwell, Wallace E and Edward H Merrill, *The New Popular History of the World, The Story of Mankind from Earliest Times to the Present Day* (Boston, MA: Sanborn & Co., 1964).

Carroll, Lewis, *Alice's Adventures in Wonderland* (Macmillan Company, 1914).

Casper, Barry M and R Noer, *Revolutions in Physics* (New York: Norton, 1972).

Cassidy, David, *Einstein and Our World* (Amherst, NY: Humanity Books, 1998).

Chartrand, Mark R and Wil Tirion, *National Audubon Society, Field Guide to the Night Sky* (New York: Chanticleer Press, Knopf, 1991).

Chesterton, G K, *St Francis of Assisi* (Garden City, NY: Image Books, 1957).

Close, Frank, Michael Marten and Christine Sutton, *The Particle Explosion* (New York: Oxford University Press, 1987).

Cohen, I Bernard, *The Birth of a New Physics* (New York: Norton, 1985).

Cooper, Leon N, *Physics: Structure and Meaning* (Hanover, NH: University Press of New England [for] Brown University, ca. 1992).

Crew, Henry and Alfonso Desalvio, *Dialogues Concerning Two New Sciences* by Galileo Galilei, Literary Licensing, LLC, 2014.

Cromer, Alan H, *Uncommon Sense/The Heretical Nature of Science* (New York: Oxford University Press, 1993).

Crooke, William, *The Popular Religion and Folk-lore of Northern India* (London: Constable, 1896).

Crowe, Michael J, *Theories of the World from Antiquity to the Copernican Revolution* (Mineola, NY: Dover, 2000).

Di Canzio, Albert, *Galileo: His Science and His Significance for the Future of Man* (Portsmouth, NH: ADASI Publishing, ca. 1996).

Dodd, J E, *The Ideas of Particle Physics, An Introduction for Scientists* (New York: Cambridge University Press, 1984).

Drake, Stillman, *Galileo Studies: Personality, Tradition, and Revolution* (Ann Arbor, MI: University of Michigan Press, 1970).

Einstein, Albert and Leopold Infeld, *The Evolution of Physics, The Growth of Ideas from Early Concepts to Relativity and Quanta* (New York: Simon & Schuster, 1938).

Everitt, CWF, *James Clerk Maxwell: Physicist and Natural Philosopher* (New York: Scribner, 1975).

Farber, Eduard, *The Evolution of Chemistry, A History of Ideas, Methods and Materials* (New York: The Ronald Press Company, 1969).

Faÿ, Bernard, *Franklin, the Apostle of Modern Times* (London: Little Brown, 1929).

Feather, Norman, *An Introduction to the Physics of Mass, Length, and Time* (Edinburgh: Edinburgh University Press, 1959).

Fermi, Laura, *Galileo and the Scientific Revolution* (New York: Basic Books, 1961).

Ferris, Timothy, *Coming of Age in the Milky Way* (New York: Morrow, 1988).

————, *The Whole Shebang* (London; New York: Touchstone Books, 1998).

Feynman, Richard, *Character of Physical Law 1965* (Modern Library, 1994).

Ficino, Marsilio, *The Letters of Marsilio Ficino*, translated by members of the Language Department of the School of Economic Science, London (London: Shepheard-Walwyn, 1975).

FitzGerald, Edward, *Rubáiyát of Omar Khayyám: A Critical Edition* (University of Virginia Press, 2008).

Fontenelle, M de (Bernard Le Bovier), *A Plurality of Worlds* translated by Mr Glanvill (London: Osbourne, 1702).

Forbes, Robert Jones, *A History of Science and Technology; Nature Obeyed and Conquered* (Harmondsworth: Penguin, 1963).

Frank, Philipp, *Einstein, His Life and Times* (New York: Knopf, 1947).

Franklin, Benjamin, *A Dissertation on Liberty and Necessity* reproduced from the 1st edition, with a bibliographical note by Lawrence C Wroth (New York: The Facsimile Text Society, 1930).

Fritzsch, Harald, *Quarks, The Stuff of Matter* (New York: Basic Books Inc., 1983).

Galilei, Galileo, *Dialogues Concerning Two New Sciences* translation by Henry Crew and Alfonso Desalvio (Literary Licensing, LLC, 2014).

Galilei, Galileo, *Sidereus nuncius, or, The Sidereal messenger* translated with introduction, conclusion, and notes by Albert van Helden (Chicago, IL: University of Chicago Press, 1989).

Gamow, G, *Biography of Physics* (New York: Harper, 1961).

Gardner, H, *Art Through the Ages* (Fort Worth, TX: Harcourt Brace, 1996).

Gauquelin, Michel, *The Scientific Basis of Astrology, Myth or Reality* (New York: Stein & Day, 1969).

Gindikin, Semen G, *Tales of Physicists and Mathematicians* (Boston, MA: Birkhäuser, 1988).

Goethe, Johann Wolfgang von, *Faust I & II, Volume 2: Goethe's Collected Works – Updated Edition* (Princeton University Press, 2014).

Goldman, Martin, *The Demon in the Aether: The Story of James Clerk Maxwell* (Edinburgh: Paul Harris Publishing, 1983).

Goldstein, Thomas, *Dawn of Modern Science: From the Arabs to Leonardo da Vinci* (Boston, MA: Houghton Mifflin, 1980).

Gombrich, E H, *The Story of Art* (London: Phaidon Press, 1995).

Grant, Edward, *Physical Science in the Middle Ages* (New York: Wiley, 1971).

Griffith, M, *The Stars and Their Stories* (New York: Holt, 1913).

Guth, Alan, *The Inflationary Universe*, Helix Books (Reading, MA: Addison-Wesley, 1997).

Hall, Alfred Rupert, *A Brief History of Science* (New York: New American Library, 1964).

————, *Isaac Newton, Adventurer in Thought* (Oxford: Blackwell; Cambridge, MA: Blackwell, 1992).

————, *The Rise of Modern Science: From Galileo to Newton* (New York: Harper & Row, 1962).

Hargittai, István, *Symmetry Through the Eyes of a Chemist* (New York: Plenum Press, 1995).

Harrison, Edward, *Darkness at Night, A Riddle of the Universe* (Cambridge, MA: Harvard University Press, 1987).

Hawking, Stephen, *The Illustrated A Brief History of Time Updated and Expanded Edition* (New York: Bantam Books, 1996).

Hawkins, Gerald S, *Stonehenge Decoded* (New York: Dell, 1966, ca. 1965).

Heath, Sir Thomas, *Aristarchus of Samos: The Ancient Copernicus* (Oxford: Clarendon Press, 1913).

Hecht, Eugene, *Physics* (Pacific Grove, CA: Brooks/Cole, 1994).

Herschel, Mrs John, *Memoir and Correspondence of Caroline Herschel* (New York: Appleton, 1876).

Hewitt, Paul G, *Conceptual Physics, Eighth Edition* (New York: Addison-Wesley, 1997).

Hobson, Art, *Physics, Concepts and Connections, Second Edition* (Englewood Cliffs, NJ: Prentice-Hall, 1999).

Hofstadter, Douglas R, *Metamagical Themas: Questing for the Essence of Mind and Pattern* (Toronto; New York: Bantam Books, 1985).

Hogben, L, *Mathematics for the Million* (New York: Norton, 1968).

Holton, Gerald James, *Introduction to Concepts and Theories in Physics* (Reading, MA: Addison-Wesley, 1952).

Hoyle, Fred, *Nicolaus Copernicus, An Essay on His Life and Work* (London: Heinemann, 1973).

Irwin, Keith Gordon, *The Romance of Physics* (New York: Scribner, 1966).

James, Jamie, *The Music of the Spheres, Music, Science and the Natural Order of the Universe* (New York: Grove Press, 1993).

Janson, H W and F Anthony, *History of Art 6th Edition* (New York: Prentice-Hall; Harry N Abrams, ca. 2001).

Jèle, Gisèle, Ollinger-Zinque and Frederik Leen (eds) *René Magritte, 1898–1967* (Ghent: Ludion Press; New York: Distributed by H N Abrams, ca. 1998).

Jesse, John Heneage, *Memoirs of King George the Third: His Life and Reign, Volume 2* (Palala Press, 2015).

Kane, Gordon, *The Particle Garden, Our Universe as Understood by Particle Physicists* (New York: Addison-Wesley, 1995).

Kaplon, Morton F, *Homage to Galileo; Papers Presented at the Galileo Quadricentennial*, University of Rochester, 8 and 9 October 1964 (Cambridge,MA: MIT Press, 1965).

Keller, Alex, *The Infancy of Atomic Physics, Hercules in His Cradle* (Oxford: Clarendon Press, 1983).

Kesten, Herman, *Copernicus and His World* (New York: Roy Publishers, 1945).

Kippax, John R, *The Call of the Stars* (New York: Putnam, 1914).

Knight, David C, *Copernicus, Titan of Modern Astronomy* (New York: Oxford University Press, 1993).

Koestler, Arthur, *The Watershed: A Biography of Johannes Kepler* (Garden City, NY: Anchor, 1960).

Kuhn, Thomas S, *The Copernican Revolution, Planetary Astronomy in the Development of Western Thought* (Cambridge, MA: Harvard University Press, 1957).

Lederman, Leon, *The God Particle: If the Universe Is the Answer, What Is the Question?* (Boston, MA: Houghton Mifflin, 1993).

Lemay, Leo J A *The Life of Benjamin Franklin, Volume 3: Soldier, Scientist, and Politician* (University of Pennsylvania Press, 2008).

Lemonick, Michael D, *The Light at the Edge of the Universe, Leading Cosmologists on the Brink of a Scientific Revolution* (New York: Villard Books, 1993).

Lerner, Lawrence S, *Modern Physics for Scientists and Engineers* (Boston, MA: Jones & Bartlett, ca. 1996).

Levi, Primo, *Collected Poems,* In the Beginning, Faber and Faber, Boston, Mass (1988).

Lindberg, David C, *The Beginnings of Western Science: The European Scientific Tradition in Philosophical, Religious, and Institutional Context, 600 BC to AD 1450* (Chicago, IL: University of Chicago Press, 1992).

Lives in science A Scientific American Book (New York: Simon & Schuster, 1957).

Lodge, Oliver, *Pioneers of Science, and the Development of their Scientific Theories* (New York: Dover, 1960).

March, Robert H, *Physics for Poets* (New York: McGraw-Hill, 1992).

Mook, Delo E, *Inside Relativity* (Princeton, NJ: Princeton University Press, 1987).

Moore, Walter, *Schroedinger: Life and Thought* (Cambridge University Press, 1990).

Morrison, Philip and Phyllis, *The Ring of Truth – An Inquiry into How We Know What We Know* (New York: Random House, 1987).

Motz, Lloyd, *The Story of Physics* (New York: Plenum Press, 1989).

Mukherjee, Siddhartha, *The Gene: An Intimate History* (Scribner, 2016).

North, John, *The Norton History of Astronomy and Cosmology* (New York: Norton, 1994).

Noyes, Alfred, *Watchers of the Sky, Volume 1* (Pinnacle Press, 2017).

Ochoa, George and Melinda Corey, *The Wilson Chronology of Science and Technology* (New York: The HW Wilson Company, 1997).

Pagels, Heinz R, *Perfect Symmetry: The Search for the Beginning of Time* (New York: Simon & Schuster, 1985).

Park, David Allen, *The How and the Why: An Essay on the Origins and Development of Physical Theory* (Princeton, NJ: Princeton University Press, 1988).

Parmenides, *Nature, from The Fragments of Parmenides* edited by David Sider and Henry W Johnstone, Jr (Bryn Mawr, PA: Thomas Library, BrynMawr College, 1986).

Pasachoff, Jay M, *Contemporary Astronomy, Third Edition* (New York: Saunders College Publishing, 1985).

Payne-Gaposchkin, Cecilia, *An Introduction to Astronomy* (Englewood Cliffs, NJ: Prentice-Hall, 1954).

Peterson, Ivars, *Newton's Clock: Chaos in the Solar System* (New York: Freeman, 1993).

Pindar, *Paeans from Pindar's Paeans: A Reading of the Fragments with a Survey of the Genre* by Ian Rutherford (New York: Oxford University Press, 2001).

Porter, Roy, *Man Masters Nature: Twenty-Five Centuries of Science* (New York: Brazilier, 1988).

Proctor, Mary, *The Romance of Comets* (New York and London: Harper & Brothers, 1926).

Riordan, Michael, *The Hunting of the Quark, A True Story of Modern Physics* (New York: Simon & Schuster, 1987).

Ronan, Colin A, *Changing Views of the Universe* (New York: Macmillan, 1961).

————, *Galileo* (New York: G P Putnam & Sons, 1974).

————, *Science; Its History and Development among the World's Cultures* (New York: Facts on File, 1982).

————, *Sir Isaac Newton* (London: International Textbook, 1969).

————, *The Astronomers* (London: Evans Brothers, 1964).

Ross, W D, *Aristotle's Metaphysics* (Oxford: Clarendon Press, 1953).

Rousseau, Pierre, *Man's Conquest of the Stars* (London: Jarrolds, 1959).

Rowan-Robinson, Michael, *Our Universe: An Armchair Guide* (New York: Freeman, 1990).

Sagan, Carl and Ann Druyan, *Comets* (New York: Random House, 1985).

Sarton, George, *A History of Science, Ancient Science Through the Golden Age of Greece* (Cambridge,MA: Harvard University Press, 1952).

————, *A History of Science, Hellenistic Science and Culture in the Last Three Centuries BC* (Cambridge, MA: Harvard University Press, 1959).

————, *The History of Science and the New Humanism* (Bloomington, IN: Indiana University Press, 1962, ca. 1937).

Seeds, Michael A, *Foundations of Astronomy, 1999 Edition* (New York: Wadsworth, 1999).

Segré, Emilio, *From Falling Bodies to Radio Waves: Classical Physicists and Their Discoveries* (New York: Freeman, 1984).

————, *From X-Rays to Quarks, Modern Physicists and Their Discoveries* (New York: Freeman, 1980).

Shapin, Steven, *The Scientific Revolution* (Chicago, IL: University of Chicago Press, 1996).

Sharlin, Harold I, *The Convergent Century; The Unification of Science in the Nineteenth Century* (London, NY: Abelard-Schuman, 1966).

Sobel, Dava, *Galileo's Daughter: A Historical Memoir of Science, Faith, and Love* (New York: Walker & Co, 1999).

————, *Longitude* (New York: Walker & Co., 1995).

Speyer, Edward, *Six Roads from Newton, Great Discoveries in Physics* (New York: Wiley, 1994).

Spielberg, Nathan and Bryon D Anderson, *Seven Ideas that Shook the Universe* (New York: Wiley, 1995).

Szabadvary, Ferenc, *Antoine Laurent Lavoisier, The Investigator and His Times* (Cincinnati, OH: University of Cincinatti Press, 1977).

Taylor, Frank Sherwood, *A Short History of Science and Scientific Thought: With Readings from the Great Scientists from the Babylonians to Einstein* (New York: Norton, 1963).

————, *An Illustrated History of Science* (Toronto: WilliamHeinemann, 1956).

Thiel, Rudoff, *And There was Light – The Discovery of the Universe* (New York: Alfred A Knopf, 1957).

Thurston, Hugh, *Early Astronomy* (Berlin: Springer Verlag, 1994).

Tilden, Sir William A, *Famous Chemists, the Men and Their Work* (Freeport, NY: Books for Libraries Press, 1968).

Trefil, James S, *Physics as a Liberal Art* (New York: Pergamon, 1978).

————, *Space Time Infinity: The Smithsonian Views the Universe* (New York: Pantheon; Washington, DC: Smithsonian, 1985).

Tyndall, John, *Faraday as a Discoverer* (New York: Crowell, 1961).

Vergara,William C, *Science, the Never Ending Quest* (New York: Harper & Row, 1965).

Watson, James D, Alexander Gann and Jan Witkowski, *The Annotated and Illustrated Double Helix* (Simon and Schuster, 1968).

Way, Barnard R and Noel D Green, *Time and Its Reckoning* (New York: Chemical Publishing Co., 1940).

Weaver, Jefferson, *The World of Physics: A Small Library of the Literature of Physics from Antiquity to the Present* (New York: Simon & Schuster, 1987).

Westfall, Richard S, *The Construction of Modern Science; Mechanisms and Mechanics* (New York: Wiley, 1971).

————, *Never at Rest: A Biography of Isaac Newton* (New York: Cambridge University Press, 1980).

————, *The Life of Isaac Newton* (Cambridge, England and New York: Cambridge University Press, 1994).

Whitfield, Peter, *Landmarks in Western Science: From Prehistory to the Atomic Age* (New York: Routledge, 1999).

Wightman, William, *The Growth of Scientific Ideas* (New Haven, CT: Yale University Press, 1964).

Wilczek, Frank, *Longing for the Harmonies: Themes and Variations From Modern Physics* (New York: Norton, 1988).

Wilkins, David G, Bernard Schultz and Katheryn M Linduff, *Art Past, Art Present* (New York: Abrams, 1994).

Williams, L Pearce, *Michael Faraday, a Biography* (New York: Basic Books, 1965).

Whitman, Walt, *Song of Myself* (Philadelphia, PA: Masterbooks, 1973).

Yang, Chen Ning, *Elementary Particles; a Short History of Some Discoveries in Atomic Physics* (Princeton, NJ: Princeton University Press, 1961).

Yeomans, Donald K, *Comets: A Chronological History of Observation, Science, Myth, and Folklore* (New York: Wiley, 1991).

Zeri, Fedrico, *Magritte: The Human Condition* (Nide Pub, 2002).

Index